# Student Solutions Manual



to accompany

# Chemistry

Third Edition

**Julia Burdge**
*University of Idaho*

**Prepared by**

**Julia Burdge**

**AccuMedia Publishing Services**

The McGraw·Hill Companies

Student Solutions Manual to accompany
CHEMISTRY, THIRD EDITION
JULIA BURDGE

Published by McGraw-Hill Higher Education, an imprint of The McGraw-Hill Companies, Inc., 1221 Avenue of the Americas, New York, NY 10020. 

This book is printed on acid-free paper.

1 2 3 4 5 6 7 8 9 0 QDB/QDB 1 0 9 8 7 6 5 4 3

ISBN: 978-0-07-757428-4
MHID: 0-07-757428-1

www.mhhe.com

# Contents

# Chapter 1

# Chemistry: The Central Science

1.5 a. **Law** – Newton's 2nd Law of Motion.

b. **Theory** – Big Bang Theory.

c. **Hypothesis** – It may be possible but we have no data to support this statement.

1.7 a. **C and O** b. **F and H** c. **N and H** d. **O**

1.13 a. **K** (potassium) d. **B** (boron) g. **S** (sulfur)

b. **Sn** (tin) e. **Ba** (barium) h. **Ar** (argon)

c. **Cr** (chromium) f. **Pu** (plutonium) i. **Hg** (mercury)

1.15 a. The sea is a heterogeneous mixture of seawater and biological matter, but seawater, with the biomass filtered out, is a **homogeneous mixture**.

b. **element**

c. **compound**

d. **homogeneous mixture**

e. **heterogeneous mixture**

f. **homogeneous mixture**

g. **heterogeneous mixture**

1.17 a. **element** b. **compound** c. **compound** d. **element**

1.23 **Strategy:** Use the density equation:

$$d = \frac{m}{V}$$

**Solution:**

$$d = \frac{m}{V} = \frac{586\ \text{g}}{188\ \text{mL}} = \mathbf{3.12\ g / mL}$$

1.25 **Strategy:** Find the appropriate equations for converting between Fahrenheit and Celsius and between Celsius and Fahrenheit given in Section 1.3 of the text. Substitute the temperature values given in the problem into the appropriate equation.

**Setup:** Conversion from Fahrenheit to Celsius:

$$°\text{C} = (°\text{F} - 32°\text{F}) \times \frac{5°\text{C}}{9°\text{F}}$$

Conversion from Celsius to Fahrenheit:

$$°\text{F} = \left(°\text{C} \times \frac{9°\text{F}}{5°\text{C}}\right) + 32°\text{F}$$

**Solution:** a. $°\text{C} = (95°\text{F} - 32°\text{F}) \times \frac{5°\text{C}}{9°\text{F}} = \mathbf{35°C}$

b. $°\text{C} = (12°\text{F} - 32°\text{F}) \times \frac{5°\text{C}}{9°\text{F}} = \mathbf{-11°C}$

c. $°\text{C} = (102°\text{F} - 32°\text{F}) \times \frac{5°\text{C}}{9°\text{F}} = \mathbf{39°C}$

d. $$°C = (1852°F - 32°F) \times \frac{5°C}{9°F} = \mathbf{1011°C}$$

e. $$°F = \left( -273.15°C \times \frac{9°F}{5°C} \right) + 32°F = \mathbf{-459.67°F}$$

1.27 **Strategy:** Use the density equation.

**Solution:**

$$\text{volume of water} = V = \frac{m}{d} = \frac{2.50 \text{ g}}{0.992 \text{ g/mL}} = \mathbf{2.52 \text{ mL}}$$

1.29 **Strategy:** Use the equation for converting °C to K. AMPS Solution Content

**Setup:** Conversion from Celsius to Kelvin:

$$K = °C + 273.15$$

**Solution:** a. $\mathbf{K} = 115.21°C + 273.15 = \mathbf{388.36 \text{ K}}$

b. $\mathbf{K} = 37°C + 273 = \mathbf{3.10 \times 10^2 \text{ K}}$

c. $\mathbf{K} = 357°C + 273 = \mathbf{6.30 \times 10^2 \text{ K}}$

Note that when there are no digits to the right of the decimal point in the original temperature, we use 273 instead of 273.15.

1.35 a. **Quantitative**. This statement involves a measurable distance.

b. **Qualitative**. This is a value judgment. There is no numerical scale of measurement for artistic excellence.

c. **Qualitative**. If the numerical values for the densities of ice and water were given, it would be a quantitative statement.

d. **Qualitative**. The statement is a value judgment.

e. **Qualitative**. Even though numbers are involved, they are not the result of measurement.

1.37 a. **Physical Change.** The material is helium regardless of whether it is located inside or outside the balloon.

b. **Chemical change** in the battery.

c. **Physical Change.** The orange juice concentrate can be regenerated by evaporation of the water.

d. **Chemical Change.** Photosynthesis changes water, carbon dioxide, etc., into complex organic matter.

e. **Physical Change.** The salt can be recovered unchanged by evaporation.

1.39 Mass is extensive and additive: 37.2 + 62.7 = **99.9 g**

Temperature is intensive: **20°C**

Density is intensive: **11.35 g/cm³**

1.45 **Strategy:** To convert an exponential number $N \times 10^n$ to a decimal number, move the decimal $n$ places to the left if $n < 0$, or move it $n$ places to the right if $n > 0$. While shifting the decimal, add place-holding zeros as needed.

**Solution:** a. $1.52 \times 10^{-2}$ = **0.0152**

b. $7.78 \times 10^{-8}$ = **0.0000000778**

c. $1 \times 10^{-6}$ = **0.000001**

d. $1.6001 \times 10^{3}$ = **1600.1**

1.47 a. Addition using scientific notation.

**Strategy:** A measurement is in *scientific notation* when it is written in the form $N \times 10^n$, where $0 \leq N < 10$ and $n$ is an integer. When adding measurements that are written in scientific notation, rewrite the quantities so that they share a common exponent. To get the "$N$ part" of the result, we simply add the "$N$ parts" of the rewritten numbers. To get the exponent of the result, we simply set it equal to the common exponent. Finally, if need be, we rewrite the result so that its value of $N$ satisfies $0 \leq N < 10$.

**Solution:** Rewrite the quantities so that they have a common exponent. In this case, choose the common exponent $n = -3$.

$$0.0095 = 9.5 \times 10^{-3}$$

Add the "$N$ parts" of the rewritten numbers and set the exponent of the result equal to the common exponent.

$$\begin{array}{r} 9.5 \times 10^{-3} \\ +\ 8.5 \times 10^{-3} \\ \hline 18.0 \times 10^{-3} \end{array}$$

Rewrite the number so that it is in scientific notation (so that $0 \leq N < 10$).

$$18.0 \times 10^{-3} = \mathbf{1.8 \times 10^{-2}}$$

b. Division using scientific notation.

**Strategy:** When dividing two numbers using scientific notation, divide the "$N$ parts" of the numbers in the usual way. To find the exponent of the result, *subtract* the exponent of the devisor from that of the dividend.

**Solution:** Make sure that all numbers are expressed in scientific notation.

$$653 = 6.53 \times 10^2$$

Divide the "$N$ parts" of the numbers in the usual way.

$$6.53 \div 5.75 = 1.14$$

*Subtract* the exponents.

$$1.14 \times 10^{+2-(-8)} = 1.14 \times 10^{+2+8} = \mathbf{1.14 \times 10^{10}}$$

c. Subtraction using scientific notation.

**Strategy:** When subtracting two measurements that are written in scientific notation, rewrite the quantities so that they share a common exponent. To get the "N part" of the result, we simply subtract the "N parts" of the rewritten numbers. To get the exponent of the result, we simply set it equal to the common exponent. Finally, if need be, we rewrite the result so that its value of N satisfies $0 \le N < 10$.

**Solution:** Rewrite the quantities sot that they have a common exponent. Rewrite 850,000 in such a way that $n = 5$.

$$850{,}000 = 8.5 \times 10^5$$

Subtract the "*N* parts" of the numbers and set the exponent of the result equal to the common exponent.

$$\begin{array}{r} 8.5\times10^5 \\ -\ 9.0\times10^5 \\ \hline -0.5\times10^5 \end{array}$$

Rewrite the number so that $0 \le N < 10$ (ignore the sign of *N* when it is negative).

$$-0.5 \times 10^5 = \mathbf{-5 \times 10^4}$$

d. Multiplication using scientific notation.

**Strategy:** When multiplying two numbers using scientific notation, multiply the "*N* parts" of the numbers in the usual way. To find the exponent of the result, *add* the exponents of the two measurements.

**Setup:**

**Solution:** Multiply the "*N* parts" of the numbers in the usual way.

$$3.6 \times 3.6 = 13$$

*Add* the exponents.

$$13 \times 10^{-4+(+6)} = 13 \times 10^2$$

Rewrite the number so that it is in scientific notation (so that $0 \le N < 10$).

$$\mathbf{13 \times 10^2 = 1.3 \times 10^3}$$

1.49 a. **one** d. **four** g. **one or two**

b. **three** e. **three**

c. **three** f. **one**

1.51 a. Division

**Strategy:** The number of significant figures in the answer is determined by the original number having the smallest number of significant figures.

**Solution:**

$$\frac{7.310 \text{ km}}{5.70 \text{ km}} = \mathbf{1.283}$$

The 3 (bolded) is a nonsignificant digit because the original number 5.70 only has three significant digits. Therefore, the answer has only three significant digits.

The correct answer rounded off to the correct number of significant figures is:

**1.28**

**Think About It:** Why are there no units?

b. Subtraction

**Strategy:** The number of significant figures to the right of the decimal point in the answer is determined by the lowest number of digits to the right of the decimal point in any of the original numbers.

**Solution:** Writing both numbers in the decimal notation, we have

$$\begin{array}{r} 0.00326 \text{ mg} \\ -\ 0.0000788 \text{ mg} \\ \hline \mathbf{0.0031812 \text{ mg}} \end{array}$$

The bold numbers are nonsignificant digits because the number 0.00326 has five digits to the right of the decimal point. Therefore, we carry five digits to the right of the decimal point in our answer.

The correct answer rounded off to the correct number of significant figures is:

$$\mathbf{0.00318\ mg = 3.18 \times 10^{-3}\ mg}$$

c. Addition

**Strategy:** The number of significant figures to the right of the decimal point in the answer is determined by the lowest number of digits to the right of the decimal point in any of the original numbers.

**Solution:** Writing both numbers with exponents = +7, we have

$$(0.402 \times 10^7\ \text{dm}) + (7.74 \times 10^7\ \text{dm}) = \mathbf{8.14 \times 10^7\ dm}$$

Since $7.74 \times 10^7$ has only two digits to the right of the decimal point, two digits are carried to the right of the decimal point in the final answer.

1.53 **Tailor Z's measurements are the most accurate. Tailor Y's measurements are the least accurate. Tailor X's measurements are the most precise. Tailor Y's measurements are the least precise.**

1.55 a. **Strategy:** The solution requires a two-step dimensional analysis because we must first convert pounds to grams and then grams to milligrams.

**Setup:** The necessary conversion factors as derived from the equalities: 1 g = 1000 mg and 1 lb = 453.6 g.

$$\frac{453.6\ \text{g}}{1\ \text{lb}} \text{ and } \frac{1\ \text{mg}}{1 \times 10^{-3}\ \text{g}}$$

**Solution:**

$$242\ \text{lb} \times \frac{453.6\ \text{g}}{1\ \text{lb}} \times \frac{1\ \text{mg}}{1 \times 10^{-3}\ \text{g}} = \mathbf{1.10 \times 10^8\ mg}$$

b. **Strategy:** We need to convert from cubic centimeters to cubic meters.

**Setup:** 1 m = 100 cm. When a unit is raised to a power, the corresponding conversion factor must also be raised to that power in order for the units to cancel.

**Solution:**

$$68.3\ \text{cm}^3 \times \left(\frac{1\text{m}}{100\ \text{cm}}\right)^3 = \mathbf{6.83 \times 10^{-5}\ m^3}$$

c. **Strategy:** In Chapter 1 of the text, a conversion is given between liters and $cm^3$ (1 L = 1000 $cm^3$). If we can convert $m^3$ to $cm^3$, we can then convert to liters. Recall that 1 cm = $1 \times 10^{-2}$ m. We need to set up two conversion factors to convert from $m^3$ to L. Arrange the appropriate conversion factors so that $m^3$ and $cm^3$ cancel, and the unit liters is obtained in your answer.

**Setup:** The sequence of conversions is $m^3 \rightarrow cm^3 \rightarrow$ L. Use the following conversion factors:

$$\left(\frac{1\ \text{cm}}{1 \times 10^{-2}\ \text{m}}\right)^3 \text{ and } \frac{1\ \text{L}}{1000\ \text{cm}^3}$$

**Solution:**

$$7.2\ \text{m}^3 \times \left(\frac{1\ \text{cm}}{1 \times 10^{-2}\ \text{m}}\right)^3 \times \frac{1\ \text{L}}{1000\ \text{cm}^3} = \mathbf{7.2 \times 10^3\ L}$$

**Think About It:** From the above conversion factors you can show that 1 $m^3$ = $1 \times 10^3$ L. Therefore, 7 $m^3$ would equal $7 \times 10^3$ L, which is close to the answer.

d. **Strategy:** A relationship between pounds and grams is given on the end sheet of your text (1 lb = 453.6 g). This relationship will allow conversion from grams to pounds. If we can convert from $\mu$g to grams, we can then convert from grams to pounds. Recall that 1 $\mu$g = $1 \times 10^{-6}$ g. Arrange the appropriate conversion factors so that $\mu$g and grams cancel, and the unit pounds is obtained in your answer.

**Setup:** The sequence of conversions is $\mu g \rightarrow g \rightarrow lb$. Use the following conversion factors:

$$\frac{1 \times 10^{-6}\ \text{g}}{1\ \mu\text{g}} \text{ and } \frac{1\ \text{lb}}{453.6\ \text{g}}$$

**Solution:**

$$28.3 \mu\text{g} \times \frac{1 \times 10^{-6}\ \text{g}}{1\ \mu\text{g}} \times \frac{1\text{lb}}{453.6\text{g}} = \mathbf{6.24 \times 10^{-8}\ lb}$$

**Think About It:** Does the answer seem reasonable? What number does the prefix $\mu$ represent? Should 28.3 $\mu$g be a very small mass?

1.57 **Strategy:** You should know conversion factors that will allow you to convert between days and hours, between hours and minutes, and between minutes and seconds. Make sure to arrange the conversion factors so that days, hours, and minutes cancel, leaving units of seconds for the answer.

**Setup:** The sequence of conversions is days → hours → minutes → seconds. Use the following conversion factors:

$$\frac{24 \text{ h}}{1 \text{ day}}, \frac{60 \text{ min}}{1 \text{ h}}, \text{ and } \frac{60 \text{ s}}{1 \text{ min}}$$

**Solution:**

$$365.24 \text{ day} \times \frac{24 \text{ h}}{1 \text{ day}} \times \frac{60 \text{ min}}{1 \text{ h}} \times \frac{60 \text{ s}}{1 \text{ min}} = \mathbf{3.1557 \times 10^7 \text{ s}}$$

**Think About It:** Does your answer seem reasonable? Should there be a very large number of seconds in 1 year?

1.59 a. **Strategy:** The measurement is given in mi/min. We are asked to convert this rate to in/s. Use conversion factors to convert mi → ft → in and to convert min → s.

**Setup:** Use the conversion factors:

$$\frac{5280 \text{ ft}}{1 \text{ mi}}, \frac{12 \text{ in}}{1 \text{ ft}}, \text{ and } \frac{1 \text{ min}}{60 \text{ s}}$$

Be sure to set the conversion factors up so that the appropriate units cancel.

**Solution:**

$$\frac{1 \text{ mi}}{13 \text{ min}} \times \frac{5280 \text{ ft}}{1 \text{ mi}} \times \frac{12 \text{ in}}{1 \text{ ft}} \times \frac{1 \text{ min}}{60 \text{ s}} = \mathbf{81 \text{ in/s}}$$

b. **Strategy:** The measurement is given in mi/min. We are asked to convert this rate to m/min. Use a conversion factor convert mi → m.

**Setup:** Use the conversion factor:

$$\frac{1609 \text{ m}}{1 \text{ mi}}$$

**Solution:**

$$\frac{1 \text{ mi}}{13 \text{ min}} \times \frac{1609 \text{ m}}{1 \text{ mi}} = \mathbf{1.2 \times 10^2 \text{ m / min}}$$

c. **Strategy:** The measurement is given in mi/min. We are asked to convert this rate to km/h. Use conversion factors to convert mi → m →km and convert min →h.

**Setup:** Use the conversion factors:

$$\frac{1609 \text{ m}}{1 \text{ mi}}, \frac{1 \text{ km}}{1000 \text{ m}}, \text{ and } \frac{60 \text{ min}}{1 \text{ h}}$$

**Solution:**

$$\frac{1 \text{ mi}}{13 \text{ min}} \times \frac{1609 \text{ m}}{1 \text{ mi}} \times \frac{1 \text{ km}}{1000 \text{ m}} \times \frac{60 \text{ min}}{1 \text{ h}} = \mathbf{7.4 \text{ km / h}}$$

1.61 **Strategy:** The rate is given in the units mi/h. The desired units are km/h. Use conversion factors to convert mi → m → km.

**Setup:** Use the conversion factors:

$$\frac{1609 \text{ m}}{1 \text{ mi}} \text{ and } \frac{1 \text{ km}}{1000 \text{ m}}$$

**Solution:**

$$\frac{55 \text{ mi}}{1 \text{ h}} \times \frac{1609 \text{ m}}{1 \text{ mi}} \times \frac{1 \text{ km}}{1000 \text{ m}} = \mathbf{88 \text{ km / h}}$$

1.63 **Strategy:** We seek to calculate the mass of Pb in a $6.0 \times 10^3$ g sample of blood. Lead is present in the blood at the rate of $0.62 \text{ ppm} = \frac{0.62 \text{ g Pb}}{1 \times 10^6 \text{ g blood}}$. Use the rate to convert g blood → g Pb.

**Setup:** Be sure to set the conversion factor up so that g blood cancels.

**Solution:**

$$6.0 \times 10^3 \text{ g of blood} \times \frac{0.62 \text{ g Pb}}{1 \times 10^6 \text{ g blood}} = \mathbf{3.7 \times 10^{-3} \text{ g Pb}}$$

1.65 a. **Strategy:** The given unit is nm and the desired unit is m. Use a conversion factor to convert nm → m.

**Setup:** Use the conversion factor:

$$\frac{1 \times 10^{-9} \text{ m}}{1 \text{ nm}}$$

**Solution:**

$$185 \text{ nm} \times \frac{1 \times 10^{-9} \text{ m}}{1 \text{ nm}} = \mathbf{1.85 \times 10^{-7} \text{ m}}$$

b. **Strategy:** The given unit is yr and the desired unit is s. Use conversion factors to convert yr → d → h → s.

**Setup:** Use the conversion factors:

$$\frac{365 \text{ d}}{1 \text{ yr}}, \frac{24 \text{ h}}{1 \text{ d}}, \text{ and } \frac{3600 \text{ s}}{1 \text{ h}}$$

**Solution:**

$$(4.5 \times 10^9 \text{ yr}) \times \frac{365 \text{ day}}{1 \text{ yr}} \times \frac{24 \text{ h}}{1 \text{ day}} \times \frac{3600 \text{ s}}{1 \text{ h}} = \mathbf{1.4 \times 10^{17} \text{ s}}$$

c. **Strategy:** The given unit is $cm^3$ and the desired unit is $m^3$. Use a conversion factor to convert $cm^3 \rightarrow m^3$.

**Setup:** Use the conversion factor:

$$\left(\frac{0.01 \text{ m}}{1 \text{ cm}}\right)^3$$

**Solution:**

$$71.2\ \text{cm}^3 \times \left(\frac{0.01\ \text{m}}{1\ \text{cm}}\right)^3 = \mathbf{7.12 \times 10^{-5}\ m^3}$$

d. **Strategy:** The given unit is $m^3$ and the desired unit is L. Use conversion factors to convert $m^3 \rightarrow cm^3 \rightarrow L$.

**Setup:** Use the conversion factors:

$$\left(\frac{1\ \text{cm}}{1 \times 10^{-2}\ \text{m}}\right)^3 \text{ and } \frac{1\ \text{L}}{1000\ \text{cm}^3}$$

**Solution:**

$$88.6\ \text{m}^3 \times \left(\frac{1\ \text{cm}}{1 \times 10^{-2}\ \text{m}}\right)^3 \times \frac{1\ \text{L}}{1000\ \text{cm}^3} = \mathbf{8.86 \times 10^4\ L}$$

1.67 **Strategy:** The given rate has units g/L and the desired units are g/$cm^3$. Use a conversion factor to convert L $\rightarrow$ $cm^3$.

**Setup:** Use the conversion factor:

$$\frac{1\ \text{L}}{1000\ \text{cm}^3}$$

**Solution:**

$$\frac{0.625\ \text{g}}{1\ \text{L}} \times \frac{1\ \text{L}}{1000\ \text{cm}^3} = \mathbf{6.25 \times 10^{-4}\ g/cm^3}$$

1.69 **Strategy:** Use the equation $t = \frac{x^2}{2D}$ to convert the given distance ($x = 10\ \mu\text{m}$) to time $t$ in seconds. Notice that both $x$ and $D$ contain distance units. But, for the given values, $x = 10\ \mu\text{m}$ and $D = 5.7 \times 10^{-7}\ \text{cm}^2/\text{s}$, the distance units are dissimilar and will not cancel. So, before calculating, convert the units of $x$ from μm to cm.

**Setup:** Use the conversion factors:

$$\frac{1\times10^{-2}\ \text{m}}{1\ \text{cm}} \text{ and } \frac{1\ \mu\text{m}}{1\times10^{-6}\,\text{m}}$$

**Solution:**

$$10\ \mu\text{m} \times \frac{1\times10^{-6}\ \text{m}}{1\,\mu\text{m}} \times \frac{1\ \text{cm}}{1\times10^{-2}\ \text{m}} = 1\times10^{-3}\ \text{cm} = x \text{ in the equation.}$$

$$t = \left[\frac{\left(10^{-3}\ \text{cm}\right)^2}{2\times5.7\times10^{-7}\ \text{cm}^2/\text{s}}\right] = \mathbf{0.88\ s}$$

1.71 a. **Upper ruler: 2.5 cm** b. **Lower ruler: 2.55 cm**

1.73 a. **chemical** b. **chemical** c. **physical** d. **physical** e. **chemical**

1.75 a. **Strategy:** Calculate the volume of the sphere using:

$$V = \frac{4}{3}\pi r^3$$

Then use the density equation, $d = \frac{m}{V}$, to find the mass.

**Setup:** Solve the density equation for $m$ to get $m = dV$. Find the volume and substitute in into the equation for $m$.

**Solution:**

$$V = \left(\frac{4}{3}\right)(3.14159)(10.0\ \text{cm})^3 = 4189\ \text{cm}^3$$

$$m = dV = \frac{19.3\ \text{g}}{1\ \text{cm}^3} \times 4189\ \text{cm}^3 = \mathbf{8.08\times10^4\ g}$$

b. **Strategy:** Compute the volume of the cube using:

$$V = s^3$$

Then, find the mass using the density equation:

$$d = \frac{m}{V}$$

**Setup:** Solve the density equation for $m$ to get $m = dV$. Find the volume and substitute in into the equation for $m$.

**Solution:** The edge of the cube is $s = 0.040$ mm = 0.0040 cm, and $V = (0.0040 \text{ cm})^3 = 6.4 \times 10^{-8} \text{ cm}^3$.

$$m = dV = \frac{21.4 \text{ g}}{1 \text{ cm}^3} \times \left(6.4 \times 10^{-8} \text{ cm}^3\right) = \mathbf{1.4 \times 10^{-6} \text{ g}}$$

c. **Strategy:** Use the density equation:

$$d = \frac{m}{V}$$

**Setup:** Solve the density equation for $m$ to get $m = dV$.

**Solution:**

$$50.0 \text{ mL} \times \frac{0.798 \text{ g}}{1 \text{ mL}} = \mathbf{39.9 \text{ g}}$$

1.77 **Strategy:** The difference between the masses of the empty and filled flasks is the mass of the water in the flask. The volume of the water (and the flask) can be found using the density equation.

**Setup:** Solve the density equation for $V$:

$$V = \frac{m}{d}$$

**Solution:**

$$87.39 - 56.12 = 31.27 \text{ g water}$$

$$V = \frac{m}{d} = \frac{31.27 \text{ g}}{0.9976 \text{ g/cm}^3} = \mathbf{31.35 \text{ cm}^3}$$

1.79 **Strategy:** The volume of the piece of silver is the same as the volume of water it displaces. Once the volume is found, use the density equation to compute the density.

**Setup:**

$$V = 260.5 - 242.0 = 18.50 \text{ mL}$$

**Solution:**

$$d = \left(\frac{194.3 \text{ g}}{18.50 \text{ mL}}\right) = \mathbf{10.50\ g\,/\,mL}$$

The density of a solid is generally reported in g/cm$^3$. (1 mL = 1 cm$^3$) Therefore, the density is reported as **10.50 g/cm$^3$**.

**Think About It:** The volume of the water displaced must equal the volume of the piece of silver. If the silver did not sink, would you have been able to determine the volume of the piece of silver?

1.81 **Strategy:** Use the density equation.

**Setup:**

$$d = \frac{m}{V}$$

**Solution:**

$$d = \frac{m}{V} = \frac{1.20 \times 10^4 \text{ g}}{1.05 \times 10^3 \text{ cm}^3} = \mathbf{11.4\ g\,/\,cm^3}$$

1.83 **Strategy:** Use the conversion equation °F → °C or the conversion equation °C → °F. The solution below uses the conversion equation °C → °F.

**Setup:** Let $t$ represent the common temperature. Substitute $t$ into the conversion equation $°\text{F} = \left(°\text{C} \times \frac{9°\text{F}}{5°\text{C}}\right) + 32°\text{F}$ and solve for $t$.

**Solution:**

$$t = \frac{9}{5}t + 32°\text{F}$$

$$t - \frac{9}{5}t = 32°\text{F}$$

$$-\frac{4}{5}t = 32°\text{F}$$

$$t = -40°F = -40°C$$

1.85 **Strategy:** The volume of seawater is given. The strategy is to use the given conversion factors to convert L seawater → g seawater → g NaCl. This result can then be converted to kg NaCl and to tons NaCl. Note that 3.1% NaCl by weight means 100 g seawater = 3.1 g NaCl.

**Setup:** Use the conversion factors:

$$\frac{1000 \text{ mL seawater}}{1 \text{ L seawater}}, \ \frac{1.03 \text{ g seawater}}{1 \text{ mL seawater}}, \text{ and } \frac{3.1 \text{ g NaCl}}{100 \text{ g seawater}}$$

**Solution:**

$$1.5\times10^{21} \text{ L seawater}\times\frac{1000 \text{ mL seawater}}{1 \text{ L seawater}}\times\frac{1.03 \text{ g seawater}}{1 \text{ mL seawater}}\times\frac{3.1 \text{ g NaCl}}{100 \text{ g seawater}} = 4.8\times10^{22} \text{ g NaCl}$$

$$\textbf{mass NaCl (kg)} = 4.8\times10^{22} \text{ g NaCl}\times\frac{1 \text{ kg}}{1000 \text{ g}} = \mathbf{4.8\times10^{19} \textbf{ kg NaCl}}$$

$$\textbf{mass NaCl (tons)} = 4.8\times10^{22} \text{ g NaCl}\times\frac{1 \text{ lb}}{453.6 \text{ g}}\times\frac{1 \text{ ton}}{2000 \text{ lb}} = \mathbf{5.3\times10^{16} \textbf{ tons NaCl}}$$

1.87 **Strategy:** Assume that the crucible is platinum. Calculate the volume of the crucible and then compare that to the volume of water that the crucible displaces.

**Setup:**

$$\text{volume} = \frac{\text{mass}}{\text{density}}$$

**Solution:**

$$\text{Volume of crucible} = \frac{860.2 \text{ g}}{21.45 \text{ g/cm}^3} = \mathbf{40.10 \textbf{ cm}^3}$$

$$\text{Volume of water displaced} = \frac{(860.2 - 820.2)\text{g}}{0.9986 \text{ g/cm}^3} = \mathbf{40.1 \textbf{ cm}^3}$$

The volumes are the same (within experimental error), so **the density of the crucible is equal to the density of pure platinum**. Therefore, the crucible is probably made of platinum

1.89 a. **Strategy:** Use the given conversion factor to convert troy oz → g.

**Setup:** Conversion factor:

$$\frac{31.103\text{ g Au}}{1\text{ troy oz Au}}$$

**Solution:**

$$2.41\text{ troy oz Au} \times \frac{31.103\text{ g Au}}{1\text{ troy oz Au}} = \mathbf{75.0\text{ g Au}}$$

b. **Strategy:** Use the given conversion factors to convert 1 troy oz → g → lb → oz.

**Setup:** Conversion factors:

$$\frac{31.103\text{ g}}{1\text{ troy oz}}, \frac{1\text{ lb}}{453.6\text{ g}}, \text{ and } \frac{16\text{ oz}}{1\text{ lb}}$$

**Solution:**

$$1\text{ troy oz} \times \frac{31.103\text{ g}}{1\text{ troy oz}} \times \frac{1\text{ lb}}{453.6\text{ g}} \times \frac{16\text{ oz}}{1\text{ lb}} = 1.097\text{ oz}$$

**1 troy oz = 1.097 oz**

**A troy ounce is heavier than an ounce.**

1.91 a. **Strategy:** Use the percent error equation.

**Setup:** The percent error of a measurement is given by:

$$\frac{|\text{true value} - \text{experimental value}|}{\text{true value}} \times 100\%$$

**Solution:**

$$\frac{|0.798\text{ g/mL} - 0.802\text{ g/mL}|}{0.798\text{ g/mL}} \times 100\% = \mathbf{0.5\%}$$

b. **Strategy:** Use the percent error equation.

**Setup:** The percent error of a measurement is given by:

$$\frac{|\text{true value} - \text{experimental value}|}{\text{true value}} \times 100\%$$

**Solution:**

$$\frac{|0.864\text{ g} - 0.837\text{ g}|}{0.864\text{ g}} \times 100\% = \mathbf{3.1\%}$$

1.93 **Gently heat the liquid to see if any solid remains after the liquid evaporates. Also, collect the vapor and then compare the densities of the condensed liquid with the original liquid. The composition of a mixed liquid frequently changes with evaporation along with its density.**

1.95 **Strategy:** As water freezes, it expands. First, calculate the mass of the water at 20°C. Then, determine the volume that this mass of water would occupy at –5°C.

**Solution:**

$$\text{Mass of water} = 242\text{ mL} \times \frac{0.998\text{ g}}{1\text{ mL}} = 241.5\text{ g}$$

$$\text{Volume of ice at } -5°\text{C} = 241.5\text{ g} \times \frac{1\text{ mL}}{0.916\text{ g}} = 264\text{ mL}$$

**The volume occupied by the ice is larger than the volume of the glass bottle. The glass bottle would break.**

1.97 **Strategy:** We are given a distance (1500 m) and a rate (3 minutes 43.13 seconds per mile). Use the rate as a conversion factor to convert m → mi → s.

**Setup:** Convert the time to run 1 mile to seconds:

$$3\text{ min } 43.13\text{ s} = 180\text{ s} + 43.13\text{ s} = 223.13\text{ s}$$

Use the conversion factors: $\frac{1\text{ mi}}{1609\text{ m}}$ $\frac{223.13\text{ s}}{1\text{ mi}}$

**Solution:**

$$1500\text{ m} \times \frac{1\text{ mi}}{1609\text{ m}} \times \frac{223.13\text{ s}}{1\text{ mi}} = \mathbf{208.0\text{ s}} = \mathbf{3\text{ min } 28.0\text{ s}}$$

1.99 a. **homogeneous**

b. **heterogeneous**. The air will contain particulate matter, clouds, etc. This mixture is not homogeneous.

1.101 **Strategy:** Step 1: Use conversion factors to convert L seawater → mL seawater → g Au.

Step 2: Use conversion factors to convert g Au → dollars

**Setup:**

Step 1: Use the conversion factors: $\dfrac{1 \text{ mL seawater}}{0.001 \text{ L seawater}}$ $\dfrac{4.0 \times 10^{-12} \text{ g Au}}{1 \text{ mL seawater}}$

Step 2: Use the conversion factor: $\dfrac{\$1350}{31.103 \text{ g Au}}$

**Solution:**

$$(1.5 \times 10^{21} \text{ L seawater}) \times \frac{1 \text{ mL seawater}}{0.001 \text{ L seawater}} \times \frac{4.0 \times 10^{-12} \text{ g Au}}{1 \text{ mL seawater}} = \mathbf{6.0 \times 10^{12} \text{ g Au}}$$

$$\textbf{value of gold} = 6.0\times10^{12} \text{ g Au} \times \frac{\$1350}{31.103 \text{ oz}} = \mathbf{\$2.6 \times 10^{14}}$$

**Think About It:** No one has become rich mining gold from the ocean, because the cost of recovering the gold would outweigh the price of the gold.

1.103 **Strategy:** Use conversion factors to convert tons of earth → kg Si. Note that 0.50% crust by mass means 100 tons earth = 0.50 tons crust and that 27.2% Si by mass means 100 tons crust = 27.2 tons Si.

**Setup:** Conversion factors:

$$\frac{0.50 \text{ ton crust}}{100 \text{ ton earth}} \quad \frac{27.2 \text{ ton Si}}{100 \text{ ton crust}} \quad \frac{2000 \text{ lb Si}}{1 \text{ ton Si}} \quad \frac{453.6 \text{ g Si}}{1 \text{ lb Si}} \quad \frac{1 \text{ kg Si}}{1000 \text{ g Si}}$$

**Solution:**

$$5.9\times10^{21} \text{ ton earth} \times \frac{0.50 \text{ ton crust}}{100 \text{ ton earth}} \times \frac{27.2 \text{ ton Si}}{100 \text{ ton crust}} \times \frac{2000 \text{ lb Si}}{1 \text{ ton Si}} \times \frac{453.6 \text{ g Si}}{1 \text{ lb Si}} \times \frac{1 \text{ kg Si}}{1000 \text{ g Si}}$$

$$= 7.3\times10^{21} \text{ kg Si}$$

$$\textbf{mass of silicon in crust} = \mathbf{7.3 \times 10^{21} \text{ kg Si}}$$

1.105 **Strategy:** We wish to calculate the density and radius of the ball bearing. For both calculations, we need the volume of the ball bearing. The data from the first experiment can be used to calculate the density of the mineral oil. In the second experiment, the density of the mineral oil can then be used to determine what part of the 40.00 mL volume is due to the mineral oil and what part is due to the ball bearing. Once the volume of the ball bearing is determined, we can calculate its density and radius.

**Solution:** From experiment one:

$$\text{Mass of oil} = 159.446\text{ g} - 124.966\text{ g} = 34.480\text{ g}$$

$$\text{Density of oil} = \frac{34.480\text{ g}}{40.00\text{ mL}} = 0.8620\text{ g/mL}$$

From the second experiment:

$$\text{Mass of oil} = 50.952\text{ g} - 18.713\text{ g} = 32.239\text{ g}$$

$$\text{Volume of oil} = 32.239\text{ g} \times \frac{1\text{ mL}}{0.8620\text{ g}} = 37.40\text{ mL}$$

The volume of the ball bearing is obtained by difference.

$$\text{Volume of ball bearing} = 40.00\text{ mL} - 37.40\text{ mL} = 2.60\text{ mL} = 2.60\text{ cm}^3$$

Now that we have the volume of the ball bearing, we can calculate its density and radius.

$$\text{Density of ball bearing} = \frac{18.713\text{ g}}{2.60\text{ cm}^3} = \mathbf{7.20\ g/cm^3}$$

Using the formula for the volume of a sphere, we can solve for the radius of the ball bearing.

$$V = \frac{4}{3}\pi r^3$$

$$2.60\text{ cm}^3 = \frac{4}{3}\pi r^3$$

$$r^3 = 0.621\text{ cm}^3$$

$$\boldsymbol{r} = \mathbf{0.853\ cm}$$

1.107 **It would be more difficult to prove that the unknown substance is an element. Most compounds would decompose on heating, making them easy to identify. On heating, the compound HgO decomposes to elemental mercury (Hg) and oxygen gas ($O_2$).**

1.109 **Strategy:** Use the given rate to convert J → yr.

**Setup:** Conversion factor:

$$\frac{1\ \text{yr}}{1.8 \times 10^{20}\ \text{J}}$$

**Solution:**

$$(2.0 \times 10^{22}\ \text{J}) \times \frac{1\ \text{yr}}{1.8 \times 10^{20}\ \text{J}} = \mathbf{1.1 \times 10^{2}\ yr}$$

1.111 **Strategy:** Use the percent composition measurement to convert kg ore → g Cu. Note that 34.63% Cu by mass means 100 g ore = 34.63 g Cu.

**Setup:** Use the conversion factors:

$$\frac{34.63\ \text{g Cu}}{100\ \text{g ore}} \quad \frac{1000\ \text{g}}{1\ \text{kg}}$$

**Solution:**

$$(5.11 \times 10^{3}\ \text{kg ore}) \times \frac{34.63\ \text{g Cu}}{100\ \text{g ore}} \times \frac{1000\ \text{g}}{1\ \text{kg}} = \mathbf{1.77 \times 10^{6}\ g\ Cu}$$

1.113 **Strategy:** Use the given rates to convert cars → kg $CO_2$.

**Setup:** Conversion factors:

$$\frac{5000\ \text{mi}}{1\ \text{car}}, \frac{1\ \text{gal gas}}{20\ \text{mi}}, \text{ and } \frac{9.5\ \text{kg CO}_2}{1\ \text{gal gas}}$$

**Solution:**

$$(40 \times 10^{6}\ \text{cars}) \times \frac{5000\ \text{mi}}{1\ \text{car}} \times \frac{1\ \text{gal gas}}{20\ \text{mi}} \times \frac{9.5\ \text{kg CO}_2}{1\ \text{gal gas}} = \mathbf{9.5 \times 10^{10}\ kg\ CO_2}$$

1.115 **Strategy:** Use dimensional analysis. The conversions should convert the units "people" to the units "kg NaF". Since the number of conversion steps is large, divide the calculation into smaller steps.

**Setup:** Use the given conversion factors and also any others you may need from the inside back cover of the text.

**Solution:** The mass of water used by 50,000 people in 1 year is:

$$50{,}000 \text{ people} \times \frac{150 \text{ gal water}}{1 \text{ person each day}} \times \frac{3.79 \text{ L}}{1 \text{ gal}} \times \frac{1000 \text{ mL}}{1 \text{ L}} \times \frac{1.0 \text{ g H}_2\text{O}}{1 \text{ mL H}_2\text{O}} \times \frac{365 \text{ days}}{1 \text{ yr}}$$
$$= 1.04 \times 10^{13} \text{ g H}_2\text{O/yr}$$

A concentration of 1 ppm of fluorine is needed. In other words, 1 g of fluorine is needed per million grams of water. NaF is 45.0% fluorine by mass. The amount of NaF needed per year in kg is:

$$(1.04 \times 10^{13} \text{ g H}_2\text{O}) \times \frac{1 \text{ g F}}{10^6 \text{ g H}_2\text{O}} \times \frac{100\% \text{ NaF}}{45\% \text{ F}} \times \frac{1 \text{ kg}}{1000 \text{ g}} = \mathbf{2.3 \times 10^4 \text{ kg NaF}}$$

An average person uses 150 gallons of water per day. This is equal to 569 L of water. If only 6 L of water is used for drinking and cooking, 563 L of water is used for purposes in which NaF is not necessary. Therefore the amount of NaF wasted is:

$$\frac{563 \text{ L}}{569 \text{ L}} \times 100\% = \mathbf{99\%}$$

1.117 **Strategy:** The key to solving this problem is to realize that all the oxygen needed must come from the 4% difference (20% – 16%) between inhaled and exhaled air. The 240 mL of pure oxygen/min requirement comes from the 4% of inhaled air that is oxygen.

**Setup:**

$$240 \text{ mL of pure oxygen/min} = (0.04)\,(\text{volume of inhaled air/min})$$

**Solution:**

$$\text{Volume of inhaled air/min} = \frac{240 \text{ mL of oxygen/min}}{0.04} = 6000 \text{ mL of inhaled air/min}$$

Since there are 12 breaths per min,

$$\textbf{volume of air / breath} = \frac{6000 \text{ mL of inhaled air}}{1 \text{ min}} \times \frac{1 \text{ min}}{12 \text{ breaths}} = \mathbf{5 \times 10^2 \text{ mL / breath}}$$

1.119 **Strategy:** For the Fahrenheit thermometer, we must convert the possible error of 0.1°F to °C. For each thermometer, use the percent error equation to find the percent error for the measurement.

**Setup:**

$$0.1°\text{F} \times \frac{5°\text{C}}{9°\text{F}} = 0.056°\text{C}\,.$$

For the Fahrenheit thermometer, we expect:

$$|\text{true value} - \text{experimental value}| = 0.056°\text{C}$$

For the Celsius thermometer, we expect:

$$|\text{true value} - \text{experimental value}| = 0.1°\text{C}$$

$$\text{Percent error} = \frac{|\text{true value} - \text{experimental value}|}{\text{true value}} \times 100\%$$

**Solution:** For the Fahrenheit thermometer, **percent error** $= \dfrac{0.056°\text{C}}{38.9°\text{C}} \times 100\% = \mathbf{0.1\%}$

For the Celsius thermometer, **percent error** $= \dfrac{0.1°\text{C}}{38.9°\text{C}} \times 100\% = \mathbf{0.3\%}$

**Think About It:** Which thermometer is more precise?

1.121 **Strategy:** To calculate the density of the pheromone, you need the mass of the pheromone, and the volume that it occupies. The mass is given in the problem.

**Setup:**

$$\text{volume of a cylinder} = \text{area} \times \text{height} = \pi r^2 \times h$$

Converting the radius and height to cm gives:

$$0.50\text{ mi} \times \frac{1609\text{ m}}{1\text{ mi}} \times \frac{1\text{ cm}}{0.01\text{ m}} = 8.05 \times 10^4\text{ cm}$$

$$40\text{ ft} \times \frac{12\text{ in}}{1\text{ ft}} \times \frac{2.54\text{ cm}}{1\text{ in}} = 1.22 \times 10^3\text{ cm}$$

**Solution:**

$$\text{volume} = \pi(8.05 \times 10^4\text{ cm})^2 \times (1.22 \times 10^3\text{ cm}) = 2.48 \times 10^{13}\text{ cm}^3$$

Density of gases is usually expressed in g/L. Let's convert the volume to liters.

$$(2.48 \times 10^{13}\ \text{cm}^3) \times \frac{1\ \text{mL}}{1\ \text{cm}^3} \times \frac{1\ \text{L}}{1000\ \text{mL}} = 2.48 \times 10^{10}\ \text{L}$$

$$\textbf{density} = \frac{\text{mass}}{\text{volume}} = \frac{1.0 \times 10^{-8}\ \text{g}}{2.48 \times 10^{10}\ \text{L}} = \mathbf{4.0 \times 10^{-19}\ g/L}$$

# Chapter 2

# Atoms, Molecules, and Ions

2.3

$$\frac{\text{ratio of N to O in compound 1}}{\text{ratio of N to O in compound 2}} = \frac{0.8756}{0.4378} \approx \mathbf{2:1}$$

2.5

$$\frac{\text{ratio of F to S in } S_2F_{10}}{\text{ratio of F to S in } SF_4} = \frac{2.962}{2.370} \approx 1.250$$

$$\frac{\text{ratio of F to S in } SF_6}{\text{ratio of F to S in } SF_4} = \frac{3.555}{2.370} \approx 1.5$$

$$\frac{\text{ratio of F to S in } SF_4}{\text{ratio of F to S in } SF_4} = 1$$

$$\text{ratio in } SF_6 : \text{ratio in } S_2F_{10} : \text{ratio in } SF_4 = 1.5:1.25:1$$

Multiply through to get all whole numbers. $4\cdot(1.5:1.25:1) = \mathbf{6:5:4}$

2.7

$$\frac{\text{g blue: 1.00 g red (right)}}{\text{g blue: 1.00 g red (left)}} = \frac{2/3}{1/1} \approx \mathbf{0.667} \approx \mathbf{2:3}$$

2.15 Note that you are given information to set up the conversion factor relating meters and miles.

$$\boldsymbol{r}_{\textbf{atom}} = 10^4 r_{\text{nucleus}} = 10^4 \times 2.0 \text{ cm} \times \frac{1 \text{ m}}{100 \text{ cm}} \times \frac{1 \text{ mi}}{1609 \text{ m}} = \mathbf{0.12\ mi}$$

2.21 **Strategy:** The 239 in Pu-239 is the mass number. The **mass number (*A*)** is the total number of neutrons and protons present in the nucleus of an atom of an element. You can look up the atomic number (number of protons) on the periodic table.

**Solution:**

$$\text{mass number} = \text{number of protons} + \text{number of neutrons}$$

$$\textbf{number of neutrons} = \text{mass number} - \text{number of protons} = 239 - 94 = \mathbf{145}$$

2.23

| Isotope | ${}^{15}_{7}N$ | ${}^{33}_{16}S$ | ${}^{63}_{29}Cu$ | ${}^{84}_{38}Sr$ | ${}^{130}_{56}Ba$ | ${}^{186}_{74}W$ | ${}^{202}_{80}Hg$ |
|---|---|---|---|---|---|---|---|
| No. Protons | 7 | 16 | 29 | 38 | 56 | 74 | 80 |
| No. Electrons | 7 | 16 | 29 | 38 | 56 | 74 | 80 |
| No. Neutrons | 8 | 17 | 34 | 46 | 74 | 112 | 122 |

2.25 The accepted way to denote the atomic number and mass number of an element X is ${}^{A}_{Z}X$ where $\boldsymbol{A}$ = mass number and $\boldsymbol{Z}$ = atomic number.

a. ${}^{186}_{74}W$ b. ${}^{201}_{80}Hg$ c. ${}^{76}_{34}Se$ d. ${}^{239}_{94}Pu$

2.27 a. 19 b. 34 c. 75 d. 192

2.35 a. Metallic character increases as you progress down a group of the periodic table. For example, moving down Group 4A, the nonmetal carbon is at the top and the metal lead is at the bottom of the group.

b. Metallic character decreases from the left side of the table (where the metals are located) to the right side of the table (where the nonmetals are located).

2.37 Na and K are both Group 1A elements; they should have similar chemical properties. N and P are both Group 5A elements; they should have similar chemical properties. F and Cl are Group 7A elements; they should have similar chemical properties.

2.39

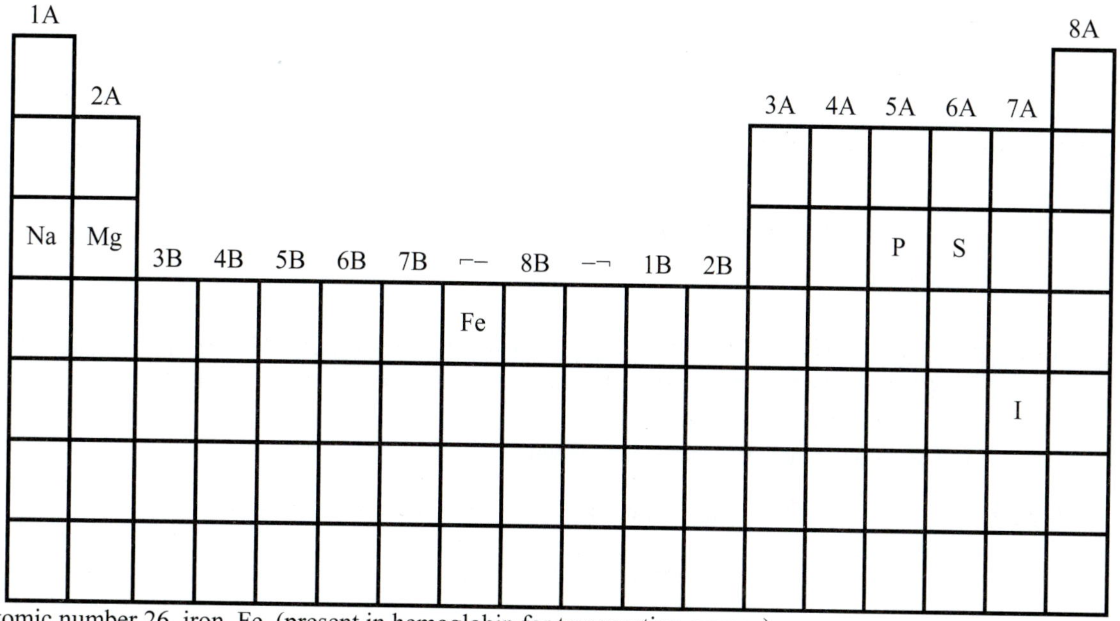

1A 8A
2A 3A 4A 5A 6A 7A
Na Mg 3B 4B 5B 6B 7B 8B 1B 2B P S
Fe
I

Atomic number 26, iron, Fe, (present in hemoglobin for transporting oxygen)

Atomic number 53, iodine, I, (present in the thyroid gland)

Atomic number 11, sodium, Na, (present in intra- and extra-cellular fluids)

Atomic number 15, phosphorus, P, (present in bones and teeth)

Atomic number 16, sulfur, S, (present in proteins)

Atomic number 12, magnesium, Mg, (present in chlorophyll molecules)

2.45 (203.973020 amu)(0.014) + (205.974440 amu)(0.241) +(206.975872 amu)(0.221)

+ (207.976627 amu)(0.524) = **207.2 amu**

2.47 **Strategy:** Each isotope contributes to the average atomic mass based on its relative abundance. Multiplying the mass of an isotope by its fractional abundance (not percent) will give the contribution to the average atomic mass of that particular isotope.

It would seem that there are two unknowns in this problem, the fractional abundance of $^{6}$Li and the fractional abundance of $^{7}$Li. However, these two quantities are not independent of each other; they are related by the fact that they must sum to 1. Start by letting $x$ be the fractional abundance of $^{6}$Li. Since the sum of the two fractional abundances must be 1, we can write

$$(6.0151\text{ amu})(x) + (7.0160\text{ amu})(1-x) = 6.941\text{ amu}$$

**Solution:** Solving for $x$ gives 0.075, which corresponds to the fractional abundance of $^{6}$Li. The fractional abundance of $^{7}$Li is $(1 - x) = 0.925$. Therefore, the natural abundances of $^{6}$Li and $^{7}$Li are **7.5%** and **92.5%**, respectively.

2.49 The conversion factor required is $\frac{6.022\times 10^{23}\text{ amu}}{1\text{ g}}$

$$8.4\text{ g}\times\frac{6.022\times 10^{23}\text{ amu}}{1\text{ g}} = \mathbf{5.1\times 10^{24}\text{ amu}}$$

2.59 a. This is a polyatomic molecule that is an elemental form of the substance. It is not a compound.

b. This is a polyatomic molecule that is a compound.

c. This is a diatomic molecule that is a compound.

2.61 **Elements:** $N_2$, $S_8$, $H_2$

**Compounds:** $NH_3$, NO, CO, $CO_2$, $SO_2$

2.63 **Strategy:** An *empirical formula* tells us which elements are present and the *simplest* whole-number ratio of their atoms. Can you divide the subscripts in the formula by a common factor to end up with smaller whole-number subscripts?

**Solution:** a. Dividing both subscripts by 2, the simplest whole number ratio of the atoms in $C_2N_2$ is **CN**.

b. Dividing all subscripts by 6, the simplest whole number ratio of the atoms in $C_6H_6$ is **CH**.

c. The molecular formula as written, $\mathbf{C_9H_{20}}$, contains the simplest whole number ratio of the atoms present. In this case, the molecular formula and the empirical formula are the same.

d. Dividing all subscripts by 2, the simplest whole number ratio of the atoms in $P_4O_{10}$ is $\mathbf{P_2O_5}$.

e. Dividing all subscripts by 2, the simplest whole number ratio of the atoms in $B_2H_6$ is $\mathbf{BH_3}$.

2.65 $C_3H_7NO_2$

2.67 a. nitrogen trichloride

b. iodine heptafluoride

c. tetraphosphorus hexoxide

d. disulfur dichloride

2.69 All of these are molecular compounds. We use prefixes to express the number of each atom in the molecule. The molecular formulas and names are:

a. $NF_3$: nitrogen trifluoride

b. $PBr_5$: phosphorus pentabromide

c. $SCl_2$: sulfur dichloride

2.75 The **atomic number (*Z*)** is the number of protons in the nucleus of each atom of an element. You can find this on a periodic table. The number of **electrons** in an *ion* is equal to the number of protons minus the charge on the ion.

**number of electrons (ion)** = number of protons − charge on the ion

| Ion | $Na^+$ | $Ca^{2+}$ | $Al^{3+}$ | $Fe^{2+}$ | $I^-$ | $F^-$ | $S^{2-}$ | $O^{2-}$ | $N^{3-}$ |
|---|---|---|---|---|---|---|---|---|---|
| No. protons | 11 | 20 | 13 | 26 | 53 | 9 | 16 | 8 | 7 |
| No. electrons | 10 | 18 | 10 | 24 | 54 | 10 | 18 | 10 | 10 |

2.77 a. Sodium ion has a +1 charge and oxide has a −2 charge. The correct formula is $\mathbf{Na_2O}$.

b. The iron ion has a +2 charge and sulfide has a −2 charge. The correct formula is **FeS**.

c. The correct formula is $\mathbf{Co_2(SO_4)_3}$.

d. Barium ion has a +2 charge and fluoride has a −1 charge. The correct formula is $\mathbf{BaF_2}$.

2.79 Compounds of metals with nonmetals are usually ionic. Nonmetal-nonmetal compounds are usually molecular.

**Ionic:** LiF, $BaCl_2$, KCl

**Molecular:** $SiCl_4$, $B_2H_6$, $C_2H_4$

2.81 **Strategy:** When naming ionic compounds, our reference for the names of cations and anions are Tables 2.8 and 2.9 of the text. Keep in mind that if a metal can form cations of different charges, we need to use the Stock system. In the Stock system, Roman numerals are used to specify the charge of the cation. The metals that have only one charge in ionic compounds are the alkali metals (+1), the alkaline earth metals (+2), $Ag^+$, $Zn^{2+}$, $Cd^{2+}$, and $Al^{3+}$.

When naming acids, binary acids are named differently than oxoacids. For binary acids, the name is based on the nonmetal. For oxoacids, the name is based on the polyatomic anion. For more detail, see Section 2.7 of the text.

**Solution:** a. This is an ionic compound in which the metal cation ($K^+$) has only one charge. The correct name is **potassium dihydrogen phosphate**.

b. This is an ionic compound in which the metal cation ($K^+$) has only one charge. The correct name is **potassium hydrogen phosphate**

c. This is molecular compound. In the gas phase, the correct name is **hydrogen bromide**.

d. The correct name of this compound in water is **hydrobromic acid**.

e. This is an ionic compound in which the metal cation ($Li^+$) has only one charge. The correct name is **lithium carbonate**.

f. This is an ionic compound in which the metal cation ($K^+$) has only one charge. The correct name is **potassium dichromate**.

g. This is an ionic compound in which the cation is a polyatomic ion with a charge of +1. The anion is an oxoanion with one less O atom than the corresponding –ate ion (nitrate). The correct name is **ammonium nitrite**.

h. The oxoanion in this acid is analogous to the chlorate ion. The correct name of this compound is **hydrogen iodate (in water, iodic acid)**

i. This is a molecular compound. We use a prefix to denote how many F atoms it contains. The correct name is **phosphorus pentafluoride**.

j. This is a molecular compound. We use prefixes to denote the numbers of both types of atom. The correct name is **tetraphosphorus hexoxide**.

k. This is an ionic compound in which the metal cation ($Cd^{2+}$) has only one charge. The correct name is **cadmium iodide**.

l. This is an ionic compound in which the metal cation ($Sr^{2+}$) has only one charge. The correct name is **strontium sulfate**.

m. This is an ionic compound in which the metal cation ($Al^{3+}$) has only one charge. The correct name is **aluminum hydroxide**.

2.83 **Strategy:** When writing formulas of molecular compounds, the prefixes specify the number of each type of atom in the compound.

When writing formulas of ionic compounds, the subscript of the cation is numerically equal to the charge of the anion, and the subscript of the anion is numerically equal to the charge on the cation. If the charges of the cation and anion are numerically equal, then no subscripts are necessary. Charges of common cations and anions are listed in Tables 2.8 and 2.9 of the text. Keep in mind that Roman numerals specify the charge of the cation, *not* the number of metal atoms. Remember that a Roman numeral is not needed for some metal cations, because the charge is known. These metals are the alkali metals (+1), the alkaline earth metals (+2), $Ag^+$, $Zn^{2+}$, $Cd^{2+}$, and $Al^{3+}$.

When writing formulas of oxoacids, you must know the names and formulas of polyatomic anions (see Table 2.9 of the text).

**Solution:**

a. Rubidium is an alkali metal. It only forms a +1 cation. The polyatomic ion nitrite, $NO_2^-$, has a −1 charge. Because the charges on the cation and anion are numerically equal, the ions combine in a one-to-one ratio. The correct formula is $\mathbf{RbNO_2}$.

b. Potassium is an alkali metal. It only forms a +1 cation. The anion, sulfide, has a charge of −2. Because the charges on the cation and anion are numerically different, the subscript of the cation is numerically equal to the charge on the anion, and the subscript of the anion is numerically equal to the charge on the cation. The correct formula is $\mathbf{K_2S}$.

c. Sodium is an alkali metal. It only forms a +1 cation. The anion is the *hydrogen sulfide* ion (the sulfide ion plus one hydrogen), $HS^-$. Because the charges are numerically the same, the ions combine in a one-to-one ratio.The correct formula is **NaHS**.

d. Magnesium is an alkaline earth metal. It only forms a +2 cation. The polyatomic phosphate anion has a charge of −3, $PO_4^{3-}$. Because the charges on the cation and anion are numerically different, the subscript of the cation is numerically equal to the charge on the anion, and the subscript of the anion is numerically equal to the charge on the cation. The correct formula is $\mathbf{Mg_3(PO_4)_2}$. Note that for its subscript to be changed, a polyatomic ion must be enclosed in parentheses.

e. Calcium is an alkaline earth metal. It only forms a +2 cation. The polyatomic ion hydrogen phosphate, $HPO_4^{2-}$, has a −2 charge. Because the charges are numerically the same, the ions combine in a one-to-one ratio. The correct formula is $\mathbf{CaHPO_4}$.

f. Lead (II), $Pb^{2+}$, is a cation with a charge +2. The polyatomic ion carbonate, $CO_3^{2-}$, has a −2 charge. Because the charges on the cation and anion are numerically equal, the ions combine in a one-to-one ratio. The correct formula is $\mathbf{PbCO_3}$.

g. Tin (II), $Sn^{2+}$, is a cation with a charge of +2. The anion, fluoride, has a change of –1. Because the charges on the cation and anion are numerically different, the subscript of the cation is numerically equal to the charge on the anion, and the subscript of the anion is numerically equal to the charge on the cation. The correct formula is **$SnF_2$**.

h. The polyatomic ion ammonium, $NH_4^+$, has a +1 charge and the polyatomic ion sulfate, $SO_4^{2-}$, has a –2 charge. To balance the charge, we need 2 $NH_4^+$ cations. The correct formula is **$(NH_4)_2SO_4$**.

i. Silver forms only a +1 ion. The perchlorate ion, $ClO_4^-$, has a charge of –1. Because the charges are numerically the same, the ions combine in a one-to-one ratio. The correct formula is **$AgClO_4$**.

j. This is a molecular compound. The Greek prefixes tell you the number of each type of atom in the molecule: no prefix indicates 1 and tri- indicates 3. The correct formula is **$BCl_3$**.

2.85 a. $Mg(NO_3)_2$ b. $Al_2O_3$ b. LiH b. $Na_2S$

2.87 acid: compound that produces $H^+$; base: compound that produces $OH^-$; oxoacids: acids that contain oxygen; oxoanions: the anions that remain when oxoacids lose $H^+$ ions; hydrates: ionic solids that have water molecules in their formulas.

2.89 **(c)** Changing the electrical charge of an atom usually has a major effect on its chemical properties. The two electrically neutral carbon isotopes should have nearly identical chemical properties.

2.91 Atomic number = 127 – 74 = 53. This anion has 53 protons, so it is an iodide ion. Since there is one more electron than protons, the ion has a –1 charge. The correct symbol is **$I^-$**.

2.93 NaCl is an ionic compound; it doesn't consist of molecules.

2.95 The species and their identification are as follows:

a. $SO_2$ molecule and compound

g. $O_3$ element and molecule

b. $S_8$ element and molecule

c. Cs element

d. $N_2O_5$ molecule and compound

e. O element

f. $O_2$ element and molecule

h. $CH_4$ molecule and compound

i. KBr compound, not molecule

j. S element

k. $P_4$ element and molecule

l. LiF compound, not molecule

2.97 All masses are relative, which means that the mass of every object is compared to the mass of a standard object (such as the piece of metal in Paris called the "standard kilogram"). The mass of the standard object is determined by an international committee, and that mass is an arbitrary number to which everyone in the scientific community agrees.

Atoms are so small it is hard to compare their masses to the standard kilogram. Instead, we compare atomic masses to the mass of one specific atom. In the 19th century the atom was $^{1}H$, and for a good part of the 20th century it was $^{16}O$. Now it is $^{12}C$, which is given the arbitrary mass of 12 amu exactly. All other isotopic masses (and therefore average atomic masses) are measured relative to the assigned mass of $^{12}C$.

2.99

| Symbol | $^{11}_{5}B$ | $^{54}_{26}Fe^{2+}$ | $^{31}_{15}P^{3-}$ | $^{196}_{79}Au$ | $^{222}_{86}Rn$ |
|---|---|---|---|---|---|
| Protons | 5 | 26 | 15 | 79 | 86 |
| Neutrons | 6 | 28 | 16 | 117 | 136 |
| Electrons | 5 | 24 | 18 | 79 | 86 |
| Net Charge | 0 | +2 | −3 | 0 | 0 |

2.101 a. $Li^+$, alkali metals always have a +1 charge in ionic compounds

b. $S^{2-}$

c. $I^-$, halogens have a −1 charge in ionic compounds

d. $N^{3-}$

e. $Al^{3+}$, aluminum always has a +3 charge in ionic compounds

f. $Cs^{+}$, alkali metals always have a +1 charge in ionic compounds

g. $Mg^{2+}$, alkaline earth metals always have a +2 charge in ionic compounds.

2.103 The binary Group 7A element acids are: HF, hydrofluoric acid; HCl, hydrochloric acid; HBr, hydrobromic acid; HI, hydroiodic acid. Oxoacids containing Group 7A elements (using the specific examples for chlorine) are: $HClO_4$, perchloric acid; $HClO_3$, chloric acid; $HClO_2$, chlorous acid: HClO, hypochlorous acid.

Examples of oxoacids containing other Group A-block elements are: $H_3BO_3$, boric acid (Group 3A); $H_2CO_3$, carbonic acid (Group 4A); $HNO_3$, nitric acid and $H_3PO_4$, phosphoric acid (Group 5A); and $H_2SO_4$, sulfuric acid (Group 6A). Hydrosulfuric acid, $H_2S$, is an example of a binary Group 6A acid while HCN, hydrocyanic acid, contains both a Group 4A and 5A element.

2.105 a.

| Isotope | ${}^{4}_{2}He$ | ${}^{20}_{10}Ne$ | ${}^{40}_{18}Ar$ | ${}^{84}_{36}Kr$ | ${}^{132}_{54}Xe$ |
|---|---|---|---|---|---|
| No. Protons | 2 | 10 | 18 | 36 | 54 |
| No. Neutrons | 2 | 10 | 22 | 48 | 78 |

b.

| neutron/proton ratio | 1.00 | 1.00 | 1.22 | 1.33 | 1.44 |
|---|---|---|---|---|---|

The neutron/proton ratio increases with increasing atomic number.

2.107 Cu, Ag, and Au are fairly chemically unreactive. This makes them especially suitable for making coins and jewelry that you want to last a very long time.

2.109 Magnesium and strontium are also alkaline earth metals. You should expect the charge of the metal to be the same (+2). **MgO** and **SrO**.

2.111 a. $\dfrac{2\text{ red} : 1\text{ blue}}{1\text{ red} : 1\text{ blue}} = \mathbf{2:1}$

b. $\dfrac{1\text{ red} : 2\text{ blue}}{1\text{ red} : 1\text{ blue}} = \mathbf{1:2}$

c. $\dfrac{4\text{ red} : 2\text{ blue}}{1\text{ red} : 1\text{ blue}} = 4:2 = \mathbf{2:1}$

d. $\dfrac{5\text{ red} : 2\text{ blue}}{1\text{ red} : 1\text{ blue}} = \mathbf{5:2}$

2.113 The mass of fluorine reacting with hydrogen and deuterium would be the same. The ratio of F atoms to hydrogen (or deuterium) atoms is 1:1 in both compounds. This does not violate the law of definite proportions. When the law of definite proportions was formulated, scientists did not know of the existence of isotopes.

2.115 a. Br a. Rn a. Se a. Rb a. Pb

2.117

| Cation | Anion | Formula | Name |
|---|---|---|---|
| $\mathbf{Mg^{2+}}$ | $\mathbf{HCO_3^-}$ | $\mathbf{Mg(HCO_3)_2}$ | Magnesium bicarbonate |
| $\mathbf{Sr^{2+}}$ | $\mathbf{Cl^-}$ | $SrCl_2$ | **Strontium chloride** |
| $Fe^{3+}$ | $NO_2^-$ | $\mathbf{Fe(NO_2)_3}$ | **Iron(III) nitrite** |
| $\mathbf{Mn^{2+}}$ | $\mathbf{ClO_3^-}$ | $\mathbf{Mn(ClO_3)_2}$ | Manganese(II) chlorate |
| $\mathbf{Sn^{4+}}$ | $\mathbf{Br^-}$ | $SnBr_4$ | **Tin(IV) bromide** |
| $Co^{2+}$ | $PO_4^{3-}$ | $\mathbf{Co_3(PO_4)_2}$ | **Cobalt(II) phosphate** |
| $Hg_2^{2+}$ | $I^-$ | $\mathbf{Hg_2I_2}$ | **Mercury(I) iodide** |
| $\mathbf{Cu^+}$ | $\mathbf{CO_3^{2-}}$ | $Cu_2CO_3$ | **Copper(I) carbonate** |
| $\mathbf{Li^+}$ | $\mathbf{N^{3-}}$ | $\mathbf{Li_3N}$ | Lithium nitride |
| $Al^{3+}$ | $S^{2-}$ | $\mathbf{Al_2S_3}$ | **Aluminum sulfide** |

2.119 The change in energy is equal to the energy released. We call this $\Delta E$. Similarly, $\Delta m$ is the change in mass.

Because $m = \frac{E}{c^2}$, we have

$$\Delta m = \frac{\Delta E}{c^2} = \frac{(1.715\times10^3\ \text{kJ})\left(\frac{1000\ \text{J}}{1\ \text{kJ}}\right)}{(2.998\times10^8\ \text{m/s})^2} = 1.908\times10^{-11}\ \text{kg} = \mathbf{1.908\times10^{-8}\ g}$$

Note that we need to convert kJ to J so that we end up with units of kg for the mass. $\left(1\ \text{J} = \frac{1\ \text{kg}\cdot\text{m}^2}{\text{s}^2}\right)$

We can add together the masses of hydrogen and oxygen to calculate the mass of water that should be formed.

$$12.096\ \text{g} + 96.000 = 108.096\ \text{g}$$

The predicted change (loss) in mass is only $1.908 \times 10^{-8}$ g which is too small a quantity to measure. Therefore, for all practical purposes, the law of conservation of mass is assumed to hold for ordinary chemical processes.

2.121 The acids, from left to right, are chloric acid, nitrous acid, hydrocyanic acid, and sulfuric acid.

2.123 a.

| Ethane | Acetylene |
|---|---|
| 2.65 g C | 4.56 g C |
| 0.665 g H | 0.383 g H |

Let's compare the ratio of the hydrogen masses in the two compounds. To do this, we need to start with the same mass of carbon. If we were to start with 4.56 g of C in ethane, how much hydrogen would combine with 4.56 g of carbon?

$$0.665\ \text{g H} \times \frac{4.56\ \text{g C}}{2.65\ \text{g C}} = 1.14\ \text{g H}$$

We can calculate the ratio of H in the two compounds.

$$\frac{1.14\ \text{g}}{0.383\ \text{g}} \approx 3$$

This is consistent with the Law of Multiple Proportions which states that if two elements combine to form more than one compound, the masses of one element that combine with a fixed mass of the other element are in ratios of small whole numbers. In this case, the ratio of the masses of hydrogen in the two compounds is 3:1.

b. For a given amount of carbon, there is 3 times the amount of hydrogen in ethane compared to acetylene. Reasonable formulas would be:

| Ethane | Acetylene |
|---|---|
| $CH_3$ | CH |
| $C_2H_6$ | $C_2H_2$ |

2.125 a. Assume that the nucleons (protons and neutrons) are hard objects of fixed size. Then the volume of the nucleus is well-approximated by the direct proportion $V = kA$, where $A$ is the number of nucleons (mass number of the atom). For a spherical nucleus, then $V = kA = \frac{4}{3}\pi r^3$. Solving for $r$:

$$kA = \frac{4}{3}\pi r^3$$

$$\left(\frac{3}{4\pi}\right)kA = r^3$$

$$\left[\left(\frac{3}{4\pi}\right)kA\right]^{1/3} = r$$

$$\left[\left(\frac{3}{4\pi}\right)k\right]^{1/3}\left(A^{1/3}\right) = r$$

$$cA^{1/3} = r \quad (c \text{ is a constant})$$

b. For the volume calculation, use lithium-7 ($A = 7$).

$$V = \frac{4}{3}\pi r^3 = \frac{4}{3}\pi\left(r_0 A^{1/3}\right)^3 = \left(\frac{4}{3}\pi r_0^3\right)(A) = \left[\frac{4}{3}\pi\left(1.2\times10^{-15}\text{ m}\right)^3\right](7) \approx 5.1\times10^{-44}\text{ m}^3$$

c. Use $r = 152$ pm $= 152 \times 10^{-12}$ m for the atomic radius. Then, the atomic volume of lithium-7 is:

$$V = \frac{4}{3}\pi r^3 = \frac{4}{3}\pi\left(152\times10^{-12}\text{ m}\right)^3 \approx 1.5\times10^{-29}\text{ m}^3$$

The fraction of the atomic radius occupied by the nucleus is $\frac{5.1\times10^{-44}}{1.5\times10^{-29}} \approx 3.4\times10^{-15}$. This is consistent with Rutherford's discovery that the nucleus occupies a very small region within the atom.

# Chapter 3

# Stoichiometry: Ratios of Combination

3.3 Add the masses of all atoms in the formula. Remember that the *absence* of a subscript means that there is *one* atom of that element present.

a. $CH_3Cl$ 1(12.01 amu) + 3(1.008 amu) + 1(35.45 amu) = **50.48 amu**

b. $N_2O_4$ 2(14.01 amu) + 4(16.00 amu) = **92.02 amu**

c. $SO_2$ 1(32.07 amu) + 2(16.00 amu) = **64.07 amu**

d. $C_6H_{12}$ 6(12.01 amu) + 12(1.008 amu) = **84.16 amu**

e. $H_2O_2$ 2(1.008 amu) + 2(16.00 amu) = **34.02 amu**

f. $C_{12}H_{22}O_{11}$ 12(12.01 amu) + 22(1.008 amu) + 11(16.00 amu) = **342.3 amu**

g. $NH_3$ 1(14.01 amu) + 3(1.008 amu) = **17.03 amu**

3.5 Using the appropriate atomic masses,

a. $CH_4$ 1(12.01 amu) + 4(1.008 amu) = **16.04 amu**

b. $NO_2$ 1(14.01 amu) + 2(16.00 amu) = **46.01 amu**

c. $SO_3$ 1(32.07 amu) + 3(16.00 amu) = **80.07 amu**

d. $C_6H_6$ 6(12.01 amu) + 6(1.008 amu) = **78.11 amu**

e. NaI 1(22.99 amu) + 1(126.9 amu) = **149.9 amu**

f. $K_2SO_4$ $2(39.10\text{ amu}) + 1(32.07\text{ amu}) + 4(16.00\text{ amu}) = \mathbf{174.27\text{ amu}}$

g. $Ca_3(PO_4)_2$ $3(40.08\text{ amu}) + 2(30.97\text{ amu}) + 8(16.00\text{ amu}) = \mathbf{310.2\text{ amu}}$

3.9 **Strategy:** Recall the procedure for calculating a percentage. Assume that we have 1 mole of $SnO_2$. The percent by mass of each element (Sn and O) is given by the mass of that element in 1 mole of $SnO_2$ divided by the molar mass of $SnO_2$, then multiplied by 100 to convert from a fractional number to a percentage.

**Solution:** Molar mass of $SnO_2 = (118.7\text{ g}) + 2(16.00\text{ g}) = 150.7\text{ g}$

$$\mathbf{\%Sn} = \frac{118.7\text{ g/mol}}{150.7\text{ g/mol}} \times 100\% = \mathbf{78.77\%}$$

$$\mathbf{\%O} = \frac{(2)(16.00\text{ g/mol})}{150.7\text{ g/mol}} \times 100\% = \mathbf{21.23\%}$$

**Think About It:** Do the percentages add to 100%? The sum of the percentages is (78.77% + 21.23%) = 100.00%.

3.11

| | Compound | Molar mass (g) | N% by mass |
|---|---|---|---|
| a. | $(NH_2)_2CO$ | 60.06 | $\frac{2(14.01\text{ g})}{60.06\text{ g}} \times 100\% = 46.65\%$ |
| b. | $NH_4NO_3$ | 80.05 | $\frac{2(14.01\text{ g})}{80.05\text{ g}} \times 100\% = 35.00\%$ |
| c. | $HNC(NH_2)_2$ | 59.08 | $\frac{3(14.01\text{ g})}{59.08\text{ g}} \times 100\% = 71.14\%$ |
| d. | $NH_3$ | 17.03 | $\frac{14.01\text{ g}}{17.03\text{ g}} \times 100\% = 82.27\%$ |

3.13 In a formula this complicated, it is easy to count the atoms incorrectly, so rewrite the formula by gathering atoms of the same kind together: $Ca_5(PO_4)_3(OH) = Ca_5(P_3O_{12})(OH) = Ca_5P_3O_{13}H$

Formula mass of $Ca_5P_3O_{13}H$ = 5(40.08) + 3(30.97) + 13(16.00) + (1.008) = 502.32 amu

$$\% \text{ Ca} = \frac{(5)\times(40.08 \text{ amu})}{(502.32 \text{ amu})}\times(100) = \mathbf{39.89\%}$$

$$\% \text{ P} = \frac{(3)\times(30.97 \text{ amu})}{(502.32 \text{ amu})}\times(100) = \mathbf{18.50\%}$$

$$\% \text{ O} = \frac{(13)\times(16.00 \text{ amu})}{(502.32 \text{ amu})}\times(100) = \mathbf{41.41\%}$$

$$\% \text{ H} = \frac{(1)\times(1.008 \text{ amu})}{(502.32 \text{ amu})}\times(100) = \mathbf{0.20\%}$$

**Check**: 39.89 + 18.50 + 41.41 + 0.20 = 100.0

Note that in problems such as this, rounding error may result in the final sum of percentages not being exactly equal to 100.

3.15 a. There are 3 vegetable noodle packs in each variety pack.

$$20 \text{ variety packs} \times \frac{3 \text{ vegetable noodle packs}}{\text{variety pack}} = 60 \text{ vegetable noodle packs}$$

$$4.667 \text{ variety packs} \times \frac{3 \text{ vegetable noodle packs}}{\text{variety pack}} = 14 \text{ vegetable noodle packs}$$

$$0.25 \text{ variety packs} \times \frac{3 \text{ vegetable noodle packs}}{\text{variety pack}} = 0.75 \text{ vegetable noodle packs}$$

b. There are 3 beef noodle packs in each variety pack.

$$72 \text{ beef noodle packs} \times \frac{1 \text{ variety pack}}{3 \text{ beef noodle packs}} = 24 \text{ variety packs}$$

$$3 \text{ beef noodle packs} \times \frac{1 \text{ variety pack}}{3 \text{ beef noodle packs}} = 1 \text{ variety pack}$$

$$10 \text{ beef noodle packs} \times \frac{1 \text{ variety pack}}{3 \text{ beef noodle packs}} = 3.333 \text{ variety packs}$$

c. Each variety pack contains 6 chicken noodle packs, 3 beef noodle packs, and 3 vegetable noodle packs.

$$30 \text{ chicken noodle packs} \times \frac{3 \text{ vegetable noodle packs}}{6 \text{ chicken noodle packs}} = 15 \text{ vegetable noodle packs}$$

$$2 \text{ chicken noodle packs} \times \frac{3 \text{ vegetable noodle packs}}{6 \text{ chicken noodle packs}} = 1 \text{ vegetable noodle pack}$$

$$25 \text{ beef noodle packs} \times \frac{3 \text{ vegetable noodle packs}}{3 \text{ beef noodle packs}} = 25 \text{ vegetable noodle pack}$$

3.21 **Strategy:** (1) translate each compound name into the correct chemical formula;

(2) translate "and" into "+"; this applies to compound phrases separated by commas;

(3) translate "react to form" into a reaction arrow "→"

Translation of a compound name into the correct chemical formula is the most difficult part of the solution. If the compound name is a **systematic name**, it will have Greek prefixes such as "di-" (2) which tell you what the subscript of a particular element is in the formula. However, in most cases the name used is a **common name** and you must simply look up (or learn and remember) the name/formula translation.

**Solution:** a. $KOH + H_3PO_4 \rightarrow K_3PO_4 + H_2O$

b. $Zn + AgCl \rightarrow ZnCl_2 + Ag$

c. $NaHCO_3 \rightarrow Na_2CO_3 + H_2O + CO_2$

d. $NH_4NO_2 \rightarrow N_2 + H_2O$

e. $CO_2 + KOH \rightarrow K_2CO_3 + H_2O$

f. $3KOH + H_3PO_4 \rightarrow K_3PO_4 + 3H_2O$

$Zn + 2AgCl \rightarrow ZnCl_2 + 2Ag$

$2NaHCO_3 \rightarrow Na_2CO_3 + H_2O + CO_2$

$NH_4NO_2 \rightarrow N_2 + 2H_2O$

$CO_2 + 2KOH \rightarrow K_2CO_3 + H_2O$

3.23 **Strategy:** (1) translate each chemical formula into the correct compound name;

(2) translate "+" into "and";

(3) translate the reaction arrow "→" into the phrase "react to form".

Translation of a chemical formula into the correct compound name is the most difficult part of the solution. Watch your spelling! In many cases you must simply look up (or learn and remember) the formula/name translation. In general, these formulas and their corresponding names are in your text.

**Solution:** a. potassium and water react to form potassium hydroxide and hydrogen.

b. barium hydroxide and hydrochloric acid react to form barium chloride and water.

c. copper and nitric acid react to form copper nitrate, nitrogen monoxide and water.

d. aluminum and sulfuric acid react to form aluminum sulfate and hydrogen.

e. hydrogen iodide reacts to form hydrogen and iodine.

3.25 **Strategy:** The goal is to have the same number of atoms of a given element on both sides of the equation. You do this by adjusting the **stoichiometric coefficients** in front of each formula (remember that if no coefficient is written, it is understood to be "1"). Remember also that you **cannot** change the **subscripts** in the formulas!

Everybody balances chemical equations the same way: we **guess** what the coefficients must be. Sometimes there are clues about what the numbers must be (see equation (a) below). Don't be afraid to guess wrong - just erase (or better yet, cross out) the wrong answer and try again. Each time you guess, count all of the atoms again very carefully. There is no "one way" or "best way" to balance chemical equations, but with practice, you will learn how to do it - and make it look as easy as your instructor does. Remember – your instructor is guessing too, but is also probably doing a lot of mental arithmetic along the way!

**Solution:** a. $N_2O_5 \rightarrow N_2O_4 + O_2$

First, count **carefully** the atoms of each element on both sides of the reaction arrow:

$$N_2O_5 \rightarrow N_2O_4 + O_2$$

2 – N – 2

5 – O – 6

In order to maintain the nitrogen atom balance, the coefficients of $N_2O_5$and $N_2O_4$ must be the same (we will leave them at 1 and 1 for now). By using only half of the oxygen molecule we can write a balanced equation:

$$N_2O_5 \rightarrow N_2O_4 + \frac{1}{2}O_2$$

2 – N – 2

5 – O – 5

Although this equation is balanced, very often we are asked to balance chemical equations with the smallest whole number stoichiometric coefficients. To clear the fraction, multiply all three coefficients by two (the denominator of the fraction):

$$\mathbf{2\,N_2O_5 \rightarrow 2N_2O_4 + O_2}$$

4 – N – 4

10 – O – 10

b. $2KNO_3 \rightarrow 2KNO_2 + O_2$

c. $NH_4NO_3 \rightarrow N_2O + 2H_2O$

d. $NH_4NO_2 \rightarrow N_2 + 2H_2O$

e. $2NaHCO_3 \rightarrow Na_2CO_3 + H_2O + CO_2$

f. $P_4O_{10} + 6H_2O \rightarrow 4H_3PO_4$

g. $2HCl + CaCO_3 \rightarrow CaCl_2 + H_2O + CO_2$

h. $2Al + 3H_2SO_4 \rightarrow Al_2(SO_4)_3 + 3H_2$

i. $CO_2 + 2KOH \rightarrow K_2CO_3 + H_2O$

j. $CH_4 + 2O_2 \rightarrow CO_2 + 2H_2O$

k. $Be_2C + 4H_2O \rightarrow 2Be(OH)_2 + CH_4$

l. $3Cu + 8HNO_3 \rightarrow 3Cu(NO_3)_2 + 2NO + 4H_2O$

m. $S + 6HNO_3 \rightarrow H_2SO_4 + 6NO_2 + 2H_2O$

n. $2NH_3 + 3CuO \rightarrow 3Cu + N_2 + 3H_2O$

3.27 On the reactants side there are 6 A atoms and 4 B atoms. On the products side, there are 4 C atoms and 2 D atoms. Writing an equation,

$$6A + 4B \rightarrow 4C + 2D$$

Chemical equations are typically written with the smallest set of whole number coefficients. Dividing the equation by two gives,

$$\mathbf{3A + 2B \rightarrow 2C + D}$$

The correct answer is choice **(d)**.

3.33 The thickness of the book in miles would be:

$$\frac{0.0036\text{ in}}{1\text{ page}} \times \frac{1\text{ ft}}{12\text{ in}} \times \frac{1\text{ mi}}{5280\text{ ft}} \times (6.022 \times 10^{23}\text{ pages}) = 3.42 \times 10^{16}\text{ mi}$$

The distance, in miles, traveled by light in one year is:

$$1\text{ yr} \times \frac{365\text{ day}}{1\text{ yr}} \times \frac{24\text{ h}}{1\text{ day}} \times \frac{3600\text{ s}}{1\text{ h}} \times \frac{3.00 \times 10^8\text{ m}}{1\text{ s}} \times \frac{1\text{ mi}}{1609\text{ m}} = 5.88 \times 10^{12}\text{ mi}$$

The thickness of the book in light-years is:

$$(3.42 \times 10^{16}\text{ mi}) \times \frac{1\text{ light-yr}}{5.88 \times 10^{12}\text{ mi}} = \mathbf{5.8 \times 10^3\text{ light-yr}}$$

It will take light $5.8 \times 10^3$ years to travel from the first page to the last one!

3.35

$$(6.00 \times 10^9\text{ Co atoms}) \times \frac{1\text{ mol Co}}{6.022 \times 10^{23}\text{ Co atoms}} = \mathbf{9.96 \times 10^{-15}\text{ mol Co}}$$

3.37 **Strategy:** We are given moles of gold and asked to solve for grams of gold. What conversion factor do we need to convert between moles and grams? Arrange the appropriate conversion factor so moles cancel, and the unit grams is obtained for the answer.

**Solution:** The conversion factor needed to covert between moles and grams is the molar mass. In the periodic table (see inside front cover of the text), we see that the molar mass of Au is 197.0 g. This can be expressed as

$$1 \text{ mol Au} = 197.0 \text{ g Au}$$

From this equality, we can write two conversion factors.

$$\frac{1 \text{ mol Au}}{197.0 \text{ g Au}} \quad \text{and} \quad \frac{197.0 \text{ g Au}}{1 \text{ mol Au}}$$

The conversion factor on the right is the correct one. Moles will cancel, leaving the unit grams for the answer.

We write

$$\textbf{? g Au} = 15.3 \text{ mol Au} \times \frac{197.0 \text{ g Au}}{1 \text{ mol Au}} = \mathbf{3.01 \times 10^3 \text{ g Au}}$$

**Think About It:** Does a mass of 3010 g for 15.3 moles of Au seem reasonable? What is the mass of 1 mole of Au?

3.39 a. **Strategy:** We can look up the molar mass of silicon (Si) on the periodic table (28.09 g/mol). We want to find the mass of a single atom of silicon (unit of g/atom). Therefore, we need to convert from the unit mole in the denominator to the unit atom in the denominator. What conversion factor is needed to convert between moles and atoms? Arrange the appropriate conversion factor so mole in the denominator cancels, and the unit atom is obtained in the denominator.

**Setup:** The conversion factor needed is Avogadro's number. We have:

$$1 \text{ mol} = 6.022 \times 10^{23} \text{ particles (atoms)}$$

From this equality, we can write two conversion factors:

$$\frac{1 \text{ mol Si}}{6.022 \times 10^{23} \text{ Si atoms}} \quad \text{and} \quad \frac{6.022 \times 10^{23} \text{ Si atoms}}{1 \text{ mol Si}}$$

The conversion factor on the left is the correct one. Moles will cancel, leaving the unit atoms in the denominator of the answer.

**Solution:** We write:

$$\textbf{? g / Si atom} = \frac{28.09\text{ g Si}}{1\text{ mol Si}} \times \frac{1\text{ mol Si}}{6.022 \times 10^{23}\text{ Si atoms}} = \mathbf{4.665 \times 10^{-23}\ g / Si\ atom}$$

b. Follow same method as part (a).

$$\textbf{? g / Fe atom} = \frac{55.85\text{ g Fe}}{1\text{ mol Fe}} \times \frac{1\text{ mol Fe}}{6.022 \times 10^{23}\text{ Fe atoms}} = \mathbf{9.274 \times 10^{-23}\ g / Fe\ atom}$$

**Strategy:** Should the mass of a single atom of Si or Fe be a very small mass?

3.41 **Strategy:** The question asks for atoms of Cu. We cannot convert directly from grams to atoms of copper. What unit do we need to convert grams of Cu to in order to convert to atoms? What does Avogadro's number represent?

**Setup:** To calculate the number of Cu atoms, we first must convert grams of Cu to moles of Cu. We use the molar mass of copper as a conversion factor. Once moles of Cu are obtained, we can use Avogadro's number to convert from moles of copper to atoms of copper.

$$1\text{ mol Cu} = 63.55\text{ g Cu}$$

The conversion factor needed is

$$\frac{1\text{ mol Cu}}{63.55\text{ g Cu}}$$

Avogadro's number is the key to the second conversion. We have

$$1\text{ mol} = 6.022 \times 10^{23}\text{ particles (atoms)}$$

From this equality, we can write two conversion factors.

$$\frac{1\text{ mol Cu}}{6.022 \times 10^{23}\text{ Cu atoms}} \quad \text{and} \quad \frac{6.022 \times 10^{23}\text{ Cu atoms}}{1\text{ mol Cu}}$$

The conversion factor on the right is the one we need because it has number of Cu atoms in the numerator, which is the unit we want for the answer.

**Solution:** Let's complete the two conversions in one step.

grams of Cu $\rightarrow$ moles of Cu $\rightarrow$ number of Cu atoms

$$\textbf{? atoms of Cu} = 25.85 \text{ g Cu} \times \frac{1 \text{ mol Cu}}{63.55 \text{ g Cu}} \times \frac{6.022 \times 10^{23} \text{ Cu atoms}}{1 \text{ mol Cu}}$$

$$= \mathbf{2.450 \times 10^{23} \text{ Cu atoms}}$$

**Think About It:** Should 25.85 g of Cu contain fewer than Avogadro's number of atoms? What mass of Cu would contain Avogadro's number of atoms?

3.43

$$2 \text{ Pb atoms} \times \frac{1 \text{ mol Pb}}{6.022 \times 10^{23} \text{ Pb atoms}} \times \frac{207.2 \text{ g Pb}}{1 \text{ mol Pb}} = 6.881 \times 10^{-22} \text{ g Pb}$$

$$(5.1 \times 10^{-23} \text{ mol He}) \times \frac{4.003 \text{ g He}}{1 \text{ mol He}} = 2.0 \times 10^{-22} \text{ g He}$$

**2 atoms of lead** have a greater mass than $5.1 \times 10^{-23}$ mol of helium.

3.45 To find the molar mass (g/mol), we simply divide the mass (in g) by the number of moles.

$$\frac{152 \text{ g}}{0.372 \text{ mol}} = \mathbf{409 \text{ g / mol}}$$

3.47 **Strategy:** We are given grams of glucose and asked to solve for atoms of C, H, and O. We cannot convert directly from grams glucose to atoms of each element. What conversions must we do to convert from grams of a compound to atoms of its constituent elements? How should Avogadro's number be used here?

**Setup:** To calculate number of atoms, we first must convert grams of glucose to moles of glucose. We use the molar mass of glucose as a conversion factor. Once moles of glucose are obtained, we can use Avogadro's number to convert from moles of glucose to molecules of glucose. From molecules of glucose we can determine atoms of individual elements using the molecular formula of glucose.

$$\text{molar mass of } C_6H_{12}O_6 = 6(12.01 \text{ g}) + 12(1.008 \text{ g}) + 6(16.00 \text{ g}) = 180.156 \text{ g}$$

The conversion factor needed is

$$\frac{1 \text{ mol } C_6H_{12}O_6}{180.2 \text{ g } C_6H_{12}O_6}$$

Avogadro's number is the key to the second conversion. We have

$$1 \text{ mol} = 6.022 \times 10^{23} \text{ particles (molecules)}$$

From this equality, we can write the conversion factor:

$$\frac{6.022\times10^{23} \text{ molecules } C_6H_{12}O_6}{1 \text{ mol } C_6H_{12}O_6}$$

The subscript for each element in the molecular formula is the key to the third conversion.

**Solution:** Let's complete these three conversions in one step to determine the number of C atoms.

grams of glucose $\rightarrow$ moles of glucose $\rightarrow$ number of glucose molecules $\rightarrow$ number of carbon atoms

$$1.50 \text{ g glucose} \times \frac{1 \text{ mol glucose}}{180.2 \text{ g glucose}} \times \frac{6.022 \times 10^{23} \text{ molecules glucose}}{1 \text{ mol glucose}} \times \frac{6 \text{ C atoms}}{1 \text{ molecule glucose}}$$

$$= \mathbf{3.01 \times 10^{22} \text{ C atoms}}$$

The ratio of H atoms to C atoms in glucose is 2:1. Therefore, there are twice as many H atoms in glucose as C atoms, so the number of H atoms = $2(3.01 \times 10^{22} \text{ atoms})$ = **$6.02 \times 10^{22}$ H atoms.**

The ratio of O atoms to C atoms in glucose is 1:1. Therefore, there are the same number of O atoms in glucose as C atoms, so the number of O atoms = **$3.01 \times 10^{22}$ O atoms.**

**Think About It:** Should 1.50 g of glucose contain fewer than Avogadro's number of C, H, and O atoms?

3.49 Using appropriate conversion factors we convert:

$$\text{g of Hg} \rightarrow \text{mol Hg} \rightarrow \text{mol S} \rightarrow \text{g S}$$

$$\mathbf{?\ g\ S} = 246 \text{ g Hg} \times \frac{1 \text{ mol Hg}}{200.6 \text{ g Hg}} \times \frac{1 \text{ mol S}}{1 \text{ mol Hg}} \times \frac{32.07 \text{ g S}}{1 \text{ mol S}} = \mathbf{39.3\ g\ S}$$

3.51 **Strategy:** Tin(II) fluoride is composed of Sn and F. The mass due to F is based on its percentage by mass in the compound. How do we calculate mass percent of an element?

**Solution:** First, we must find the mass % of fluorine in $SnF_2$. Then, we convert this percentage to a fraction and multiply by the mass of the compound (24.6 g), to find the mass of fluorine in 24.6 g of $SnF_2$.

The percent by mass of fluorine in tin(II) fluoride, is calculated as follows:

$$\text{mass \% F} = \frac{\text{mass of F in 1 mol SnF}_2}{\text{molar mass of SnF}_2} \times 100\%$$

$$= \frac{2(19.00 \text{ g})}{156.7 \text{ g}} \times 100\% = 24.25\% \text{ F}$$

Converting this percentage to a fraction, we obtain 24.25/100 = 0.2425.

Next, multiply the fraction by the total mass of the compound.

$$\mathbf{? \text{ g F in } 24.6 \text{ g SnF}_2} = (0.2425)(24.6 \text{ g}) = \mathbf{5.97 \text{ g F}}$$

**Think About It:** As a ball-park estimate, note that the mass percent of F is roughly 25 percent, so that a quarter of the mass should be F. One quarter of approximately 24 g is 6 g, which is close to the answer.

3.53 a. **Strategy:** In a chemical formula, the subscripts represent the ratio of the number of moles of each element that combine to form the compound. Therefore, we need to convert from mass percent to moles in order to determine the empirical formula. If we assume an exactly 100 g sample of the compound, do we know the mass of each element in the compound? How do we then convert from grams to moles?

**Solution:** If we have 100 g of the compound, then each percentage can be converted directly to grams. In this sample, there will be 40.1 g of C, 6.6 g of H, and 53.3 g of O. Because the subscripts in the formula represent a mole ratio, we need to convert the grams of each element to moles. The conversion factor needed is the molar mass of each element. Let $n$ represent the number of moles of each element so that

$$n_C = 40.1 \text{ g C} \times \frac{1 \text{ mol C}}{12.01 \text{ g C}} = 3.339 \text{ mol C}$$

$$n_H = 6.6 \text{ g H} \times \frac{1 \text{ mol H}}{1.008 \text{ g H}} = 6.55 \text{ mol H}$$

$$n_O = 53.3 \text{ g O} \times \frac{1 \text{ mol O}}{16.00 \text{ g O}} = 3.331 \text{ mol O}$$

Thus, we arrive at the formula $C_{3.339}H_{6.55}O_{3.331}$, which gives the identity and the mole ratios of atoms present. However, chemical formulas are written with whole numbers. Try to convert to whole numbers by dividing all the subscripts by the smallest subscript (3.331).

$$\textbf{C:} \ \frac{3.339}{3.331} \approx 1 \qquad \textbf{H:} \ \frac{6.55}{3.331} \approx 2 \qquad \textbf{O:} \ \frac{3.331}{3.331} = 1$$

This gives the empirical formula, $\mathbf{CH_2O}$.

**Think About It:** Are the subscripts in $CH_2O$ reduced to the smallest whole numbers?

b. Following the same procedure as part (a), we find:

$$n_C = 18.4 \text{ g C} \times \frac{1 \text{ mol C}}{12.01 \text{ g C}} = 1.532 \text{ mol C}$$

$$n_N = 21.5 \text{ g N} \times \frac{1 \text{ mol N}}{14.01 \text{ g N}} = 1.535 \text{ mol N}$$

$$n_K = 60.1 \text{ g K} \times \frac{1 \text{ mol K}}{39.10 \text{ g K}} = 1.537 \text{ mol K}$$

Dividing by the smallest number of moles (1.532 mol) gives the empirical formula, **KCN**.

3.55 Find the molar mass corresponding to each formula.

For $C_4H_5N_2O$: $4(12.01 \text{ g}) + 5(1.008 \text{ g}) + 2(14.01 \text{ g}) + (16.00 \text{ g}) = 97.10 \text{ g}$

For $C_8H_{10}N_4O_2$: $8(12.01 \text{ g}) + 10(1.008 \text{ g}) + 4(14.01 \text{ g}) + 2(16.00 \text{ g}) = 194.20 \text{ g}$

The molecular formula is $\mathbf{C_8H_{10}N_4O_2}$.

3.57 First, calculate the number of grams of arsenic(VI) oxide corresponding to the $LD_{50}$, noting that the units of $LD_{50}$ are g/kg (grams of poison per kg of body weight). The formula for arsenic(VI) oxide is $AsO_3$.

$$? \text{ g } AsO_3 = 184 \text{ lb} \times \frac{1 \text{ kg}}{2.20 \text{ lb}} \times \frac{0.015 \text{ g } AsO_3}{1 \text{ kg}} = \mathbf{1.25 \text{ g } AsO_3}$$

Next, convert grams to number of molecules using the molar mass of $AsO_3$ and Avogadro's number.

$$? \text{ } AsO_3 = 1.25 \text{ g} \times \frac{\text{mol } AsO_3}{122.92 \text{ g}} \times \frac{6.022 \times 10^{23} \text{ } AsO_3 \text{ molecules}}{\text{mol } AsO_3} = \mathbf{6.12 \times 10^{21} \text{ molecules}}$$

3.59 Molecular formula: $C_9H_{16}O_4$

Empirical formula: $C_9H_{16}O_4$

$$\% \text{ C} = \frac{9 \text{ mol C} \times \left(\frac{12.01 \text{ g C}}{\text{mol C}}\right)}{1 \text{ mol } C_9H_{16}O_4 \times \left(\frac{188.22 \text{ g } C_9H_{16}O_4}{\text{mol } C_9H_{16}O_4}\right)} = 57.43\% \text{ C}$$

$$\% \text{ H} = \frac{16 \text{ mol H} \times \left(\frac{1.008 \text{ g H}}{\text{mol H}}\right)}{1 \text{ mol } C_9H_{16}O_4 \times \left(\frac{188.22 \text{ g } C_9H_{16}O_4}{\text{mol } C_9H_{16}O_4}\right)} = 8.57\% \text{ H}$$

$$\% \text{ O} = \frac{4 \text{ mol O} \times \left(\frac{16.00 \text{ g O}}{\text{mol O}}\right)}{1 \text{ mol } C_9H_{16}O_4 \times \left(\frac{188.22 \text{ g } C_9H_{16}O_4}{\text{mol } C_9H_{16}O_4}\right)} = 34.00\% \text{ O}$$

3.63 The process of combustion analysis involves the following steps:

(1) Convert the mass of $CO_2$ to moles of $CO_2$ - this is the same as the number of moles of C in the sample because every C atom in the sample becomes the C atom in a $CO_2$ molecule;

(2) Convert moles of C to grams of C;

(3) Convert the mass of $H_2O$ to moles of $H_2O$. Twice this number is the number of moles of H in the sample (1 mol $H_2O$ = 2 mol H) because for every *two* H atoms in the sample, *one* water molecule is produced;

(4) Convert moles of H to grams of H;

(5) Subtract the combined mass of C and H from the sample mass; if the result is zero, then only C and H are present in the sample; if the result is greater than zero, then this is the mass of the other element in the sample (in this example, oxygen); convert the mass of this element to moles;

(6) The mole ratio leads to the empirical formula as in percent composition problems.

$$? \text{ mol C} = 28.16 \text{ mg } CO_2 \times \frac{1 \text{g}}{1000 \text{mg}} \times \frac{1 \text{ mol } CO_2}{44.01 \text{ g } CO_2} \times \frac{1 \text{ mol C}}{1 \text{ mol } CO_2} = 0.0006399 \text{ mol C}$$

and

$$0.0006399 \text{ mol C} \times \frac{12.01 \text{g C}}{1 \text{mol C}} = 0.007685 \text{ g C} = 7.685 \text{ mg C}$$

$$? \text{ mol H} = 11.53 \text{ mg } H_2O \ \frac{1 \text{g}}{1000 \text{ mg}} \times \frac{1 \text{mol } H_2O}{18.02 \text{g } H_2O} \times \frac{2 \text{mol H}}{1 \text{mol } H_2O} = 0.001280 \text{ mol H}$$

and

$$0.001280 \text{ mol H} \times \frac{1.008 \text{ g H}}{1 \text{ mol H}} = 0.001290 \text{ g H} = 1.290 \text{ mg H}$$

$$\text{mg O} = 10.00 - (7.685+1.290) = 1.025 \text{ mg O} \times \frac{1 \text{ g}}{1000 \text{ mg}} \times \frac{1 \text{ mol O}}{16.00 \text{ g O}} = 0.0000641 \text{ mol O}$$

$$\text{C:H:O} = 0.0006399 : 0.001280 : 0.0000641 = 9.98 : 19.97 : 1.00 \approx 10 : 20 : 1$$

The empirical formula for menthol is $\mathbf{C_{10}H_{20}O}$

3.65 Begin by assuming a mass equal to the molar mass.

$$? \text{ mol C} = 0.2974 \times 121 \text{ g} = 35.99 \text{ g C} \times \frac{1 \text{ mol C}}{12.01 \text{ g C}} = 2.996 \text{ mol C}$$

$$? \text{ mol H} = 0.0582 \times 121 \text{ g} = 7.042 \text{ g H} \times \frac{1 \text{ mol H}}{1.008 \text{ g H}} = 6.986 \text{ mol H}$$

$$? \text{ mol O} = 0.2641 \times 121 \text{ g} = 31.96 \text{ g O} \times \frac{1 \text{ mol O}}{16.00 \text{ g O}} = 1.997 \text{ mol O}$$

$$? \text{ mol N} = 0.1156 \times 121 \text{ g} = 13.99 \text{ g N} \times \frac{1 \text{ mol N}}{14.01 \text{ g N}} = 0.998 \text{ mol N}$$

$$? \text{ mol S} = 0.2647 \times 121 \text{ g} = 32.03 \text{ g S} \times \frac{1 \text{ mol S}}{32.07 \text{ g S}} = 0.999 \text{ mol S}$$

Molecular formula $\mathbf{C_3H_7O_2NS}$

3.67 a. All the C and H in the combustion products come from the acetylene. In acetylene, the ratio of C to H is 1:1. Thus, the ratio of C to H in the combustion products must also be 1:1. Since each water molecule contains 2 atoms of H and each carbon dioxide molecule contains 1 atom of C, there must be twice the number of carbon dioxide molecules as water molecules. The answer is diagram (b).

b. In ethylene, the ratio of C to H is 1:2. Since this is the same ratio as the number of C atoms in each carbon dioxide molecule to the number of H atoms in each water molecule, combustion would produce equal numbers of $CO_2$ and $H_2O$ molecules. The answer is diagram (a).

3.71 $Si(s) + 2Cl_2(g) \longrightarrow SiCl_4(l)$

**Strategy:** Looking at the balanced equation, how do we compare the amounts of $Cl_2$ and $SiCl_4$? We can compare them based on the mole ratio from the balanced equation.

**Solution:** Because the balanced equation is given in the problem, the mole ratio between $Cl_2$ and $SiCl_4$ is known: 2 moles $Cl_2 \simeq$ 1 mole $SiCl_4$. From this relationship, we have two conversion factors.

$$\frac{2\text{ mol Cl}_2}{1\text{ mol SiCl}_4} \quad\text{and}\quad \frac{1\text{ mol SiCl}_4}{2\text{ mol Cl}_2}$$

Which conversion factor is needed to convert from moles of $SiCl_4$ to moles of $Cl_2$? The conversion factor on the left is the correct one. Moles of $SiCl_4$ will cancel, leaving units of "mol $Cl_2$" for the answer. We calculate moles of $Cl_2$ reacted as follows:

$$\textbf{? mol Cl}_\mathbf{2}\text{ reacted} = 0.507\ \textit{mol SiCl}_4 \times \frac{2\text{ mol Cl}_2}{1\text{ mol SiCl}_4} = \mathbf{1.01\ mol\ Cl_2}$$

**Think About It:** Does the answer seem reasonable? Should the moles of $Cl_2$ consumed be *double* the moles of $SiCl_4$ produced?

3.73 Starting with the 5.0 moles of $C_4H_{10}$, we can use the mole ratio from the balanced equation to calculate the moles of $CO_2$ formed.

$$2C_4H_{10}(g) + 13O_2(g) \rightarrow 8CO_2(g) + 10H_2O(l)$$

$$?\text{ mol CO}_2 = 5.0\text{ mol C}_4\text{H}_{10} \times \frac{8\text{ mol CO}_2}{2\text{ mol C}_4\text{H}_{10}} = 20\text{ mol CO}_2 = 2.0\times10^1\text{ mol CO}_2$$

3.75 a. $2NaHCO_3 \longrightarrow Na_2CO_3 + CO_2 + H_2O$

b.

$$\text{Molar mass NaHCO}_3 = 22.99\text{ g} + 1.008\text{ g} + 12.01\text{ g} + 3(16.00\text{ g}) = 84.008\text{ g}$$

$$\text{Molar mass CO}_2 = 12.01\text{ g} + 2(16.00\text{ g}) = 44.01\text{ g}$$

The balanced equation shows one mole of $CO_2$ formed from two moles of $NaHCO_3$.

$$\textbf{mass NaHCO}_\mathbf{3} = 20.5\text{ g CO}_2 \times \frac{1\text{ mol CO}_2}{44.01\text{ g CO}_2} \times \frac{2\text{ mol NaHCO}_3}{1\text{ mol CO}_2} \times \frac{84.008\text{ g NaHCO}_3}{1\text{ mol NaHCO}_3}$$

$$= \mathbf{78.3\ g\ NaHCO_3}$$

3.77 $C_6H_{12}O_6 \longrightarrow 2C_2H_5OH + 2CO_2$

glucose ethanol

**Strategy:** We compare glucose and ethanol based on the *mole ratio* in the balanced equation. Before we can determine moles of ethanol produced, we need to convert to moles of glucose. What conversion factor is needed to convert from grams of glucose to moles of glucose? Once moles of ethanol are obtained, another conversion factor is needed to convert from moles of ethanol to grams of ethanol.

**Solution:** The molar mass of glucose will allow us to convert from grams of glucose to moles of glucose. The molar mass of glucose = 6(12.01 g) + 12(1.008 g) + 6(16.00 g) = 180.16 g. The balanced equation is given, so the mole ratio between glucose and ethanol is known; that is 1 mole glucose ≏ 2 moles ethanol. Finally, the molar mass of ethanol will convert moles of ethanol to grams of ethanol. This sequence of three conversions is summarized as follows:

grams of glucose → moles of glucose → moles of ethanol → grams of ethanol

$$\textbf{? g C}_2\textbf{H}_5\textbf{OH} = 500.4\text{ g C}_6\text{H}_{12}\text{O}_6 \times \frac{1\text{ mol C}_6\text{H}_{12}\text{O}_6}{180.16\text{ g C}_6\text{H}_{12}\text{O}_6} \times \frac{2\text{ mol C}_2\text{H}_5\text{OH}}{1\text{ mol C}_6\text{H}_{12}\text{O}_6} \times \frac{46.068\text{ g C}_2\text{H}_5\text{OH}}{1\text{ mol C}_2\text{H}_5\text{OH}}$$

$$= \mathbf{255.9\ g\ C_2H_5OH}$$

**Think About It:** Does the answer seem reasonable? Should the mass of ethanol produced be approximately half the mass of glucose reacted? Twice as many moles of ethanol are produced compared to the moles of glucose reacted, but the molar mass of ethanol is about one-fourth that of glucose.

The liters of ethanol can be calculated from the density and the mass of ethanol.

$$\text{volume} = \frac{\text{mass}}{\text{density}}$$

$$\textbf{Volume of ethanol obtained} = \frac{255.9\text{ g}}{0.789\text{ g/mL}} = 324\text{ mL} = \mathbf{0.324\ L}$$

3.79 The balanced equation shows that eight moles of KCN are needed to combine with four moles of Au.

$$\textbf{? mol KCN} = 29.0\text{ g Au} \times \frac{1\text{ mol Au}}{197.0\text{ g Au}} \times \frac{8\text{ mol KCN}}{4\text{ mol Au}} = \mathbf{0.294\ mol\ KCN}$$

3.81 a. $NH_4NO_3(s) \longrightarrow N_2O(g) + 2H_2O(g)$

b. Starting with moles of $NH_4NO_3$, we can use the mole ratio from the balanced equation to find moles of $N_2O$. Once we have moles of $N_2O$, we can use the molar mass of $N_2O$ to convert to grams of $N_2O$.

Combining the two conversions into one calculation, we have:

$$\text{mol } NH_4NO_3 \rightarrow \text{mol } N_2O \rightarrow \text{g } N_2O$$

$$\textbf{? g N}_2\textbf{O} = 0.46 \text{ mol } NH_4NO_3 \times \frac{1 \text{ mol } N_2O}{1 \text{ mol } NH_4NO_3} \times \frac{44.02 \text{ g } N_2O}{1 \text{ mol } N_2O} = \mathbf{2.0 \times 10^1 \text{ g } N_2O}$$

3.83 The balanced equation for the decomposition is :

$$2KClO_3(s) \longrightarrow 2KCl(s) + 3O_2(g)$$

$$\textbf{? g O}_2 = 46.0 \text{ g } KClO_3 \times \frac{1 \text{ mol } KClO_3}{122.55 \text{ g } KClO_3} \times \frac{3 \text{ mol } O_2}{2 \text{ mol } KClO_3} \times \frac{32.00 \text{ g } O_2}{1 \text{ mol } O_2} = \mathbf{18.0 \text{ g } O_2}$$

3.89 This is a limiting reactant problem. Let's calculate the moles of $Cl_2$ produced assuming complete reaction for each reactant.

$$0.86 \text{ mol } MnO_2 \times \frac{1 \text{ mol } Cl_2}{1 \text{ mol } MnO_2} = 0.86 \text{ mol } Cl_2$$

$$48.2 \text{ g HCl} \times \frac{1 \text{ mol HCl}}{36.458 \text{ g HCl}} \times \frac{1 \text{ mol } Cl_2}{4 \text{ mol HCl}} = 0.3305 \text{ mol } Cl_2$$

**HCl** is the limiting reactant; it limits the amount of product produced. It will be used up first. The amount of product produced is 0.3305 mole $Cl_2$. Let's convert this to grams.

$$\textbf{? g Cl}_2 = 0.3305 \text{ mol } Cl_2 \times \frac{70.90 \text{ g } Cl_2}{1 \text{ mol } Cl_2} = \mathbf{23.4 \text{ g } Cl_2}$$

3.91 a. Start with a balanced chemical equation. It's given in the problem. We use NG as an abbreviation for nitroglycerin. The molar mass of NG = 227.1 g/mol.

$$4C_3H_5N_3O_9 \longrightarrow 6N_2 + 12CO_2 + 10H_2O + O_2$$

Map out the following strategy to solve this problem.

$$\text{g NG} \rightarrow \text{mol NG} \rightarrow \text{mol } O_2 \rightarrow \text{g } O_2$$

Calculate the grams of $O_2$ using the strategy above.

$$\textbf{? g O}_2 = 2.00 \times 10^2 \text{ g NG} \times \frac{1 \text{ mol NG}}{227.1 \text{ g NG}} \times \frac{1 \text{ mol } O_2}{4 \text{ mol NG}} \times \frac{32.00 \text{ g } O_2}{1 \text{ mol } O_2} = \mathbf{7.05 \text{ g } O_2}$$

b. The theoretical yield was calculated in part (a), and the actual yield is given in the problem (6.55 g). The percent yield is:

$$\% \text{ yield} = \frac{\text{actual yield}}{\text{theoretical yield}} \times 100\%$$

$$\textbf{\% yield} = \frac{6.55 \text{ g } O_2}{7.05 \text{ g } O_2} \times 100\% = \textbf{92.9\%}$$

3.93 The actual yield of ethylene is 481 g. Let's calculate the yield of ethylene if the reaction is 100 percent efficient. We can calculate this from the definition of percent yield. We can then calculate the mass of hexane needed.

$$\% \text{ yield} = \frac{\text{actual yield}}{\text{theoretical yield}} \times 100\%$$

$$42.5\% \text{ yield} = \frac{481 \text{ g } C_2H_4}{\text{theoretical yield}} \times 100\%$$

$$\text{theoretical yield } C_2H_4 = 1.132 \times 10^3 \text{ g } C_2H_4$$

The mass of hexane needed is:

$$(1.132 \times 10^3 \text{ g } C_2H_4) \times \frac{1 \text{ mol } C_2H_4}{28.052 \text{ g } C_2H_4} \times \frac{1 \text{ mol } C_6H_{14}}{1 \text{ mol } C_2H_4} \times \frac{86.172 \text{ g } C_6H_{14}}{1 \text{ mol } C_6H_{14}} = \mathbf{3.48 \times 10^3 \text{ g } C_6H_{14}}$$

3.95 This is a limiting reactant problem. Let's calculate the moles of $S_2Cl_2$ produced assuming complete reaction for each reactant.

$$S_8(l) + 4Cl_2(g) \rightarrow 4S_2Cl_2(l)$$

$$4.06 \text{ g } S_8 \times \frac{1 \text{ mol } S_8}{256.56 \text{ g } S_8} \times \frac{4 \text{ mol } S_2Cl_2}{1 \text{ mol } S_8} = 0.0633 \text{ mol } S_2Cl_2$$

$$6.24 \text{ g } Cl_2 \times \frac{1 \text{ mol } Cl_2}{70.90 \text{ g } Cl_2} \times \frac{4 \text{ mol } S_2Cl_2}{4 \text{ mol } Cl_2} = 0.0880 \text{ mol } S_2Cl_2$$

$S_8$ is the limiting reactant; it limits the amount of product produced. The amount of product produced is 0.0633 mole $S_2Cl_2$. Let's convert this to grams.

$$? \text{ g } S_2Cl_2 = 0.0633 \text{ mol } S_2Cl_2 \times \frac{135.04 \text{ g } S_2Cl_2}{1 \text{ mol } S_2Cl_2} = 8.55 \text{ g } S_2Cl_2$$

This is the theoretical yield of $S_2Cl_2$. The actual yield is given in the problem (6.55 g). The percent yield is:

$$\% \text{ yield} = \frac{\text{actual yield}}{\text{theoretical yield}} \times 100\% = \frac{6.55 \text{ g}}{8.55 \text{ g}} \times 100\% = \mathbf{76.6\%}$$

3.97 According to the diagram, the reaction consumes one $O_2$ molecule and four $NO_2$ molecules, and two $N_2O_5$ molecules are produced. The balanced equation is $\mathbf{O_2 + 4NO_2 \rightarrow 2N_2O_5}$. The limiting reagent is $\mathbf{NO_2}$ since it was consumed fully with no leftover molecules remaining. There was some $O_2$ left over, so $O_2$ is in excess.

3.99

$$N_2 + 3H_2 \rightarrow 2NH_3$$

The number of $N_2$ molecules shown in the diagram is 3. The balanced equation shows 3 moles $H_2$ are stoichiometrically equivalent to 1 mole $N_2$. Therefore, we need 9 molecules of $H_2$ to react completely with 3 molecules of $N_2$. There are 10 molecules of $H_2$ present in the diagram. $H_2$ is in excess. $N_2$ is the limiting reactant.

9 molecules of $H_2$ will react with 3 molecules of $N_2$, leaving 1 molecule of $H_2$ in excess. The mole ratio between $N_2$ and $NH_3$ is 1:2. When 3 molecules of $N_2$ react, 6 molecules of $NH_3$ will be produced.

3.101 a. combustion b. combination c. decomposition

3.103

$$2H_2(g) + O_2(g) \rightarrow 2H_2O(g)$$

We start with 8 molecules of $H_2$ and 3 molecules of $O_2$. The balanced equation shows 2 moles $H_2$ are stoichiometrically equivalent to 1 mole $O_2$. If 3 molecules of $O_2$ react, 6 molecules of $H_2$ will react, leaving 2 molecules of $H_2$ in excess. The balanced equation also shows 1 mole $O_2$ is stoichiometrically equivalent to 2 moles $H_2O$. If 3 molecules of $O_2$ react, 6 molecules of $H_2O$ will be produced.

After complete reaction, there will be **2 molecules of $H_2$** and **6 molecules of $H_2O$**. The correct diagram is choice **(b)**.

3.105 The symbol "O" refers to moles of oxygen atoms, not oxygen molecule ($O_2$). Look at the molecular formulas given in parts (a) and (b). What do they tell you about the relative amounts of carbon and oxygen?

a.
$$0.212 \text{ mol C} \times \frac{1 \text{ mol O}}{1 \text{ mol C}} = \mathbf{0.212 \text{ mol O}}$$

b.
$$0.212 \text{ mol C} \times \frac{2 \text{ mol O}}{1 \text{ mol C}} = \mathbf{0.424 \text{ mol O}}$$

3.107 We assume that all the Cl in the compound ends up as HCl and all the O ends up as $H_2O$. Therefore, we need to find the number of moles of Cl in HCl and the number of moles of O in $H_2O$.

$$\text{mol Cl} = 0.233 \text{ g HCl} \times \frac{1 \text{ mol HCl}}{36.458 \text{ g HCl}} \times \frac{1 \text{ mol Cl}}{1 \text{ mol HCl}} = 0.006391 \text{ mol Cl}$$

$$\text{mol O} = 0.403 \text{ g H}_2\text{O} \times \frac{1 \text{ mol H}_2\text{O}}{18.016 \text{ g H}_2\text{O}} \times \frac{1 \text{ mol O}}{1 \text{ mol H}_2\text{O}} = 0.02237 \text{ mol O}$$

Dividing by the smallest number of moles (0.006391 mole) gives the formula, $ClO_{3.5}$. Multiplying both subscripts by two gives the empirical formula, $\mathbf{Cl_2O_7}$.

3.109 The amount of Fe that reacted is:

$$\frac{1}{8} \times 664 \text{ g} = 83.0 \text{ g}$$

The amount of Fe remaining is:

$$664 \text{ g} - 83.0 \text{ g} = \mathbf{581 \text{ g}}$$

Thus, 83.0 g of Fe reacts to form the compound $Fe_2O_3$, which has two moles of Fe atoms per 1 mole of compound. The mass of $Fe_2O_3$ produced is:

$$83.0 \text{ g Fe} \times \frac{1 \text{ mol Fe}}{55.85 \text{ g Fe}} \times \frac{1 \text{ mol Fe}_2\text{O}_3}{2 \text{ mol Fe}} \times \frac{159.7 \text{ g Fe}_2\text{O}_3}{1 \text{ mol Fe}_2\text{O}_3} = \mathbf{119 \text{ g Fe}_2\text{O}_3}$$

**The final mass of the iron bar and rust is**: 581 g Fe + 119 g $Fe_2O_3$ = **700 g**

3.111 a. The following strategy can be used to convert from the volume of the Mg cube to the number of Mg atoms.

$$\text{cm}^3 \rightarrow \text{grams} \rightarrow \text{moles} \rightarrow \text{atoms}$$

$$1.0 \text{ cm}^3 \times \frac{1.74 \text{ g Mg}}{1 \text{ cm}^3} \times \frac{1 \text{ mol Mg}}{24.31 \text{ g Mg}} \times \frac{6.022 \times 10^{23} \text{ Mg atoms}}{1 \text{ mol Mg}} = \mathbf{4.3 \times 10^{22} \text{ Mg atoms}}$$

b. Since 74 percent of the available space is taken up by Mg atoms, $4.3 \times 10^{22}$ atoms occupy the following volume:

$$0.74 \times 1.0 \text{ cm}^3 = 0.74 \text{ cm}^3$$

We are trying to calculate the radius of a single Mg atom, so we need the volume occupied by a single Mg atom.

$$\text{volume Mg atom} = \frac{0.74 \text{ cm}^3}{4.3 \times 10^{22} \text{ Mg atoms}} = 1.7 \times 10^{-23} \text{ cm}^3/\text{Mg atom}$$

The volume of a sphere is $\frac{4}{3}\pi r^3$. Solving for the radius:

$$V = 1.7 \times 10^{-23} \text{ cm}^3 = \frac{4}{3}\pi r^3$$

$$r^3 = 4.1 \times 10^{-24} \text{ cm}^3$$

$$r = 1.6 \times 10^{-8} \text{ cm}$$

Converting to picometers:

$$\textbf{radius Mg atom} = (1.6 \times 10^{-8} \text{ cm}) \times \frac{0.01 \text{ m}}{1 \text{ cm}} \times \frac{1 \text{ pm}}{1 \times 10^{-12} \text{ m}} = \mathbf{1.6 \times 10^2 \text{ pm}}$$

3.113 The molar mass of chlorophyll is 893.48 g/mol. Finding the mass of a 0.0011-mol sample:

$$0.0011 \text{ mol chlorophyll} \times \frac{893.48 \text{ g chlorophyll}}{1 \text{ mol chlorophyll}} = 0.98 \text{ g chlorophyll}$$

The chlorophyll sample has the greater mass.

3.115 a.

$$8.38 \text{ g KBr} \times \frac{1 \text{ mol KBr}}{119.0 \text{ g KBr}} \times \frac{6.022 \times 10^{23} \text{ KBr}}{1 \text{ mol KBr}} \times \frac{1 \text{ K}^+ \text{ ion}}{1 \text{ KBr}} = \mathbf{4.24 \times 10^{22} \text{ K}^+ \text{ ions}}$$

Since there is one $Br^-$ for every one $K^+$, the number of $Br^-$ ions = **$4.24 \times 10^{22}$ $Br^-$ ions**

b.

$$5.40 \text{ g Na}_2\text{SO}_4 \times \frac{1 \text{ mol Na}_2\text{SO}_4}{142.05 \text{ g Na}_2\text{SO}_4} \times \frac{6.022 \times 10^{23} \text{ Na}_2\text{SO}_4}{1 \text{ mol Na}_2\text{SO}_4} \times \frac{2 \text{ Na}^+ \text{ ions}}{1 \text{ Na}_2\text{SO}_4} = \mathbf{4.58 \times 10^{22} \text{ Na}^+ \text{ ions}}$$

Since there are two $Na^+$ for every one $SO_4^{2-}$, the number of $SO_4^{2-}$ ions = **$2.29 \times 10^{22}$ $SO_4^{2-}$ ions**

c.

$$7.45 \text{ g Ca}_3(\text{PO}_4)_2 \times \frac{1 \text{ mol Ca}_3(\text{PO}_4)_2}{310.18 \text{ g Ca}_3(\text{PO}_4)_2} \times \frac{6.022 \times 10^{23} \text{ Ca}_3(\text{PO}_4)_2}{1 \text{ mol Ca}_3(\text{PO}_4)_2} \times \frac{3 \text{ Ca}^{2+} \text{ ions}}{1 \text{ Ca}_3(\text{PO}_4)_2}$$

$$= \mathbf{4.34 \times 10^{22} \text{ Ca}^{2+} \text{ ions}}$$

Since there are three $Ca^{2+}$ for every two $PO_4^{3-}$, the number of $PO_4^{3-}$ ions is:

$$4.34 \times 10^{22}\ \text{Ca}^{2+}\ \text{ions} \times \frac{2\ \text{PO}_4^{3-}\ \text{ions}}{3\ \text{Ca}^{2+}\ \text{ions}} = \mathbf{2.89 \times 10^{22}\ PO_4^{3-}\ ions}$$

3.117 The mass of one fluorine atom is 19.00 amu. The mass of one mole of fluorine atoms is 19.00 g. Multiplying the mass of one atom by Avogadro's number gives the mass of one mole of atoms. We can write:

$$\frac{19.00\ \text{amu}}{1\ \text{F atom}} \times (6.022 \times 10^{23}\ \text{F atoms}) = 19.00\ \text{g F}$$

or

$$\mathbf{6.022 \times 10^{23}\ amu = 1\ g}$$

This is why Avogadro's numbers has sometimes been described as a conversion factor between amu and grams.

3.119 First, we can calculate the moles of oxygen.

$$2.445\ \text{g C} \times \frac{1\ \text{mol C}}{12.01\ \text{g C}} \times \frac{1\ \text{mol O}}{1\ \text{mol C}} = 0.2036\ \text{mol O}$$

Next, we can calculate the molar mass of oxygen.

$$\text{molar mass O} = \frac{3.257\ \text{g O}}{0.2036\ \text{mol O}} = 16.00\ \text{g/mol}$$

If 1 mole of oxygen atoms has a mass of 16.00 g, then 1 atom of oxygen has an **atomic mass of 16.00 amu**.

3.121 a. The mass of chlorine is **5.0 g**.

b. From the percent by mass of Cl, we can calculate the mass of chlorine in 60.0 g of $NaClO_3$.

$$\text{mass \% Cl} = \frac{35.45\ \text{g Cl}}{106.44\ \text{g compound}} \times 100\% = 33.31\%\ \text{Cl}$$

$$\mathbf{mass\ Cl} = 60.0\ \text{g} \times 0.3331 = \mathbf{20.0\ g\ Cl}$$

c. 0.10 mol of KCl contains 0.10 mol of Cl.

$$0.10\ \text{mol Cl} \times \frac{35.45\ \text{g Cl}}{1\ \text{mol Cl}} = \mathbf{3.5\ g\ Cl}$$

d. From the percent by mass of Cl, we can calculate the mass of chlorine in 30.0 g of $MgCl_2$.

$$\text{mass \% Cl} = \frac{(2)(35.45\text{ g Cl})}{95.21\text{ g compound}} \times 100\% = 74.47\%\text{ Cl}$$

$$\textbf{mass Cl} = 30.0\text{ g} \times 0.7447 = \textbf{22.3 g Cl}$$

e. The mass of Cl can be calculated from the molar mass of $Cl_2$.

$$0.50\text{ mol Cl}_2 \times \frac{70.90\text{ g Cl}}{1\text{ mol Cl}_2} = \textbf{35.45 g Cl}$$

Thus, **(e) 0.50 mol $Cl_2$** contains the greatest mass of chlorine.

3.123 Both compounds contain only Pt and Cl. The percent by mass of Pt can be calculated by subtracting the percent Cl from 100 percent.

**Compound A:** Assume 100 g of compound.

$$26.7\text{ g Cl} \times \frac{1\text{ mol Cl}}{35.45\text{ g Cl}} = 0.753\text{ mol Cl}$$

$$73.3\text{ g Pt} \times \frac{1\text{ mol Pt}}{195.1\text{ g Pt}} = 0.376\text{ mol Pt}$$

Dividing by the smallest number of moles (0.376 mole) gives the empirical formula, **$PtCl_2$**.

**Compound B:** Assume 100 g of compound.

$$42.1\text{ g Cl} \times \frac{1\text{ mol Cl}}{35.45\text{ g Cl}} = 1.19\text{ mol Cl}$$

$$57.9\text{ g Pt} \times \frac{1\text{ mol Pt}}{195.1\text{ g Pt}} = 0.297\text{ mol Pt}$$

Dividing by the smallest number of moles (0.297 mole) gives the empirical formula, **$PtCl_4$**.

3.125 Both compounds contain only Mn and O. When the first compound is heated, oxygen gas is evolved. Let's calculate the empirical formulas for the two compounds, then we can write a balanced equation.

a. **Compound X:** Assume 100 g of compound.

$$63.3\text{ g Mn} \times \frac{1\text{ mol Mn}}{54.94\text{ g Mn}} = 1.15\text{ mol Mn}$$

$$36.7\text{ g O} \times \frac{1\text{ mol O}}{16.00\text{ g O}} = 2.29\text{ mol O}$$

Dividing by the smallest number of moles (1.15 moles) gives the empirical formula, $\mathbf{MnO_2}$.

**Compound Y:** Assume 100 g of compound.

$$72.0\text{ g Mn} \times \frac{1\text{ mol Mn}}{54.94\text{ g Mn}} = 1.31\text{ mol Mn}$$

$$28.0\text{ g O} \times \frac{1\text{ mol O}}{16.00\text{ g O}} = 1.75\text{ mol O}$$

Dividing by the smallest number of moles gives $MnO_{1.33}$. Recall that an empirical formula must have whole number coefficients. Multiplying by a factor of 3 gives the empirical formula $\mathbf{Mn_3O_4}$.

b. The unbalanced equation is:

$$MnO_2 \longrightarrow Mn_3O_4 + O_2$$

Balancing by inspection gives:

$$\mathbf{3MnO_2 \longrightarrow Mn_3O_4 + O_2}$$

3.127 We assume that the increase in mass results from the element nitrogen. The mass of nitrogen is:

$$0.378\text{ g} - 0.273\text{ g} = 0.105\text{ g N}$$

The empirical formula can now be calculated. Convert to moles of each element.

$$0.273\text{ g Mg} \times \frac{1\text{ mol Mg}}{24.31\text{ g Mg}} = 0.0112\text{ mol Mg}$$

$$0.105\text{ g N} \times \frac{1\text{ mol N}}{14.01\text{ g N}} = 0.00749\text{ mol N}$$

Dividing by the smallest number of moles gives $Mg_{1.5}N$. Recall that an empirical formula must have whole number coefficients. Multiplying by a factor of 2 gives the empirical formula $\mathbf{Mg_3N_2}$. The name of this compound is **magnesium nitride**.

3.129 The molar mass of air can be calculated by multiplying the mass of each component by its abundance and adding them together. Recall that nitrogen gas and oxygen gas are diatomic.

**molar mass air** = (0.7808)(28.02 g/mol) + (0.2095)(32.00 g/mol) + (0.0097)(39.95 g/mol) = **28.97 g/mol**

3.131 Possible formulas for the metal bromide could be MBr, $MBr_2$, $MBr_3$, etc. Assuming 100 g of compound, the moles of Br in the compound can be determined. From the mass and moles of the metal for each possible formula, we can calculate a molar mass for the metal. The molar mass that matches a metal on the periodic table would indicate the correct formula.

Assuming 100 g of compound, we have 53.79 g Br and 46.21 g of the metal (M). The moles of Br in the compound are:

$$53.79 \text{ g Br} \times \frac{1 \text{ mol Br}}{79.90 \text{ g Br}} = 0.67322 \text{ mol Br}$$

If the formula is MBr, the moles of M are also 0.67322 mole. If the formula is $MBr_2$, the moles of M are 0.67322/2 = 0.33661 mole, and so on. For each formula (MBr, $MBr_2$, and $MBr_3$), we calculate the molar mass of the metal.

$$\text{MBr:} \ \frac{46.21 \text{ g M}}{0.67322 \text{ mol M}} = 68.64 \text{ g/mol (no such metal)}$$

$$\text{MBr}_2\text{:} \ \frac{46.21 \text{ g M}}{0.33661 \text{ mol M}} = 137.3 \text{ g/mol (The metal is Ba. The formula is BaBr}_2\text{)}$$

$$\text{MBr}_3\text{:} \ \frac{46.21 \text{ g M}}{0.22441 \text{ mol M}} = 205.9 \text{ g/mol (no such metal)}$$

3.133 Assume 100 g of sample. Then,

$$\text{mol Na} = 32.08 \text{ g Na} \times \frac{1 \text{ mol Na}}{22.99 \text{ g Na}} = 1.395 \text{ mol Na}$$

$$\text{mol O} = 36.01 \text{ g O} \times \frac{1 \text{ mol O}}{16.00 \text{ g O}} = 2.251 \text{ mol O}$$

$$\text{mol Cl} = 19.51 \text{ g Cl} \times \frac{1 \text{ mol Cl}}{35.45 \text{ g Cl}} = 0.5504 \text{ mol Cl}$$

Since Cl is only contained in NaCl, the moles of Cl equals the moles of Na contained in NaCl.

$$\text{mol Na (in NaCl)} = 0.5504 \text{ mol}$$

The number of moles of Na in the remaining two compounds is: 1.395 mol – 0.5504 mol = 0.8446 mol Na.

To solve for moles of the remaining two compounds, let

$$x = \text{moles of Na}_2\text{SO}_4$$

$$y = \text{moles of NaNO}_3$$

Then, from the mole ratio of Na and O in each compound, we can write

$$2x + y = \text{mol Na} = 0.8446 \text{ mol}$$

$$4x + 3y = \text{mol O} = 2.251 \text{ mol}$$

Solving two equations with two unknowns gives

$$x = 0.1414 = \text{mol } Na_2SO_4 \quad \text{and} \quad y = 0.5618 = \text{mol } NaNO_3$$

Finally, we convert to mass of each compound to calculate the mass percent of each compound in the sample. Remember, the sample size is 100 g.

$$\textbf{mass \% NaCl} = 0.5504 \text{ mol NaCl} \times \frac{58.44 \text{ g NaCl}}{1 \text{ mol NaCl}} \times \frac{1}{100 \text{ g sample}} \times 100\% = \textbf{32.17\% NaCl}$$

$$\textbf{mass \% Na}_2\textbf{SO}_4 = 0.1414 \text{ mol Na}_2\text{SO}_4 \times \frac{142.1 \text{ g Na}_2\text{SO}_4}{1 \text{ mol Na}_2\text{SO}_4} \times \frac{1}{100 \text{ g sample}} \times 100\% = \textbf{20.09\% Na}_2\textbf{SO}_4$$

$$\textbf{mass \% NaNO}_3 = 0.5618 \text{ mol NaNO}_3 \times \frac{85.00 \text{ g NaNO}_3}{1 \text{ mol NaNO}_3} \times \frac{1}{100 \text{ g sample}} \times 100\% = \textbf{47.75\% NaNO}_3$$

3.135 a. The balanced equation is:

$$C_3H_8(g) + 5O_2(g) \longrightarrow 3CO_2(g) + 4H_2O(l)$$

b. The balanced equation shows a mole ratio of 3 moles $CO_2$ : 1 mole $C_3H_8$. The mass of $CO_2$ produced is:

$$3.65 \text{ mol C}_3\text{H}_8 \times \frac{3 \text{ mol CO}_2}{1 \text{ mol C}_3\text{H}_8} \times \frac{44.01 \text{ g CO}_2}{1 \text{ mol CO}_2} = \textbf{482 g CO}_2$$

3.137 a. $Zn(s) + H_2SO_4(aq) \longrightarrow ZnSO_4(aq) + H_2(g)$

b. We assume that a pure sample would produce the theoretical yield of $H_2$. The balanced equation shows a mole ratio of 1 mole $H_2$ : 1 mole Zn. The theoretical yield of $H_2$ is:

$$3.86 \text{ g Zn} \times \frac{1 \text{ mol Zn}}{65.41 \text{ g Zn}} \times \frac{1 \text{ mol H}_2}{1 \text{ mol Zn}} \times \frac{2.016 \text{ g H}_2}{1 \text{ mol H}_2} = 0.119 \text{ g H}_2$$

$$\textbf{percent purity} = \frac{0.0764 \ g\ H_2}{0.119 \ g\ H_2} \times 100\% = \textbf{64.2\%}$$

c. We assume that the impurities are inert and do not react with the sulfuric acid to produce hydrogen.

3.139 a. The balanced chemical equation is:

$$C_3H_8(g) + 3H_2O(g) \longrightarrow 3CO(g) + 7H_2(g)$$

b. You should come up with the following strategy to solve this problem. In this problem, we use kg-mol to save a couple of steps.

$$\text{kg } C_3H_8 \rightarrow \text{mol } C_3H_8 \rightarrow \text{mol } H_2 \rightarrow \text{kg } H_2$$

$$\textbf{? kg H}_2 = (2.84 \times 10^3 \text{ kg } C_3H_8) \times \frac{1 \text{ kg-mol } C_3H_8}{44.09 \text{ kg } C_3H_8} \times \frac{7 \text{ kg-mol } H_2}{1 \text{ kg-mol } C_3H_8} \times \frac{2.016 \text{ kg } H_2}{1 \text{ kg-mol } H_2}$$

$$= \mathbf{909\ kg\ H_2}$$

3.141 The balanced equations for the combustion of sulfur and the reaction of $SO_2$ with CaO are:

$$S(s) + O_2(g) \longrightarrow SO_2(g) \qquad\qquad SO_2(g) + CaO(s) \longrightarrow CaSO_3(s)$$

First, find the amount of sulfur present in the daily coal consumption.

$$(6.60 \times 10^6 \text{ kg coal}) \times \frac{1.6\% \text{ S}}{100\%} = 1.06 \times 10^5 \text{ kg S} = 1.06 \times 10^8 \text{ g S}$$

The daily amount of CaO needed is:

$$(1.06 \times 10^8 \text{ g S}) \times \frac{1 \text{ mol S}}{32.07 \text{ g S}} \times \frac{1 \text{ mol } SO_2}{1 \text{ mol S}} \times \frac{1 \text{ mol CaO}}{1 \text{ mol } SO_2} \times \frac{56.08 \text{ g CaO}}{1 \text{ mol CaO}} \times \frac{1 \text{ kg}}{1000 \text{ g}} = \mathbf{1.85 \times 10^5\ kg\ CaO}$$

3.143

$$\text{? mol C} = 14.7 \text{ g } CO_2 \times \frac{1 \text{ mol } CO_2}{44.01 \text{ g } CO_2} \times \frac{1 \text{ mol C}}{1 \text{ mol } CO_2} = 0.3340 \text{ mol C} \times \frac{12.01 \text{ g C}}{1 \text{ mol C}} = 4.012 \text{ g C}$$

$$\text{? mol H} = 6.00 \text{ g } H_2O \times \frac{1 \text{ mol } H_2O}{18.02 \text{ g } H_2O} \times \frac{2 \text{ mol H}}{1 \text{ mol } H_2O} = 0.6659 \text{ mol H} \times \frac{1.008 \text{ g H}}{1 \text{ mol H}} = 0.6712 \text{ g H}$$

$$\text{Mass O} = \text{mass Sample} - (\text{Mass C} + \text{Mass H}) = 5.3168 \text{ g O} \times \frac{1 \text{ mol O}}{16.00 \text{ g O}} = 0.332 \text{ mol O}$$

$$\text{Molar ratios: C : H : O} = 0.334 : 0.666 : 0.332 = \frac{0.334}{0.332} : \frac{0.666}{0.332} : \frac{0.332}{0.332} = 1.00 : 2.00 : 1.00$$

**The empirical formula is $CH_2O$**

3.145 a. First, calculate the mass of C in $CO_2$, the mass of H in $H_2O$, and the mass of N in $NH_3$. For now, we will carry more than 3 significant figures and then round to the correct number at the end of the problem.

$$? \text{ g C} = 3.94 \text{ g CO}_2 \times \frac{1 \text{ mol CO}_2}{44.01 \text{ g CO}_2} \times \frac{1 \text{ mol C}}{1 \text{ mol CO}_2} \times \frac{12.01 \text{ g C}}{1 \text{ mol C}} = 1.075 \text{ g C}$$

$$? \text{ g H} = 1.89 \text{ g H}_2\text{O} \times \frac{1 \text{ mol H}_2\text{O}}{18.02 \text{ g H}_2\text{O}} \times \frac{2 \text{ mol H}}{1 \text{ mol H}_2\text{O}} \times \frac{1.008 \text{ g H}}{1 \text{ mol H}} = 0.2114 \text{ g H}$$

$$? \text{ g N} = 0.436 \text{ g NH}_3 \times \frac{1 \text{ mol NH}_3}{17.03 \text{ g NH}_3} \times \frac{1 \text{ mol N}}{1 \text{ mol NH}_3} \times \frac{14.01 \text{ g N}}{1 \text{ mol N}} = 0.3587 \text{ g N}$$

Next, we can calculate the % C, % H, and the % N in each sample, then we can calculate the % O by difference.

$$\% \text{C} = \frac{1.075 \text{ g C}}{2.175 \text{ g sample}} \times 100\% = 49.43\% \text{ C}$$

$$\% \text{H} = \frac{0.2114 \text{ g H}}{2.175 \text{ g sample}} \times 100\% = 9.720\% \text{ H}$$

$$\% \text{N} = \frac{0.3587 \text{ g N}}{1.873 \text{ g sample}} \times 100\% = 19.15\% \text{ N}$$

The $\% \text{O} = 100\% - (49.43\% + 9.720\% + 19.15\%) = 21.70\% \text{ O}$

Assuming 100 g of compound, calculate the moles of each element.

$$? \text{ mol C} = 49.43 \text{ g C} \times \frac{1 \text{ mol C}}{12.01 \text{ g C}} = 4.116 \text{ mol C}$$

$$? \text{ mol H} = 9.720 \text{ g H} \times \frac{1 \text{ mol H}}{1.008 \text{ g H}} = 9.643 \text{ mol H}$$

$$? \text{ mol N} = 19.15 \text{ g N} \times \frac{1 \text{ mol N}}{14.01 \text{ g N}} = 1.367 \text{ mol N}$$

$$? \text{ mol O} = 21.70 \text{ g O} \times \frac{1 \text{ mol O}}{16.00 \text{ g O}} = 1.356 \text{ mol O}$$

Thus, we arrive at the formula $C_{4.116}H_{9.643}N_{1.367}O_{1.356}$. Dividing by 1.356 gives the empirical formula, **$C_3H_7NO$**.

b. The empirical molar mass is 73.10 g. Since the approximate molar mass of lysine is 150 g, we have:

$$\frac{150\text{ g}}{73.10\text{ g}} \approx 2$$

Therefore, the molecular formula is $(C_3H_7NO)_2$ or $\mathbf{C_6H_{14}N_2O_2}$.

3.147 Molar mass of $C_4H_8Cl_2S$ = 4(12.01 g) + 8(1.008 g) + 2(35.45 g) + 32.07 g = 159.07 g

$$\mathbf{\%C} = \frac{4(12.01\text{ g/mol})}{159.07\text{ g/mol}} \times 100\% = \mathbf{30.20\%}$$

$$\mathbf{\%H} = \frac{8(1.008\text{ g/mol})}{159.07\text{ g/mol}} \times 100\% = \mathbf{5.069\%}$$

$$\mathbf{\%Cl} = \frac{2(35.45\text{ g/mol})}{159.07\text{ g/mol}} \times 100\% = \mathbf{44.57\%}$$

$$\mathbf{\%S} = \frac{32.07\text{ g/mol}}{159.07\text{ g/mol}} \times 100\% = \mathbf{20.16\%}$$

3.149 a. The molar mass of hemoglobin is:

$$2952(12.01\text{ g}) + 4664(1.008\text{ g}) + 812(14.01\text{ g}) + 832(16.00\text{ g}) + 8(32.07\text{ g}) + 4(55.85\text{ g})$$

$$= \mathbf{6.532 \times 10^4\ g}$$

b. To solve this problem, the following conversions need to be completed:

L → mL → red blood cells → hemoglobin molecules → mol hemoglobin → mass hemoglobin

We will use the following abbreviations: RBC = red blood cells, HG = hemoglobin

$$5.00\text{ L} \times \frac{1\text{ mL}}{0.001\text{ L}} \times \frac{5.0 \times 10^9\text{ RBC}}{1\text{ mL}} \times \frac{2.8 \times 10^8\text{ HG molecules}}{1\text{ RBC}} \times \frac{1\text{ mol HG}}{6.022 \times 10^{23}\text{ molecules HG}} \times \frac{6.532 \times 10^4\text{ g HG}}{1\text{ mol HG}}$$

$$= \mathbf{7.6 \times 10^2\ g\ HG}$$

3.151 $\mathbf{C_3H_2ClF_5O}$, 3(12.01 g) +2(1.008 g) + 1(35.45 g) + 5(19.00 g) + 1(16.00 g) = **184.50 g/mol**

3.153 $SO_2$ is converted to $H_2SO_4$ by reaction with water. The mole ratio between $SO_2$ and $H_2SO_4$ is 1:1.

This is a unit conversion problem. You should come up with the following strategy to solve the problem.

$$\text{tons SO}_2 \rightarrow \text{ton-mol SO}_2 \rightarrow \text{ton-mol H}_2\text{SO}_4 \rightarrow \text{tons H}_2\text{SO}_4$$

$$\textbf{? tons H}_2\textbf{SO}_4 = (4.0 \times 10^5 \text{ tons SO}_2) \times \frac{1 \text{ ton-mol SO}_2}{64.07 \text{ tons SO}_2} \times \frac{1 \text{ ton-mol H}_2\text{SO}_4}{1 \text{ ton-mol SO}_2} \times \frac{98.09 \text{ tons H}_2\text{SO}_4}{1 \text{ ton-mol H}_2\text{SO}_4}$$

$$= \mathbf{6.1 \times 10^5 \text{ tons H}_2\text{SO}_4}$$

3.155 **METHOD 1:**

*Step 1:* Assume you have exactly 100 g of substance. 100 g is a convenient amount, because all the percentages sum to 100%. The percentage of oxygen is found by difference:

$$100\% - (19.8\% + 2.50\% + 11.6\%) = 66.1\%$$

In 100 g of PAN there will be 19.8 g C, 2.50 g H, 11.6 g N, and 66.1 g O.

*Step 2:* Calculate the number of moles of each element in the compound. Remember, an *empirical formula* tells us which elements are present and the simplest whole-number ratio of their atoms. This ratio is also a mole ratio. Use the molar masses of these elements as conversion factors to convert to moles.

$$n_\text{C} = 19.8 \text{ g C} \times \frac{1 \text{ mol C}}{12.01 \text{ g C}} = 1.649 \text{ mol C}$$

$$n_\text{H} = 2.50 \text{ g H} \times \frac{1 \text{ mol H}}{1.008 \text{ g H}} = 2.480 \text{ mol H}$$

$$n_\text{N} = 11.6 \text{ g N} \times \frac{1 \text{ mol N}}{14.01 \text{ g N}} = 0.8280 \text{ mol N}$$

$$n_\text{O} = 66.1 \text{ g O} \times \frac{1 \text{ mol O}}{16.00 \text{ g O}} = 4.131 \text{ mol O}$$

*Step 3:* Try to convert to whole numbers by dividing all the subscripts by the smallest subscript. The formula is $C_{1.649}H_{2.480}N_{0.8280}O_{4.131}$. Dividing the subscripts by 0.8280 gives the empirical formula, $\mathbf{C_2H_3NO_5}$.

To determine the molecular formula, remember that the molar mass/empirical mass will be an integer greater than or equal to one.

$$\frac{\text{molar mass}}{\text{empirical molar mass}} \geq 1 \text{ (integer values)}$$

In this case,

$$\frac{\text{molar mass}}{\text{empirical molar mass}} = \frac{120 \text{ g}}{121.05 \text{ g}} \approx 1$$

Hence, the molecular formula and the empirical formula are the same, $\mathbf{C_2H_3NO_5}$.

**METHOD 2:**

*Step 1:* Multiply the mass % (converted to a decimal) of each element by the molar mass to convert to grams of each element. Then, use the molar mass to convert to moles of each element.

$$n_{\text{C}} = (0.198) \times (120 \text{ g}) \times \frac{1 \text{ mol C}}{12.01 \text{ g C}} = 1.98 \text{ mol C} \approx 2 \text{ mol C}$$

$$n_{\text{H}} = (0.0250) \times (120 \text{ g}) \times \frac{1 \text{ mol H}}{1.008 \text{ g H}} = 2.98 \text{ mol H} \approx 3 \text{ mol H}$$

$$n_{\text{N}} = (0.116) \times (120 \text{ g}) \times \frac{1 \text{ mol N}}{14.01 \text{ g N}} = 0.994 \text{ mol N} \approx 1 \text{ mol N}$$

$$n_{\text{O}} = (0.661) \times (120 \text{ g}) \times \frac{1 \text{ mol O}}{16.00 \text{ g O}} = 4.96 \text{ mol O} \approx 5 \text{ mol O}$$

*Step 2:* Since we used the molar mass to calculate the moles of each element present in the compound, this method directly gives the molecular formula. The formula is $\mathbf{C_2H_3NO_5}$.

*Step 3:* Try to reduce the molecular formula to a simpler whole number ratio to determine the empirical formula. The formula is already in its simplest whole number ratio. The molecular and empirical formulas are the same. The empirical formula is $\mathbf{C_2H_3NO_5}$.

3.157 a. We need to compare the mass % of K in both KCl and $K_2SO_4$.

$$\%\text{K in KCl} = \frac{39.10 \text{ g}}{74.55 \text{ g}} \times 100\% = 52.45\% \text{ K}$$

$$\%\text{K in K}_2\text{SO}_4 = \frac{2(39.10 \text{ g})}{174.27 \text{ g}} \times 100\% = 44.87\% \text{ K}$$

The price depends on the %K.

$$\frac{\text{Price of K}_2\text{SO}_4}{\text{Price of KCl}} = \frac{\%\text{K in K}_2\text{SO}_4}{\%\text{K in KCl}}$$

$$\text{Price of K}_2\text{SO}_4 = \text{Price of KCl} \times \frac{\%\text{K in K}_2\text{SO}_4}{\%\text{K in KCl}}$$

$$\textbf{Price of } \mathbf{K_2SO_4} = \frac{\$0.55}{\text{kg}} \times \frac{44.87\%}{52.45\%} = \mathbf{\$0.47 / kg}$$

b. First, calculate the number of moles of K in 1.00 kg of KCl.

$$(1.00 \times 10^3 \text{ g KCl}) \times \frac{1 \text{ mol KCl}}{74.55 \text{ g KCl}} \times \frac{1 \text{ mol K}}{1 \text{ mol KCl}} = 13.4 \text{ mol K}$$

Next, calculate the amount of $K_2O$ needed to supply 13.4 mol K.

$$13.4 \text{ mol K} \times \frac{1 \text{ mol K}_2\text{O}}{2 \text{ mol K}} \times \frac{94.20 \text{ g K}_2\text{O}}{1 \text{ mol K}_2\text{O}} \times \frac{1 \text{ kg}}{1000 \text{ g}} = \mathbf{0.631 \text{ kg } K_2O}$$

3.159 The surface area of the water can be calculated assuming that the dish is circular.

$$\text{surface area of water} = \pi r^2 = \pi(10 \text{ cm})^2 = 3.1 \times 10^2 \text{ cm}^2$$

The cross-sectional area of one stearic acid molecule in $cm^2$ is:

$$0.21 \text{ nm}^2 \times \left(\frac{1 \times 10^{-9} \text{ m}}{1 \text{ nm}}\right)^2 \times \left(\frac{1 \text{ cm}}{0.01 \text{ m}}\right)^2 = 2.1 \times 10^{-15} \text{ cm}^2\text{/molecule}$$

Assuming that there is no empty space between molecules, we can calculate the number of stearic acid molecules that will fit in an area of $3.1 \times 10^2$ cm$^2$.

$$(3.1 \times 10^2 \text{ cm}^2) \times \frac{1 \text{ molecule}}{2.1 \times 10^{-15} \text{ cm}^2} = 1.5 \times 10^{17} \text{ molecules}$$

Next, we can calculate the moles of stearic acid in the $1.4 \times 10^{-4}$ g sample. Then, we can calculate Avogadro's number (the number of molecules per mole).

$$1.4 \times 10^{-4} \text{ g stearic acid} \times \frac{1 \text{ mol stearic acid}}{284.5 \text{ g stearic acid}} = 4.9 \times 10^{-7} \text{ mol stearic acid}$$

$$\textbf{Avogadro's number } (N_A) = \frac{1.5 \times 10^{17} \text{ molecules}}{4.9 \times 10^{-7} \text{ mol}} = \mathbf{3.1 \times 10^{23} \text{ molecules / mol}}$$

# Chapter 4

# Reactions in Aqueous Solutions

4.7 When NaCl dissolves in water it dissociates into $Na^+$ and $Cl^-$ ions. When the ions are hydrated, the water molecules will be oriented so that the negative end of the water dipole interacts with the positive sodium ion, and the positive end of the water dipole interacts with the negative chloride ion. The negative end of the water dipole is near the oxygen atom, and the positive end of the water dipole is near the hydrogen atoms. The diagram that best represents the hydration of NaCl when dissolved in water is **diagram (c)**.

4.9 Ionic compounds, strong acids, and strong bases (metal hydroxides) are strong electrolytes (completely broken up into ions of the compound). Weak acids and weak bases are weak electrolytes. Molecular substances other than acids or bases are nonelectrolytes.

a. strong electrolyte (ionic)

b. nonelectrolyte

c. weak electrolyte (weak base)

d. strong electrolyte (strong base)

4.11 a. Solid NaCl does not conduct. The ions are locked in a rigid lattice structure. **Non-conducting.**

b. Molten NaCl conducts. The ions can move around in the liquid state. **Conducting.**

c. Aqueous NaCl conducts. NaCl dissociates completely to $Na^+(aq)$ and $Cl^-(aq)$ in water. **Conducting.**

4.13 Since HCl dissolved in water conducts electricity, then HCl(*aq*) must actually exists as $H^+(aq)$ cations and $Cl^-(aq)$ anions. Since HCl dissolved in benzene solvent does not conduct electricity, then we must assume that the HCl molecules in benzene solvent do not ionize, but rather exist as un-ionized molecules.

4.17 Refer to Table 4.2 of the text to solve this problem. AgCl is insoluble in water. It will precipitate from solution. $NaNO_3$ is soluble in water and will remain as $Na^+$ and $NO_3^-$ ions in solution. The mixture is best represented by **diagram (c)**.

4.19 **Strategy:** Although it is not necessary to memorize the solubilities of compounds, you should keep in mind the following useful rules: all ionic compounds containing alkali metal cations, the ammonium ion, and the nitrate, bicarbonate, and chlorate ions are soluble. For other compounds, refer to Tables 4.2 and 4.3 of the text.

**Solution:** a. $Ca_3(PO_4)_2$ is ***insoluble***. Most phosphate compounds are insoluble.

b. $Mn(OH)_2$ is ***insoluble***. Most hydroxide compounds are insoluble.

c. $AgClO_3$ is ***soluble***. All chlorate compounds are soluble.

d. $K_2S$ is ***soluble***. All compounds containing alkali metal cations are soluble.

4.21 **Strategy:** Recall that an *ionic equation* shows dissolved ionic compounds in terms of their free ions. What ions do the aqueous solutions contain before they are combined? A *net ionic equation* shows only the species that actually take part in the reaction.

**Solution:** a. In solution, $AgNO_3$ dissociates into $Ag^+$ and $NO_3^-$ ions and $Na_2SO_4$ dissociates into $Na^+$ and $SO_4^{2-}$ ions. According to Tables 4.2 and 4.3 of the text, silver ions ($Ag^+$) and sulfate ions ($SO_4^{2-}$) will form an insoluble compound, silver sulfate ($Ag_2SO_4$), while the other product, $NaNO_3$, is soluble and remains in solution. This is a precipitation reaction. The balanced molecular equation is:

$$2AgNO_3(aq) + Na_2SO_4(aq) \rightarrow Ag_2SO_4(s) + 2NaNO_3(aq)$$

The ionic and net ionic equations are:

**Ionic:**

$$\mathbf{2Ag^+(aq) + 2NO_3^-(aq) + 2Na^+(aq) + SO_4^{2-}(aq) \rightarrow Ag_2SO_4(s) + 2Na^+(aq) + 2NO_3^-(aq)}$$

**Net ionic:**

$$\mathbf{2Ag^+(aq) + SO_4^{2-}(aq) \rightarrow Ag_2SO_4(s)}$$

b. In solution, $BaCl_2$ dissociates into $Ba^{2+}$ and $Cl^-$ ions and $ZnSO_4$ dissociates into $Zn^{2+}$ and

$SO_4^{2-}$ ions. According to Tables 4.2 and 4.3 of the text, barium ions ($Ba^{2+}$) and sulfate ions ($SO_4^{2-}$) will form an insoluble compound, barium sulfate ($BaSO_4$), while the other product, $ZnCl_2$, is soluble and remains in solution. This is a precipitation reaction. The balanced molecular equation is:

$$BaCl_2(aq) + ZnSO_4(aq) \rightarrow BaSO_4(s) + ZnCl_2(aq)$$

The ionic and net ionic equations are:

**Ionic:**

$$\mathbf{Ba^{2+}(aq) + 2Cl^-(aq) + Zn^{2+}(aq) + SO_4^{2-}(aq) \rightarrow BaSO_4(s) + Zn^{2+}(aq) + 2Cl^-(aq)}$$

**Net ionic:**

$$\mathbf{Ba^{2+}(aq) + SO_4^{2-}(aq) \rightarrow BaSO_4(s)}$$

c. In solution, $(NH_4)_2CO_3$ dissociates into $NH_4^+$ and $CO_3^{2-}$ ions and $CaCl_2$ dissociates into $Ca^{2+}$ and $Cl^-$ ions. According to Tables 4.2 and 4.3 of the text, calcium ions ($Ca^{2+}$) and carbonate ions ($CO_3^{2-}$) will form an insoluble compound, calcium carbonate ($CaCO_3$), while the other product, $NH_4Cl$, is soluble and remains in solution. This is a precipitation reaction. The balanced molecular equation is:

$$(NH_4)_2CO_3(aq) + CaCl_2(aq) \rightarrow CaCO_3(s) + 2NH_4Cl(aq)$$

The ionic and net ionic equations are:

**Ionic:**

$$\mathbf{2NH_4^+(aq) + CO_3^{2-}(aq) + Ca^{2+}(aq) + 2Cl^-(aq) \rightarrow CaCO_3(s) + 2NH_4^+(aq) + 2Cl^-(aq)}$$

**Net ionic:**

$$\mathbf{Ca^{2+}(aq) + CO_3^{2-}(aq) \rightarrow CaCO_3(s)}$$

**Think About It:** Note that because we balanced the molecular equations first, the net ionic equations are balanced both as to the number of *atoms* on each side, and as to the number of positive and negative *charges* on each side.

4.23 Only the combination in (b) will result in a precipitation reaction.

a. Both reactants are soluble ionic compounds. The other possible ion combinations, $Na_2SO_4$ and

$Cu(NO_3)_2$, are also soluble. **No precipitate forms.**

b. Both reactants are soluble. Of the other two possible ion combinations, KCl is soluble, but $BaSO_4$ is insoluble and will precipitate. The net ionic equation for the precipitation is:

$$\mathbf{Ba^{2+}(\mathit{aq}) + SO_4^{2-}(\mathit{aq}) \rightarrow BaSO_4(\mathit{s})}$$

4.31 **Strategy:** What are the characteristics of a Brønsted acid? Does it contain at least an H atom? With the exception of ammonia, most Brønsted bases that you will encounter at this stage are anions.

**Solution:** a. $PO_4^{3-}$ in water can accept a proton to become $HPO_4^{2-}$, and is thus a **Brønsted base**.

b. $ClO_2^-$ in water can accept a proton to become $HClO_2$, and is thus a **Brønsted base**.

c. $NH_4^+$ dissolved in water can donate a proton $H^+$, thus behaving as a **Brønsted acid**.

d. $HCO_3^-$ can either donate a proton to yield $H^+$ and $CO_3^{2-}$, thus behaving as a **Brønsted acid**. Or, $HCO_3^-$ can accept a proton to become $H_2CO_3$, thus behaving as a **Brønsted base**.

**Think About It:** The $HCO_3^-$ species is said to be *amphoteric* because it possesses both acidic and basic properties.

4.33 **Strategy:** Recall that strong acids and strong bases are strong electrolytes. They are completely ionized in solution. An *ionic equation* will show strong electrolytes, including strong acids and strong bases as separate ions. Weak acids and weak bases are weak electrolytes. They only ionize to a small extent in solution. Weak acids and weak bases are shown as molecules in ionic and net ionic equations. A *net ionic equation* shows only the species that actually take part in the reaction.

**Solution:** a. $HC_2H_3O_2$ is a weak acid. It will be shown as a molecule in the ionic equation. KOH is a strong base. It completely dissociates to $K^+$ and $OH^-$ ions. Since $HC_2H_3O_2$ is an acid, it donates an $H^+$ to the base, $OH^-$, producing water. The other product is the salt, $KC_2H_3O_2$, which is soluble and remains in solution. The balanced molecular equation is:

$$HC_2H_3O_2(aq) + KOH(aq) \rightarrow KC_2H_3O_2(aq) + H_2O(l)$$

The ionic and net ionic equations are:

**Ionic:**

$$HC_2H_3O_2(aq) + K^+(aq) + OH^-(aq) \rightarrow C_2H_3O_2^-(aq) + K^+(aq) + H_2O(l)$$

**Net ionic:**

$$HC_2H_3O_2(aq) + OH^-(aq) \rightarrow C_2H_3O_2^-(aq) + H_2O(l)$$

b. $H_2CO_3$ is a weak acid. It will be shown as a molecule in the ionic equation. NaOH is a strong base. It completely dissociates to $Na^+$ and $OH^-$ ions. Since $H_2CO_3$ is an acid, it donates an $H^+$ to the base, $OH^-$, producing water. The other product is the salt, $Na_2CO_3$, which is soluble and remains in solution. The balanced molecular equation is:

$$H_2CO_3(aq) + 2NaOH(aq) \rightarrow Na_2CO_3(aq) + 2H_2O(l)$$

The ionic and net ionic equations are:

**Ionic:**

$$H_2CO_3(aq) + 2Na^+(aq) + 2OH^-(aq) \rightarrow 2Na^+(aq) + CO_3^{2-}(aq) + 2H_2O(l)$$

**Net ionic:**

$$H_2CO_3(aq) + 2OH^-(aq) \rightarrow CO_3^{2-}(aq) + 2H_2O(l)$$

c. $HNO_3$ is a strong acid. It completely ionizes to $H^+$ and $NO_3^-$ ions. $Ba(OH)_2$ is a strong base. It completely dissociates to $Ba^{2+}$ and $OH^-$ ions. Since $HNO_3$ is an acid, it donates an $H^+$ to the base, $OH^-$, producing water. The other product is the salt, $Ba(NO_3)_2$, which is soluble and remains in solution. The balanced molecular equation is:

$$2HNO_3(aq) + Ba(OH)_2(aq) \rightarrow Ba(NO_3)_2(aq) + 2H_2O(l)$$

The ionic and net ionic equations are:

**Ionic:**

$$2H^+(aq) + 2NO_3^-(aq) + Ba^{2+}(aq) + 2OH^-(aq) \rightarrow Ba^{2+}(aq) + 2NO_3^-(aq) + 2H_2O(l)$$

**Net ionic:**

$$2H^+(aq) + 2OH^-(aq) \rightarrow 2H_2O(l) \text{ or } H^+(aq) + OH^-(aq) \rightarrow H_2O(l)$$

4.41 **Strategy:** In order to break a redox reaction down into an oxidation half-reaction and a reduction half-reaction, you should first assign oxidation numbers to all the atoms in the reaction. In this way, you can determine which element is oxidized (loses electrons) and which element is reduced (gains electrons).

**Solution:** In each part, the reducing agent is the reactant in the first half-reaction and the oxidizing agent is the reactant in the second half-reaction.

a. The product is an ionic compound whose ions are $Sr^{2+}$ and $O^{2-}$. The half-reactions are

$$\mathbf{2Sr \rightarrow 2Sr^{2+} + 4e^- \quad and \quad O_2 + 4e^- \rightarrow 2O^{2-}}$$

**Sr is the reducing agent. $O_2$ is the oxidizing agent.**

b. The product is an ionic compound whose ions are $Li^+$ and $H^-$. The half-reactions are

$$\mathbf{2Li \rightarrow 2Li^+ + 2e^- \quad and \quad H_2 + 2e^- \rightarrow 2H^-}$$

**Li is the reducing agent. $H_2$ is the oxidizing agent.**

c. The product is an ionic compound whose ions are $Cs^+$ and $Br^-$. The half-reactions are

$$\mathbf{2Cs \rightarrow 2Cs^+ + 2e^- \quad and \quad Br_2 + 2e^- \rightarrow 2Br^-}$$

**Cs is the reducing agent. $Br_2$ is the oxidizing agent.**

d. The product is an ionic compound whose ions are $Mg^{2+}$ and $N^{3-}$. The half-reactions are

$$\mathbf{3Mg \rightarrow 3Mg^{2+} + 6e^- \quad and \quad N_2 + 6e^- \rightarrow 2N^{3-}}$$

**Mg is the reducing agent. $N_2$ is the oxidizing agent.**

4.43 **Strategy:** In general, we follow the rules listed in Section 4.4 of the text for assigning oxidation numbers.

**Solution:** The oxidation number for hydrogen is +1, and for oxygen is –2. The oxidation number for sulfur in $S_8$ is zero (rule 1). Remember that in a molecule, the sum of the oxidation numbers of all the atoms must be zero, and in a polyatomic ion the sum of oxidation numbers of all elements in the ion must equal the net charge of the ion (rule 2).

$$\mathbf{H_2S\ (-2),\ S^{2-}\ (-2),\ HS^-\ (-2) < S_8\ (0) < SO_2\ (+4) < SO_3\ (+6),\ H_2SO_4\ (+6)}$$

The number in parentheses denotes the oxidation number of sulfur.

**Think About It:** In each case, does the sum of the oxidation numbers of all the atoms equal the net charge on the species?

4.45 See the guidelines for assigning oxidation numbers in Section 4.4 of the text.

a. $\underline{Cl}F$: F −1, **Cl +1**

b. $\underline{I}F_7$: F −1, **I +7**

c. $\underline{C}H_4$: H +1, **C −4**

d. $\underline{C_2}H_2$: H +1, **C −1**

e. $\underline{C_2}H_4$: H +1, **C −2**

f. $K_2\underline{Cr}O_4$: K +1, O −2, **Cr +6**

g. $K_2\underline{Cr_2}O_7$: K +1, O −2, **Cr +6**

h. $K\underline{Mn}O_4$: K +1, O −2, **Mn +7**

i. $NaH\underline{C}O_3$: Na +1, H +1, O −2, **C +4**

j. $\underline{Li_2}$: **Li 0** (rule 1)

k. $Na\underline{I}O_3$: Na +1, O −2, **I +5**

l. $K\underline{O_2}$: K +1, **O −1/2**

m. $\underline{P}F_6^-$: F −1, **P +5**

n. $K\underline{Au}Cl_4$: K +1, Cl −1, **Au +3**

4.47 a. $\underline{Cs_2}O$, **+1**

b. $Ca\underline{I}_2$, **–1**

c. $\underline{Al}_2O_3$, **+3**

d. $H_3\underline{As}O_3$, **+3**

e. $\underline{Ti}O_2$, **+4**

f. $\underline{Mo}O_4^{2-}$, **+6**

g. $\underline{Pt}Cl_4^{2-}$, **+2**

h. $\underline{Pt}Cl_6^{2-}$, **+4**

i. $\underline{Sn}F_2$, **+2**

j. $\underline{Cl}F_3$, **+3**

k. $\underline{Sb}F_6^-$, **+5**

4.49 **If nitric acid is a strong oxidizing agent and zinc is a strong reducing agent, then zinc metal will probably reduce nitric acid when the two react; that is, N will gain electrons and the oxidation number of N must decrease. Since the oxidation number of nitrogen in nitric acid is +5 (verify!), then the nitrogen-containing product must have a smaller oxidation number for nitrogen. The only compound in the list that doesn't have a nitrogen oxidation number less than +5 is $N_2O_5$, (what is the oxidation number of N in $N_2O_5$?). This is never a product of the reduction of nitric acid.**

4.51 In order to work this problem, you need to assign the oxidation numbers to all the elements in the compounds. In each case oxygen has an oxidation number of –2. These oxidation numbers should then be compared to the range of possible oxidation numbers that each element can have. (See Figure 4.7 in the text.) **Molecular oxygen is a powerful oxidizing agent. In $SO_3$, the oxidation number of the element bound to oxygen (S) is at its maximum value (+6); the sulfur cannot be oxidized further. The other elements bound to oxygen in this problem have less than their maximum oxidation number and can undergo further oxidation. Only $SO_3$ does not react with molecular oxygen.**

4.53 a. **Decomposition** (This is a special type of decomposition reaction called a "**disproportionation**," in which a single species undergoes both oxidation and reduction.)

b. **Displacement**

c. **Decomposition**

d. **Combination**

4.59 **Strategy:** First calculate the moles of KI needed to prepare the solution, and then convert from moles to grams. How many moles of KI does $5.00 \times 10^2$ mL of a 2.80 *M* solution contain?

**Solution:** From the molarity (2.80 *M*) and the volume ($5.00 \times 10^2$ mL), we can calculate the moles of KI needed to prepare the solution.

First, calculate the moles of KI needed to prepare the solution.

$$\text{mol of KI} = \frac{2.80 \text{ mol KI}}{1000 \text{ mL soln}}(5.00 \times 10^2 \text{ mL soln}) = 1.40 \text{ mol KI}$$

Converting to grams of KI:

$$1.40 \text{ mol KI} \times \frac{166.0 \text{ g KI}}{1 \text{ mol KI}} = \mathbf{232 \text{ g KI}}$$

**Think About It:** As a ball-park estimate, the mass should be given by [molarity (mol/L) × volume (L) = moles × molar mass (g/mol) = grams]. Let's round the molarity to 3 *M* and the molar mass to 150 g, because we are simply making an estimate. This gives: [3 mol/L × (1/2)L × 150 g = 225 g]. This is close to our answer of 232 g.

4.61 Since the problem asks for moles of solute ($MgCl_2$), you should be thinking that you can calculate moles of solute from the molarity and volume of solution.

$$\text{mol} = M \times \text{L}$$

$$60.0 \text{ mL} = 0.0600 \text{ L}$$

$$\text{mol MgCl}_2 = \frac{0.100 \text{ mol MgCl}_2}{1 \text{ L soln}} \times 0.0600 \text{ L soln} = \mathbf{6.00 \times 10^{-3} \text{ mol MgCl}_2}$$

4.63 Molar mass of $C_2H_5OH$ = 46.068 g/mol; molar mass of $C_{12}H_{22}O_{11}$ = 342.3 g/mol; molar mass of NaCl = 58.44 g/mol.

a.

$$? \text{ mol } C_2H_5OH = 29.0 \text{ g } C_2H_5OH \times \frac{1 \text{ mol } C_2H_5OH}{46.068 \text{ g } C_2H_5OH} = 0.6295 \text{ mol } C_2H_5OH$$

$$\text{Molarity} = \frac{\text{mol solute}}{\text{L soln}} = \frac{0.6295 \text{ mol } C_2H_5OH}{0.545 \text{ L soln}} = \mathbf{1.16}\ \boldsymbol{M}$$

b.

$$? \text{ mol } C_{12}H_{22}O_{11} = 15.4 \text{ g } C_{12}H_{22}O_{11} \times \frac{1 \text{ mol } C_{12}H_{22}O_{11}}{342.3 \text{ g } C_{12}H_{22}O_{11}} = 0.04499 \text{ mol } C_{12}H_{22}O_{11}$$

$$\text{Molarity} = \frac{\text{mol solute}}{\text{L soln}} = \frac{0.04499 \text{ mol } C_{12}H_{22}O_{11}}{74.0 \times 10^{-3} \text{ L soln}} = \mathbf{0.608}\ \boldsymbol{M}$$

c.

$$? \text{ mol NaCl} = 9.00 \text{ g NaCl} \times \frac{1 \text{ mol NaCl}}{58.44 \text{ g NaCl}} = 0.154 \text{ mol NaCl}$$

$$\text{Molarity} = \frac{\text{mol solute}}{\text{L soln}} = \frac{0.154 \text{ mol NaCl}}{86.4 \times 10^{-3} \text{ L soln}} = \mathbf{1.78}\ \boldsymbol{M}$$

4.65 First, calculate the moles of each solute. Then, you can calculate the volume (in L) from the molarity and the number of moles of solute.

a.

$$? \text{ mol NaCl} = 2.14 \text{ g NaCl} \times \frac{1 \text{ mol NaCl}}{58.44 \text{ g NaCl}} = 0.03662 \text{ mol NaCl}$$

$$\text{L soln} = \frac{\text{mol solute}}{\text{Molarity}} = \frac{0.03662 \text{ mol NaCl}}{0.270 \text{ mol/L}} = 0.136 \text{ L} = \mathbf{136 \text{ mL soln}}$$

b.

$$? \text{ mol } C_2H_5OH = 4.30 \text{ g } C_2H_5OH \times \frac{1 \text{ mol } C_2H_5OH}{46.068 \text{ g } C_2H_5OH} = 0.09334 \text{ mol } C_2H_5OH$$

$$\text{L soln} = \frac{\text{mol solute}}{\text{Molarity}} = \frac{0.09334 \text{ mol } C_2H_5OH}{1.50 \text{ mol/L}} = 0.0622 \text{ L} = \mathbf{62.2 \text{ mL soln}}$$

c.

$$? \text{ mol } HC_2H_3O_2 = 0.85 \text{ g } HC_2H_3O_2 \times \frac{1 \text{ mol } HC_2H_3O_2}{60.052 \text{ g } HC_2H_3O_2} = 0.0142 \text{ mol } HC_2H_3O_2$$

$$\text{L soln} = \frac{\text{mol solute}}{\text{Molarity}} = \frac{0.0142 \text{ mol } HC_2H_3O_2}{0.30 \text{ mol/L}} = 0.047 \text{ L} = \mathbf{47\ mL\ soln}$$

4.67 **Strategy:** Because the volume of the final solution is greater than the original solution, this is a dilution process. Keep in mind that in a dilution, the concentration of the solution decreases, but the number of moles of the solute remains the same (Equation 4.2).

**Solution:** We prepare for the calculation by tabulating our data. The subscripts c and d stand for concentrated and dilute.

$$M_c = 2.00\ M \qquad M_d = 0.646\ M$$

$$V_c = ? \qquad V_d = 1.00 \text{ L}$$

We substitute the data into Equation (4.3) of the text.

$$M_cV_c = M_dV_d$$

ou can solve the equation algebraically for $V_c$. Then substitute in the given quantities to solve for the volume of 2.00 $M$ HCl needed to prepare 1.00 L of a 0.646 $M$ HCl solution.

$$\mathbf{Vc} = \frac{M_d \times M_d}{M_c} = \frac{0.646\ M \times 1.00 \text{ L}}{2.00\ M} = \mathbf{0.323\ L = 323\ mL}$$

To prepare the 0.646 $M$ solution, you would **dilute 323 mL of the 2.00 *M* HCl solution to a final volume of 1.00 L.**

4.69 You can solve the equation algebraically for $V_c$. Then substitute in the given quantities to solve the for the volume of 4.00 $M$ $HNO_3$ needed to prepare 60.0 mL of a 0.200 $M$ $HNO_3$ solution.

$$V_c = \frac{M_d \times V_d}{M_c} = \frac{0.200\ M \times 60.00 \text{ mL}}{4.00\ M} = \mathbf{3.00\ mL}$$

To prepare the 0.200 $M$ solution, you would **dilute 3.00 mL of the 4.00 *M* $HNO_3$ solution to a final volume of 60.0 mL.**

4.71 a.

$$\left(\frac{0.150 \text{ mol } BaCl_2}{1 \text{ L}}\right)\left(\frac{2 \text{ mol } Cl^-}{1 \text{ mol } BaCl_2}\right) = \mathbf{0.300\ \mathit{M}\ Cl^-}$$

$$\left(\frac{0.566 \text{ mol NaCl}}{1 \text{ L}}\right)\left(\frac{1 \text{ mol } Cl^-}{1 \text{ mol NaCl}}\right) = \mathbf{0.566\ \mathit{M}\ Cl^-}$$

$$\left(\frac{1.202 \text{ mol AlCl}_3}{1 \text{ L}}\right)\left(\frac{3 \text{ mol Cl}^-}{1 \text{ mol AlCl}_3}\right) = \mathbf{3.606}\ \boldsymbol{M}\ \mathbf{Cl^-}$$

b.

$$\left(2.55\frac{\text{mol NO}_3^{2-}}{\text{L}}\right)\left(\frac{1 \text{ mol Sr(NO}_3)_2}{2 \text{ mol NO}_3^{2-}}\right) = \mathbf{1.28}\ \boldsymbol{M}\ \mathbf{Sr(NO_3)_2}$$

4.73 Moles of nitrate in the 0.992 *M* solution:

$$(0.0950 \text{ L})\left(\frac{0.992 \text{ mol KNO}_3}{1 \text{ L}}\right)\left(\frac{1 \text{ mol NO}_3^-}{1 \text{ mol KNO}_3}\right) = 0.0942 \text{ mol NO}_3^-$$

Moles of nitrate in the 1.570 *M* solution:

$$(0.1555 \text{ L})\left(\frac{1.570 \text{ mol Ca(NO}_3)_2}{1 \text{ L}}\right)\left(\frac{2 \text{ mol NO}_3^-}{1 \text{ mol Ca(NO}_3)_2}\right) = 0.4883 \text{ mol NO}_3^-$$

$$\text{Total nitrate} = 0.0942 \text{ mol} + 0.4883 \text{ mol} = 0.5825 \text{ mol}$$

Assume the volumes are additive. Then, the volume of the final solution is 0.0950 L + 0.1555 L = 0.2505 L. The concentration of nitrate in the final solution is 0.5825 mol / 0.2505 L = **2.325 *M*.**

4.81 The balanced equation is:

$$CaCl_2(aq) + 2AgNO_3(aq) \rightarrow Ca(NO_3)_2(aq) + 2AgCl(s)$$

We need to determine the limiting reactant. $Ag^+$ and $Cl^-$ combine in a 1:1 mole ratio to produce AgCl. Let's calculate the amount of $Ag^+$ and $Cl^-$ in solution.

$$\text{mol Ag}^+ = \frac{0.100 \text{ mol Ag}^+}{1000 \text{ mL soln}} \times 15.0 \text{ mL soln} = 1.50 \times 10^{-3} \text{ mol Ag}^+$$

$$\text{mol Cl}^- = \frac{0.150 \text{ mol CaCl}_2}{1000 \text{ mL soln}} \times \frac{2 \text{ mol Cl}^-}{1 \text{ mol CaCl}_2} \times 30.0 \text{ mL soln} = 9.00 \times 10^{-3} \text{ mol Cl}^-$$

Since $Ag^+$ and $Cl^-$ combine in a 1:1 mole ratio, $AgNO_3$ is the limiting reactant. Only $1.50 \times 10^{-3}$ mole of AgCl can form. Converting to grams of AgCl:

$$1.50 \times 10^{-3} \text{ mol AgCl} \times \frac{143.35 \text{ g AgCl}}{1 \text{ mol AgCl}} = \mathbf{0.215\ g\ AgCl}$$

4.83 The net ionic equation is:

$$\mathbf{Ag^+(aq) + Cl^-(aq) \rightarrow AgCl(s)}$$

One mole of $Cl^-$ is required per mole of $Ag^+$. First, find the number of moles of $Ag^+$.

$$\text{mol } Ag^+ = \frac{0.0113 \text{ mol } Ag^+}{1000 \text{ mL soln}} \times (2.50 \times 10^2 \text{ mL soln}) = 2.825 \times 10^{-3} \text{ mol } Ag^+$$

Now, calculate the mass of NaCl using the mole ratio from the balanced equation.

$$(2.825 \times 10^{-3} \text{ mol } Ag^+) \times \frac{1 \text{ mol } Cl^-}{1 \text{ mol } Ag^+} \times \frac{1 \text{ mol NaCl}}{1 \text{ mol } Cl^-} \times \frac{58.44 \text{ g NaCl}}{1 \text{ mol NaCl}} = \mathbf{0.165 \text{ g NaCl}}$$

4.85 a. In order to have the correct mole ratio to solve the problem, you must start with a balanced chemical equation.

$$HCl(aq) + NaOH(aq) \rightarrow NaCl(aq) + H_2O(l)$$

From the molarity and volume of the HCl solution, you can calculate moles of HCl. Then, using the mole ratio from the balanced equation above, you can calculate moles of NaOH.

$$? \text{ mol NaOH} = 25.00 \text{ mL} \times \frac{2.430 \text{ mol HCl}}{1000 \text{ mL soln}} \times \frac{1 \text{ mol NaOH}}{1 \text{ mol HCl}} = 6.075 \times 10^{-2} \text{ mol NaOH}$$

Solving for the volume of NaOH:

$$\text{liters of solution} = \frac{\text{moles of solute}}{M}$$

$$\text{volume of NaOH} = \frac{6.075 \times 10^{-2} \text{ mol NaOH}}{1.420 \text{ mol/L}} = 4.278 \times 10^{-2} \text{ L} = \mathbf{42.78 \text{ mL}}$$

b. This problem is similar to part (a). The difference is that the mole ratio between base and acid is 2:1.

$$H_2SO_4(aq) + 2NaOH(aq) \rightarrow Na_2SO_4(aq) + 2H_2O(l)$$

$$? \text{ mol NaOH} = 25.00 \text{ mL} \times \frac{4.500 \text{ mol } H_2SO_4}{1000 \text{ mL soln}} \times \frac{2 \text{ mol NaOH}}{1 \text{ mol } H_2SO_4} = 0.2250 \text{ mol NaOH}$$

$$\text{volume of NaOH} = \frac{0.2250 \text{ mol NaOH}}{1.420 \text{ mol/L}} = 0.1585 \text{ L} = \mathbf{158.5 \text{ mL}}$$

c. This problem is similar to parts (a) and (b). The difference is that the mole ratio between base and acid is 3:1.

$$H_3PO_4(aq) + 3NaOH(aq) \longrightarrow Na_3PO_4(aq) + 3H_2O(l)$$

$$? \text{ mol NaOH} = 25.00 \text{ mL} \times \frac{1.500 \text{ mol } H_3PO_4}{1000 \text{ mL soln}} \times \frac{3 \text{ mol NaOH}}{1 \text{ mol } H_3PO_4} = 0.1125 \text{ mol NaOH}$$

$$\text{volume of NaOH} = \frac{0.1125 \text{ mol NaOH}}{1.420 \text{ mol/L}} = 0.07923 \text{ L} = \mathbf{79.23\ mL}$$

4.87 Diagram (b) corresponds to $H_3PO_4$. Diagram (c) corresponds to HCl. Diagram (d) corresponds to $H_2SO_4$.

4.89 In redox reactions the oxidation numbers of elements change. To test whether an equation represents a redox process, assign the oxidation numbers to each of the elements in the reactants and products. If oxidation numbers change, it is a redox reaction.

a. On the left the oxidation number of chlorine in $Cl_2$ is zero (rule 1). On the right it is $-1$ in $Cl^-$ (rule 2) and $+1$ in $OCl^-$. Since chlorine is both oxidized and reduced, this is a disproportionation **redox reaction**.

b. The oxidation numbers of calcium and carbon do not change. This is not a redox reaction; it is a **precipitation reaction**.

c. The oxidation numbers of nitrogen and hydrogen do not change. This is not a redox reaction; it is an **acid-base reaction**.

d. The oxidation numbers of carbon, chlorine, chromium, and oxygen do not change. This is not a redox reaction; it doesn't fit easily into any category, but could be considered as a type of **combination reaction**.

e. The oxidation number of calcium changes from 0 to +2, and the oxidation number of fluorine changes from 0 to $-1$. This is a combination **redox reaction**.

The remaining parts (f) through (j) can be worked the same way.

f. **Redox**

g. **Precipitation**

h. **Redox**

i. **Redox**

j. **Redox**

4.91 Choice **(d)**, 0.20 *M* $Mg(NO_3)_2$, should be the best conductor of electricity; the total ion concentration in this solution is 0.60 *M*. The total ion concentrations for solutions (a) and (c) are 0.40 *M* and 0.50 *M*, respectively. We can rule out choice (b), because acetic acid is a weak electrolyte.

4.93 The balanced equation for the displacement reaction is:

$$Zn(s) + CuSO_4(aq) \rightarrow ZnSO_4(aq) + Cu(s)$$

The moles of $CuSO_4$ that react with 7.89 g of zinc are:

$$7.89 \text{ g Zn} \times \frac{1 \text{ mol Zn}}{65.41 \text{ g Zn}} \times \frac{1 \text{ mol CuSO}_4}{1 \text{ mol Zn}} = 0.1206 \text{ mol CuSO}_4$$

The volume of the 0.156 *M* $CuSO_4$ solution needed to react with 7.89 g Zn is:

$$\text{L of soln} = \frac{\text{mole solute}}{M} = \frac{0.1206 \text{ mol CuSO}_4}{0.156 \text{ mol/L}} = 0.773 \text{ L} = \mathbf{773 \text{ mL}}$$

**Think About It:** Would you expect Zn to displace $Cu^{2+}$ from solution, as shown in the equation?

4.95 a. **Weak electrolyte**. Ethanolamine ($C_2H_5ONH_2$) is a weak base.

b. **Strong electrolyte**. Potassium fluoride (KF) is a soluble ionic compound.

c. **Strong electrolyte**. Ammonium nitrate ($NH_4NO_3$) is a soluble ionic compound.

d. **Nonelectrolyte**. Isopropanol ($C_3H_7OH$) is a molecular compound that is neither an acid nor a base.

4.97 a. Weak electrolytes exist predominantly as molecules in solution. **$C_2H_5ONH_2$ molecules**.

b. Strong electrolytes exist predominantly as ions in solution. **$K^+$ and $F^-$ ions**.

c. Strong electrolytes exist predominantly as ions in solution. **$NH_4^+$ and $NO_3^-$ ions.**

d. Nonelectrolytes exist entirely as molecules in solution. **$C_3H_7OH$ molecules.**

4.99 The neutralization reaction is: $HA(aq) + NaOH(aq) \rightarrow NaA(aq) + H_2O(l)$

The mole ratio between the acid and NaOH is 1:1. The moles of HA that react with NaOH are:

$$20.27 \text{ mL soln} \times \frac{0.1578 \text{ mol NaOH}}{1000 \text{ mL soln}} \times \frac{1 \text{ mol HA}}{1 \text{ mol NaOH}} = 3.1986 \times 10^{-3} \text{ mol HA}$$

3.664 g of the acid reacted with the base. The molar mass of the acid is:

$$\text{Molar mass} = \frac{3.664 \text{ g HA}}{3.1986 \times 10^{-3} \text{ mol HA}} = \mathbf{1146\ g/mol}$$

4.101 Let's call the original solution, soln 1; the first dilution, soln 2; and the second dilution, soln 3. Start with the concentration of soln 3, 0.00383 *M*. From the concentration and volume of soln 3, we can find the concentration of soln 2. Then, from the concentration and volume of soln 2, we can find the concentration of soln 1, the original solution.

$$M_2V_2 = M_3V_3$$

$$M_2 = \frac{M_3V_3}{V_2} = \frac{(0.00383\ M)(1.000 \times 10^3 \text{ mL})}{25.00 \text{ mL}} = 0.1532\ M$$

$$M_1V_1 = M_2V_2$$

$$\boldsymbol{M_1} = \frac{M_2V_2}{V_1} = \frac{(0.1532\ M)(125.0 \text{ mL})}{15.00 \text{ mL}} = \mathbf{1.28}\ \boldsymbol{M}$$

4.103 The balanced equation is:

$$Ba(OH)_2(aq) + Na_2SO_4(aq) \rightarrow BaSO_4(s) + 2NaOH(aq)$$

moles $Ba(OH)_2$: $(2.27 \text{ L})(0.0820 \text{ mol/L}) = 0.1861 \text{ mol } Ba(OH)_2$

moles $Na_2SO_4$: $(3.06 \text{ L})(0.0664 \text{ mol/L}) = 0.2032 \text{ mol } Na_2SO_4$

Since the mole ratio between $Ba(OH)_2$ and $Na_2SO_4$ is 1:1, $Ba(OH)_2$ is the limiting reactant. The mass of $BaSO_4$ formed is:

$$0.1861 \text{ mol } Ba(OH)_2 \times \frac{1 \text{ mol } BaSO_4}{1 \text{ mol } Ba(OH)_2} \times \frac{233.37 \text{ g } BaSO_4}{1 \text{ mol } BaSO_4} = \mathbf{43.4\ g\ BaSO_4}$$

4.105 First, calculate the number of moles of glucose present.

$$\frac{0.513 \text{ mol glucose}}{1000 \text{ mL soln}} \times 60.0 \text{ mL} = 0.03078 \text{ mol glucose}$$

$$\frac{2.33 \text{ mol glucose}}{1000 \text{ mL soln}} \times 120.0 \text{ mL} = 0.2796 \text{ mol glucose}$$

Add the moles of glucose, then divide by the total volume of the combined solutions to calculate the molarity.

$$60 \text{ mL} + 120.0 \text{ mL} = 180.0 \text{ mL} = 0.180 \text{ L}$$

$$\text{Molarity of final solution} = \frac{(0.03078 + 0.2796) \text{ mol glucose}}{0.180 \text{ L}} = 1.72 \text{ mol/L} = \mathbf{1.72}\ \boldsymbol{M}$$

4.107 The three chemical tests might include:

**(1)** **Electrolysis to ascertain if hydrogen and oxygen were produced,**

**(2)** **The reaction with an alkali metal to see if a base and hydrogen gas were produced, and**

**(3)** **The dissolution of a metal oxide to see if a base was produced (or a nonmetal oxide to see if an acid was produced).**

4.109 Moles of $Ca(NO_3)_2$ from the 0.568 *M* solution:

$$(0.0462 \text{ L})\left(\frac{0.568 \text{ mol Ca}(NO_3)_2}{1 \text{ L}}\right) = 0.0262 \text{ mol Ca}(NO_3)_2$$

Moles of $Ca(NO_3)_2$ from the 1.396 *M* solution:

$$(0.0805 \text{ L})\left(\frac{1.396 \text{ mol Ca}(NO_3)_2}{1 \text{ L}}\right) = 0.112 \text{ mol Ca}(NO_3)_2$$

Total moles of $Ca(NO_3)_2$ = 0.0262 + 0.112 = 0.138 mol.

Assume the volumes are additive. Then, the volume of the final solution will be 0.0462 + 0.0805 = 0.1267 L. The concentration is 0.138 mol / 0.1267 L = 1.09 *M* $Ca(NO_3)_2$.

4.111 Diagram (a) showing $Ag^+$ and $NO_3^-$ ions. The balanced molecular equation for the reaction is

$$AgOH(aq) + HNO_3(aq) \longrightarrow H_2O(l) + AgNO_3(aq)$$

4.113 a. **Check with litmus paper, combine with carbonate or bicarbonate to see if $CO_2$ gas is produced, combine with a base and check for neutralization with an indicator.**

b. **Titrate a known quantity of acid with a standard NaOH solution. Since it is a monoprotic acid, the moles of NaOH reacted equals the moles of the acid. Dividing the mass of acid by the number of moles gives the molar mass of the acid.**

c. **Visually compare the conductivity of the acid with a standard NaCl solution of the same molar concentration. A strong acid will have a similar conductivity to the NaCl solution. The conductivity of a weak acid will be considerably less than the NaCl solution.**

4.115 **No.** In a redox reaction, electrons must be transferred between reacting species. In other words, oxidation numbers must change in a redox reation. In both $O_2$ (molecular oxygen) and $O_3$ (ozone), the oxidation number of oxygen is zero. This is ***not*** a redox reaction.

4.117 a. An acid and a base react to form water and a salt. Potassium iodide is a salt; therefore, the acid and base are chosen to produce this salt.

$$\mathbf{HI}(aq) + \mathbf{KOH}(aq) \rightarrow \mathbf{KI}(aq) + \mathbf{H_2O}(l)$$

**The resulting solution could be evaporated to dryness to isolate the KI.**

b. Acids react with carbonates to form carbon dioxide gas. Again, chose the acid and carbonate salt so that KI is produced.

$$\mathbf{2HI}(aq) + \mathbf{K_2CO_3}(aq) \rightarrow \mathbf{2KI}(aq) + \mathbf{CO_2}(g) + \mathbf{H_2O}(l)$$

Again, **the resulting solution could be evaporated to dryness to isolate the KI.**

4.119 All three products are water insoluble. Use this information in formulating your answer.

a. **Combine any soluble magnesium salt with a soluble hydroxide, filter the precipitate.**

Example:

$$MgCl_2(aq) + 2NaOH(aq) \rightarrow \mathbf{Mg(OH)_2}(s) + 2NaCl(aq)$$

b. **Combine any soluble silver salt with any soluble iodide salt, filter the precipitate.**

Example:

$$AgNO_3(aq) + NaI(aq) \rightarrow \mathbf{AgI}(s) + NaNO_3(aq)$$

c. **Combine any soluble barium salt with any soluble phosphate salt, filter the precipitate.**

Example:

$$3Ba(OH)_2(aq) + 2H_3PO_4(aq) \rightarrow \mathbf{Ba_3(PO_4)_2(s)} + 6H_2O(l)$$

4.121 a. **Add $Na_2SO_4$**, or any soluble sulfate salt. Barium sulfate would precipitate leaving sodium ions in solution.

b. **Add KOH**, or any soluble compound containing hydroxide, carbonate, phosphate, or sulfide ion to precipitate magnesium cations. In the case of adding KOH, $Mg(OH)_2$ would precipitate leaving potassium cations in solution.

c. **Add $AgNO_3$**, or any soluble silver salt. AgBr would precipitate, leaving nitrate ions in solution.

d. **Add $Ca(NO_3)_2$**, or any soluble compound containing a cation other than ammonium or a Group 1A cation to precipitate the phosphate ions; the nitrate ions will remain in solution.

e. **Add $Mg(NO_3)_2$**, or any solution containing a cation other than ammonium or a Group 1A cation to precipitate the carbonate ions; the nitrate ions will remain in solution.

4.123 Reaction 1:

$$\mathbf{SO_3^{2-}(aq) + H_2O_2(aq) \rightarrow SO_4^{2-}(aq) + H_2O(l)}$$

Reaction 2:

$$\mathbf{SO_4^{2-}(aq) + Ba^{2+}(aq) \rightarrow BaSO_4(s)}$$

4.125 We assume that O has an oxidation state of –2. From this, we determine the ratio of combination of chlorine and oxygen necessary to make a neutral formula for each of the specified oxidation states of Cl.

**$Cl_2O$ (Cl = +1)** **$Cl_2O_3$ (Cl = +3)** **$ClO_2$ (Cl = +4)** **$Cl_2O_6$ (Cl = +6)** **$Cl_2O_7$ (Cl = +7)**

4.127 The reaction between $Mg(NO_3)_2$ and NaOH is:

$$Mg(NO_3)_2(aq) + 2NaOH(aq) \rightarrow Mg(OH)_2(s) + 2NaNO_3(aq)$$

Magnesium hydroxide, $Mg(OH)_2$, precipitates from solution. $Na^+$ and $NO_3^-$ are spectator ions. This is most likely a limiting reactant problem as the amounts of both reactants are given. Let's first determine which reactant is the limiting reactant before we try to determine the concentration of ions remaining in the solution.

$$1.615\text{ g Mg(NO}_3)_2 \times \frac{1\text{ mol Mg(NO}_3)_2}{148.33\text{ g Mg(NO}_3)_2} = 0.010888\text{ mol Mg(NO}_3)_2$$

$$1.073\text{ g NaOH} \times \frac{1\text{ mol NaOH}}{39.998\text{ g NaOH}} = 0.026826\text{ mol NaOH}$$

From the balanced equation, we need twice as many moles of NaOH compared to $Mg(NO_3)_2$. We have more than twice as much NaOH (2 × 0.010888 mol = 0.021776 mol) and therefore $Mg(NO_3)_2$ is the limiting reactant. NaOH is in excess and ions of $Na^+$, $OH^-$, and $NO_3^-$ will remain in solution. Because $Na^+$ and $NO_3^-$ are spectator ions, the number of moles after reaction will equal the initial number of moles. The excess moles of $OH^-$ need to be calculated based on the amount that reacts with $Mg^{2+}$. The combined volume of the two solutions is: 22.02 mL + 28.64 mL = 50.66 mL = 0.05066 L.

$$[\text{Na}^+] = 0.026826\text{ mol NaOH} \times \frac{1\text{ mol Na}^+}{1\text{ mol NaOH}} \times \frac{1}{0.05066\text{ L}} = \mathbf{0.5295}\ \boldsymbol{M}$$

$$[\text{NO}_3^-] = 0.010888\text{ mol Mg(NO}_3)_2 \times \frac{2\text{ mol NO}_3^-}{1\text{ mol Mg(NO}_3)_2} \times \frac{1}{0.05066\text{ L}} = \mathbf{0.4298}\ \boldsymbol{M}$$

The moles of $OH^-$ reacted are:

$$0.010888\text{ mol Mg}^{2+} \times \frac{2\text{ mol OH}^-}{1\text{ mol Mg}^{2+}} = 0.021776\text{ mol OH}^-\text{ reacted}$$

The moles of excess $OH^-$ are:

$$0.026826\text{ mol} - 0.021776\text{ mol} = 0.005050\text{ mol OH}^-$$

$$[\text{OH}^-] = \frac{0.005050\text{ mol}}{0.05066\text{ L}} = \mathbf{0.09968}\ \boldsymbol{M}$$

$$[\mathbf{Mg^{2+}}] \approx \mathbf{0}\ \boldsymbol{M}$$

(The concentration of $Mg^{2+}$ is approximately zero as almost all of it will precipitate as $Mg(OH)_2$.)

4.129 Moles of $KMnO_4$ from the 1.66 *M* solution:

$$(0.0352\text{ L})\left(\frac{1.66\text{ mol KMnO}_4}{1\text{ L}}\right) = 0.0584\text{ mol KMnO}_4$$

Moles of $KMnO_4$ from the 0.892 *M* solution:

$$(0.0167\text{ L})\left(\frac{0.892\text{ mol KMnO}_4}{1\text{ L}}\right) = 0.0149\text{ mol KMnO}_4$$

Total moles of $KMnO_4$ = 0.0584 + 0.0149 = 0.0733 mol.

Assume the volumes are additive. Then, the volume of the final solution will be 0.0352 + 0.0167 = 0.0519 L. The concentration is 0.0733 mol / 0.0519 L = 1.41 *M* $KMnO_4$.

4.131 a. **The precipitate $CaSO_4$ formed over Ca preventing the Ca from reacting with the sulfuric acid.**

b. **Aluminum is protected by a tenacious oxide layer with the composition $Al_2O_3$.**

c. **These metals react more readily with water.**

$$2Na(s) + 2H_2O(l) \rightarrow 2NaOH(aq) + H_2(g)$$

d. **The metal should be placed below Fe and above H.**

e. **Any metal above Al in the activity series will react with $Al^{3+}$. Metals from Mg to Li will work.**

4.133 The precipitation reaction is:

$$Ag^+(aq) + Br^-(aq) \rightarrow AgBr(s)$$

In this problem, the relative amounts of NaBr and $CaBr_2$ are not known. However, the total amount of $Br^-$ in the mixture can be determined from the amount of AgBr produced. Find the number of moles of $Br^-$.

$$1.6930\text{ g AgBr} \times \frac{1\text{ mol AgBr}}{187.8\text{ g AgBr}} \times \frac{1\text{ mol Br}^-}{1\text{ mol AgBr}} = 9.0149 \times 10^{-3}\text{ mol Br}^-$$

The amount of $Br^-$ comes from both NaBr and $CaBr_2$. Let $x$ = number of moles NaBr. Then, the number of moles of $CaBr_2 = \frac{9.0149 \times 10^{-3}\text{ mol} - x}{2}$. The moles of $CaBr_2$ are divided by 2, because 1 mol of $CaBr_2$ produces 2 moles of $Br^-$. The sum of the NaBr and $CaBr_2$ masses must equal the mass of the mixture, 0.9157 g. We can write:

$$\text{mass NaBr} + \text{mass CaBr}_2 = 0.9157\text{ g}$$

$$\left[x\text{ mol NaBr} \times \frac{102.89\text{ g NaBr}}{1\text{ mol NaBr}}\right] + \left[\left(\frac{9.0149 \times 10^{-3} - x}{2}\right)\text{mol CaBr}_2 \times \frac{199.88\text{ g CaBr}_2}{1\text{ mol CaBr}_2}\right] = 0.9157\text{ g}$$

$$2.95x = 0.014751$$

$$x = 5.0003 \times 10^{-3} = \text{moles NaBr}$$

Converting moles to grams:

$$\text{mass NaBr} = (5.0003 \times 10^{-3} \text{ mol NaBr}) \times \frac{102.89 \text{ g NaBr}}{1 \text{ mol NaBr}} = 0.51448 \text{ g NaBr}$$

The percentage by mass of NaBr in the mixture is:

$$\% \text{ NaBr} = \frac{0.51448 \ g}{0.9157 \ g} \times 100\% = \mathbf{56.2\% \ NaBr}$$

4.135 There are two moles of $Cl^-$ per one mole of $CaCl_2$.

a.

$$25.3 \text{ g CaCl}_2 \times \frac{1 \text{ mol CaCl}_2}{110.98 \text{ g CaCl}_2} \times \frac{2 \text{ mol Cl}^-}{1 \text{ mol CaCl}_2} = 0.4559 \text{ mol Cl}^-$$

$$\text{Molarity Cl}^- = \frac{0.4559 \text{ mol Cl}^-}{0.325 \text{ L soln}} = 1.40 \text{ mol/L} = \mathbf{1.40} \ \boldsymbol{M}$$

b. We need to convert from mol/L to grams in 0.100 L.

$$\frac{1.40 \text{ mol Cl}^-}{1 \text{ L soln}} \times \frac{35.45 \text{ g Cl}^-}{1 \text{ mol Cl}^-} \times 0.100 \text{ L soln} = \mathbf{4.96 \ g \ Cl^-}$$

4.137 a. **acid: $H_3O^+$, base: $OH^-$**

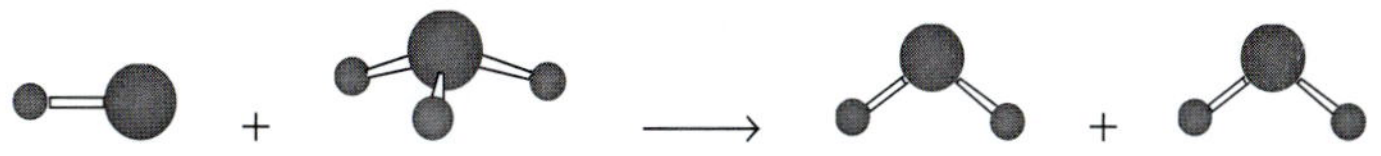

$OH^-$ + $H_3O^+$ ⟶ $H_2O$ + $H_2O$

b. **acid: $NH_4^+$, base: $NH_2^-$**

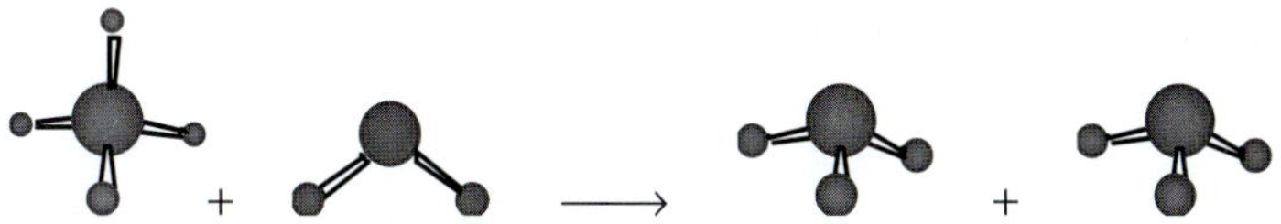

$NH_4^+$ + $NH_2^-$ ⟶ $NH_3$ + $NH_3$

4.139 **When a solid dissolves in solution, the volume of the solution usually changes.**

4.141 **Electric furnace method:**

$$P_4(s) + 5O_2(g) \longrightarrow P_4O_{10}(s) \qquad \text{(redox)}$$

$$P_4O_{10}(s) + 6H_2O(l) \longrightarrow 4H_3PO_4(aq) \qquad \text{(acid-base)}$$

**Wet process:**

$$Ca_5(PO_4)_3F(s) + 5H_2SO_4(aq) \longrightarrow 3H_3PO_4(aq) + HF(aq) + 5CaSO_4(s)$$

(acid-base reaction and precipitation)

4.143 a.

$$\mathbf{CaF_2(s) + H_2SO_4(aq) \rightarrow CaSO_4(s) + 2HF(g)}$$

$$\mathbf{2NaCl(s) + H_2SO_4(aq) \rightarrow Na_2SO_4(aq) + 2HCl(g)}$$

b. HBr and HI ***cannot*** be prepared similarly, because **the sulfuric acid would oxidize $Br^-$ and $I^-$ ions to $Br_2$ and $I_2$**.

c.

$$\mathbf{PBr_3(l) + 3H_2O(l) \rightarrow 3HBr(g) + H_3PO_3(aq)}$$

4.145 a.

$$\mathbf{4KO_2(s) + 2CO_2(g) \rightarrow 2K_2CO_3(s) + 3O_2(g)}$$

b. The oxidation state of O in the $O_2^-$ ion is −1/2. **$O_2^-$ (−1/2).**

c.

$$? \text{ L air} = 7.00 \text{ g KO}_2 \times \frac{1 \text{ mol KO}_2}{71.10 \text{ g KO}_2} \times \frac{2 \text{ mol CO}_2}{4 \text{ mol KO}_2} \times \frac{44.01 \text{ g CO}_2}{1 \text{ mol CO}_2} \times \frac{1 \text{ L air}}{0.063 \text{ g CO}_2} = \mathbf{34.4 \text{ L air}}$$

4.147 Since aspirin is a monoprotic acid, it will react with NaOH in a 1:1 mole ratio.

First, calculate the moles of aspirin in the tablet.

$$12.25 \text{ mL soln} \times \frac{0.1466 \text{ mol NaOH}}{1000 \text{ mL soln}} \times \frac{1 \text{ mol aspirin}}{1 \text{ mol NaOH}} = 1.7959 \times 10^{-3} \text{ mol aspirin}$$

Next, convert from moles of aspirin to grains of aspirin.

$$1.7959 \times 10^{-3} \text{ mol aspirin} \times \frac{180.15 \text{ g aspirin}}{1 \text{ mol aspirin}} \times \frac{1 \text{ grain}}{0.0648 \text{ g}} = \mathbf{4.99 \text{ grains aspirin in one tablet}}$$

4.149 a.

$$\mathbf{Pb(NO_3)_2(\mathit{aq}) + Na_2SO_4(\mathit{aq}) \rightarrow PbSO_4(s) + 2NaNO_3(\mathit{aq})}$$

$$\mathbf{Pb^{2+}(\mathit{aq}) + SO_4^{2-}(\mathit{aq}) \rightarrow PbSO_4(\mathit{s})}$$

b. First, calculate the moles of $Pb^{2+}$ in the polluted water.

$$0.00450 \text{ g } Na_2SO_4 \times \frac{1 \text{ mol } Na_2SO_4}{142.05 \text{ g } Na_2SO_4} \times \frac{1 \text{ mol } Pb(NO_3)_2}{1 \text{ mol } Na_2SO_4} \times \frac{1 \text{ mol } Pb^{2+}}{1 \text{ mol } Pb(NO_3)_2} = 3.168 \times 10^{-5} \text{ mol } Pb^{2+}$$

The volume of the polluted water sample is 500 mL (0.500 L). The molar concentration of $Pb^{2+}$ is:

$$[Pb^{2+}] = \frac{\text{mol } Pb^{2+}}{\text{L soln}} = \frac{3.168 \times 10^{-5} \text{ mol } Pb^{2+}}{0.500 \text{ L soln}} = \mathbf{6.34 \times 10^{-5}} \; \boldsymbol{M}$$

4.151 Write the ionic equation:

$$2Na^+(aq) + S^{2-}(aq) + Cu^{2+}(aq) + SO_4^{2-}(aq) \rightarrow 2Na^+(aq) + SO_4^{2-}(aq) + CuS(s)$$

Cancel the spectator ions and write the net ionic equation:

$$Cu^{2+}(aq) + S^{2-}(aq) \rightarrow CuS(s)$$

Calculate the number of moles of $Cu^{2+}$ in 0.0177 g of CuS:

$$(0.0177 \text{ g CuS})\left(\frac{1 \text{ mol CuS}}{95.62 \text{ g CuS}}\right)\left(\frac{1 \text{ mol } Cu^{2+}}{1 \text{ mol CuS}}\right) = 1.85 \times 10^{-4} \text{ mol } Cu^{2+}$$

The concentration of $Cu^{2+}$ in the discharge is $\left(1.85 \times 10^{-4} \text{ mol } Cu^{2+}\right)/\left(0.800 \text{ L}\right) = 2.31 \times 10^{-4} \; M \; Cu^{2+}$.

4.153 a. $CH_3CH_2OH$ is a nonelectrolyte, $K_2Cr_2O_7$ is a strong electrolyte, and $H_2SO_4$ is a strong electrolyte. $HC_2H_3O_2$ is a weak electrolyte, $Cr_2(SO_4)_3$ is a strong electrolyte, $K_2SO_4$ is a strong electrolyte, and $H_2O$ is a nonelectolyte.

b. The ionic equation is:

$$3CH_3CH_2OH(g) + 4K^+(aq) + 2Cr_2O_7^{2-}(aq) + 8H^+(aq) + 8HSO_4^-(aq) \rightarrow$$
$$3HC_2H_3O_2(aq) + 4Cr^{3+}(aq) + 6SO_4^{2-}(aq) + 4K^+(aq) + 2SO_4^{2-}(aq) + 11H_2O(l)$$

The net ionic equation is:

$$3CH_3CH_2OH(g) + 2Cr_2O_7^{2-}(aq) + 8H^+(aq) + 8HSO_4^-(aq) \rightarrow$$
$$3HC_2H_3O_2(aq) + 4Cr^{3+}(aq) + 8SO_4^{2-}(aq) + 11H_2O(l)$$

c. $$3CH_3CH_2OH(g) + 2K_2Cr_2O_7(aq) + 8H_2SO_4(aq) \rightarrow 3HC_2H_3O_2(aq) + 2Cr_2(SO_4)_3(aq) + 2K_2SO_4(aq) + 11H_2O(l)$$

| | $CH_3CH_2OH$ | $K_2Cr_2O_7$ | $H_2SO_4$ | $HC_2H_3O_2$ | $Cr_2(SO_4)_3$ | $K_2SO_4$ | $H_2O$ |
|---|---|---|---|---|---|---|---|
| Oxidation number (per atom) | −2, +1, −2, +1, −2, +1 | +1, +6, −2 | +1, +6, −2 | +1, 0, +1, −2 | +3, +6, −2 | +1, +6, −2 | +1, −2 |
| Total | −2, +3, −2, +2, −2, +1 | +2, +12, −14 | +2, +6, −8 | +1, 0, +3, −4 | +6, +18, −24 | +2, +6, −8 | +2, −2 |

d. $$\frac{0.025 \text{ g } K_2Cr_2O_7}{100 \text{ mL solution}} \times \frac{1 \text{ mol } K_2Cr_2O_7}{294.2 \text{ g } K_2Cr_2O_7} \times \frac{1000 \text{ mL solution}}{\text{L solution}} = 8.5 \times 10^{-4}\ M$$

e. $$(0.014\ M)x = (8.5 \times 10^{-4}\ M)(250 \text{ mL})$$
$$x = 15 \text{ mL}$$

f. $$[K^+] = (2)(8.5 \times 10^{-4}\ M) = 1.7 \times 10^{-3}\ M$$
$$[Cr_2O_7^{2-}] = 8.5 \times 10^{-4}\ M$$

# Chapter 5

# Thermochemistry

5.7 Law of Conservation of Energy

5.9 Energy is needed to break chemical bonds, while energy is released when bonds are formed.

5.13 Using Equation 5.3 and the sign conventions for $q$ and $w$,

$$\Delta U = q + w$$

$$\mathbf{\Delta U} = (-93\text{ J}) + (47\text{ J}) = \mathbf{-46\ J}$$

5.15 Using Equation 5.3 and the sign conventions for $q$ and $w$,

$$\Delta U = q + w$$

$$510\text{ J} = (-415) + w$$

$$\mathbf{w = 925\ J\ (work\ done\ \mathit{on}\ the\ system)}$$

5.17 Diagram (i) shows no change in volume. Therefore, $w = 0$.

Diagram (ii) shows an increase in volume. Therefore, $w < 0$.

Diagram (iii) shows a decrease in volume. Therefore, $w > 0$.

a. Rearrange equation 5.3 to solve for work.

$$\Delta U = q + w$$

$$w = \Delta U - q$$

Since $\Delta U < 0$ and $q > 0$, w must be less than 0. This corresponds to diagram (ii).

b. Since $w < 0$, this corresponds to diagram (ii).

c. Since $w < 0$, this corresponds to diagram (ii).

5.25 a. Because the external pressure is zero, no work is done in the expansion.

$$w = -P\Delta V = -(0)(89.3 - 26.7)\text{mL}$$

$$\boldsymbol{w = 0\ \text{J}}$$

b. The external, opposing pressure is 1.5 atm, so:

$$w = -P\Delta V = -(1.5\ \text{atm})(89.3 - 26.7)\text{mL}$$

$$w = -94\ \text{mL}\cdot\text{atm} \times \frac{0.001\ \text{L}}{1\ \text{mL}} = -0.094\ \text{L}\cdot\text{atm}$$

To convert the answer to joules, we write:

$$\boldsymbol{w} = -0.094\ \text{L}\cdot\text{atm} \times \frac{101.3\ \text{J}}{1\ \text{L}\cdot\text{atm}} = \mathbf{-9.5\ J}$$

c. The external, opposing pressure is 2.8 atm, so

$$w = -P\Delta V = -(2.8\ \text{atm})(89.3 - 26.7)\text{mL}$$

$$w = (-1.8 \times 10^2\ \text{mL}\cdot\text{atm}) \times \frac{0.001\ \text{L}}{1\ \text{mL}} = -0.18\ \text{L}\cdot\text{atm}$$

To convert the answer to joules, we write:

$$\boldsymbol{w} = -0.18\ \text{L}\cdot\text{atm} \times \frac{101.3\ \text{J}}{1\ \text{L}\cdot\text{atm}} = \mathbf{-18\ J}$$

5.27 **Strategy:** The thermochemical equation shows that for every 2 moles of ZnS roasted, 879 kJ of heat are given off (note the negative sign). We can write a conversion factor from this information.

$$\frac{-879\ \text{kJ}}{2\ \text{mol ZnS}}$$

How many moles of ZnS are in 1 g of ZnS? What conversion factor is needed to convert between grams and moles?

**Solution:** We need to first calculate the number of moles of ZnS in 1 g of the compound. Then, we can convert to the number of kilojoules produced from the exothermic reaction. The sequence of conversions is:

grams of ZnS $\rightarrow$ moles of ZnS $\rightarrow$ kilojoules of heat generated

Therefore, the enthalpy change for the roasting of 1 g of ZnS is:

$$\Delta H = \frac{-879\text{ kJ}}{2\text{ mol ZnS}} \times \frac{1\text{ mol ZnS}}{97.48\text{ g ZnS}} = -4.51\text{ kJ/g ZnS}$$

Because the question asked *how much heat* is evolved, it is not necessary to include the negative sign in the answer. The roasting of ZnS produces **4.51 kJ/g**.

5.29 We can calculate $\Delta U$ by rearranging Equation 5.10 of the text.

$$\Delta U = \Delta H - P\Delta V$$

Substituting into the above equation:

$$\Delta U = 483.6\text{ kJ/mol} - (1.00\text{ atm})(32.7\text{ L})\left(\frac{101.3\text{ J}}{\text{L}\cdot\text{atm}}\right)\left(\frac{1\text{ kJ}}{1000\text{ J}}\right)$$

$$\mathbf{\Delta U = 4.80 \times 10^2\ kJ}$$

5.31 The second reaction is the reverse of the first reaction. Consequently, the magnitude of $\Delta$H remains the same, but its sign changes.

$$\mathbf{\Delta H = 595.8\ kJ/mol}$$

5.35 $q = m_{Cu}s_{Cu}\Delta t = (6.22 \times 10^3\text{ g})(0.385\text{ J/g}\cdot{}^\circ\text{C})(324.3^\circ\text{C} - 20.5^\circ\text{C}) = 7.28 \times 10^5\text{ J} = \mathbf{728\ kJ}$

5.37 **Strategy:** We know the masses of gold and iron as well as the initial temperatures of each. We can look up the specific heats of gold and iron in Table 10.2 of the text. Assuming no heat is lost to the surroundings, we can equate the heat lost by the iron sheet to the heat gained by the gold sheet. With this information, we can solve for the final temperature of the combined metals.

**Solution:** Treating the calorimeter as an isolated system (no heat lost to the surroundings), we can write:

$$q_{Au} + q_{Fe} = 0$$

or

$$q_{Au} = -q_{Fe}$$

The heat gained by the gold sheet is given by:

$$q_{Au} = m_{Au}s_{Au}\Delta T = (10.0\text{ g})(0.129\text{ J/g}\cdot{}^\circ\text{C})(T_f - 18.0)^\circ\text{C}$$

where $m$ and $s$ are the mass and specific heat, and $\Delta T = T_{\text{final}} - T_{\text{initial}}$.

The heat lost by the iron sheet is given by:

$$q_{\text{Fe}} = m_{\text{Fe}}s_{\text{Fe}}\Delta T = (20.0\text{ g})(0.444\text{ J/g}\cdot{}^\circ\text{C})(T_f - 55.6)^\circ\text{C}$$

Substituting into the equation derived above, we can solve for $T_f$.

$$q_{\text{Au}} = -q_{\text{Fe}}$$

$$(10.0\text{ g})(0.129\text{ J/g}\cdot{}^\circ\text{C})(T_f - 18.0)^\circ\text{C} = -(20.0\text{ g})(0.444\text{ J/g}\cdot{}^\circ\text{C})(T_f - 55.6)^\circ\text{C}$$

$$1.29\,T_f - 23.2 = -8.88\,T_f + 494$$

$$10.2\,T_f = 517$$

$$\mathbf{T_f = 50.7^\circ C}$$

**Think About It:** Must the final temperature be between the two starting values?

5.39 **Strategy:** The neutralization reaction is exothermic. 56.2 kJ of heat are released when 1 mole of $H^+$ reacts with 1 mole of $OH^-$. Assuming no heat is lost to the surroundings, we can equate the heat lost by the reaction to the heat gained by the combined solution. How do we calculate the heat released during the reaction? Are we combining 1 mole of $H^+$ with 1 mole of $OH^-$? How do we calculate the heat absorbed by the combined solution?

**Solution:** Assuming no heat is lost to the surroundings, we can write:

$$q_{\text{soln}} + q_{\text{rxn}} = 0$$

or

$$q_{\text{soln}} = -q_{\text{rxn}}$$

First, let's set up how we would calculate the heat gained by the solution,

$$q_{\text{soln}} = m_{\text{soln}}s_{\text{soln}}\Delta T$$

where $m$ and $s$ are the mass and specific heat of the solution and $\Delta T = T_f - T_i$.

We assume that the specific heat of the solution is the same as the specific heat of water, and we assume that the density of the solution is the same as the density of water (1.00 g/mL). Since the density is 1.00 g/mL, the mass of 400 mL of solution (200 mL + 200 mL) is 400 g.

Substituting into the equation above, the heat gained by the solution can be represented as:

$$q_{\text{soln}} = (4.00 \times 10^2\ \text{g})(4.184\ \text{J/g·°C})(T_f - 20.48°\text{C})$$

Next, let's calculate $q_{\text{rxn}}$, the heat released when 200 mL of 0.862 *M* HCl are mixed with 200 mL of 0.431 *M* $Ba(OH)_2$. The equation for the neutralization is:

$$HCl(aq) + Ba(OH)_2(aq) \longrightarrow 2H_2O(l) + BaCl_2(aq)$$

There is exactly enough $Ba(OH)_2$ to neutralize all the HCl. Note that 2 mole HCl is stoichiometrically equivalent to 1 mole $Ba(OH)_2$, and that the concentration of HCl is double the concentration of $Ba(OH)_2$. The number of moles of HCl is:

$$(2.00 \times 10^2\ \text{mL}) \times \frac{0.862\ \text{mol HCl}}{1000\ \text{mL}} = 0.172\ \text{mol HCl}$$

The amount of heat released when 1 mole of $H^+$ is neutralized is given in the problem (−56.2 kJ/mol). The amount of heat liberated when 0.172 mole of $H^+$ is neutralized is:

$$q_{\text{rxn}} = 0.172\ \text{mol} \times \frac{-56.2 \times 10^3\ \text{J}}{1\ \text{mol}} = -9.67 \times 10^3\ \text{J}$$

Finally, knowing that the heat lost by the reaction equals the heat gained by the solution, we can solve for the final temperature of the mixed solution.

$$q_{\text{soln}} = -q_{\text{rxn}}$$

$$(4.00 \times 10^2\ \text{g})(4.184\ \text{J/g·°C})(T_f - 20.48°\text{C}) = -(-9.67 \times 10^3\ \text{J})$$

$$(1.67 \times 10^3)T_f - (3.43 \times 10^4) = 9.67 \times 10^3\ \text{J}$$

$$T_f = \mathbf{26.3°C}$$

5.41 **Strategy:** The first law of thermodynamics dictates that the energy given off by a system must be absorbed by the surroundings. In this case, the surroundings consist of the ethanol and the calorimeter, both of which absorb the heat given off by the methanol, which we will call the system. Because $q_{\text{surr}} = -q_{\text{sys}}$, we know that the sum of $q_{\text{ethanol}}$ and $q_{\text{calorimeter}}$ must equal $-q_{methanol}$. According to Equation 10.11, $q = ms\Delta T$.

**Setup:** The specific heat of ethanol, from Table 10.2, is 2.46 J/g · °C. $\Delta T_{\text{methanol}}$ is 28.5°C – 35.6°C = −7.1°C. $\Delta T_{\text{ethanol}}$ and $\Delta T_{\text{calorimeter}}$ are both 28.5°C – 24.7°C = 3.8°C.

**Solution:** Since $q_{\text{ethanol}} + q_{\text{calorimeter}} = -q_{methanol}$, we can write:

$$38.65\ \text{g} \times \frac{2.46\ \text{J}}{\text{g}^\circ\text{C}} \times 3.8^\circ\text{C} + \frac{19.3\ \text{J}}{^\circ\text{C}} \times 3.8^\circ\text{C} = -\left[25.95\ \text{g} \times s \times (-7.1^\circ\text{C})\right]$$

$$361 \text{ J} + 73.3 \text{ J} = (184 \text{ g} \cdot {}^\circ\text{C})s$$

$$s = \mathbf{2.36\ J/g \cdot {}^\circ C}$$

5.43 The specific heat of a substance is the amount of heat required to increase the temperature of 1 g of the substance by 1°C. Because metal A has a higher specific heat than metal B, it will take more heat to increase the temperature of metal A. Under the same heating conditions, this means it will take longer for the temperature of **metal A** to increase.

5.47 **Strategy:** Our goal is to calculate the enthalpy change for the formation of monoclinic sulfur from rhombic sulfur. To do so, we must arrange the equations that are given in the problem in such a way that they will sum to the desired overall equation. This requires reversing the second equation and changing the sign of its $\Delta H^\circ$ value.

**Solution:**

| Reaction | $\Delta H^\circ$ (kJ/mol) |
|---|---|
| $\text{S(rhombic)} + O_2(g) \rightarrow SO_2(g)$ | −296.06 |
| $SO_2(g) \rightarrow \text{S(monoclinic)} + O_2(g)$ | 296.36 |
| $\text{S(rhombic)} \rightarrow \text{S(monoclinic)}$ | $\Delta H^\circ_{\text{rxn}}$ = **0.30 kJ/mol** |

**Think About It:** Which is the more stable allotropic form of sulfur?

5.49 Making the necessary changes to the reactions given in the problem, and making the corresponding changes to the $\Delta H^\circ$ values:

| Reaction | $\Delta H^\circ$ (kJ/mol) |
|---|---|
| $CO_2(g) + 2H_2O(l) \rightarrow CH_3OH(l) + \frac{3}{2}O_2(g)$ | 726.4 |
| $\text{C(graphite)} + O_2(g) \rightarrow CO_2(g)$ | −393.5 |
| $2H_2(g) + O_2(g) \rightarrow 2H_2O(l)$ | 2(−285.8) |
| $\text{C(graphite)} + 2H_2(g) + \frac{1}{2}O_2(g) \rightarrow CH_3OH(l)$ | $\Delta H^\circ_{\text{rxn}}$ = **−238.7 kJ/mol** |

We have just calculated an enthalpy at standard conditions, which we abbreviate $\Delta H^\circ_{\text{rxn}}$. In this case, the reaction in question was for the formation of *one* mole of $CH_3OH$ *from its elements* in their standard state.

Therefore, the $\Delta H^\circ_{rxn}$ that we calculated is also, by definition, the standard heat of formation $\Delta H^\circ_f$ of $CH_3OH$ (**−238.7 kJ/mol**).

5.51 The reaction can be written as:

$$6 \text{ green} + 8 \text{ grey} \rightarrow 4 \text{ blue}$$

Arrange the given equations so that their sum is the desired equation.

| | |
|---|---|
| 2 green + 8 purple → 4 blue | $\Delta H = -100$ kJ/mol |
| 8 grey + 4 green → 8 purple | $\Delta H = \frac{4}{3}(-60 \text{ kJ/mol}) = -80$ kJ/mol |
| 6 green + 8 grey → 4 blue | $\Delta H = -180$ kJ/mol |

5.57 **$CH_4(g)$** and **$H(g)$**. All the other choices are elements in their most stable form ($\Delta H^\circ_f = 0$). The most stable form of hydrogen is $H_2(g)$.

5.59 We know that heat is required to convert liquid water to water vapor:

$$H_2O(l) \rightarrow H_2O(g) \qquad \text{Endothermic}$$

Therefore, we know that

$$\Delta H^\circ_{rxn} = \Delta H^\circ_f[H_2O(g)] - \Delta H^\circ_f[H_2O(l)] > 0$$

and

$$\mathbf{\Delta H^\circ_f\ [H_2O(l)]} \text{ is more negative since } \Delta H^\circ_{rxn} > 0.$$

5.61 a.

$$\Delta H^\circ = 2\Delta H^\circ_f(H_2O) - 2\Delta H^\circ_f(H_2) - \Delta H^\circ_f(O_2)$$

$$\mathbf{\Delta H^\circ} = (2)(-285.8 \text{ kJ/mol}) - (2)(0) - (1)(0) = \mathbf{-571.6 \text{ kJ/mol}}$$

b.

$$\Delta H^\circ = 4\Delta H^\circ_f(CO_2) + 2\Delta H^\circ_f(H_2O) - 2\Delta H^\circ_f(C_2H_2) - 5\Delta H^\circ_f(O_2)$$

$$\mathbf{\Delta H^\circ} = (4)(-393.5 \text{ kJ/mol}) + (2)(-285.8 \text{ kJ/mol}) - (2)(226.6 \text{ kJ/mol}) - (5)(0) = \mathbf{-2599 \text{ kJ/mol}}$$

5.63 The given enthalpies are in units of kJ/g. We must convert them to units of kJ/mol.

a.

$$\frac{-22.6 \text{ kJ}}{1 \text{ g}} \times \frac{32.04 \text{ g}}{1 \text{ mol}} = \mathbf{-724 \text{ kJ/mol}}$$

b.

$$\frac{-29.7 \text{ kJ}}{1 \text{ g}} \times \frac{46.07 \text{ g}}{1 \text{ mol}} = \mathbf{-1.37 \times 10^3 \text{ kJ/mol}}$$

c.

$$\frac{-33.4 \text{ kJ}}{1 \text{ g}} \times \frac{60.09 \text{ g}}{1 \text{ mol}} = \mathbf{-2.01 \times 10^3 \text{ kJ/mol}}$$

5.65

$$\Delta H° = 6\Delta H_f°(CO_2) + 6\Delta H_f°(H_2O) - [\Delta H_f°(C_6H_{12}) + 9\Delta H_f°(O_2)]$$

$$\mathbf{\Delta H°} = (6)(-393.5 \text{ kJ/mol}) + (6)(-285.8 \text{ kJ/mol}) - (1)(-151.9 \text{ kJ/mol}) - (1)(0) = \mathbf{-3924 \text{ kJ/mol}}$$

**Think About It:** Why is the standard heat of formation of oxygen zero?

5.67

$$w = -P\Delta V = -(1 \text{ atm})(-98 \text{ L}) = 98 \text{ L} \cdot \text{atm} \times \frac{101.3 \text{ J}}{1 \text{ L} \cdot \text{atm}} = \mathbf{9.9 \times 10^3 \text{ J} = 9.9 \text{ kJ}}$$

$$\Delta H = \Delta U + P\Delta V \quad \text{or} \quad \Delta U = \Delta H - P\Delta V$$

Using $\Delta H$ as $-185.2$ kJ = ($2 \times -92.6$ kJ), (because the question involves the formation of 4 moles of ammonia, not 2 moles of ammonia for which the standard enthalpy is given in the question), and $-P\Delta V$ as 9.9 kJ (for which we just solved):

$$\mathbf{\Delta U} = -185.2 \text{ kJ} + 9.9 \text{ kJ} = \mathbf{-175.3 \text{ kJ}}$$

5.69 Using the $\Delta H_f°$ values in Appendix 2 and Equation 5.18 of the text, we write

$$\Delta H_{rxn}° = [5\Delta H_f°(B_2O_3) + 9\Delta H_f°(H_2O)] - [2\Delta H_f°(B_5H_9) + 12\Delta H_f°(O_2)]$$

$$\Delta H° = [(5)(-1263.6 \text{ kJ/mol}) + (9)(-285.8 \text{ kJ/mol})] - [(2)(73.2 \text{ kJ/mol}) + (12)(0 \text{ kJ/mol})]$$

$$\mathbf{\Delta H° = -9036.6 \text{ kJ/mol}}$$

Looking at the balanced equation, this is the amount of heat released for every 2 moles of $B_5H_9$ consumed. We can use the following ratio

$$\frac{-9036.6 \text{ kJ}}{2 \text{ mol } B_5H_9}$$

to convert to kJ/g $B_5H_9$. The molar mass of $B_5H_9$ is 63.12 g, so

$$\text{heat released per gram } B_5H_9 = \frac{-9036.6 \text{ kJ}}{2 \text{ mol } B_5H_9} \times \frac{1 \text{ mol } B_5H_9}{63.12 \text{ g } B_5H_9} = \mathbf{-71.58\ kJ/g\ B_5H_9}$$

5.71 a. $Br_2(l)$ is the most stable form of bromine at 25°C; therefore, $\Delta H_f^\circ[Br_2(l)] = 0$. Since $Br_2(g)$ is less stable than $Br_2(l)$, $\Delta H_f^\circ[Br_2(g)] > 0$.

b. $I_2(s)$ is the most stable form of iodine at 25°C; therefore, $\Delta H_f^\circ[I_2(s)] = 0$. Since $I_2(g)$ is less stable than $I_2(s)$, $\Delta H_f^\circ[I_2(g)] > 0$.

5.73 The balanced equation showing the formation of $Ag_2O(s)$ from its elements is:

$$2Ag(s) + \tfrac{1}{2}O_2(g) \longrightarrow Ag_2O(s)$$

Knowing that the standard enthalpy of formation of any element in its most stable form is zero, and using Equation 5.18 of the text, we write:

$$\Delta H^\circ_{rxn} = \Sigma n\Delta H_f^\circ(\text{products}) - \Sigma m\Delta H_f^\circ(\text{reactants})$$

$$\Delta H^\circ_{rxn} = [\Delta H_f^\circ(Ag_2O)] - \left[2\Delta H_f^\circ(Ag) + \frac{1}{2}\Delta H_f^\circ(O_2)\right]$$

$$\Delta H^\circ_{rxn} = [\Delta H_f^\circ(Ag_2O)] - [0+0]$$

$$\mathbf{\Delta H_f^\circ(Ag_2O) = \Delta H^\circ_{rxn}}$$

In a similar manner, you should be able to show that $\mathbf{\Delta H_f^\circ(CaCl_2) = \Delta H^\circ_{rxn}}$ for the reaction

$$Ca(s) + Cl_2(g) \longrightarrow CaCl_2(s)$$

In both cases, calorimetry could be used to measure the enthalpy changes.

5.75 In a chemical reaction the same elements and the same numbers of atoms are always on both sides of the equation. This provides a consistent reference which allows the energy change in the reaction to be interpreted in terms of the chemical or physical changes that have occurred. In a nuclear reaction the same elements are not always on both sides of the equation and no common reference point exists.

5.77 The reaction corresponding to standard enthalpy of formation, $\Delta H_f^\circ$, of $AgNO_2(s)$ is:

$$Ag(s) + \frac{1}{2}N_2(g) + O_2(g) \rightarrow AgNO_2(s)$$

Rather than measuring the enthalpy directly, we can use the enthalpy of formation of $AgNO_3(s)$ and the $\Delta H^\circ_{rxn}$ provided.

$$AgNO_3(s) \rightarrow AgNO_2(s) + \frac{1}{2}O_2(g)$$

$$\Delta H^\circ_{rxn} = \Delta H^\circ_f(AgNO_2) + \frac{1}{2}\Delta H^\circ_f(O_2) - \Delta H^\circ_f(AgNO_3)$$

$$78.67 \text{ kJ/mol} = \Delta H^\circ_f(AgNO_2) + 0 - (-123.02 \text{ kJ/mol})$$

$$\mathbf{\Delta H^\circ_f(AgNO_2) = -44.35\ kJ/mol}$$

5.79

$$q_{sys} = q_{metal} + q_{water} + q_{calorimeter} = 0$$

$$m_{metal}s_{metal}(T_{final} - T_{initial}) + m_{water}s_{water}(T_{final} - T_{initial}) + C_{calorimeter}(T_{final} - T_{initial}) = 0$$

All the needed values are given in the problem. Plug in the values and solve for $s_{metal}$.

$$(44.0\text{ g})(s_{metal})(28.4 - 99.0)°\text{C} + (80.0\text{ g})(4.184\text{ J/g·°C})(28.4 - 24.0)°\text{C} + (12.4\text{ J/°C})(28.4 - 24.0)°\text{C} = 0$$

$$(-3.11 \times 10^3)s_{metal}\ (\text{g·°C}) = -1.53 \times 10^3\text{ J}$$

$$\mathbf{s_{metal} = 0.492\ J/g·°C}$$

5.81

$$H(g) + Br(g) \longrightarrow HBr(g) \qquad \Delta H^\circ_{rxn} = ?$$

Rearrange the equations as necessary so they can be added to yield the desired equation.

$$H(g) \longrightarrow \frac{1}{2}H_2(g) \qquad \Delta H^\circ_{rxn} = \frac{1}{2}(-436.4\text{ kJ/mol}) = -218.2\text{ kJ/mol}$$

$$Br(g) \longrightarrow \frac{1}{2}Br_2(g) \qquad \Delta H^\circ_{rxn} = \frac{1}{2}(-192.5\text{ kJ/mol}) = -96.25\text{ kJ/mol}$$

$$\frac{1}{2}H_2(g) + \frac{1}{2}Br_2(g) \longrightarrow HBr(g) \qquad \Delta H^\circ_{rxn} = \frac{1}{2}(-72.4\text{ kJ/mol}) = -36.2\text{ kJ/mol}$$

---

$$H(g) + Br(g) \longrightarrow HBr(g) \qquad \mathbf{\Delta H^\circ = -350.7\ kJ/mol}$$

5.83 The reaction for the combustion of octane is:

$$C_8H_{18}(l) + \frac{25}{2}O_2(g) \rightarrow 8CO_2(g) + 9H_2O(l)$$

$\Delta H^\circ_{rxn}$ for this reaction was calculated in problem 5.83 from standard enthalpies of formation.

$$\Delta H^\circ_{rxn} = -5470 \text{ kJ/mol}$$

$$\text{Heat/gal of octane} = \frac{5470 \text{ kJ}}{1 \text{ mol } C_8H_{18}} \times \frac{1 \text{ mol } C_8H_{18}}{114.2 \text{ g}} \times \frac{0.7025 \text{ g}}{1 \text{ mL}} \times \frac{3785 \text{ mL}}{1 \text{ gal}}$$

$$\text{Heat/gal of octane} = 1.27 \times 10^5 \text{ kJ/gal gasoline}$$

The reaction for the combustion of ethanol is:

$$C_2H_5OH(l) + 3O_2(g) \rightarrow 2CO_2(g) + 3H_2O(l)$$

$$\Delta H^\circ_{rxn} = 2\Delta H^\circ_f(CO_2) + 3\Delta H^\circ_f(H_2O) - \Delta H^\circ_f(C_2H_5OH) - 3\Delta H^\circ_f(O_2)$$

$$\Delta H^\circ_{rxn} = (2)(-393.5 \text{ kJ/mol}) + (3)(-285.8 \text{ kJ/mol}) - (1)(-277.0 \text{ kJ/mol}) = -1367 \text{ kJ/mol}$$

$$\text{Heat/gal of ethanol} = \frac{1367 \text{ kJ}}{1 \text{ mol } C_2H_5OH} \times \frac{1 \text{ mol } C_2H_5OH}{46.07 \text{ g}} \times \frac{0.7894 \text{ g}}{1 \text{ mL}} \times \frac{3785 \text{ mL}}{1 \text{ gal}}$$

$$\text{Heat/gal of ethanol} = 8.87 \times 10^4 \text{ kJ/gal ethanol}$$

For ethanol, what would the cost have to be to supply the same amount of heat per dollar of gasoline? For gasoline, it cost \$2.20 to provide $1.27 \times 10^5$ kJ of heat.

$$\frac{\$2.20}{1.27\times10^5 \text{ kJ}} \times \frac{8.87\times10^4 \text{ kJ}}{1 \text{ gal ethanol}} = \mathbf{\$1.54/gal\ ethanol}$$

5.85 The heat gained by the liquid nitrogen must be equal to the heat lost by the water.

$$q_{N_2} = -q_{H_2O}$$

If we can calculate the heat lost by the water, we can calculate the heat gained by 60.0 g of the nitrogen.

$$\text{Heat lost by the water} = q_{H_2O} = m_{H_2O}s_{H_2O}\Delta T$$

$$q_{H_2O} = (2.00 \times 10^2 \text{ g})(4.184 \text{ J/g}\cdot{}^\circ\text{C})(41.0 - 55.3)^\circ\text{C} = -1.20 \times 10^4 \text{ J}$$

The heat gained by 60.0 g nitrogen is the opposite sign of the heat lost by the water.

$$q_{N_2} = -q_{H_2O}$$

$$q_{N_2} = 1.20 \times 10^4 \text{ J}$$

The problem asks for the molar heat of vaporization of liquid nitrogen. Above, we calculated the amount of heat necessary to vaporize 60.0 g of liquid nitrogen. We need to convert from J/60.0 g $N_2$ to J/mol $N_2$.

$$\Delta H_{\text{vap}} = \frac{1.20 \times 10^4 \text{ J}}{60.0 \text{ g } N_2} \times \frac{28.02 \text{ g } N_2}{1 \text{ mol } N_2} = \mathbf{5.60 \times 10^3 \text{ J/mol}} = \mathbf{5.60 \text{ kJ/mol}}$$

5.87 Recall that the standard enthalpy of formation ( $\Delta H_f^\circ$ ) is defined as the heat change that results when 1 mole of a compound is formed from its elements at a pressure of 1 atm. Only in choice **(a)** does $\Delta H_{\text{rxn}}^\circ = \Delta H_f^\circ$. In choice (b), C(diamond) is *not* the most stable form of elemental carbon under standard conditions; C(graphite) is the most stable form.

5.89 a. **No work is done by a gas expanding in a vacuum**, because the pressure exerted on the gas is zero.

b.

$$w = -P\Delta V$$

$$w = -(0.20 \text{ atm})(0.50 - 0.050)\text{L} = -0.090 \text{ L·atm}$$

Converting to units of joules:

$$w = -0.090 \text{ L·atm} \times \frac{101.3 \text{ J}}{\text{L·atm}} = \mathbf{-9.1 \text{ J}}$$

5.91 The heat of the reaction (combustion) is absorbed by both the water and the calorimeter.

$$q_{\text{rxn}} = -(q_{\text{water}} + q_{\text{cal}})$$

If we can calculate both $q_{\text{water}}$ and $q_{\text{rxn}}$, then we can calculate $q_{\text{cal}}$. First, let's calculate the heat absorbed by the water.

$$q_{\text{water}} = m_{\text{water}} s_{\text{water}} \Delta T$$

$$q_{\text{water}} = (2000 \text{ g})(4.184 \text{ J/g·°C})(25.67 - 21.84)\text{°C} = 3.20 \times 10^4 \text{ J} = 32.0 \text{ kJ}$$

Next, let's calculate the heat released ($q_{\text{rxn}}$) when 1.9862 g of benzoic acid are burned. $\Delta H_{\text{rxn}}$ is given in units of kJ/mol. Convert to $q_{\text{rxn}}$ in kJ.

$$q_{\text{rxn}} = 1.9862 \text{ g benzoic acid} \times \frac{1 \text{ mol benzoic acid}}{122.1 \text{ g benzoic acid}} \times \frac{-3226.7 \text{ kJ}}{1 \text{ mol benzoic acid}} = -52.49 \text{ kJ}$$

And,

$$q_{\text{cal}} = -q_{\text{rxn}} - q_{\text{water}}$$

$$q_{\text{cal}} = 52.49 \text{ kJ} - 32.0 \text{ kJ} = 20.5 \text{ kJ}$$

To calculate the heat capacity of the bomb calorimeter, we can use the following equation:

$$q_{cal} = C_{cal}\Delta T$$

$$\mathbf{C_{cal}} = \frac{q_{cal}}{\Delta T} = \frac{20.5\ \text{kJ}}{(25.67 - 21.84)°\text{C}} = \mathbf{5.35\ kJ\,/\,°C}$$

5.93 The equation we are interested in is the formation of CO from its elements.

$$\text{C(graphite)} + \tfrac{1}{2}\text{O}_2(g) \longrightarrow \text{CO}(g) \qquad \Delta H° = ?$$

Try to add the given equations together to end up with the equation above.

$$\text{C(graphite)} + \text{O}_2(g) \longrightarrow \text{CO}_2(g) \qquad \Delta H° = -393.5\ \text{kJ/mol}$$

$$\text{CO}_2(g) \longrightarrow \text{CO}(g) + \tfrac{1}{2}\text{O}_2(g) \qquad \Delta H° = +283.0\ \text{kJ/mol}$$

---

$$\mathbf{C(graphite)} + \tfrac{1}{2}\mathbf{O_2}(g) \longrightarrow \mathbf{CO}(g) \qquad \mathbf{\Delta H° = -110.5\ kJ/mol}$$

We cannot obtain $\Delta H_f^\circ$ for CO directly, because burning graphite in oxygen will form both CO and $CO_2$.

5.95 The reaction we are interested in is the formation of ethanol from its elements.

$$2\text{C(graphite)} + \tfrac{1}{2}\text{O}_2(g) + 3\text{H}_2(g) \longrightarrow \text{C}_2\text{H}_5\text{OH}(l)$$

Along with the reaction for the combustion of ethanol, we can add other reactions together to end up with the above reaction.

Reversing the reaction representing the combustion of ethanol gives:

$$2\text{CO}_2(g) + 3\text{H}_2\text{O}(l) \longrightarrow \text{C}_2\text{H}_5\text{OH}(l) + 3\text{O}_2\,(g) \qquad \Delta H° = +1367.4\ \text{kJ/mol}$$

We need to add equations to add C (graphite) and remove $H_2O$ from the reactants side of the equation. We write:

$$2\text{CO}_2(g) + 3\text{H}_2\text{O}(l) \longrightarrow \text{C}_2\text{H}_5\text{OH}(l) + 3\text{O}_2(g) \qquad \Delta H° = +1367.4\ \text{kJ/mol}$$

$$2\text{C(graphite)} + 2\text{O}_2(g) \longrightarrow 2\text{CO}_2(g) \qquad \Delta H° = 2(-393.5\ \text{kJ/mol})$$

$$3\text{H}_2(g) + \tfrac{3}{2}\text{O}_2(g) \longrightarrow 3\text{H}_2\text{O}(l) \qquad \Delta H° = 3(-285.8\ \text{kJ/mol})$$

---

$$\mathbf{2C(graphite)} + \tfrac{1}{2}\mathbf{O_2}(g) + \mathbf{3H_2}(g) \longrightarrow \mathbf{C_2H_5OH}(l) \qquad \mathbf{\Delta H_f^\circ = -277.0\ kJ/mol}$$

5.97 Heat gained by ice = Heat lost by the soft drink

$$m_{ice} \times 334\ \text{J/g} = -m_{sd}s_{sd}\Delta T$$

$$m_{ice} \times 334\ \text{J/g} = -(361\ \text{g})(4.184\ \text{J/g}\cdot{}^\circ\text{C})(0 - 23)^\circ\text{C}$$

$$\boldsymbol{m_{ice} = 104\ \text{g}}$$

5.99 When 1.034 g of naphthalene are burned, 41.56 kJ of heat are evolved. Convert this to the amount of heat evolved on a molar basis. The molar mass of naphthalene is 128.2 g/mol.

$$q = \frac{-41.56\ \text{kJ}}{1.034\ \text{g}\ C_{10}H_8} \times \frac{128.2\ \text{g}\ C_{10}H_8}{1\ \text{mol}\ C_{10}H_8} = -5153\ \text{kJ/mol}$$

$q$ has a negative sign because this is an exothermic reaction.

This reaction is run at constant volume ($\Delta V = 0$); therefore, no work will result from the change.

$$w = -P\Delta V = 0$$

From Equation 5.3 of the text, it follows that the change in energy is equal to the heat change.

$$\boldsymbol{\Delta U} = q + w = q_v = \mathbf{-5153\ kJ/mol}$$

5.101 First, we calculate $\Delta H$ for the combustion of 1 mole of glucose using data in Appendix 2 of the text. We can then calculate the heat produced in the calorimeter. Using the heat produced along with $\Delta H$ for the combustion of 1 mole of glucose will allow us to calculate the mass of glucose in the sample. Finally, the mass % of glucose in the sample can be calculated.

$$C_6H_{12}O_6(s) + 6O_2(g) \rightarrow 6CO_2(g) + 6H_2O(l)$$

$$\Delta H^\circ_{rxn} = (6)(-393.5\ \text{kJ/mol}) + (6)(-285.8\ \text{kJ/mol}) - (1)(-1274.5\ \text{kJ/mol}) = -2801.3\ \text{kJ/mol}$$

The heat produced in the calorimeter is:

$$(3.134^\circ\text{C})(19.65\ \text{kJ/}^\circ\text{C}) = 61.58\ \text{kJ}$$

Let $x$ equal the mass of glucose in the sample:

$$x\ \text{g glucose} \times \frac{1\ \text{mol glucose}}{180.2\ \text{g glucose}} \times \frac{2801.3\ \text{kJ}}{1\ \text{mol glucose}} = 61.58\ \text{kJ}$$

$$x = 3.961\ \text{g}$$

$$\%\ \text{glucose} = \frac{3.961\ \text{g}}{4.117\ \text{g}} \times 100\% = \mathbf{96.21\%}$$

5.103 **Strategy:** Solve the equation $q = C\Delta T$ for $\Delta T$, then use $\Delta T = T_{\text{final}} - T_{\text{initial}}$ to compute the final temperature.

**Setup:** The water *absorbs* the heat, so $q = +21.8$ kJ. Also, since the heat capacity of the calorimeter is negligible, the heat capacity of the apparatus is simply the heat capacity of the 150 g of water it contains.

**Solution:**

$$\Delta T = \frac{q}{C} = \frac{+21.8\times10^{3}\text{ J}}{(150\text{ g H}_2\text{O})(4.184\text{ J/(g}\cdot{}^\circ\text{C}))} = 34.7^\circ\text{C}$$

$$T_{\text{final}} - T_{\text{initial}} = 34.7^\circ\text{C} \quad \text{or} \quad T_{\text{finial}} = 34.7^\circ\text{C} + 23.4^\circ\text{C} = \mathbf{58.1^\circ C}$$

5.105 a. As heat is added to water at 25°C, the temperature increases until the boiling point is reached.

b. At 1 atm of pressure, water at 100°C will remain at that temperature until the added heat has converted all the liquid to a gas.

c. The temperature of a system can change without heat being added if a chemical reaction occurs. Example: If an exothermic reaction occurs, the temperature will increase. If an endothermic reaction occurs, the temperature will decrease.

5.107 **Strategy:** For parts (a)-(d), assume that pressure is constant and non-zero for all processes. Also, it is useful to recognize that $\Delta H = q$ and $\Delta U = \Delta H - P\Delta V = \Delta H + w$ under these conditions.

a. $$C_6H_6(l) \rightarrow C_6H_6(s)$$

This process gives off heat to the surroundings, so $q$ and $\Delta H$ are negative. For most substances, the volumes of the liquid state and the solid state are nearly equal, so $\Delta V \approx 0$. Thus, $w = -P\Delta V \approx 0$ and $\Delta U = \Delta H + w \approx \Delta H$ (negative).

b. $$2Na(s) + 2H_2O(l) \rightarrow 2NaOH(aq) + H_2(g)$$

This process is exothermic, so $\Delta H$ and $q$ are negative. Since a gas is produced in the reaction, the volume of the system increases, making $\Delta V$ positive and $w = -P\Delta V$ negative. Also,

$\Delta U = \Delta H + w = (\text{negative}) + (\text{negative}) = \text{negative}$ .

c. $NH_3(l) \rightarrow NH_3(g)$

Boiling a liquid requires that heat be added to the system. It is an endothermic process. So, $q$ and $\Delta H$ are positive. Since there are more moles of gas in the products than reactants, $\Delta V$ is positive. Because $w = -P\Delta V$, $w$ is negative. Since $\Delta U = \Delta H - P\Delta V = (\text{positive}) - P(\text{positive})$, the sign of $\Delta U$ cannot be determined without more information.

d. $H_2O(s) \rightarrow H_2O(l)$

Melting a solid requires that heat be added to the system. It is an endothermic process. So, $q$ and $\Delta H$ are positive. The volume change is nearly zero, so $w = -P\Delta V \approx 0$. Because $w = -P\Delta V$, $w = 0$. Also, $\Delta U = \Delta H + w \approx (\text{positive}) + 0 \approx \text{positive}$ .

e. Assume the gas is ideal, in which case its change in energy is due solely to changes in its *kinetic energy*. If the expansion of the gas occurs at a constant temperature, the kinetic energy of the system remains the same (since temperature is constant). Thus, $\Delta U$ is zero. Recall that $\Delta H = \Delta U + \Delta(PV)$ and also recognize that, for an ideal gas at constant $n$, $\Delta(PV) = \Delta(nRT) = nR\Delta T = (nR)(0) = 0$ . So, $\Delta H = \Delta U + \Delta(PV) = 0 + 0 = 0$ . Since the gas expanded, it did work (assuming a non-zero external pressure) and $w$ will be negative. Because $\Delta U = q + w = 0$, $q$ must be positive.

| | $q$ | $w$ | $\Delta U$ | $\Delta H$ |
|---|---|---|---|---|
| Freezing of benzene | – | 0 | – | – |
| Reaction of sodium with water | – | – | – | – |
| Boiling of liquid ammonia | + | – | ? | + |
| Melting of ice | + | 0 | + | + |
| Expansion of a gas at constant temperature | + | – | 0 | 0 |

5.109 **Strategy:** First, we need to determine if the reaction uses all the reactants. Then, we use the molar heat of neutralization to determine the amount of heat produced by the reaction. Finally, we can calculate the change in temperature of the solution.

**Solution:** $OH^-(aq) + H^+(aq) \rightarrow H_2O(l)$

Find the moles of $OH^-(aq)$.

$$50.0\text{ mL Ba(OH)}_2\left(\frac{1\text{ L}}{1000\text{ mL}}\right)\left(\frac{0.200\text{ mol Ba(OH)}_2}{1\text{ L}}\right)\left(\frac{2\text{ mol OH}^-}{1\text{ mol Ba(OH)}_2}\right) = 0.0200\text{ mol OH}^-$$

Find the moles of $H^+(aq)$.

$$50.0\text{ mL HNO}_3\left(\frac{1\text{ L}}{1000\text{ mL}}\right)\left(\frac{0.400\text{ mol HNO}_3}{1\text{ L}}\right)\left(\frac{1\text{ mol H}^+}{1\text{ mol HNO}_3}\right) = 0.0200\text{ mol H}^+$$

This is a stoichiometric mixture, so all the reactants will be converted to water.

Use the molar heat of neutralization to find the heat produced by the reaction.

$$\begin{aligned} q_{\text{rxn}} &= (-56.2\text{ kJ/mol})(0.0200\text{ mol}) \\ &= -1.124\text{ kJ} \end{aligned}$$

The heat produced by the reaction is equal to the heat gained by the surroundings. In this case, the surroundings are composed of the water and the calorimeter.

$$q_{\text{surr}} = -q_{\text{rxn}}$$

$$\begin{aligned} q_{\text{surr}} &= q_{\text{H}_2\text{O}} + q_{\text{cal}} \\ &= m_{\text{H}_2\text{O}}s_{\text{H}_2\text{O}}\Delta T + C_{\text{cal}}\Delta T; \end{aligned}$$

where $m_{\text{H}_2\text{O}}$ is the mass of water,
$s_{\text{H}_2\text{O}}$ is the specific heat capacity of water,
and $C_{\text{cal}}$ is the heat capacity of the calorimeter.

Find the mass of water.

$$(100.0\text{ mL})\left(\frac{1\text{ g}}{\text{mL}}\right) + (0.0200\text{ mol})\left(\frac{18.02\text{ g}}{\text{mol}}\right) = 100.4\text{ g}$$

Calculate the final temperature.

$$\begin{aligned} 1124\text{ J} &= (100.4\text{ g})(4.184\text{ J/g}\cdot{}^\circ\text{C})\Delta T + (496\text{ J/}^\circ\text{C})\Delta T \\ 1124\text{ J} &= (916\text{ J/}^\circ\text{C})\Delta T \\ \Delta T &= 1.23^\circ\text{C} \end{aligned}$$

$$\begin{aligned} \Delta T &= T_\text{f} - T_\text{i} \\ 1.23^\circ\text{C} &= T_\text{f} - 22.4^\circ\text{C} \\ \boldsymbol{T_\text{f}} &= \mathbf{23.6^\circ C} \end{aligned}$$

5.111 A good starting point would be to calculate the standard enthalpy for both reactions.

Calculate the standard enthalpy for the reaction:

$$C(s) + \frac{1}{2}O_2(g) \longrightarrow CO(g)$$

This reaction corresponds to the standard enthalpy of formation of CO, so we use the value of −110.5 kJ/mol (see Appendix 2 of the text).

Calculate the standard enthalpy for the reaction:

$$C(s) + H_2O(g) \longrightarrow CO(g) + H_2(g)$$

$$\Delta H^\circ_{rxn} = [\Delta H^\circ_f(CO) + \Delta H^\circ_f(H_2)] - [\Delta H^\circ_f(C) + \Delta H^\circ_f(H_2O)]$$

$$\Delta H^\circ_{rxn} = [(1)(-110.5\text{ kJ/mol}) + (1)(0)] - [(1)(0) + (1)(-241.8\text{ kJ/mol})] = 131.3\text{ kJ/mol}$$

The first reaction, which is exothermic, can be used to promote the second reaction, which is endothermic. Thus, the two gases are produced alternately.

5.113 The heat produced by the reaction heats the solution and the calorimeter: $q_{rxn} = -(q_{soln} + q_{cal})$

$$q_{soln} = ms\Delta T = (50.0\text{ g})(4.184\text{ J/g}\cdot{}^\circ\text{C})(22.17^\circ\text{C} - 19.25^\circ\text{C}) = 611\text{ J}$$

$$q_{cal} = C\Delta T = (98.6\text{ J/}^\circ\text{C})(22.17^\circ\text{C} - 19.25^\circ\text{C}) = 288\text{ J}$$

$$-q_{rxn} = (q_{soln} + q_{cal}) = (611 + 288)\text{J} = 899\text{ J}$$

The 899 J produced was for 50.0 mL of a 0.100 $M$ $AgNO_3$ solution.

$$50.0\text{ mL} \times \frac{0.100\text{ mol Ag}^+}{1000\text{ mL soln}} = 5.00 \times 10^{-3}\text{ mol Ag}^+$$

On a molar basis the heat produced was:

$$\frac{899\text{ J}}{5.00 \times 10^{-3}\text{ mol Ag}^+} = 1.80 \times 10^5\text{ J/mol Ag}^+ = 180\text{ kJ/mol Ag}^+$$

The balanced equation involves 2 moles of $Ag^+$, so the heat produced is 2 mol × 180 kJ/mol = 360 kJ

Since the reaction produces heat (or by noting the sign convention above), then:

$$\mathbf{\Delta H_{rxn}} = q_{rxn} = \mathbf{-3.60 \times 10^2\ kJ/mol\ Zn}\ (\text{or } -3.60 \times 10^2\text{ kJ/2 mol Ag}^+)$$

5.115 The heat required to heat 200 g of water (assume $d = 1$ g/mL) from 20°C to 100°C is:

$$q = ms\Delta T$$

$$q = (200\text{ g})(4.184\text{ J/g}\cdot{}^\circ\text{C})(100 - 20)^\circ\text{C} = 6.69 \times 10^4\text{ J}$$

Since 50% of the heat from the combustion of methane is lost to the surroundings, twice the amount of heat needed must be produced during the combustion: $2(6.69 \times 10^4\ \text{J}) = 1.34 \times 10^5\ \text{J} = 1.34 \times 10^2\ \text{kJ}$.

Use standard enthalpies of formation (see Appendix 2) to calculate the heat of combustion of methane.

$$CH_4(g) + 2O_2(g) \rightarrow CO_2(g) + 2H_2O(l) \qquad \Delta H° = -890.3\ \text{kJ/mol}$$

The number of moles of methane needed to produce $1.34 \times 10^2$ kJ of heat is:

$$(1.34\times10^2\ \text{kJ})\times\frac{1\ \text{mol}\ CH_4}{890.3\ \text{kJ}} = 0.151\ \text{mol}\ CH_4$$

$$(0.151\ \text{mol}\ CH_4)\times\frac{\$0.27}{1\ \text{mol}\ CH_4} = \$0.041$$

The cost of the methane is about **4.1 cents**.

Assuming 100% metabolic efficiency, the energy required is

$$\boldsymbol{E} = mgh = 5.2 \times 10^{-3}\ \text{kg} \times 9.8\ \text{m/s}^2 \times 12\ \text{m} = \mathbf{0.61\ J}$$

5.117 **Strategy:** Rearrange $\Delta H = \Delta U + P\Delta V$ (constant pressure) to get $\Delta H - \Delta U = P\Delta V$. The pressure is 50,000 atm. The volume change of one mole of substance may be computed from the given densities.

**Setup:** Find the molar volume of graphite and diamond.

$$V_{\text{diamond}} = \left(\frac{12.01\ \text{g C}(s)}{1\ \text{mol C}(s)}\right)\left(\frac{1\ \text{cm}^3}{3.52\ \text{g C}(s)}\right)\left(\frac{1\ \text{L}}{1000\ \text{cm}^3}\right) = 3.41\times10^{-3}\ \text{L/mol}$$

$$V_{\text{graphite}} = \left(\frac{12.01\ \text{g C}(s)}{1\ \text{mol C}(s)}\right)\left(\frac{1\ \text{cm}^3}{2.25\text{g C}(s)}\right)\left(\frac{1\ \text{L}}{1000\ \text{cm}^3}\right) = 5.34\times10^{-3}\ \text{L/mol}$$

**Solution:**

$$\Delta H - \Delta U = P\Delta V = (50{,}000\ \text{atm})(3.41\times10^{-3}\ \text{L/mol} - 5.34\times10^{-3}\ \text{L/mol})\left(\frac{101.3\ \text{kJ}}{1\ \text{L}\cdot\text{atm}}\right)$$

$$= -9.78\times10^3\ \text{J/mol}$$

$$= \mathbf{-9.78\ kJ/mol}$$

5.119 The heat required to raise the temperature of 1 liter of water by 1°C is:

$$4.184\frac{\text{J}}{\text{g}\cdot°\text{C}}\times\frac{1\ \text{g}}{1\ \text{mL}}\times\frac{1000\ \text{mL}}{1\ \text{L}}\times1°\text{C} = 4184\ \text{J/L}$$

Next, convert the volume of the Pacific Ocean to liters.

$$(7.2 \times 10^8\ \text{km}^3) \times \left(\frac{1000\ \text{m}}{1\ \text{km}}\right)^3 \times \left(\frac{100\ \text{cm}}{1\ \text{m}}\right)^3 \times \frac{1\ \text{L}}{1000\ \text{cm}^3} = 7.2 \times 10^{20}\ \text{L}$$

The amount of heat needed to raise the temperature of $7.2 \times 10^{20}$ L of water is:

$$(7.2 \times 10^{20}\ \text{L}) \times \frac{4184\ \text{J}}{1\ \text{L}} = 3.0 \times 10^{24}\ \text{J}$$

Finally, we can calculate the number of atomic bombs needed to produce this much heat.

$$(3.0 \times 10^{24}\ \text{J}) \times \frac{1\ \text{atomic bomb}}{1.0 \times 10^{15}\ \text{J}} = \mathbf{3.0 \times 10^9\ atomic\ bombs} = \mathbf{3.0\ billion\ atomic\ bombs}$$

5.121 a. **Although we cannot measure $\Delta H^{\circ}_{\text{rxn}}$ for this reaction, the reverse process is the combustion of glucose. We could easily measure $\Delta H^{\circ}_{\text{rxn}}$ for this combustion by burning a mole of glucose in a bomb calorimeter.**

$$C_6H_{12}O_6(s) + 6O_2(g) \longrightarrow 6CO_2(g) + 6H_2O(l)$$

b. We can calculate $\Delta H^{\circ}_{\text{rxn}}$ using standard enthalpies of formation.

$$\Delta H^{\circ}_{\text{rxn}} = \Delta H^{\circ}_{\text{f}}[C_6H_{12}O_6(s)] + 6\Delta H^{\circ}_{\text{f}}[O_2(g)] - \left[6\Delta H^{\circ}_{\text{f}}[CO_2(g)] + 6\Delta H^{\circ}_{\text{f}}[H_2O(l)]\right]$$

$$\Delta H^{\circ}_{\text{rxn}} = [(1)(-1274.5\ \text{kJ/mol}) + 0] - [(6)(-393.5\ \text{kJ/mol}) + (6)(-285.8\ \text{kJ/mol})] = 2801.3\ \text{kJ/mol}$$

$\Delta H^{\circ}_{\text{rxn}}$ has units of kJ/mol glucose. We want the $\Delta H^{\circ}$ change for $7.0 \times 10^{14}$ kg glucose. We need to calculate how many moles of glucose are in $7.0 \times 10^{14}$ kg glucose. Use the following strategy to solve the problem.

$$\text{kg glucose} \rightarrow \text{g glucose} \rightarrow \text{mol glucose} \rightarrow \text{kJ}\ (\Delta H^{\circ})$$

$$\boldsymbol{\Delta H^{\circ}} = (7.0 \times 10^{14}\ \text{kg}) \times \frac{1000\ \text{g}}{1\ \text{kg}} \times \frac{1\ \text{mol}\ C_6H_{12}O_6}{180.2\ \text{g}\ C_6H_{12}O_6} \times \frac{2801.3\ \text{kJ}}{1\ \text{mol}\ C_6H_{12}O_6} = \mathbf{1.1 \times 10^{19}\ kJ}$$

5.123 Energy intake for mechanical work:

$$0.17 \times 500\ \text{g} \times \frac{3000\ \text{J}}{1\ \text{g}} = 2.6 \times 10^5\ \text{J}$$

$$2.6 \times 10^5\ \text{J} = mgh$$

$$1\ \text{J} = 1\ \text{kg·m}^2\text{s}^{-2}$$

$$2.6 \times 10^5\ \frac{\text{kg} \cdot \text{m}^2}{\text{s}^2} = (46\ \text{kg})(9.8\ \text{m/s}^2)h$$

$$\boldsymbol{h = 5.8 \times 10^2\ \text{m}}$$

5.125 Water has a larger specific heat than air. Thus cold, damp air can extract more heat from the body than cold, dry air. By the same token, hot, humid air can deliver more heat to the body.

5.127 a.

$$\mathbf{2LiOH(\mathit{aq}) + CO_2(\mathit{g}) \rightarrow Li_2CO_3(\mathit{aq}) + H_2O(\mathit{l})}$$

b. The metabolism of glucose is the same as the combustion of glucose:

$$C_6H_{12}O_6(aq) + 6O_2(g) \rightarrow 6CO_2(g) + 6H_2O(l)$$

Using Equation 5.18 from the text,

$$\Delta H^\circ_{\text{rxn}} = [6\Delta H^\circ_{\text{f}}(CO_2) + 6\Delta H^\circ_{\text{f}}(H_2O)] - \Delta H^\circ_{\text{f}}(C_6H_{12}O_6)$$

$$= [6(-393.5\ \text{kJ/mol}) + 6(-285.8\ \text{kJ/mol})] - (-1274.5\ \text{kJ/mol}) = -2801\ \text{kJ}$$

$$\textbf{? g CO}_\mathbf{2} = 1.2 \times 10^4\ \text{kJ} \times \frac{6\ \text{mol CO}_2}{2801\ \text{kJ}} \times \frac{44.01\ \text{g CO}_2}{1\ \text{mol CO}_2} = 1131\ \text{g} = \mathbf{1.1\ kg\ CO_2}$$

$$\textbf{? g LiOH} = 1131\ \text{g CO}_2 \times \frac{1\ \text{mol CO}_2}{44.01\ \text{g CO}_2} \times \frac{2\ \text{mol LiOH}}{1\ \text{mol CO}_2} \times \frac{23.9\ \text{g LiOH}}{1\ \text{mol LiOH}} = 1229\ \text{g} = \mathbf{1.2\ kg\ LiOH}$$

5.129 a. $CaC_2(s) + 2H_2O(l) \longrightarrow Ca(OH)_2(s) + C_2H_2(g)$

b. The reaction for the combustion of acetylene is:

$$2C_2H_2(g) + 5O_2(g) \longrightarrow 4CO_2(g) + 2H_2O(l)$$

We can calculate the enthalpy change for this reaction from standard enthalpy of formation values given in Appendix 2 of the text.

$$\Delta H^\circ_{\text{rxn}} = [4\Delta H^\circ_{\text{f}}(CO_2) + 2\Delta H^\circ_{\text{f}}(H_2O)] - [2\Delta H^\circ_{\text{f}}(C_2H_2) + 5\Delta H^\circ_{\text{f}}(O_2)]$$

$$\Delta H^\circ_{\text{rxn}} = [(4)(-393.5\ \text{kJ/mol}) + (2)(-285.8\ \text{kJ/mol})] - [(2)(226.6\ \text{kJ/mol}) + (5)(0)]$$

$$\Delta H^\circ_{rxn} = -2599 \text{ kJ/mol}$$

Looking at the balanced equation, this is the amount of heat released when two moles of $C_2H_2$ are reacted. The problem asks for the amount of heat that can be obtained starting with 74.6 g of $CaC_2$. From this amount of $CaC_2$, we can calculate the moles of $C_2H_2$ produced.

$$74.6 \text{ g } CaC_2 \times \frac{1 \text{ mol } CaC_2}{64.10 \text{ g } CaC_2} \times \frac{1 \text{ mol } C_2H_2}{1 \text{ mol } CaC_2} = 1.16 \text{ mol } C_2H_2$$

Now, we can use the $\Delta H^\circ_{rxn}$ calculated above as a conversion factor to determine the amount of heat obtained when 1.16 moles of $C_2H_2$ are burned.

$$1.16 \text{ mol } C_2H_2 \times \frac{2599 \text{ kJ}}{2 \text{ mol } C_2H_2} = \mathbf{1.51 \times 10^3 \text{ kJ}}$$

or

$$\mathbf{1.51 \times 10^6 \text{ J}}$$

5.131 a.

$$2.0 \text{ g glucose} \times \frac{1 \text{ mol glucose}}{180.2 \text{ g glucose}} = 0.0111 \text{ mol glucose}$$

The heat of combustion for glucose is –2801 kJ/mol. Therefore, the energy released by the combustion of a 2.0 g glucose tablet is:

$$0.0111 \text{ mol glucose} \times \frac{2801 \text{ kJ}}{1 \text{ mol glucose}} = \mathbf{31 \text{ kJ}}$$

For sucrose, $C_{12}H_{22}O_{11}$,

$$2.0 \text{ g sucrose} \times \frac{1 \text{ mol sucrose}}{342.3 \text{ g sucrose}} = 0.00584 \text{ mol sucrose}$$

The heat of combustion for sucrose (using values from Appendix 2) is –5644 kJ/mol. The energy released by combustion of a 2.0 g sucrose tablet is:

$$0.00584 \text{ mol sucrose} \times \frac{5644 \text{ kJ}}{1 \text{ mol sucrose}} = \mathbf{33 \text{ kJ}}$$

b.

$$31 \text{ kJ} \times 0.30 = 9.3 \text{ kJ} = 9.3 \times 10^3 \text{ J}$$

$$9.3 \times 10^3 \text{ J} = 65 \text{ kg} \times 9.8 \text{ m/s}^2 \times h$$

$$\mathbf{h = 15 \text{ m}}$$

For sucrose, 33 kJ × 0.30 = 9.9 kJ = 9.9 × $10^3$ J

$$9.9 \times 10^3 \text{ J} = 65 \text{ kg} \times 9.8 \text{ m/s}^2 \times h$$

$$\mathbf{h = 16\ m}$$

5.133 Begin by calculating the standard enthalpy of reaction.

$$\Delta H^\circ_{rxn} = 2\Delta H^\circ_f(CaSO_4) - [2\Delta H^\circ_f(CaO) + 2\Delta H^\circ_f(SO_2) + \Delta H^\circ_f(O_2)]$$

$$= (2)(-1432.7 \text{ kJ/mol}) - [(2)(-635.6 \text{ kJ/mol}) + (2)(-296.4 \text{ kJ/mol}) + 0]$$

$$= -1001 \text{ kJ/mol}$$

This is the enthalpy change for every 2 moles of $SO_2$ that are removed. The problem asks to calculate the enthalpy change for this process if 6.6 × $10^5$ g of $SO_2$ are removed.

$$(6.6 \times 10^5 \text{ g } SO_2) \times \frac{1 \text{ mol } SO_2}{64.07 \text{ g } SO_2} \times \frac{-1001 \text{ kJ}}{2 \text{ mol } SO_2} = \mathbf{-5.2 \times 10^6\ kJ}$$

5.135 Since the humidity is very low in deserts, there is little water vapor in the air to trap and hold the heat radiated back from the ground during the day. Once the sun goes down, the temperature drops dramatically. 40°F temperature drops between day and night are common in desert climates. Coastal regions have much higher humidity levels compared to deserts. The water vapor in the air retains heat, which keeps the temperature at a more constant level during the night. In addition, sand and rocks in the desert have small specific heats compared with water in the ocean. The water absorbs much more heat during the day compared to sand and rocks, which keeps the temperature warmer at night.

5.137 a.

$$3N_2H_4(g) \rightarrow 4NH_3(g) + N_2(g)$$

b.

$$\Delta H^\circ_{rxn} = \left[\Delta H^\circ_f(N_2) + 4H^\circ_f(NH_3)\right] - 3H^\circ_f(N_2H_4)$$
$$= \left[0 \text{ kJ/mol} + 4(-46.3 \text{ kJ/mol})\right] - 3(50.42 \text{ kJ/mol})$$
$$= -336.46 \text{ kJ/mol}$$

c.

$$N_2H_4(g) + O_2(g) \rightarrow N_2(g) + 2H_2O(l)$$

$$\Delta H^\circ_{rxn} = \left[2(-285.8 \text{ kJ/mol}) + 0 \text{ kJ/mol}\right] - \left[50.42 \text{ kJ/mol} + 0 \text{ kJ/mol}\right]$$
$$= -622.02 \text{ kJ/mol}$$

$$4NH3(g) + 3O_2(g) \rightarrow 2N_2(g) + 6H_2O(l)$$

$$\Delta H^\circ_{\text{rxn}} = \left[6(-285.8 \text{ kJ/mol}) + 0 \text{ kJ/mol}\right] - \left[4(-46.3 \text{ kJ/mol}) + 0 \text{ kJ/mol}\right]$$
$$= -1529.6 \text{ kJ/mol}$$

d.

$$(1 \text{ g } N_2H_4)\left(\frac{\text{mol } N_2H_4}{32.06 \text{ g } N_2H_4}\right)\left(\frac{-622.02 \text{ kJ}}{\text{mol } N_2H_4}\right) = -19.40 \text{ kJ}$$

$$(1 \text{ g } NH_3)\left(\frac{\text{mol } NH_3}{17.03 \text{ g } NH_3}\right)\left(\frac{-1529.6 \text{ kJ}}{\text{mol } NH_3}\right) = -22.45 \text{ kJ}$$

Ammonia would cause the greater increase in temperature.

# Chapter 6

# Quantum Theory and the Electronic Structure of Atoms

6.5 a. **Strategy:** We are given the frequency of an electromagnetic wave and asked to calculate the wavelength. Rearranging Equation 6.1 of the text to solve for wavelength gives:

$$\lambda = \frac{c}{\nu}$$

**Solution:** Substituting the frequency and the speed of light ($3.00 \times 10^8$ m/s) into the above equation, the wavelength is:

$$\lambda = \frac{c}{\nu} = \frac{3.00\times10^{8}\ \text{m/s}}{8.6\times10^{13}\ /\text{s}} = 3.5\times10^{-6}\ \text{m} = \mathbf{3.5\times10^{3}\ nm}$$

b. **Strategy:** We are given the wavelength of an electromagnetic wave and asked to calculate the frequency. Rearranging Equation 6.1 of the text to solve for frequency gives:

$$\nu = \frac{c}{\lambda}$$

**Solution:** Because the speed of light is given in meters per second, it is convenient to first convert wavelength to units of meters. Recall that 1 nm = $1 \times 10^{-9}$ m (see Table 1.3 of the text). We write:

$$566\ \text{nm}\times\frac{1\times10^{-9}\ \text{m}}{1\ \text{nm}} = 566\times10^{-9}\ \text{m or } 5.66\times10^{-7}\ \text{m}$$

Substituting in the wavelength and the speed of light ($3.00 \times 10^8$ m/s), the frequency is:

$$\nu = \frac{c}{\lambda} = \frac{3.00\times10^{8}\ \text{m/s}}{566\times10^{-9}\ \text{m}} = 5.30\times10^{14}\ /\text{s} = \mathbf{5.30\times10^{14}\ Hz}$$

6.7

$$\lambda = \frac{c}{\nu} = \frac{3.00\times10^{8}\ \text{m/s}}{9{,}192{,}631{,}770\ \text{s}^{-1}} = 3.26\times10^{-2}\ \text{m} = \mathbf{3.26\times10^{7}\ nm}$$

This radiation falls in the **microwave region** of the spectrum. (See Figure 6.1 of the text.)

6.9 Since the speed of light is $3.00 \times 10^8$ m/s, we can write:

$$(1.3\times10^{8}\ \text{mi})\times\frac{1.61\ \text{km}}{1\ \text{mi}}\times\frac{1000\ \text{m}}{1\ \text{km}}\times\frac{1\ \text{s}}{3.00\times10^{8}\ \text{m}} = \mathbf{7.0\times10^{2}\ s}$$

**Think About It:** Would the time be different for other types of electromagnetic radiation?

6.15 **Strategy:** We are given the wavelength of photon and asked to calculate the energy. Rearranging Equation 6.1 of the text to solve for frequency gives:

$$\nu = \frac{c}{\lambda}$$

**Solution:** Substituting this into Equation 6.2, the energy is:

$$E = h\nu = \frac{hc}{\nu} = \frac{\left(6.63\times10^{-34}\ \text{J}\cdot\text{s}\right)\left(3.00\times10^{8}\ \text{m/s}\right)}{7.05\times10^{-7}\ \text{m}} = \mathbf{2.82\times10^{-19}\ J}$$

6.17 a. Rearranging Equation 6.1 to solve for wavelength:

$$\lambda = \frac{c}{\nu} = \frac{3.00\times10^{8}\ \text{m/s}}{6.5\times10^{9}\ \text{s}^{-1}} = 0.046\ \text{m} = \mathbf{4.6\times10^{7}\ nm}$$

The radiation **does not fall in the visible region**; it is microwave radiation. (See Figure 6.1 of the text.)

b. Using Equation 6.2:

$$E = h\nu = (6.63\times10^{-34}\ \text{J·s})(6.5\times10^{9}\ \text{s}^{-1}) = \mathbf{4.3\times10^{-24}\ J/photon}$$

c. Using Avogadro's number to convert to J/mol:

$$E = \frac{4.3\times10^{-24}\ \text{J}}{1\ \text{photon}}\times\frac{6.022\times10^{23}\ \text{photons}}{1\ \text{mol}} = \mathbf{2.6\ J/mol}$$

6.19 We can rearrange Equation 6.1 to solve for frequency,

$$\nu = \frac{c}{\lambda}$$

and substitute the result into Equation 6.2 to solve for the energy of a photon.

$$\mathbf{E} = h\nu = \frac{hc}{\lambda} = \frac{(6.63 \times 10^{-34}\ \text{J}\cdot\text{s})(3.00 \times 10^{8}\ \text{m/s})}{(0.154 \times 10^{-9}\ \text{m})} = \mathbf{1.29 \times 10^{-15}\ J}$$

6.21 Convert wavelength to frequency using Equation 6.1 and then calculate the energy per photon using Equation 6.2:

$$\nu = \frac{c}{\lambda} = \frac{3.00\times10^{8}\ \text{m/s}}{5.85\times10^{-7}\ \text{m}} = 5.13\times10^{14}\ \text{s}^{-1}$$

$$E = h\nu = (6.63 \times 10^{-34}\ \text{J·s})(5.13 \times 10^{14}\ \text{s}^{-1}) = 3.40 \times 10^{-19}\ \text{J/photon}$$

$$\textbf{Number of photons} = \frac{4.0\times10^{-17}\ \text{J}}{3.40\times10^{-19}\ \text{J/photon}} = \mathbf{1.2 \times 10^{2}\ photons}$$

6.23 **Infrared photons have insufficient energy to cause the chemical changes.**

6.25 A "blue" photon (shorter wavelength) is higher energy than a "yellow" photon. For the same amount of energy delivered to the metal surface, there must be fewer "blue" photons than "yellow" photons. Thus, the yellow light would eject more electrons since there are more "yellow" photons. Since the "blue" photons are of higher energy, blue light will eject electrons with greater kinetic energy.

6.29 Note that we use more significant figures than we usually do for the values of *h* and *c* for this problem.

$$\mathbf{E} = \frac{hc}{\lambda} = \frac{(6.6256 \times 10^{-34}\ \text{J}\cdot\text{s})(2.998 \times 10^{8}\ \text{m/s})}{656.3 \times 10^{-9}\ \text{m}} = \mathbf{3.027 \times 10^{-19}\ J}$$

6.31 **Strategy:** We are given the initial and final states in the emission process. We can calculate the energy of the emitted photon using Equation 6.6 of the text. Then, from this energy, we can solve for the frequency of the photon, and from the frequency we can solve for the wavelength.

**Solution:** From Equation 6.6 we write:

$$\Delta E = -2.18\times10^{-18}\ \text{J}\left(\frac{1}{n_\text{f}^2} - \frac{1}{n_\text{i}^2}\right) = -2.18\times10^{-18}\ \text{J}\left(\frac{1}{3^2} - \frac{1}{4^2}\right) = -1.06\times10^{-19}\ \text{J}$$

The negative sign for $\Delta E$ indicates that this is energy associated with an emission process. To calculate the frequency, we will omit the minus sign for $\Delta E$ because the frequency of the photon must be positive. We know that:

$$\Delta E = h\nu$$

Rearranging the equation and substituting in the known values:

$$\nu = \frac{\Delta E}{h} = \frac{1.06\times10^{-19}\ \text{J}}{6.63\times10^{-34}\ \text{J}\cdot\text{s}} = 1.60\times10^{14}\ \text{s}^{-1}\ \text{or}\ \mathbf{1.60\times10^{14}\ Hz}$$

We also know that $\lambda = \frac{c}{\nu}$. Substituting the frequency calculated above into this equation gives:

$$= \frac{(3.00\times10^{8}\ \text{m/s})}{1.60\times10^{14}\ \text{s}^{-1}} = 1.88\times10^{-6}\ \text{m} = \mathbf{1.88\times10^{3}\ nm}$$

This wavelength is in the infrared region of the electromagnetic spectrum (see Figure 6.1 of the text).

6.33

$$\Delta E = -2.18\times10^{-18}\ \text{J}\left(\frac{1}{1^2}-\frac{1}{n_i^2}\right)$$

$n_f$ is given in the problem but we need to calculate $\Delta E$. The photon energy is:

$$E = \frac{hc}{\lambda} = \frac{(6.63\times10^{-34}\ \text{J}\cdot\text{s})(3.00\times10^{8}\ \text{m/s})}{94.9\times10^{-9}\ \text{m}} = 2.10\times10^{-18}\ \text{J}$$

Since this is an emission process, the energy change $\Delta E$ must be negative, or $-2.10\times10^{-18}$ J.

Substitute $\Delta E$ into the following equation, and solve for $n_i$.

$$-2.10\times10^{-18}\ \text{J} = -2.18\times10^{-18}\ \text{J}\left(\frac{1}{1^2}-\frac{1}{n_i^2}\right)$$

$$\frac{1}{n_i^2} = \left(\frac{-2.10\times10^{-18}\ \text{J}}{2.18\times10^{-18}\ \text{J}}\right) + \frac{1}{1^2} = -0.963 + 1 = 0.0367$$

$$\boldsymbol{n_i} = \frac{1}{\sqrt{0.0367}} = \mathbf{5}$$

6.35 **Analyze the emitted light by passing it through a prism.**

6.37 **Excited atoms of the chemical elements emit the same characteristic frequencies or lines in a terrestrial laboratory, in the sun, or in a star many light-years distant from earth.**

6.41 **Strategy:** We are given the mass and the speed of the neutron and asked to calculate the wavelength. We need the de Broglie equation, which is Equation 6.9 of the text. Note that because the units of Planck's constant are J·s, $m$ must be in kg and $u$ must be in m/s (1 J = 1 kg·m$^2$/s$^2$).

**Solution:** Using Equation 6.9 we write:

$$\lambda = \frac{h}{mu}$$

$$\lambda = \frac{h}{mu} = \frac{6.63\times10^{-34}\ \text{J}\cdot\text{s}}{(1.675\times10^{-27}\ \text{kg})(7.00\times10^{2}\ \text{m/s})} = 5.65\times10^{-10}\ \text{m} = \mathbf{0.565\ nm}$$

6.43 **Strategy:** We are given the mass and the speed of the honey bee and asked to calculate the de Broglie wavelength. We need the de Broglie equation, which is Equation 6.9 of the text. Note that because the units of Planck's constant are J·s, $m$ must be in kg and $u$ must be in m/s (1 J = 1 kg·m$^2$/s$^2$).

**Solution:** Because mass in this problem is given in g and speed is given in mph, we must first convert these to kg and m/s, respectively.

$$m = 12.4\ \text{g} \times \frac{1\ \text{kg}}{1000\ \text{g}} = 0.0124\ \text{kg}$$

$$u = \frac{1.20\times10^{2}\ \text{mi}}{1\ \text{h}} \times \frac{1.61\ \text{km}}{1\ \text{mi}} \times \frac{1000\ \text{m}}{1\ \text{km}} \times \frac{1\ \text{h}}{3600\ \text{s}} = 53.7\ \text{m/s}$$

Using these values in Equation 6.9 we write:

$$\lambda = \frac{h}{mu} = \frac{6.63\times10^{-34}\ \text{J}\cdot\text{s}}{(0.0124\ \text{kg})(53.7\ \text{m/s})} = 9.96\times10^{-34}\ \text{m} \times \frac{100\ \text{cm}}{1\ \text{m}} = \mathbf{9.96\times10^{-32}\ cm}$$

6.51 **Strategy:** Use Equation 6.11 to calculate $\Delta x$ :

$$\left(\Delta x \cdot m\Delta u \geq \frac{h}{4\pi}\right)$$

The uncertainty in the velocity, 2.0 km/s, is $\Delta u$. The mass of a thermal neutron is $1.675 \times 10^{-27}$ kg according to Problem 6.41.

**Setup:** Rearranging to solve for $\Delta x$, we have:

$$\Delta x = \frac{h}{4\pi \cdot m\Delta u}$$

We use the equal sign in the uncertainty equation to calculate the *minimum* uncertainty values.

The uncertainty in the speed is 2.0 km/s. In m/s, the speed is:

$$\frac{2.0\text{ km}}{\text{s}} \times \frac{1\times 10^3\text{ m}}{1\text{ km}} = 2.0\times 10^3\text{ m/s}$$

**Solution:**

$$\Delta x = \frac{h}{4\pi \cdot m\Delta u}$$

$$\Delta x = \frac{6.63\times 10^{-34}\text{ kg}\cdot\text{m}^2/\text{s}}{4\pi\left(1.675\times 10^{-27}\text{ kg}\right)\left(2.0\times 10^3\text{ m/s}\right)} = \mathbf{1.6\times 10^{-11}\text{ m}}$$

6.53 Rearranging Equation 6.11 to solve for the uncertainty in velocity, $\Delta u$, we write:

$$\Delta u \geq \frac{h}{4\pi m\Delta x} = \frac{6.63\times 10^{-34}\text{ J}\cdot\text{s}}{(4\pi)\left(2.80\times 10^{-3}\text{ kg}\right)\left(4.30\times 10^{-7}\text{ m}\right)} \geq 4.38\times 10^{-26}\text{ m/s}$$

**This uncertainty is far smaller than can be measured**. Therefore, we are able to determine the speed of a macroscopic object with great certainty using a visible wavelength of light.

6.57 **Strategy:** What are the relationships among $n$, $\ell$, and $m_\ell$?

**Solution:** The angular momentum quantum number $\ell$ can have integral (i.e. whole number) values from 0 to $n - 1$. In this case $n = 2$, so the allowed values of the angular momentum quantum number, $\boldsymbol{\ell}$, are **0**, corresponding to an $s$ orbital; and **1**, corresponding to a $p$ orbital.

Each allowed value of the angular momentum quantum number labels a subshell. Within a given subshell (label $\ell$) there are $2\ell + 1$ allowed energy states (orbitals) each labeled by a different value of the magnetic quantum number. The allowed values run from $-\ell$ through 0 to $+\ell$ (whole numbers only). For the subshell labeled by the angular momentum quantum number $\boldsymbol{\ell} = \mathbf{1}$, the allowed values of the magnetic quantum number, $m_\ell$, are **–1, 0, and 1**. For the other subshell in this problem labeled by the angular momentum quantum number $\boldsymbol{\ell} = \mathbf{0}$, the allowed value of the magnetic quantum number is **0**.

**Think About It:** If the allowed whole number values run from $-\ell$ to $+\ell$, are there always $2\ell + 1$ values? Why?

6.59 **For $n = 4$, the allowed values of $\ell$ are 0, 1, 2, and 3 [$\ell = 0$ to $(n - 1)$, integer values].** These $\ell$ values correspond to the **4*s*, 4*p*, 4*d*, and 4*f* subshells**. These subshells each have **1, 3, 5, and 7 orbitals, respectively** (number of orbitals = $2\ell + 1$).

6.65 a. **2*p*: $n = 2$, $\ell = 1$, $m_\ell = 1, 0,$ or $-1$**

b. **3*s*: $n = 3$, $\ell = 0$, $m_\ell = 0$ (only allowed value)**

c. **5*d*: $n = 5$, $\ell = 2$, $m_\ell = 2, 1, 0, -1,$ or $-2$**

An orbital in a subshell can have any of the allowed values of the magnetic quantum number for that subshell. All the orbitals in a subshell have exactly the same energy.

6.67 **A 2*s* orbital is larger than a 1*s* orbital and exhibits a node. Both have the same spherical shape. The 1*s* orbital is lower in energy than the 2*s*.**

6.69 **In H, energy depends only on *n*, but for all other atoms, energy depends on *n* and $\ell$.** This is because in the many-electron atom, the 3*p* orbital electrons are more effectively shielded by the inner electrons of the atom (that is, the 1*s*, 2*s*, and 2*p* electrons) than the 3*s* electrons. The 3*s* orbital is said to be more "penetrating" than the 3*p* and 3*d* orbitals. In the hydrogen atom there is only one electron (no shielding), so the 3*s*, 3*p*, and 3*d* orbitals have the same energy.

6.71 Equation 6.5 of the text gives the orbital energy in terms of the principal quantum number, *n*, alone (for the hydrogen atom). The energy does not depend on any of the other quantum numbers. If two orbitals in the hydrogen atom have the same value of *n*, they have equal energy.

a. **2*s*** > 1*s*

b. **3*p*** > 2*p*

c. **equal**

d. **equal**

e. **5*s*** > 4*f*

6.73 a. Based on the relative size of the orbitals, **orbital (b) has the greatest value of *n*. The principal quantum number, *n*, designates the size of the orbital.** As *n* increases, the average distance of an electron in the orbital from the nucleus increases. Therefore, the orbital is larger.

b. The angular momentum quantum number, $\ell$, describes the shape of the atomic orbital. **A value of $\ell = 1$, corresponds to a *p* orbital.** Figure 6.19 shows that each *p* orbital can be thought of as two lobes on opposite sides of the nucleus. **Orbitals (a) and (d) represent *p* orbitals with $\ell = 1$.**

c. **None.** The shape of orbital (b) is that of a *d* orbital, specifically the $d_{z^2}$ orbital (see Figure 6.20). For each value of $n \geq 3$, there is only one $d_{z^2}$ orbital. Therefore, there can be no other orbitals with the same general shape as (b) with the same value of *n*.

6.77 There can be a maximum of two electrons occupying one orbital.

a. **two** b. **six** c. **ten** d. **fourteen**

**Think About It:** What rule of nature demands a maximum of two electrons per orbital? Do they have the same energy? How are they different? Would five 4*d* orbitals hold as many electrons as five 3*d* orbitals? In other words, does the principal quantum number *n* affect the number of electrons in a given subshell?

6.79 3*s*: **two** 3*d*: **ten** 4*p*: **six** 4*f*: **fourteen** 5*f*: **fourteen**

6.81 a. **is wrong because the magnetic quantum number $m_\ell$ can have only whole number values.**

b. **is wrong because the magnetic quantum number $m_\ell$ can only have the value 0 when the angular momentum quantum number $\ell$ is 0.**

c. **is wrong because the magnetic quantum number $m_\ell$ can only have the value 0 when the angular momentum quantum number $\ell$ is 0.**

d. **is acceptable.**

e. **is wrong because the electron spin quantum number $m_s$ can have only half-integral values.**

6.83

| | | | |
|---|---|---|---|
| B: | $[He]2s^22p^1$ (1 unpaired electron) | Ne: | (0 unpaired electrons) |
| P: | $[Ne]3s^23p^3$ (3 unpaired electrons) | Sc: | $[Ar]4s^23d^1$ (1 unpaired electron) |
| Mn: | $[Ar]4s^23d^5$ (5 unpaired electrons) | Se: | $[Ar]4s^23d^{10}4p^4$ (2 unpaired electrons) |
| Kr: | (0 unpaired electrons) | Fe: | $[Ar]4s^23d^6$ (4 unpaired electrons) |

Cd: $[Kr]5s^24d^{10}$ (0 unpaired electrons) I: $[Kr]5s^24d^{10}5p^5$ (1 unpaired electron)

Pb: $[Xe]6s^24f^{14}5d^{10}6p^2$ (2 unpaired electrons)

6.85

| ↑↓ | ↑ ↑ ↑ | ↑↓ | ↑↓ ↑ ↑ | ↑↓ | ↑↓ ↑↓ ↑ |
|---|---|---|---|---|---|
| $3s^2$ | $3p^3$ | $3s^2$ | $3p^4$ | $3s^2$ | $3p^5$ |

| $S^+$ (5 valence electrons) | S (6 valence electrons) | $S^-$ (7 valence electrons) |
|---|---|---|
| 3 unpaired electrons | 2 unpaired electrons | 1 unpaired electron |

**$S^+$** has the most unpaired electrons.

6.95 The ground state electron configuration of Tc is: $[Kr]5s^24d^5$.

6.97 **Strategy:** How many electrons are in the Ge atom (Z = 32)? We start with $n = 1$ and proceed to fill orbitals in the order shown in Figure 6.23 of the text. Remember that any given orbital can hold at most 2 electrons. However, don't forget about degenerate orbitals. Starting with $n = 2$, there are three $p$ orbitals of equal energy, corresponding to $m_\ell = -1, 0, 1$. Starting with $n = 3$, there are five $d$ orbitals of equal energy, corresponding to $m_\ell = -2, -1, 0, 1, 2$. We can place electrons in the orbitals according to the Pauli exclusion principle and Hund's rule. The task is simplified if we use the noble gas core preceding Ge for the inner electrons.

**Solution:** Germanium has 32 electrons. The noble gas core in this case is [Ar]. (Ar is the noble gas in the period preceding germanium.) [Ar] represents $1s^22s^22p^63s^23p^6$. This core accounts for 18 electrons, which leaves 14 electrons to place.

See Figure 6.23 of your text to check the order of filling subshells past the Ar noble gas core. You should find that the order of filling is 4$s$, 3$d$, 4$p$. There are 14 remaining electrons to distribute among these orbitals. The 4$s$ orbital can hold two electrons. Each of the five 3$d$ orbitals can hold two electrons for a total of *10* electrons. This leaves two electrons to place in the 4$p$ orbitals.

The electrons configuration for **Ge** is:

$$\mathbf{[Ar]4s^23d^{10}4p^2}$$

You should follow the same reasoning for the remaining atoms.

**Fe: $[Ar]4s^23d^6$** **Zn: $[Ar]4s^23d^{10}$** **Ni: $[Ar]4s^23d^8$**

**W: $[Xe]6s^24f^{14}5d^4$** **Tl: $[Xe]6s^24f^{14}5d^{10}6p^1$**

6.99 **Part (b) is correct in the view of contemporary quantum theory. Bohr's explanation of emission and absorption line spectra appears to have universal validity. Parts (a) and (c) are artifacts of Bohr's early planetary model of the hydrogen atom and are *not* considered to be valid today.**

6.101 a. With $n = 2$, there are $n^2$ orbitals $= 2^2 = 4$. $m_s = +½$, specifies 1 electron per orbital, for a total of **4 electrons** (one $e^-$ in the 2*s* and 2*p* orbitals).

b. $n = 4$ and $m_\ell = +1$, specifies one orbital in each subshell with $\ell = 1, 2,$ or 3 (i.e., a 4*p*, 4*d*, and 4*f* orbital). Each of the three orbitals holds 2 electrons for a total of **6 electrons**.

c. If $n = 3$ and $\ell = 2$, $m_\ell$ has the values 2, 1, 0, –1, or –2. Each of the five orbitals can hold 2 electrons for a total of **10 electrons** (2 $e^-$ in each of the five 3*d* orbitals).

d. If $n = 2$ and $\ell = 0$, then $m_\ell$ can only be zero. $m_s = -½$ specifies 1 electron in this orbital for a total of **1 electron** (one $e^-$ in the 2*s* orbital).

e. $n = 4$, $\ell = 3$ and $m_\ell = -2$, specifies one 4*f* orbital. This orbital can hold **2 electrons**.

6.103 a. **Strategy:** Use Equation 6.3 ($h\nu = \text{KE} + W$) to determine the binding energy of each metal.

**Setup:** Solving Equation 6.3 for the binding energy, $W$ gives:

$$W = h\nu - \text{KE}$$

At the frequency values provided on the graph, KE = 0. Therefore, we write:

$$W = h\nu$$

Planck's constant, $h$, is $6.63 \times 10^{-34}$ J·s.

**Solution:** Metal A:

$$W = (6.63\times10^{-34}\ \text{J·s})(5.2\times10^{14}\ \text{s}^{-1}) = \mathbf{3.4\times10^{-19}\ J}$$

Metal B:

$$W = (6.63\times10^{-34}\ \text{J·s})(8.5\times10^{14}\ \text{s}^{-1}) = \mathbf{5.6\times10^{-19}\ J}$$

Metal C:

$$W = (6.63\times10^{-34}\ \text{J·s})(9.9\times10^{14}\ \text{s}^{-1}) = \mathbf{6.6\times10^{-19}\ J}$$

**Metal C has the highest binding energy, 6.6 × 10$^{-19}$ J.**

b. **Strategy:** In order for an electron to be ejected, the energy of the photon striking the metal surface must be at least the binding energy, *W*. Use Equation 6.1 ($c = \lambda\nu$) to convert the wavelength to frequency; then use Equation 6.2 ($E = h\nu$) to determine the energy of the photon. Compare this energy to the binding energy of each of the metals.

**Setup:** The wavelength must be converted from nanometers to meters:

$$333\ \text{nm}\times\frac{1\times10^{-9}\ \text{m}}{1\ \text{nm}} = 3.33\times10^{-7}\ \text{m}$$

Planck's constant, *h*, is 6.63 × 10$^{-34}$ J·s.

**Solution:**

$$\nu = \frac{c}{\lambda} = \frac{3.00\times10^{8}\ \text{m/s}}{3.33\times10^{-7}\ \text{m}} = 9.009\times10^{14}\ \text{s}^{-1}$$

$$E = h\nu = (6.63\times10^{-34}\ \text{J·s})(9.009\times10^{14}\ \text{s}^{-1}) = \mathbf{5.97\times10^{-19}\ J}$$

The energy of a photon with a wavelength of 333 nm is greater than the binding energies of metals A and B (3.4 × 10$^{-19}$ J and 5.6 × 10$^{-19}$, respectively). Therefore, **electrons will be ejected from metals A and B**.

6.105 The Balmer series corresponds to transitions to the $n = 2$ level.

Rearranging Equation 6.4 to solve for wavelength we write:

$$\lambda = \frac{1}{R_\infty\left(\frac{1}{n_1^2} - \frac{1}{n_2^2}\right)}$$

The Rydberg constant for $He^+$ is 4.39 × 10$^7$ m$^{-1}$. Therefore, for the transition, $n = 3 \rightarrow 2$:

$$\lambda = \frac{1}{4.39\times10^{7}\ \text{m}^{-1}\left(\frac{1}{3^2} - \frac{1}{2^2}\right)} = -1.64\times10^{-7}\ \text{m}$$

Note that the negative sign indicates the *emission* of light. Wavelengths are always positive quantities.

For the transition, $n = 3 \rightarrow 2$,

$$\lambda = \mathbf{164\ nm}$$

For the transition, $n = 4 \rightarrow 2$,

$$\lambda = \mathbf{121\ nm}$$

For the transition, $n = 5 \rightarrow 2$,

$$\lambda = \mathbf{108\ nm}$$

For the transition, $n = 6 \rightarrow 2$,

$$\lambda = \mathbf{103\ nm}$$

For H, the calculations are identical to those above, except the Rydberg constant for H is $1.097 \times 10^7\ m^{-1}$.

For the transition, $n = 3 \rightarrow 2$,

$$\lambda = \mathbf{656\ nm}$$

For the transition, $n = 4 \rightarrow 2$,

$$\lambda = \mathbf{486\ nm}$$

For the transition, $n = 5 \rightarrow 2$,

$$\lambda = \mathbf{434\ nm}$$

For the transition, $n = 6 \rightarrow 2$,

$$\lambda = \mathbf{410\ nm}$$

All the Balmer transitions for $He^+$ are in the ultraviolet region; whereas, the transitions for H are all in the visible region.

6.107 Applying the Pauli exclusion principle and Hund's rule:

a. $\underline{\uparrow\downarrow}\ \underline{\uparrow\downarrow}\ \underline{\uparrow\downarrow}\,\underline{\uparrow\downarrow}\,\underline{\uparrow\ }$ — $\mathbf{1s^2}$ $\mathbf{2s^2}$ $\mathbf{2p^5}$

b. **[Ne]** $\underline{\uparrow\downarrow}\ \underline{\uparrow\ }\,\underline{\uparrow\ }\,\underline{\uparrow\ }$ — $\mathbf{3s^2}$ $\mathbf{3p^3}$

c. **[Ar]** $\underline{\uparrow\downarrow}\ \underline{\uparrow\downarrow}\,\underline{\uparrow\downarrow}\,\underline{\uparrow\ }\,\underline{\uparrow\ }\,\underline{\uparrow\ }$ — $\mathbf{4s^2}$ $\mathbf{3d^7}$

6.109 a. **False**. $n = 2$ is the first excited state.

b. **False**. In the $n = 4$ state, the electron is (on average) further from the nucleus and hence easier to remove.

c. **True**.

d. **False**. The $n = 4$ to $n = 1$ transition is a higher energy transition, which corresponds to a *shorter* wavelength.

e. **True**.

6.111 The excited atoms are still neutral, so the total number of electrons is the same as the atomic number of the element.

a. **He (2 electrons), $1s^2$**

b. **N (7 electrons), $1s^22s^22p^3$**

c. **Na (11 electrons), $1s^22s^22p^63s^1$**

d. **As (33 electrons), [Ar]$4s^23d^{10}4p^3$**

e. **Cl (17 electrons), [Ne]$3s^23p^5$**

6.113 Based on the *selection rule*, which states that $\Delta\ell = \pm1$, only **(b)** and **(d)** are allowed transitions. Any of the transitions in Figure 6.11 is possible as long as $\ell$ for the final state differs from $\ell$ of the initial state by 1.

6.115 Since the energy corresponding to a photon of wavelength $\lambda_1$ equals the energy of photon of wavelength $\lambda_2$ plus the energy of photon of wavelength $\lambda_3$, then the equation must relate the wavelength to energy.

$$\text{energy of photon 1} = (\text{energy of photon 2} + \text{energy of photon 3})$$

Since $E = \frac{hc}{\lambda}$, then:

$$\frac{hc}{\lambda_1} = \frac{hc}{\lambda_2} + \frac{hc}{\lambda_3}$$

Dividing by $hc$:

$$\frac{1}{\lambda_1} = \frac{1}{\lambda_2} + \frac{1}{\lambda_3}$$

6.117 **Strategy:** Nodes occur at locations where the wavefunction is zero. Set $\Psi_{2s}$ equal to zero and solve for $\rho$. Then, calculate $r$ by solving $\rho = Z\left(\frac{r}{a_0}\right)$ for $r$ and substituting.

**Solution:**

$$0 = \frac{1}{\sqrt{2a_0^3}}\left(1 - \frac{\rho}{2}\right)e^{-\rho/2}$$

The exponential factor is non-zero for all finite values of $\rho$. The factor $1 - \frac{\rho}{2}$ is zero for $\rho = 2$.

$$r = \frac{a_0\rho}{Z} = \frac{(0.529\text{ nm})(2)}{1} = \mathbf{1.06\text{ nm}}$$

6.119 a. First, we need to calculate the moving mass of the proton, and then we can calculate its wavelength using the de Broglie equation.

$$m_{\text{moving}} = \frac{m_{\text{rest}}}{\sqrt{1 - \left(\frac{u}{c}\right)^2}} = \frac{1.673 \times 10^{-27}\text{ kg}}{\sqrt{1 - \left(\frac{(0.50)(3.00 \times 10^8\text{ m/s})}{3.00 \times 10^8\text{ m/s}}\right)^2}} = 1.93 \times 10^{-27}\text{ kg}$$

$$\lambda = \frac{h}{mu} = \frac{6.63 \times 10^{-34}\text{ J}\cdot\text{s}}{(1.93 \times 10^{-27}\text{ kg})[(0.50)(3.00 \times 10^8\text{ m/s})]}$$

$$\lambda = 2.29 \times 10^{-15}\text{ m} = \mathbf{2.29 \times 10^{-6}\text{ nm}}$$

b.

$$m_{\text{moving}} = \frac{m_{\text{rest}}}{\sqrt{1 - \left(\frac{u}{c}\right)^2}} = \frac{6.0 \times 10^{-2}\text{ kg}}{\sqrt{1 - \left(\frac{63\text{ m/s}}{3.00 \times 10^8\text{ m/s}}\right)^2}} \approx 6.0 \times 10^{-2}\text{ kg}$$

**The equation is only important for speeds close to that of light**. Note that photons have a rest mass of zero; otherwise, their moving mass would be infinite!

6.121 The energy given in the problem is the energy of 1 mole of gamma rays. We need to convert this to the energy of one gamma ray, then we can calculate the wavelength and frequency of this gamma ray.

$$\frac{3.14\times10^{11}\text{ J}}{1\text{ mol}} \times \frac{1\text{ mol}}{6.022\times10^{23}\text{ gamma particles}} = 5.21\times10^{-13}\text{ J/gamma particle}$$

Now, we can calculate the wavelength and frequency from this energy.

$$E = \frac{hc}{\lambda}$$

$$\lambda = \frac{hc}{E} = \frac{(6.63\times10^{-34}\ \text{J}\cdot\text{s})(3.00\times10^{8}\ \text{m/s})}{5.21\times10^{-13}\ \text{J}} = \mathbf{3.82\times10^{-13}\ m = 0.382\ pm}$$

and

$$E = h\nu$$

$$\nu = \frac{E}{h} = \frac{5.21\times10^{-13}\ \text{J}}{6.63\times10^{-34}\ \text{J}\cdot\text{s}} = \mathbf{7.86\times10^{20}\ s^{-1}}$$

6.123 In the photoelectric effect, light of sufficient energy shining on a metal surface causes electrons to be ejected (photoelectrons). Since the electrons are charged particles, the metal surface becomes positively charged as more electrons are lost. After a long enough period of time, the positive surface charge becomes large enough to start attracting the ejected electrons back toward the metal with the result that the kinetic energy of the departing electrons becomes smaller.

6.125 The kinetic energy acquired by the electrons is equal to the voltage times the charge on the electron. After calculating the kinetic energy, we can calculate the velocity of the electrons (KE = ½mu$^2$). Finally, we can calculate the wavelength associated with the electrons using the de Broglie equation.

$$\text{KE} = (5.00\times10^{3}\ \text{V})\times\frac{1.602\times10^{-19}\ \text{J}}{1\ \text{V}} = 8.01\times10^{-16}\ \text{J}$$

We can now calculate the velocity of the electrons.

$$\text{KE} = \frac{1}{2}mu^2$$

$$8.01\times10^{-16}\ \text{J} = \frac{1}{2}(9.1094\times10^{-31}\ \text{kg})u^2$$

$$u = 4.19\times10^{7}\ \text{m/s}$$

Finally, we can calculate the wavelength associated with the electrons using the de Broglie equation.

$$\lambda = \frac{h}{mu}$$

$$\lambda = \frac{(6.63\times10^{-34}\ \text{J}\cdot\text{s})}{(9.1094\times10^{-31}\ \text{kg})(4.19\times10^{7}\ \text{m/s})} = \mathbf{1.74\times10^{-11}\ m = 17.4\ pm}$$

6.127 $n_i = 236,\; n_f = 235$

$$\Delta E = (-2.18\times10^{-18}\text{ J})\left(\frac{1}{235^2}-\frac{1}{236^2}\right) = -3.34\times10^{-25}\text{ J}$$

$$\lambda = \frac{hc}{\Delta E} = \frac{(6.63\times10^{-34}\text{ J}\cdot\text{s})(3.00\times10^{8}\text{ m/s})}{3.34\times10^{-25}\text{ J}} = \mathbf{0.596\ m}$$

This wavelength is in the ***microwave/radio*** region. (See Figure 6.1 of the text.)

6.129 The energy needed per photon for the process is:

$$\frac{248\times10^{3}\text{ J}}{1\text{ mol}}\times\frac{1\text{ mol}}{6.022\times10^{23}\text{ photons}} = 4.12\times10^{-19}\text{ J/photon}$$

$$\lambda = \frac{hc}{E} = \frac{(6.63\times10^{-34}\text{ J}\cdot\text{s})(3.00\times10^{8}\text{ m/s})}{(4.12\times10^{-19}\text{ J})} = \mathbf{4.83\times10^{-7}\ m} = \mathbf{483\ nm}$$

Any wavelength shorter than 483 nm will also promote this reaction. Once a person goes indoors, the reverse reaction $Ag + Cl \rightarrow AgCl$ takes place.

6.131 Energy of a photon at 360 nm:

$$E = h\nu = \frac{hc}{\lambda} = \frac{(6.63\times10^{-34}\text{ J}\cdot\text{s})(3.00\times10^{8}\text{ m/s})}{360\times10^{-9}\text{ m}} = 5.53\times10^{-19}\text{ J}$$

Area of exposed body in cm$^2$:

$$0.45\text{ m}^2\times\left(\frac{1\text{ cm}}{1\times10^{-2}\text{ m}}\right)^2 = 4.5\times10^{3}\text{ cm}^2$$

The number of photons absorbed by the body in 2.5 hours is:

$$0.5\times\frac{2.0\times10^{16}\text{ photons}}{\text{cm}^2\cdot\text{s}}\times\left(4.5\times10^{3}\text{ cm}^2\right)\times\frac{9000\text{ s}}{2.5\text{ h}} = 4.1\times10^{23}\text{ photons/2.5 h}$$

The factor of 0.5 is used above because only 50% of the radiation is absorbed.

$4.1\times10^{23}$ photons with a wavelength of 360 nm correspond to an energy of:

$$4.1\times10^{23}\text{ photons}\times\frac{5.53\times10^{-19}\text{ J}}{1\text{ photon}} = \mathbf{2.2\times10^{5}\ J}$$

6.133 a. **We note that the maximum solar radiation centers around 500 nm. Thus, over billions of years, organisms have adjusted their development to capture energy at or near this wavelength. The two most notable cases are photosynthesis and vision.**

b. **Astronomers record blackbody radiation curves from stars and compare them with those obtained from objects at different temperatures in the laboratory. Because the shape of the curve and the wavelength corresponding to the maximum depend on the temperature of an object, astronomers can reliably determine the temperature at the surface of a star from the closest matching curve and wavelength.**

6.135 The heat needed to raise the temperature of 150 mL of water from 20°C to 100°C is:

$$q = ms\Delta T = (150\text{ g})(4.184\text{ J/g}\cdot°\text{C})(100 - 20)°\text{C} = 5.0 \times 10^4\text{ J}$$

The microwave will need to supply more energy than this because only 92.0% of microwave energy is converted to thermal energy of water. The energy that needs to be supplied by the microwave is:

$$\frac{5.0 \times 10^4\text{ J}}{0.920} = 5.4 \times 10^4\text{ J}$$

The energy supplied by one photon with a wavelength of $1.22 \times 10^8$ nm (0.122 m) is:

$$E = \frac{hc}{\lambda} = \frac{(6.63 \times 10^{-34}\text{ J}\cdot\text{s})(3.00 \times 10^8\text{ m/s})}{(0.122\text{ m})} = 1.63 \times 10^{-24}\text{ J}$$

The number of photons needed to supply $5.4 \times 10^4$ J of energy is:

$$(5.4 \times 10^4\text{ J}) \times \frac{1\text{ photon}}{1.63 \times 10^{-24}\text{ J}} = \mathbf{3.3 \times 10^{28}\text{ photons}}$$

6.137 The energy required to heat the water is: $ms\Delta T = (368\text{ g})(4.184\text{ J/g}\cdot°\text{C})(5.00°\text{C}) = 7.70 \times 10^3\text{ J}$

Energy of a photon with a wavelength = $1.06 \times 10^4$ nm:

$$E = h\nu = \frac{hc}{\lambda} = \frac{(6.63 \times 10^{-34}\text{ J}\cdot\text{s})(3.00 \times 10^8\text{ m/s})}{1.06 \times 10^{-5}\text{ m}} = 1.88 \times 10^{-20}\text{ J/photon}$$

The number of photons required is:

$$(7.70 \times 10^3\text{ J}) \times \frac{1\text{ photon}}{1.88 \times 10^{-20}\text{ J}} = \mathbf{4.10 \times 10^{23}\text{ photons}}$$

# Chapter 7

# Electron Configuration and the Periodic Table

7.17 **Strategy:** To write the electron configuration, we refer to the information in Section 6.9 of the text. We start with principal quantum number $n = 1$ and continue upward in energy until all electrons are accounted for.

**Setup:** We know that for $n = 1$, we have a $1s$ orbital (2 electrons). For $n = 2$, we have a $2s$ orbital (2 electrons) and three $2p$ orbitals (6 electrons). This constitutes the neon noble gas core. For $n = 3$, we have a $3s$ orbital (2 electrons) and $3p$ orbitals (6 electrons). The number of electrons left to place is 34 – 18 = 16. The electrons with the next lowest energy are in the $4s$ orbitals (2 electrons). The next electrons are in the $3d$ orbitals (10 electrons). The number of electrons left to place is 34 – 30 = 4. These 4 electrons are placed in the $4p$ orbitals. **The electron configuration is** $1s^22s^22p^63s^23p^64s^23d^{10}4p^4$ or **[Ar]$4s^23d^{10}4p^4$**. **A neutral atom with 34 electrons** (and 34 protons) **is selenium**, Se.

7.19 Elements that have the same number of valence electrons will have similarities in chemical behavior. Looking at the periodic table, elements with the same number of valence electrons are in the same group. Therefore, the pairs that would represent similar chemical properties of their atoms are: **(a) and (d), (b) and (e), (c) and (f)**.

7.21 a. **Group 1A or 1** b. **Group 5A or 15** c. **Group 5A or 15** d. **Group 8B or 8**

7.25 a. Carbon has a nuclear charge of +6. If each core electron were totally effective in screening the valence electrons from the nucleus, and the valence electrons did not shield one another, the shielding constant ($\sigma$) would be 2 and the nuclear charge ($Z_{\text{eff}}$) would be +4 for both the $2s$ and the $2p$ electrons. Therefore, **$\sigma = 2$ and $Z_{\text{eff}} = +4$.**

b. In reality, for $2s$ electrons, $Z_{\text{eff}} = +6 - 2.78 = +3.22$; for $2p$ electrons, $Z_{\text{eff}} = +6 - 2.86 = +3.14$. **$2s$, $Z_{\text{eff}} = +3.22$; $2p$, $Z_{\text{eff}} = +3.14$.**

**These values are lower than those in part (a) because the $2s$ and $2p$ electrons actually do shield each other somewhat.**

7.33 The atomic number of mercury is 80. We carry an extra significant figure throughout this calculation to avoid rounding errors.

$$\Delta E = (2.18 \times 10^{-18}\ \text{J})(80^2)\left(\frac{1}{1^2} - \frac{1}{\infty^2}\right) = 1.395 \times 10^{-14}\ \text{J/ion}$$

$$\Delta E = \frac{1.395 \times 10^{-14}\ \text{J}}{1\ \text{ion}} \times \frac{6.022 \times 10^{23}\ \text{ions}}{1\ \text{mol}} \times \frac{1\ \text{kJ}}{1000\ \text{J}} = \mathbf{8.40 \times 10^{6}\ kJ/mol}$$

7.35 **Strategy:** Recall that the general periodic trends in atomic size are:

(1) Moving from left to right across a row (period) of the periodic table, the atomic radius ***decreases*** due to an increase in effective nuclear charge.

(2) Moving down a column (group) of the periodic table, the atomic radius ***increases*** since the orbital size increases with increasing principal quantum number.

**Setup:** The atoms that we are considering are all in the same period of the periodic table. Hence, the atom furthest to the left in the row will have the largest atomic radius, and the atom furthest to the right in the row will have the smallest atomic radius. Arranged in order of decreasing atomic radius, we have:

$$\text{Na} > \text{Mg} > \text{Al} > \text{P} > \text{Cl}$$

**Solution:** See Figure 7.6 of the text to confirm that the above is the correct order of decreasing atomic radius.

7.37 **Fluorine** is the smallest atom in Group 7A. Atomic radius increases moving down a group since the orbital size increases with increasing principal quantum number, $n$.

7.39 **Left to right: S, Se, Ca, and K.**

7.41 The atomic radius is largely determined by how strongly the outer-shell electrons are held by the nucleus. The larger the effective nuclear charge, the more strongly the electrons are held and the smaller the atomic radius. For the second period, the atomic radius of Li is largest because the 2$s$ electron is well shielded by the filled 1$s$ shell. The effective nuclear charge that the outermost electrons feel increases across the period as a result of incomplete shielding by electrons in the same shell. Consequently, the orbital containing the electrons is compressed and the atomic radius decreases.

7.43 The general periodic trend for first ionization energy is that it increases across a period (row) of the periodic table and it decreases down a group (column). Of the choices, K will have the smallest ionization energy. Ca, just to the right of K, will have a higher first ionization energy. Moving to the right across the periodic table, the ionization energies will continue to increase as we move to P. Continuing across to Cl and moving up the halogen group, F will have a higher ionization energy than P. Finally, Ne is to the right of F in period two, thus it will have a higher ionization energy. The correct order of increasing first ionization energy is:

**K < Ca < P < F < Ne**

You can check the above answer by looking up the first ionization energies for these elements in Table 7.3 of the text.

7.45 The Group 3A elements (such as Al) all have a single electron in the outermost *p* subshell, which is well shielded from the nuclear charge by the inner electrons and the $ns^2$ electrons. Therefore, less energy is needed to remove a single *p* electron than to remove a paired *s* electron from the same principal energy level (such as for Mg).

7.47 **Strategy:** Removal of the outermost electron requires less energy if it is shielded by a filled inner shell.

**Setup:** The lone electron in the 3*s* orbital will be much easier to remove. This lone electron is shielded from the nuclear charge by the filled inner shell. Therefore, **the ionization energy of 496 kJ/mol is paired with the electron configuration $1s^22s^22p^63s^1$**.

**A noble gas electron configuration, such as $1s^22s^22p^6$**, is a very stable configuration, making it extremely difficult to remove an electron. The 2*p* electron is not as effectively shielded by electrons in the same energy level. **The high ionization energy of 2080 kJ/mol would be associated with the element having this noble gas electron configuration**.

**Solution:** Compare this answer to the data in Table 7.3 The electron configuration of $1s^22s^22p^63s^1$ corresponds to a Na atom, and the electron configuration of $1s^22s^22p^6$ corresponds to a Ne atom.

7.49 **Setup:** What are the trends in electron affinity in a periodic group and in a particular period. Which of the above elements are in the same group and which are in the same period?

**Solution:** One of the general periodic trends for electron affinity is that the tendency to accept electrons increases (that is, electron affinity values become more positive) as we move from left to right across a period. However, this trend does not include the noble gases. We know that noble gases are extremely stable, and they do not want to gain or lose electrons.

7.51 Alkali metals have a valence electron configuration of $ns^1$ so they can accept another electron in the ns orbital. On the other hand, alkaline earth metals have a valence electron configuration of $ns^2$. Alkaline earth metals have little tendency to accept another electron, as it would have to go into a higher energy *p* orbital.

7.57 You should realize that the metal ion in question is a transition metal ion because it has five electrons in the 3*d* subshell. Remember that in a transition metal ion, the $(n-1)d$ orbitals are more stable than the *ns* orbital.

Hence, when a cation is formed from an atom of a transition metal, electrons are *always* removed first from the *ns* orbital and then from the $(n-1)d$ orbitals if necessary. Since the metal ion has a +3 charge, three electrons have been removed. Since the 4*s* subshell is less stable than the 3*d*, two electrons would have been lost from the 4*s* and one electron from the 3*d*. Therefore, the electron configuration of the neutral atom is $[Ar]4s^23d^6$. This is the electron configuration of iron. Thus, the metal is **iron, Fe**.

7.59 Isoelectronic means that the species have the same number of electrons and the same electron configuration.

$Be^{2+}$ and He (2 $e^-$) $N^{3-}$ and $F^-$ (10 $e^-$) $Fe^{2+}$ and $Co^{3+}$ (24 $e^-$) $S^{2-}$ and Ar (18 $e^-$)

7.61 a. $Cr^{3+}$ b. $Sc^{3+}$ c. $Rh^{3+}$ d. $Ir^{3+}$

7.65 **Strategy:** In comparing ionic radii, it is useful to classify the ions into three categories: (1) isoelectronic ions (See Section 7.6 of the text), (2) ions that carry the same charges and are generated from atoms of the same periodic group, and (3) ions that carry different charges but are generated from the same atom. In case (1), ions carrying a greater negative charge are always larger; in case (2), ions from atoms having a greater atomic number are always larger; in case (3), ions that have a smaller positive charge (or a larger negative charge) are always larger.

**Solution:** a. **Cl** is smaller than $Cl^-$. An atom gets bigger when more electrons are added.

b. $\mathbf{Na^+}$ is smaller than Na. An atom gets smaller when electrons are removed.

c. $\mathbf{O^{2-}}$ is smaller than $S^{2-}$. Both elements belong to the same group, and ionic radius increases going down a group.

d. $\mathbf{Al^{3+}}$ is smaller than $Mg^{2+}$. The two ions are isoelectronic and in such cases the radius gets smaller as the charge becomes more positive.

e. $\mathbf{Au^{3+}}$ is smaller than $Au^+$ for the same reason as part (b).

**Think About It:** In each of the above cases from which atom (or ion) would it be harder to remove an electron?

7.67 The $Cu^+$ ion is larger than $Cu^{2+}$ because it has one more electron.

7.73 We assume the approximate boiling point of argon is the mean of the boiling points of neon and krypton, based on its position in the periodic table being between Ne and Kr in Group 8A.

$$\textbf{b.p.} = \frac{-246.1°\text{C} + (-153.2°\text{C})}{2} = \mathbf{-199.7°C}$$

The actual boiling point of argon is −185.7°C.

7.75 Since ionization energies decrease going down a column in the periodic table, francium should have the lowest first ionization energy of all the alkali metals. As a result, Fr should be the most reactive of all the Group 1A elements toward water and oxygen. The reaction with oxygen would probably be similar to that of K, Rb, or Cs.

7.77 The Group 1B elements are much less reactive than the Group 1A elements. The 1B elements are more stable because they have much higher ionization energies resulting from incomplete shielding of the nuclear charge by the inner *d* electrons. The $ns^1$ electron of a Group 1A element is shielded from the nucleus more effectively by the completely filled noble gas core. Consequently, the outer *s* electrons of 1B elements are more strongly attracted by the nucleus.

7.79 a. Lithium oxide is a basic oxide. It reacts with water to form the metal hydroxide:

$$Li_2O(s) + H_2O(l) \longrightarrow 2LiOH(aq)$$

b. Calcium oxide is a basic oxide. It reacts with water to form the metal hydroxide:

$$CaO(s) + H_2O(l) \longrightarrow Ca(OH)_2(aq)$$

c. Sulfur trioxide is an acidic oxide. It reacts with water to form sulfuric acid:

$$SO_3(g) + H_2O(l) \longrightarrow H_2SO_4(aq)$$

7.81 As we move down a column, the metallic character of the elements increases. Since magnesium and barium are both Group 2A elements, we expect barium to be more metallic than magnesium and **BaO** to be more basic than MgO.

7.83 a. bromine, Br b. nitrogen, N c. rubidium, Rb d. magnesium, Mg

7.85 This is an isoelectronic series with ten electrons in each species. The species with the smallest nuclear charge will be the largest. Recall that the largest species will be the *easiest* to ionize.

increasing ionization energy: $O^{2-} < F^- < Na^+ < Mg^{2+}$

7.87 $O^+$ and N $S^{2-}$ and Ar $N^{3-}$ and Ne $As^{3+}$ and Zn $Cs^+$ and Xe

7.89 **(a)** and **(d)**

7.91 Fluorine is a yellow-green gas that attacks glass; chlorine is a pale yellow gas; bromine is a fuming red liquid; and iodine is a dark, metallic-looking solid.

7.93 Fluorine, F

7.95 $H^-$ and He are isoelectronic species with two electrons. Since $H^-$ has only one proton compared to two protons for He, the nucleus of $H^-$ will attract the two electrons less strongly compared to He. Therefore, $\mathbf{H^-}$ is larger.

7.97

| Oxide | Name | Property |
|---|---|---|
| $Li_2O$ | lithium oxide | basic |
| $BeO$ | beryllium oxide | amphoteric |
| $B_2O_3$ | diboron trioxide | acidic |
| $CO_2$ | carbon dioxide | acidic |
| $N_2O_5$ | dinitrogen pentoxide | acidic |

7.99 Replacing Z in the equation given in Problem 7.46 with $(Z - \sigma)$ gives:

$$E_n = \left(2.18\times10^{-18}\ \text{J}\right)\left(Z-\sigma\right)^2\left(\frac{1}{n^2}\right)$$

For helium, the atomic number ($Z$) is 2, and in the ground state, its two electrons are in the first energy level, so $n = 1$. Substitute $Z$, $n$, and the first ionization energy into the above equation to solve for $\sigma$.

$$E_1 = 3.94\times10^{-18}\ \text{J} = (2.18\times10^{-18}\ \text{J})(2-\sigma)^2\left(\frac{1}{1^2}\right)$$

$$(2-\sigma)^2 = \frac{3.94\times10^{-18}\ \text{J}}{2.18\times10^{-18}\ \text{J}}$$

$$2-\sigma = \sqrt{1.81}$$

$$\sigma = 2-1.34 = \mathbf{0.66}$$

7.101 The volume of a sphere is $4/3\pi r^3$. The percentage of volume occupied by $K^+$ compared to K is:

$$\frac{\text{volume of K}^+\text{ ion}}{\text{volume of K atom}}\times 100\% = \frac{\frac{4}{3}\pi(138\text{ pm})^3}{\frac{4}{3}\pi(227\text{ pm})^3}\times 100\% = 22.5\%$$

Therefore, there is a **decrease in atomic volume** of (100 – 22.5)% = **77.5%** when $K^+$ is formed from K.

7.103 a. matches bromine ($Br_2$)

c. matches calcium (Ca)

e. matches argon (Ar)

b. matches hydrogen ($H_2$)

d. matches gold (Au)

7.105 X must belong to Group 4A; it is probably **Sn** or **Pb** because it is not a very reactive metal (it is certainly not reactive like an alkali metal).

Y is a nonmetal since it does *not* conduct electricity. Since it is a light yellow solid, it is probably **phosphorus** (Group 5A).

Z is an **alkali metal** since it reacts with air to form a basic oxide or peroxide.

7.107 The plot is:

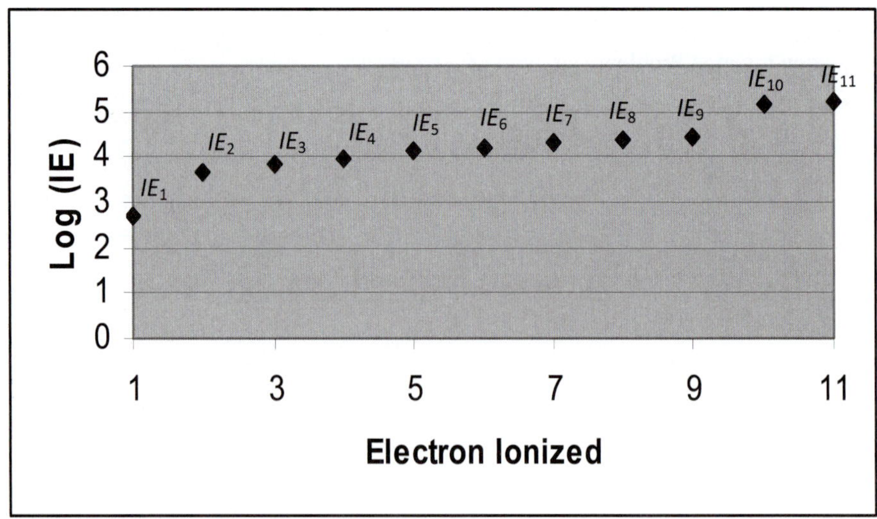


a. $IE_1$ corresponds to the electron in $3s^1$

$IE_7$ corresponds to the electron in $2p^1$

$IE_2$ corresponds to the first electron in $2p^6$

$IE_3$ corresponds to the first electron in $2p^5$

$IE_4$ corresponds to the first electron in $2p^4$

$IE_5$ corresponds to the first electron in $2p^3$

$IE_6$ corresponds to the first electron in $2p^2$

$IE_8$ corresponds to the first electron in $2s^2$

$IE_9$ corresponds to the electron in $2s^1$

$IE_{10}$ corresponds to the first electron in $1s^2$

$IE_{11}$ corresponds to the electron in $1s^1$

b. Each break ($IE_1 \rightarrow IE_2$ and $IE_9 \rightarrow IE_{10}$) represents the transition to another shell ($n = 1 \rightarrow 2$ and $n = 2 \rightarrow 3$).

7.109 The hydrides are: LiH (lithium hydride), $CH_4$ (methane), $NH_3$ (ammonia), $H_2O$ (water), and HF (hydrogen fluoride).

The reactions with water:

$$LiH + H_2O \rightarrow LiOH + H_2$$

$$CH_4 + H_2O \rightarrow \text{no reaction at room temperature.}$$

$$NH_3 + H_2O \rightarrow NH_4^+ + OH^-$$

$$H_2O + H_2O \rightarrow H_3O^+ + OH^-$$

$$HF + H_2O \rightarrow H_3O^+ + F^-$$

The last three reactions involve *equilibria* that will be discussed in later chapters.

7.111 a.

$$2KClO_3(s) \rightarrow 2KCl(s) + 3O_2(g)$$

b.

$$N_2(g) + 3H_2(g) \rightarrow 2NH_3(g) \qquad \text{(industrial)}$$

$$NH_4Cl(s) + NaOH(aq) \rightarrow NH_3(g) + NaCl(aq) + H_2O(l)$$

c.

$$CaCO_3(s) \rightarrow CaO(s) + CO_2(g) \qquad \text{(industrial)}$$

$$CaCO_3(s) + 2HCl(aq) \rightarrow CaCl_2(aq) + H_2O(l) + CO_2(g)$$

d.

$$Zn(s) + H_2SO_4(aq) \rightarrow ZnSO_4(aq) + H_2(g)$$

e. Same as **(c)**, (first equation)

7.113 Examine a solution of $Na_2SO_4$ which is colorless. This shows that the $SO_4^{2-}$ ion is colorless. Thus the blue color is due to $Cu^{2+}(aq)$.

7.115 $Z_{eff}$ increases from left to right across the table, so electrons are held more tightly. (This explains the electron affinity values of C and O.) Nitrogen has a zero value of electron affinity because of the stability of the half-filled 2*p* subshell (that is, N has little tendency to accept another electron).

7.117 Once an atom gains an electron forming a negative ion, adding additional electrons is typically an unfavorable process due to electron-electron repulsions. 2nd and 3rd electron affinities do not occur spontaneously and are therefore difficult to measure.

7.119 There is a large jump from the second to the third ionization energy, indicating a change in the principal quantum number *n*. In other words, the third electron removed is an inner, noble gas core electron, which is difficult to remove. Therefore, the element is in **Group 2A**.

7.121 a. $SiH_4$, $GeH_4$, $SnH_4$, $PbH_4$

b. Metallic character increases going down a family of the periodic table. Therefore, **RbH should be more ionic than NaH**.

c. Since Ra is in Group 2A, we would expect the reaction to be the same as other alkaline earth metals with water.

$$\mathbf{Ra}(s) + \mathbf{2H_2O}(l) \rightarrow \mathbf{Ra(OH)_2}(aq) + \mathbf{H_2}(g)$$

d. Beryllium (diagonal relationship)

7.123 We rearrange the equation given in the problem to solve for $Z_{eff}$.

$$Z_{eff} = n\sqrt{\frac{I_1}{1312\ \text{kJ/mol}}}$$

Li:

$$Z_{eff} = (2)\sqrt{\frac{520\ \text{kJ/mol}}{1312\ \text{kJ/mol}}} = 1.26$$

Na:

$$Z_{\text{eff}} = (3)\sqrt{\frac{495.9 \text{ kJ/mol}}{1312 \text{ kJ/mol}}} = 1.84$$

K:

$$Z_{\text{eff}} = (4)\sqrt{\frac{418.7 \text{ kJ/mol}}{1312 \text{ kJ/mol}}} = 2.26$$

As we move down a group, $Z_{\text{eff}}$ increases. This is what we would expect because shells with larger $n$ values are less effective at shielding the outer electrons from the nuclear charge.

Li:

$$\frac{Z_{\text{eff}}}{n} = \frac{1.26}{2} = 0.630$$

Na:

$$\frac{Z_{\text{eff}}}{n} = \frac{1.84}{3} = 0.613$$

K:

$$\frac{Z_{\text{eff}}}{n} = \frac{2.26}{4} = 0.565$$

The $Z_{\text{eff}}/n$ values are fairly constant, meaning that the screening per shell is about the same.

7.125 Lithium, like magnesium, forms a stable nitride ($Li_3N$), so **nitrogen** should not be used for lithium storage.

7.127 The first statement that an allotropic form of the element is a colorless crystalline solid, might lead you to think about diamond, a form of carbon. When carbon is reacted with excess oxygen, the colorless gas, carbon dioxide is produced.

$$C(s) + O_2(g) \rightarrow CO_2(g)$$

When $CO_2(g)$ is dissolved in water, carbonic acid is produced.

$$CO_2(g) + H_2O(l) \rightarrow H_2CO_3(aq)$$

Element X is most likely **carbon**, choice **(c)**.

7.129 Rearrange the given equation to solve for ionization energy.

$$IE = h\nu - \frac{1}{2}mu^2$$

or,

$$IE = \frac{hc}{\lambda} - KE$$

The kinetic energy of the ejected electron is given in the problem. Substitute $h$, $c$, and $\lambda$ into the above equation to solve for the ionization energy.

$$IE = \frac{(6.63 \times 10^{-34}\ \text{J}\cdot\text{s})(3.00 \times 10^{8}\ \text{m/s})}{162 \times 10^{-9}\ \text{m}} - (5.34 \times 10^{-19}\ \text{J})$$

$$\mathbf{IE = 6.94 \times 10^{-19}\ J/electron}$$

If there are no other electrons with lower kinetic energy, then this is the electron from the valence shell.

To ensure that the ejected electron is the valence electron, UV light of the *longest* wavelength (lowest energy) should be used that can still eject electrons.

7.131 **Strategy:** In the formation of a **cation** from the neutral atom of a representative element, one or more electrons are *removed* from the highest occupied $n$ shell. In the formation of an **anion** from the neutral atom of a representative element, one or more electrons are *added* to the highest partially filled $n$ shell. Representative elements typically gain or lose electrons to achieve a stable noble gas electron configuration. When a cation is formed from an atom of a transition metal, electrons are *always* removed first from the $ns$ orbital and then from the $(n-1)d$ orbitals if necessary.

**Solution:**

a. [Ne]

b. same as (a). Do you see why?

c. [Ar]

d. Same as (c). Do you see why?

e. Same as (c)

f. $[Ar]3d^6$. Why isn't it $[Ar]4s^23d^4$?

g. $[Ar]3d^9$. Why not $[Ar]4s^23d^7$?

h. $[Ar]3d^{10}$. Why not $[Ar]4s^23d^8$?

7.133 The binding of a cation to an anion results from electrostatic attraction. As the +2 cation gets smaller (from $Ba^{2+}$ to $Mg^{2+}$), the distance between the opposite charges decreases and the electrostatic attraction increases.

7.135 a. It was determined that the periodic table was based on atomic number, not atomic mass.

b. Argon:

$$(0.00337 \times 35.9675 \text{ amu}) + (0.00063 \times 37.9627 \text{ amu}) + (0.9960 \times 39.9624 \text{ amu}) = \mathbf{39.95 \text{ amu}}$$

Potassium:

$$(0.93258 \times 38.9637 \text{ amu}) + (0.000117 \times 39.9640 \text{ amu}) + (0.0673 \times 40.9618 \text{ amu}) = \mathbf{39.10 \text{ amu}}$$

7.137 The electron configuration of titanium is: $[\text{Ar}]4s^2 3d^2$. Titanium has four valence electrons, so the maximum oxidation number it is likely to have in a compound is +4. The compounds followed by the oxidation state of titanium are: $K_3TiF_6$, +3; $K_2Ti_2O_5$, +4; $TiCl_3$, +3; $K_2TiO_4$, +6; and $K_2TiF_6$, +4. **$K_2TiO_4$** is unlikely to exist because of the oxidation state of Ti of +6. Ti in an oxidation state greater than +4 is unlikely because of the very high ionization energies needed to remove the fifth and sixth electrons.

7.139 The reaction representing the electron affinity of chlorine is:

$$\text{Cl}(g) + e^- \longrightarrow \text{Cl}^-(g) \qquad \Delta H° = +349 \text{ kJ/mol}$$

It follows that the photon energy needed for the reverse process is also +349 kJ/mol.

$$\text{Cl}^-(g) + \text{photon} \longrightarrow \text{Cl}(g) + e^- \qquad h\nu = +349 \text{ kJ/mol}$$

The energy above is the energy of one mole of photons. We need to convert to the energy of one photon in order to calculate the wavelength of the photon.

$$\frac{349 \text{ kJ}}{1 \text{ mol photons}} \times \frac{1 \text{ mol photons}}{6.022 \times 10^{23} \text{ photons}} \times \frac{1000 \text{ J}}{1 \text{ kJ}} = 5.80 \times 10^{-19} \text{ J/photon}$$

Now, we can calculate the wavelength of a photon with this energy.

$$\lambda = \frac{hc}{E} = \frac{(6.63 \times 10^{-34} \text{ J}\cdot\text{s})(3.00 \times 10^8 \text{ m/s})}{5.80 \times 10^{-19} \text{ J}} = 3.43 \times 10^{-7} \text{ m} = \mathbf{343 \text{ nm}}$$

The radiation is in the **ultraviolet** region of the electromagnetic spectrum.

# Chapter 8

# Chemical Bonding I: Basic Concepts

8.3 a. ·Be· b. ·K c. ·Ca· d. ·Ga· (with dot above) e. ·O· (with lone pairs) f. :Br· (with lone pairs) g. ·N· h. :I· i. ·As· j. :F·

8.5 a. :I· b. [:I:]$^-$ c. ·S· d. [:S:]$^{2-}$ e. ·P· f. [:P:]$^{3-}$ g. ·Na h. $Na^+$ i. ·Mg· j. $Mg^{2+}$ k. ·Al· l. [:As]$^{3+}$ m. ·Pb· n. [:Pb]$^{2+}$

8.19 Using the method outlined in Figure 8.3, we determine the lattice energy as follows:

$$\text{Lattice energy} = \Delta H_f^\circ[\text{Li}(g)] + \Delta H_f^\circ[\text{Cl}(g)] + IE_1(\text{Li}) + \left|\Delta H_f^\circ[\text{LiCl}(s)]\right| - EA(\text{Cl})$$

$\text{Li}(s) \rightarrow \text{Li}(g)$ $\Delta H_f^\circ[\text{Li}(g)] = 159.3$ kJ/mol

$\frac{1}{2}\text{Cl}_2(g) \rightarrow \text{Cl}(g)$ $\Delta H_f^\circ[\text{Cl}(g)] = 121.7$ kJ/mol

$\text{Li}(g) \rightarrow \text{Li}^+(g) + e^-$ $IE_1(\text{Li}) = 520$ kJ/mol

$\text{Li}(s) + \frac{1}{2}\text{Cl}_2(g) \rightarrow \text{LiCl}(s)$ $\Delta H_f^\circ[\text{LiCl}(s)] = -408.3$ kJ/mol

$\left|\Delta H_f^\circ[\text{LiCl}(s)]\right| = 408.3$ kJ/mol

$\text{Cl}(g) + e^- \rightarrow \text{Cl}^-(g)$ $EA(\text{Cl}) = 349$ kJ/mol

---

(159.3 kJ/mol) + (121.7 kJ/mol) + (520 kJ/mol) + (408.3 kJ/mol) – (349 kJ/mol) = 860 kJ/mol

The lattice energy of LiCl is **860 kJ/mol**.

8.21 We use Coulomb's law to answer this question: $E \propto \dfrac{Q_1 \times Q_2}{d}$ where $Q_1$ is the charge on the cation, $A^+$, and $Q_2$ is the charge on the anion, $B^-$.

a. Doubling the radius of the cation would increase the distance, *d*, between the centers of the ions. A larger value of *d* results in a smaller energy, *E*, of the ionic bond. **Doubling the radius of the cation decreases the ionic bond energy**. Is it possible to say how much smaller *E* will be?

b. **Tripling the charge on $A^+$ triples the ionic bond energy**, since the energy of the bond is directly proportional to the charge on the cation, $Q_1$.

c. **Doubling the charge on both the cation and anion increases the bond energy by a factor of 4.**

d. Decreasing the radius of both the cation and the anion to half of their original values is the same as halving the distance, *d*, between the centers of the ions. **Decreasing the radii of $A^+$ and $B^-$ to half their original values increases the bond energy by a factor of 2.**

8.23 Lewis representations for the ionic reactions are as follows.

a. $Na\cdot \ + \ :\ddot{\underset{\cdot\cdot}{F}}\cdot \longrightarrow Na^+ :\ddot{\underset{\cdot\cdot}{F}}:^-$

b. $2K\cdot \ + \ \cdot\ddot{\underset{\cdot\cdot}{S}}\cdot \longrightarrow 2K^+ :\ddot{\underset{\cdot\cdot}{S}}:^{2-}$

c. $\dot{\underset{\cdot}{Ba}} \ + \ \cdot\ddot{\underset{\cdot\cdot}{O}}\cdot \longrightarrow Ba^{2+} :\ddot{\underset{\cdot\cdot}{O}}:^{2-}$

d. $\dot{\underset{\cdot}{Al}}\cdot \ + \ \cdot\ddot{\underset{\cdot}{N}}\cdot \longrightarrow Al^{3+} :\ddot{\underset{\cdot\cdot}{N}}:^{3-}$

8.33 a. $BF_3$, boron trifluoride, covalent

b. KBr, potassium bromide, ionic

8.37 The order of increasing ionic character is:

Cl–Cl (zero difference in electronegativity) < Br–Cl (difference 0.2) < Si–C (difference 0.7) < Cs–F (difference 3.3).

8.39 a. The two silicon atoms are the same. The bond is **covalent.**

b. The electronegativity difference between Cl and Si is 3.0 – 1.8 = 1.2. The bond is **polar covalent**.

c. The electronegativity difference between F and Ca is 4.0 – 1.0 = 3.0. The bond is **ionic**.

d. The electronegativity difference between N and H is 3.0 – 2.1 = 0.9. The bond is **polar covalent**.

8.41 **Strategy:** We can look up electronegativity values in Figure 8.6 of the text. The amount of ionic character is based on the electronegativity difference between the two atoms. The larger the electronegativity difference, the greater the ionic character.

**Solution:** Let ΔEN = electronegativity difference. The bonds arranged in order of increasing ionic character are:

C–H (ΔEN = 0.4) < Br–H (ΔEN = 0.7) < F–H (ΔEN = 1.9) < Li–Cl (ΔEN = 2.0) < Na–Cl (ΔEN = 2.1) < K–F (ΔEN = 3.2)

8.43 The Lewis structures are:

a. :F̤̈—Ö̤—F̤̈:

b. :F̤̈—N̈=N̈—F̤̈:

c.
```
    H   H
    |   |
H—Si—Si—H
    |   |
    H   H
```

d. $^-$:Ö̤—H

e.
```
   H  :O:
   |   ||    ..
H—C—C—O:⁻
   |         ..
  :Cl:
```

f.
```
   H  H
   |  |+
H—C—N—H
   |  |
   H  H
```

8.45 a. O=C(Br)(Br) c. H—N(H)—O—H e. H—C(H)(H)—C(H)(H)—Br

b. H—Se—H d. O=P(Cl)(Cl)(Cl) f. Cl—N(Cl)—Cl

8.47 **Strategy:** We follow the procedure for drawing Lewis structures outlined in Section 8.5 of the text.

**Solution:** a. *Step 1:* The skeletal structure is: O N O

*Step 2:* The outer-shell electron configuration of O is $2s^22p^4$. Each O has six valence electrons. The outer-shell electron configuration of N is $2s^22p^3$. N has five valence electrons. Also, we must subtract an electron to account for the positive charge. Thus, there are

$$(2 \times 6) + 5 - 1 = 16 \text{ valence electrons}$$

*Step 3:* We draw single covalent bonds between the atoms, and then attempt to complete the octets—first for the O atoms.

:Ö—N—Ö:

Because this uses up all of the available electrons, step 5 outlined in the text is not required, but step 6 requires us to move electron pairs to satisfy the octet for nitrogen.

:Ö=N=O:

We determine the formal charges on the atoms using the method outlined in Section 8.6 and Equation 8.2 in the text. For each oxygen atom, formal charge = 6 – 6 = 0. For the nitrogen atom, formal charge = 5 – 4 = +1.

O=N⁺=O

**Check:** As a final check, we verify that there are 16 valence electrons in the Lewis structure of $NO_2^+$.

b. $\ddot{S}=C=\ddot{N}^-$ or $^-:\ddot{S}-C\equiv N:$

c. $^-:\ddot{S}-\ddot{S}:^-$

d. $:\ddot{F}-\ddot{Cl}^+-\ddot{F}:$

8.49 a. Neither oxygen atom has a complete octet, and the left-most hydrogen atom shows two bonds (4 electrons). Hydrogen can hold only two electrons in its valence shell.

b. The correct structure is:

$H_3C-C(=\ddot{O})-\ddot{O}-H$

8.53 **Strategy:** We follow the procedure for drawing Lewis structures outlined in Section 8.5 of the text. After we complete the Lewis structure, we draw the resonance structures.

**Solution:** Following the procedure in Sections 8.5 and 8.7 of the text, we come up with the following resonance structures:

$H-C(-\ddot{O}:^-)=\ddot{O}: \longleftrightarrow H-C(=\ddot{O}:)-\ddot{O}:^-$

$H_2C=N^+(-\ddot{O}:^-)-\ddot{O}:^- \longleftrightarrow H_2\ddot{C}^--N^+(-\ddot{O}:^-)=\ddot{O}: \longleftrightarrow H_2\ddot{C}^--N^+(=\ddot{O}:)-\ddot{O}:^-$

8.55 The structures of the most important resonance forms are:

$H-\ddot{N}=\overset{+}{N}=\ddot{N}:^{-} \longleftrightarrow H-\ddot{\underset{..}{N}}^{-}-\overset{+}{N}\equiv N: \longleftrightarrow H-\overset{+}{N}\equiv\overset{+}{N}-\ddot{\underset{..}{N}}:^{2-}$

8.57 Three reasonable resonance structures for $OCN^-$ are:

$\ddot{\underset{..}{O}}=C=\ddot{\underset{..}{N}}^{-} \longleftrightarrow {}^{-}:\ddot{\underset{..}{O}}-C\equiv N: \longleftrightarrow :\overset{+}{O}\equiv C-\ddot{\underset{..}{N}}:^{2-}$

8.59 The two resonance structures are:

H—N̈—H | C ... (adenine resonance structures) ↔

All of the atoms have formal charges of zero in both resonance structures.

8.67 **Strategy:** We follow the procedure outlined in Section 8.5 of the text for drawing Lewis structures. We assign formal charges as discussed in Section 8.6 of the text.

**Solution:** Drawing the structure with single bonds between Be and each of the Cl atoms, we see that Be is "electron deficient." The octet rule is not followed for Be. The Lewis structure is:

$:\ddot{\underset{..}{Cl}}-Be-\ddot{\underset{..}{Cl}}:$

An octet of electrons on Be can only be formed by making two double bonds as shown below:

$\overset{+}{\ddot{\underset{..}{Cl}}}=\overset{2-}{Be}=\overset{+}{\ddot{\underset{..}{Cl}}}$

This places a high negative formal charge on Be and positive formal charges on the Cl atoms. This structure distributes the formal charges counter to the electronegativities of the elements. It is not a plausible Lewis structure.

8.69 The outer electron configuration of antimony is $5s^25p^3$. The Lewis structure is shown below. All five valence electrons are shared in the five covalent bonds. **No**, the octet rule is not obeyed. (The electrons on the chlorine atoms have been omitted for clarity.)

Cl
Cl—Sb—Cl
Cl Cl

8.71 The reaction can be represented as:

:Cl—Al—Cl: + :Cl:⁻ → :Cl—Al⁻—Cl: (with :Cl: above and below)

The new bond formed is called a **coordinate covalent bond.**

8.73 Completed octet on S: $[:\ddot{O}-\ddot{S}(-\ddot{O}:)-\ddot{O}:]^{2-}$ ; zero formal charge on S: $[:\ddot{O}-\ddot{S}(=\ddot{O})-\ddot{O}:]^{2-}$

8.77 **Strategy:** Keep in mind that bond breaking is an energy absorbing (endothermic) process and bond making is an energy releasing (exothermic) process. Therefore, the overall energy change is the difference between these two opposing processes, as described in Equation 8.3 of the text.

**Solution:** There are two oxygen-to-oxygen bonds in ozone. We will represent these bonds as O–O. However, these bonds might not be true oxygen-to-oxygen single bonds. Using Equation 8.3 of the text, we write:

$$\Delta H^\circ = \Sigma BE(\text{reactants}) - \Sigma BE(\text{products})$$

$$\Delta H^\circ = BE(\text{O=O}) - 2BE(\text{O–O})$$

In the problem, we are given $\Delta H^\circ$ for the reaction, and we can look up the O=O bond enthalpy in Table 8.6 of the text. Solving for the average bond enthalpy in ozone,

$$-2BE(\text{O–O}) = \Delta H^\circ - BE(\text{O=O})$$

$$BE(\text{O}-\text{O}) = \frac{BE(\text{O=O}) - \Delta H^\circ}{2} = \frac{498.7\text{ kJ/mol} + 107.2\text{ kJ/mol}}{2} = \mathbf{303.0\ kJ/mol}$$

Considering the resonance structures for ozone, is it expected that the O–O bond enthalpy in ozone is between the single O–O bond enthalpy (142 kJ) and the double O=O bond enthalpy (498.7 kJ)?

8.79 a.

| *Bonds Broken* | *Number Broken* | *Bond Enthalpy* (kJ/mol) | *Enthalpy Change* (kJ) |
|---|---|---|---|
| C – H | 12 | 414 | 4968 |
| C – C | 2 | 347 | 694 |
| O = O | 7 | 498.7 | 3491 |

| *Bonds Formed* | *Number Formed* | *Bond Enthalpy* (kJ/mol) | *Enthalpy Change* (kJ) |
|---|---|---|---|
| C = O | 8 | 799 | 6392 |
| O – H | 12 | 460 | 5520 |

$\Delta H°$ = total energy input – total energy released

= (4968 + 694 + 3491) – (6392 + 5520) = **–2759 kJ/mol**

b.

$$\Delta H° = 4\Delta H_f^\circ(CO_2) + 6\Delta H_f^\circ(H_2O) - [2\Delta H_f^\circ(C_2H_6) + 7\Delta H_f^\circ(O_2)]$$

$$\boldsymbol{\Delta H°} = (4)(-393.5\text{ kJ/mol}) + (6)(-241.8\text{ kJ/mol}) - [(2)(-84.7\text{ kJ/mol}) + (7)(0)] = \mathbf{-2855.4\text{ kJ/mol}}$$

The answers for part (a) and (b) are different, because *average* bond enthalpies are used for part (a).

8.81 **Strategy:** Draw Lewis structures for the reactants and products in order to determine the types and numbers of bonds. The balanced equation for the reaction is given below. We use this equation and Table 8.6 to estimate the enthalpy of reaction.

C≡O + F—F + F—F + F—F ⟶ $CF_4$ (F—C—F with F above and below C) + F—O—F

**Solution:**

$$\begin{aligned}\Delta H_{rxn} &= BE(C\equiv O)+3BE(F\text{–}F)-[4BE(C\text{–}F)+2BE(O\text{–}F)]\\ &= 1070\text{ kJ/mol}+(3)(156.9\text{ kJ/mol})-[(4)(453\text{ kJ/mol})+(2)(190\text{ kJ/mol})]\\ &= \mathbf{-651\text{ kJ/mol}}\end{aligned}$$

8.83 Typically, ionic compounds are composed of a metal cation and a nonmetal anion. RbCl and $KO_2$ are ionic compounds.

Typically, covalent compounds are composed of nonmetals. $PF_5$, $BrF_3$, and $CI_4$ are covalent compounds.

8.85 Recall that you can classify bonds as ionic or covalent based on electronegativity difference.

The melting points (°C) are shown in parentheses following the formulas.

**Ionic:** NaF (993) $MgF_2$ (1261) $AlF_3$ (1291)

**Covalent:** $SiF_4$ (−90.2) $PF_5$ (−83) $SF_6$ (−50) $ClF_3$ (−83)

8.87 KF is an ionic compound. It is a solid at room temperature made up of $K^+$ and $F^-$ ions. It has a high melting point, and it is a strong electrolyte. Benzene, $C_6H_6$, is a covalent compound that exists as discrete molecules. It is a liquid at room temperature. It has a low melting point, is insoluble in water, and is a nonelectrolyte.

8.89 The resonance structures are:

$$^{-}\ddot{\underset{..}{N}}=\overset{+}{N}=\ddot{\underset{..}{N}}^{-} \longleftrightarrow :N\equiv\overset{+}{N}-\ddot{\underset{..}{N}}:^{2-} \longleftrightarrow {}^{2-}:\ddot{\underset{..}{N}}-\overset{+}{N}\equiv N:$$

**Think About It:** Which is the most plausible structure based on a formal charge argument?

8.91 a. An example of an aluminum species that satisfies the octet rule is the anion $AlCl_4^-$. The Lewis dot structure is drawn in Problem 8.71.

b. An example of an aluminum species containing an expanded octet is anion $AlF_6^{3-}$. (How many pairs of electrons surround the central atom?)

c. An aluminum species that has an incomplete octet is the compound $AlCl_3$. The dot structure is given in Problem 8.71.

8.93 $CF_2$ would be very unstable because carbon does not have an octet. (How many electrons does it have?)

$LiO_2$ would not be stable because the lattice energy between $Li^+$ and superoxide $O_2^-$ would be too low to stabilize the solid.

$CsCl_2$ requires a $Cs^{2+}$ cation. The second ionization energy is too large to be compensated by the increase in lattice energy.

$PI_5$ appears to be a reasonable species (compared to $PF_5$ in Practice Problem 8.12A of the text). However, the iodine atoms are too large to have five of them "fit" around a single P atom.

8.95 a. False b. true c. false d. false

For question (c), what is an example of a second-period species that violates the octet rule?

8.97 The formation of $CH_4$ from its elements is:

$$C(s) + 2H_2(g) \longrightarrow CH_4(g)$$

The reaction could take place in two steps:

Step 1: $C(s) + 2H_2(g) \longrightarrow C(g) + 4H(g)$ $\Delta H^\circ_{rxn} = (716 + 872.8)\text{kJ/mol} = 1589\text{ kJ/mol}$

Step 2: $C(g) + 4H(g) \longrightarrow CH_4(g)$ $\Delta H^\circ_{rxn} \approx -4 \times (\text{bond energy of C}-\text{H bond})$

$$= -4 \times 414\text{ kJ/mol} = -1656\text{ kJ/mol}$$

Therefore, $\Delta H^\circ_f(CH_4)$ would be approximately the sum of the enthalpy changes for the two steps. See Section 5.5 of the text (Hess's law).

$$\Delta H^\circ_f(CH_4) = \Delta H^\circ_{rxn}(1) + \Delta H^\circ_{rxn}(2)$$

$$\Delta H^\circ_f(CH_4) = (1589 - 1656)\text{kJ/mol} = \mathbf{-67\text{ kJ/mol}}$$

The actual value of $\Delta H^\circ_f(CH_4) = -74.85$ kJ/mol.

8.99 Only $\mathbf{N_2}$ has a triple bond. Therefore, it has the shortest bond length.

8.101 To be isoelectronic, molecules must have the same number and arrangement of valence electrons. $CH_4$ and $NH_4^+$ are isoelectronic (8 valence electrons), as are $N_2$ and CO (10 valence electrons), as are $C_6H_6$ and $B_3N_3H_6$ (30 valence electrons). Draw Lewis structures to convince yourself that the electron arrangements are the same in each isoelectronic pair.

8.103

```
 [      ..         ]3-   [       . .        ]3-
 [     :O:         ]     [       .O.        ]
 [  ..  |   ..     ]     [  ..   ||   ..    ]
 [ :O — P — O:     ]     [ :O —  P  — O:    ]
 [  ..  |   ..     ]     [  ..   |    ..    ]
 [     :O:         ]     [      :O:         ]
 [      ..         ]     [       ..         ]
```

H—O—Cl(—O)—O    H—O—Cl(=O)=O

O—S(=O)—O    O=S(=O)=O

O=S—O    O=S=O

8.105 The reaction can be represented as:

$$H-\ddot{N}^{-}(-H) + H-O(-H) \rightarrow H-N(-H)-H + {}^{-}O-H$$

8.107 The central iodine atom in $I_3^-$ has *ten* electrons surrounding it: two bonding pairs and three lone pairs. The central iodine has an expanded octet. Elements in the second period such as fluorine cannot have an expanded octet as would be required for $F_3^-$.

8.109 :N≡N⁺—N=N⁺=N⁻ ⟷ ⁻N=N⁺=N—N⁺≡N: ⟷ :N≡N⁺—N⁻—N⁺≡N:

8.111 Form (a) is the most important resonance structure with no formal charges and all satisfied octets. (b) is likely not as important as (a) because of the positive formal charge on O. Forms (c) and (d) do not satisfy the octet rule for all atoms and are likely not important.

8.113

Cl₂Al(μ-Cl)₂AlCl₂ (each bridging Cl bonds covalently to one Al and coordinately, shown by an arrow, to the other Al)

The arrows indicate coordinate covalent bonds. This dimer does not possess a dipole moment.

8.115 a. Using Equation 8.3 of the text,

$$\Delta H = \sum BE(\text{reactants}) - \sum BE(\text{products})$$

$$\mathbf{\Delta H} = [(436.4 + 151.0) - 2(298.3)] = \mathbf{-9.2\ kJ/mol}$$

b. Using Equation 5.19 of the text,

$$\Delta H^\circ = 2\Delta H^\circ_f[\text{HI}(g)] - \left[\Delta H^\circ_f[\text{H}_2(g)] + \Delta H^\circ_f[\text{I}_2(g)]\right]$$

$$\boldsymbol{\Delta H^\circ} = (2)(25.9 \text{ kJ/mol}) - [(0) + (1)(61.0 \text{ kJ/mol})] = \mathbf{-9.2 \text{ kJ/mol}}$$

8.117 The Lewis structures are:

a. $:\overset{-}{C}\equiv\overset{+}{O}:$

b. $:N\equiv\overset{+}{O}:$

c. $:\overset{-}{C}\equiv N:$

d. $:N\equiv N:$

8.119 True. Each noble gas atom already has completely filled *ns* and *np* subshells.

8.121 a. The bond enthalpy of $F_2^-$ is the energy required to break up $F_2^-$ into an F atom and an $F^-$ ion.

$$F_2^-(g) \longrightarrow F(g) + F^-(g)$$

We can arrange the equations given in the problem so that they add up to the above equation. See Section 5.5 of the text (Hess's law).

$$F_2^-(g) \longrightarrow F_2(g) + e^- \qquad \Delta H^\circ = 290 \text{ kJ/mol}$$

$$F_2(g) \longrightarrow 2F(g) \qquad \Delta H^\circ = 156.9 \text{ kJ/mol}$$

$$F(g) + e^- \longrightarrow F^-(g) \qquad \Delta H^\circ = -333 \text{ kJ/mol}$$

---

$$F_2^-(g) \longrightarrow F(g) + F^-(g)$$

The bond enthalpy of $F_2^-$ is the sum of the enthalpies of reaction.

$$\boldsymbol{BE(F_2^-)} = [290 + 156.9 + (-333 \text{ kJ})]\text{kJ/mol} = \mathbf{114 \text{ kJ/mol}}$$

b. **The bond in $F_2^-$ is weaker** (114 kJ/mol) than the bond in $F_2$ (156.9 kJ/mol), because the extra electron increases repulsion between the F atoms.

8.123 a.

$$:\dot{N}{=}\ddot{O} \longleftrightarrow \bar{:}\ddot{N}{=}\dot{O}^{+}$$

The first structure is the most important. Both N and O have formal charges of zero. In the second structure, the more electronegative oxygen atom has a formal charge of +1. Having a positive formal charge on an highly electronegative atom is not favorable. In addition, both structures leave one atom with an incomplete octet. This cannot be avoided due to the odd number of electrons.

b. It is not possible to draw a structure with a triple bond between N and O.

$$:N{\equiv}\dot{O}$$

Any structure drawn with a triple bond will lead to an expanded octet. Elements in the second row of the period table cannot exceed the octet rule.

8.125 There are four C–H bonds in $CH_4$, so the average bond enthalpy of a C–H bond is:

$$\frac{1656\ \text{kJ/mol}}{4} = 414\ \text{kJ/mol}$$

The Lewis structure of propane is:

```
    H   H   H
    |   |   |
H — C — C — C — H
    |   |   |
    H   H   H
```

There are eight C–H bonds and two C–C bonds. We write:

$$8(\text{C–H}) + 2(\text{C–C}) = 4006\ \text{kJ/mol}$$

$$8(414\ \text{kJ/mol}) + 2(\text{C–C}) = 4006\ \text{kJ/mol}$$

$$2(\text{C–C}) = 694\ \text{kJ/mol}$$

So, the average bond enthalpy of a C–C bond is: $\frac{694}{2}\ \text{kJ/mol} = \mathbf{347\ kJ/mol}$

8.127

$$\text{EN(O)} = \frac{1314 + 141}{2} = 727.5 \qquad \text{EN(F)} = \frac{1680 + 328}{2} = 1004$$

$$\text{EN(Cl)} = \frac{1256 + 349}{2} = 803$$

Using Mulliken's definition, the electronegativity of chlorine is greater than that of oxygen, and fluorine is still the most electronegative element. We can convert to the Pauling scale by dividing each of the above by 230 kJ/mol.

$$\text{EN(O)} = \frac{727.5}{230} = \mathbf{3.16} \text{ or } \mathbf{3.2} \qquad \text{EN(F)} = \frac{1004}{230} = \mathbf{4.37} \text{ or } \mathbf{4.4} \qquad \text{EN(Cl)} = \frac{803}{230} = \mathbf{3.49} \text{ or } \mathbf{3.5}$$

These values compare to the Pauling values for oxygen of 3.5, fluorine of 4.0, and chlorine of 3.0.

8.129 C—C: 347 kJ/mol; N—N: 193 kJ/mol; O—O: 142 kJ/mol. **Lone pairs appear to weaken the bond.**

8.131

$$\text{Work done} = \text{force} \times \text{distance}$$

$$= (2.0 \times 10^{-9}\text{ N}) \times (2 \times 10^{-10}\text{ m})$$

$$= 4 \times 10^{-19}\text{ N·m}$$

$$= 4 \times 10^{-19}\text{ J to break one bond}$$

Expressing the bond enthalpy in kJ/mol:

$$\frac{4 \times 10^{-19}\text{ J}}{1\text{ bond}} \times \frac{1\text{ kJ}}{1000\text{ J}} \times \frac{6.022 \times 10^{23}\text{ bonds}}{1\text{ mol}} = \mathbf{2 \times 10^2\ kJ/mol}$$

8.133 (1) You could estimate the lattice energy of the solid by trying to measure its melting point. $Mg^+O^-$ would have a lattice energy (and, therefore, a melting point) similar to that of $Na^+Cl^-$. This lattice energy and melting point are much lower than those of $Mg^{2+}O^{2-}$.

(2) You could determine the magnetic properties of the solid. An $Mg^+O^-$ solid would be paramagnetic while $Mg^{2+}O^{2-}$ solid is diamagnetic. See Chapter 9 of the text.

8.135 The complete Lewis structure is:

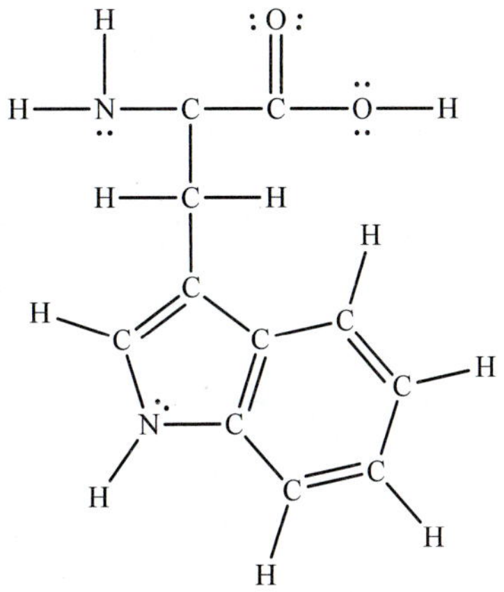

8.137 The skeletal structure is:

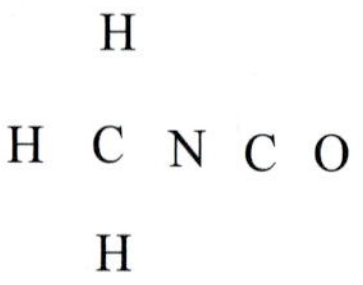

H
H C N C O
H

The number of valence electron is: $(1 \times 3) + (2 \times 4) + 5 + 6 = 22$ valence electrons

We can draw two resonance structures for methyl isocyanate.

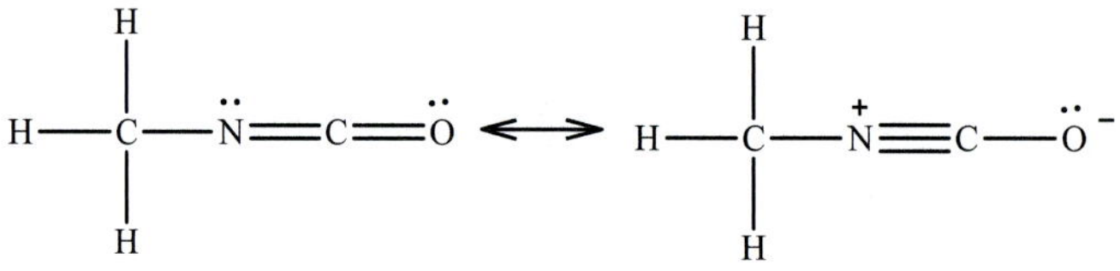

$H_3C{-}\ddot{N}{=}C{=}\ddot{O} \longleftrightarrow H_3C{-}\overset{+}{N}{\equiv}C{-}\ddot{O}^{-}$

8.139 a. $CFCl_3$ (C bonded to :F:, :Cl:, :Cl:, :Cl:)

c. $CHF_2Cl$ (C bonded to H, :F:, :F:, :Cl:)

b. $CF_2Cl_2$ (C bonded to :F:, :F:, :Cl:, :Cl:)

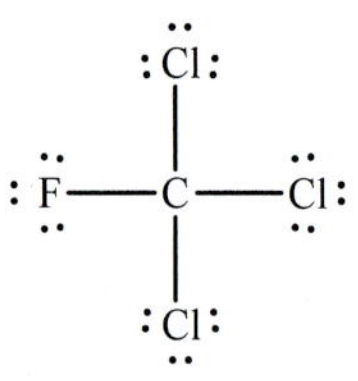

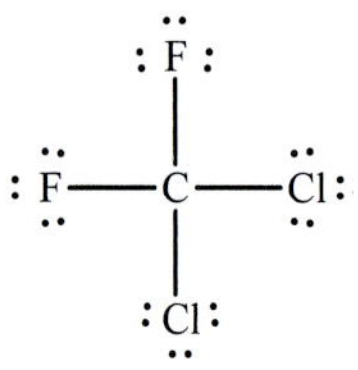

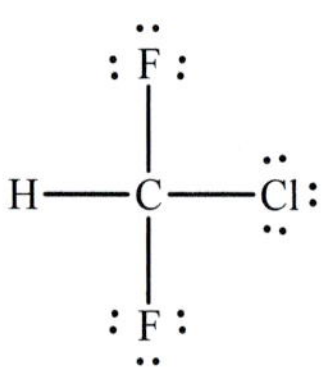

d. $CF_3{-}CHF_2$ (F–C(F)(F)–C(F)(F)–H)

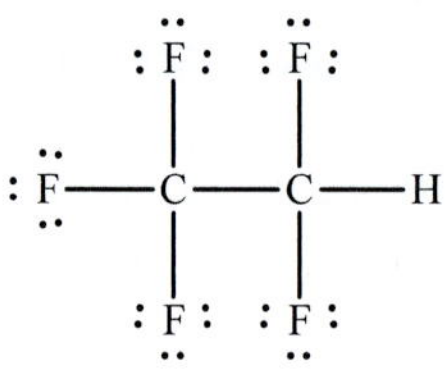

8.141 A reasonable Lewis structure is:

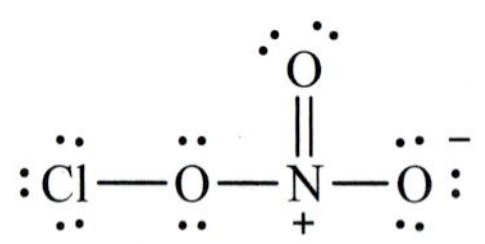

$:\ddot{Cl}{-}\ddot{O}{-}\overset{+}{N}(=\ddot{O}){-}\ddot{O}:^{-}$

8.143 a. $H{-}H\ \ H^{+} \longleftrightarrow H^{+}\ \ H{-}H \longleftrightarrow H{-}H^{+}{-}H$

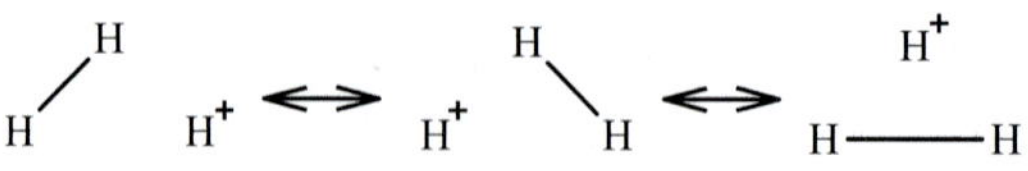

b. This is an application of Hess's Law.

$2H + H^+ \rightarrow H_3^+$ $\Delta H = -849$ kJ/mol

$H_2 \rightarrow 2H$ $\Delta H = 436.4$ kJ/mol

---

$H^+ + H_2 \rightarrow H_3^+$ **$\Delta H = -413$ kJ/mol**

The energy released in forming $H_3^+$ from $H^+$ and $H_2$ is almost as large as the formation of $H_2$ from 2 H atoms.

# Chapter 9

# Chemical Bonding II: Molecular Geometry and Bonding Theories

9.7 **Strategy:** The sequence of steps in determining molecular geometry is as follows:

draw Lewis structure ⟶ count electron-domains around the central atom ⟶ find electron-domain geometry ⟶ determine molecular geometry based on position of atoms

**Solution:** a. The Lewis structure of $PCl_3$ is shown below. In the VSEPR method, it is the number of bonding pairs and lone pairs of electrons around the *central atom* (phosphorus, in this case) that is important in determining the structure, the lone pairs of electrons around the chlorine atoms have been omitted for simplicity. There are four electron domains around the central atom. The electron-domain geometry is tetrahedral. There is one lone pair around the central atom, phosphorus, which makes this an $AB_3$ case with one lone pair. The information in Table 9.2 shows that the structure is a trigonal pyramid like ammonia.

```
Cl—P̈—Cl
   |
   Cl
```

b. The Lewis structure of $CHCl_3$ is shown below. There are four electron domains, all of which are bonds, around carbon which makes this an $AB_4$ case with no lone pairs. The molecule should be tetrahedral like the $AB_4$ example shown in Figure 9.2.

```
    H
    |
Cl—C—Cl
    |
    Cl
```

c. The Lewis structure of $SiH_4$ is shown below. Like part (b), it is a tetrahedral $AB_4$ molecule.

H
|
H—Si—H
|
H

d. The Lewis structure of $TeCl_4$ is shown below. There are five electron domains, four of which are bonds, which make this an $AB_4$ case with one lone pair. Consulting Table 9.2 shows that the structure should be that of a see-saw shape like $SF_4$.

Cl—Te—Cl (with lone pair on Te)
Cl / \ Cl

9.9

| | Lewis Structure | Electron-domain geometry | Molecular geometry |
|---|---|---|---|
| **(a)** | $CBr_4$: :Br: above and below C, :Br— C —Br: (each Br with three lone pairs) | tetrahedral | tetrahedral |
| **(b)** | $BCl_3$: :Cl—B—Cl: with :Cl: below B (each Cl with three lone pairs) | trigonal planar | trigonal planar |
| **(c)** | $NF_3$: :F—N—F: with :F: below N (N with one lone pair; each F with three lone pairs) | tetrahedral | trigonal pyramidal |
| **(d)** | H—Se—H (Se with two lone pairs) | tetrahedral | bent |
| **(e)** | $[O{=}N{-}O]^-$ (N with one lone pair) | trigonal planar | bent |

9.11 a. $AB_2$ (no lone pairs on the central atom) linear

b. $AB_4$ (no lone pairs on the central atom) tetrahedral

c. $AB_5$ (no lone pairs on the central atom) trigonal bipyramidal

d. $AB_3$ (one lone pair on the central atom) trigonal pyramidal

e. $AB_4$ (no lone pairs on the central atom) tetrahedral

9.13 The Lewis structure is:

```
     H   :O:
     |   ||   ..
H —— C —— C —— O —— H
     |        ..
     H
```

Carbon at the center of $H_3C$— electron-domain geometry = tetrahedral

molecular geometry = tetrahedral

```
   :O:
   ||   ..
 — C —  O — H
        ..
```

Carbon at the center of the above: electron-domain geometry = trigonal planar

molecular geometry = trigonal planar

```
   ..
 — O — H
   ..
```

Oxygen in the above: electron-domain geometry = tetrahedral
molecular geometry = bent

9.17 **Strategy:** The sequence of steps in determining molecular polarity is as follows:

draw Lewis structure ⟶ count electron-domains around the central atom ⟶ find electron-domain geometry ⟶ determine molecular geometry based on position of atoms

Once the molecular geometry is known, the polarity of individual bonds is determined. Bonds connecting atoms of different electronegativities are polar. Molecules containing polar bonds may be polar or nonpolar depending on the arrangement of polar bonds. In general, a symmetrical arrangement of identical bonds will result in a nonpolar molecule. Conversely, an asymmetrical distribution of identical bonds—or a distribution of bonds with different dipole magnitudes, will result in a polar molecule. Each molecule must be considered individually to determine its polarity. In order for a molecule to be nonpolar, its bond dipoles must sum to zero.

**Solution:** a. $BrF_5$ has the Lewis structure (The lone pairs on the F atoms are not shown.):

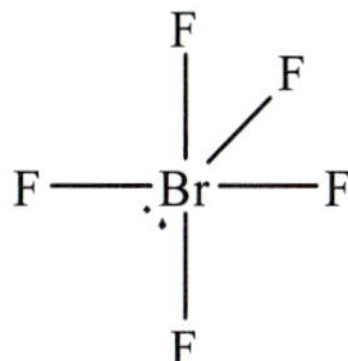


With six electron domains around the central atom, the electron-domain geometry is octahedral. Because one of the electron domains is a lone pair, we expect the molecular geometry to be a square pyramid (See Figure 9.2). The bond dipoles in this molecule do not sum to zero. Therefore, the molecule is **polar**.

b. $BCl_3$ has the Lewis structure (The lone pairs on the Cl atoms are not shown.):

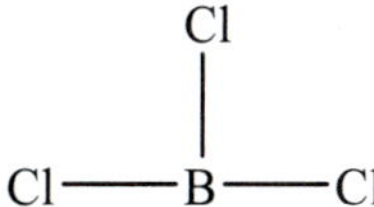


Recall that boron is one of the elements that does not necessarily obey the octet rule (See Section 8.8 in the text). With three electron domains around the central atom, the electron-domain geometry is trigonal planar. Because none of the three electron domains on the central atom is a lone pair, the molecular geometry is also trigonal planar. The bonds are symmetrically distributed and the individual bond dipoles sum to zero. Therefore, the molecule is **nonpolar**.

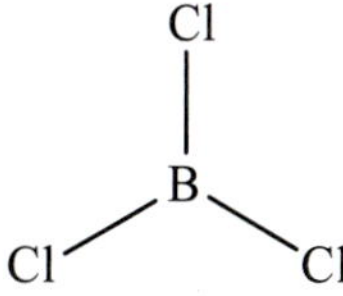


9.19 **Only (c) is polar**. The others are symmetrical distributions of identical bonds, with the bond dipoles summing to zero.

9.29 **Strategy:** The steps for determining the hybridization of the central atom in a molecule are:

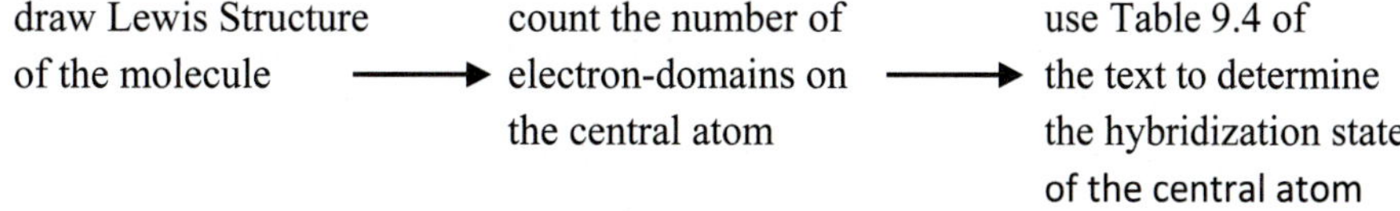

**Solution:** a. Draw the Lewis structure of the molecule.

$$\begin{array}{c} \text{H} \\ | \\ \text{H}-\text{Si}-\text{H} \\ | \\ \text{H} \end{array}$$

Count the number of electron domains around the central atom. Since there are four electron domains around Si, the electron-domain geometry is tetrahedral and we conclude that Si is ***sp*$^3$ hybridized**.

b. Draw the Lewis structure of the molecule.

$$\begin{array}{c} \text{H}\quad\text{H} \\ |\quad\;\; | \\ \text{H}-\text{Si}-\text{Si}-\text{H} \\ |\quad\;\; | \\ \text{H}\quad\text{H} \end{array}$$

Count the number of electron domains around the "central atoms". Since there are four electron domains around each Si, the electron-domain geometry for each Si is tetrahedral and we conclude that each Si is ***sp*$^3$ hybridized**.

9.31 Draw the Lewis structures. Before the reaction, boron is $sp^2$ hybridized in $BF_3$ (three electron domains around the central atom, trigonal planar electron-domain geometry) and nitrogen is $sp^3$ hybridized in $NH_3$ (four electron domains around the central atom, tetrahedral electron-domain geometry). After the reaction, boron and nitrogen are both $sp^3$ hybridized (tetrahedral electron-domain geometry).

9.33 **Strategy:** The steps for determining the hybridization of the central atom in a molecule are:

draw Lewis Structure of the molecule ⟶ count the number of electron-domains on the central atom ⟶ use Table 9.4 of the text to determine the hybridization state of the central atom

**Solution:** Draw the Lewis structure of the molecule.

$$\begin{array}{c} :\ddot{\text{F}}: \\ | \\ :\ddot{\text{F}}-\text{P}\begin{array}{l}\diagup :\ddot{\text{F}}: \\ \diagdown :\ddot{\text{F}}:\end{array} \\ | \\ :\ddot{\text{F}}: \end{array}$$

Count the number of electron domains around the central atom. Since there are five electron domains around P, the electron-domain geometry is trigonal bipyramidal ($AB_5$ w/no lone pairs) and we conclude that P is ***sp*$^3$*d* hybridized**.

9.37 a. *sp* b. *sp* c. *sp*

9.39 **Strategy:** The steps for determining the hybridization of the central atom in a molecule are:

draw Lewis Structure of the molecule ⟶ count the number of electron-domains on the central atom ⟶ use Table 9.4 of the text to determine the hybridization state of the central atom

**Solution:** Draw the Lewis structure of the molecule. Several resonance forms with formal charges are shown.

$$\left[{}^{-}\ddot{\underset{..}{N}}{=}\overset{+}{N}{=}\ddot{\underset{..}{N}}{}^{-}\right]^{-} \longleftrightarrow \left[:N{\equiv}\overset{+}{N}{-}\ddot{\underset{..}{N}}:^{2-}\right]^{-} \longleftrightarrow \left[{}^{2-}:\ddot{\underset{..}{N}}{-}\overset{+}{N}{\equiv}N:\right]^{-}$$

Count the number of electron domains around the central atom. Since there are two electron domains around N, the electron-domain geometry is linear ($AB_2$ w/no lone pairs) and we conclude that N is ***sp* hybridized**.

Remember, a multiple bond is just *one* electron domain.

9.41 A single bond is usually a sigma bond, a double bond is usually a sigma bond and a pi bond, and a triple bond is always a sigma bond and two pi bonds. Therefore, there are **nine $\sigma$ bonds** and **nine $\pi$ bonds** in the molecule.

9.43 The benzo(*a*)pyrene molecule contains 36 $\sigma$ bonds (some hydrogen atoms are omitted from the figure) and 10 $\pi$ bonds.

9.49 In order for the two hydrogen atoms to combine to form a $H_2$ molecule, the electrons must have opposite spins. If two H atoms collide and their electron spins are parallel, no bond will form.

9.51 The electron configurations are listed. Refer to Figure 9.18 of the text for the molecular orbital diagram.

$Li_2$: $(\sigma_{1s})^2(\sigma^{\star}_{1s})^2(\sigma_{2s})^2$ bond order = 1

$Li_2^+$: $(\sigma_{1s})^2(\sigma^{\star}_{1s})^2(\sigma_{2s})^1$ bond order = $\frac{1}{2}$

$$Li_2^- : \quad (\sigma_{1s})^2(\sigma_{1s}^\star)^2(\sigma_{2s})^2(\sigma_{2s}^\star)^1 \qquad \text{bond order} = \frac{1}{2}$$

Order of increasing stability: $\mathbf{Li_2^-} = \mathbf{Li_2^+} < \mathbf{Li_2}$

In reality, $Li_2^+$ is more stable than $Li_2^-$ because there is less electrostatic repulsion in $Li_2^+$.

9.53 See Figure 9.18 of the text. Removing an electron from $B_2$ gives $B_2^+$, which has a bond order of (1/2). Therefore, $\mathbf{B_2^+}$, with a bond order of 1/2, has the longer B—B bond. ($B_2$ has a bond order of 1.)

9.55 The Lewis diagram has all electrons paired (incorrect) and a double bond (correct). The MO diagram has two unpaired electrons (correct), and a bond order of 2 (correct).

9.57 We refer to Figure 9.18 of the text.

$O_2$ has a bond order of 2 and is paramagnetic (two unpaired electrons).

$O_2^+$ has a bond order of 2.5 and is paramagnetic (one unpaired electron).

$O_2^-$ has a bond order of 1.5 and is paramagnetic (one unpaired electron).

$O_2^{2-}$ has a bond order of 1 and is diamagnetic.

9.59 As shown in the text (see Figure 9.18), the two shared electrons that make up the single bond in $B_2$ both lie in pi molecular orbitals and constitute a pi bond. The four shared electrons that make up the double bond in $C_2$ all lie in pi molecular orbitals and constitute two pi bonds.

9.63 The left symbol shows three delocalized double bonds (correct). The right symbol shows three localized double bonds and three single bonds (incorrect).

9.65 a. Two Lewis resonance forms are shown below. (Formal charges different than zero are indicated.)

:F: (single bond to N)
O=N⁺—O:⁻ ⟷ ⁻:O—N⁺=O

b. There are no lone pairs on the nitrogen atom; it should have a trigonal planar electron-domain geometry and therefore use ***sp***$^2$ hybrid orbitals.

c. The bonding consists of sigma bonds joining the nitrogen atom to the fluorine and oxygen atoms. In addition there is a pi molecular orbital delocalized over the N and O atoms. Is nitryl fluoride isoelectronic with the carbonate ion?

9.67 The Lewis structures of ozone are:

$$\ddot{\underset{\cdot\cdot}{O}}=\ddot{O}-\ddot{\underset{\cdot\cdot}{O}}: \longleftrightarrow :\ddot{\underset{\cdot\cdot}{O}}-\ddot{O}=\ddot{\underset{\cdot\cdot}{O}}$$

The central oxygen atom is ***sp***$^2$ hybridized ($AB_2$ w/one lone pair). The unhybridized $2p_z$ orbital on the central oxygen overlaps with the $2p_z$ orbitals on the two terminal atoms.

9.69 **Strategy:** The sequence of steps in determining molecular geometry is as follows:

draw Lewis structure ⟶ count electron-domains around ⟶ find electron-domain geometry the central atom ⟶ determine molecular geometry based on position of atoms

**Solution:** Write the Lewis structure of the molecule.

$$:\ddot{\underset{\cdot\cdot}{Br}}-Hg-\ddot{\underset{\cdot\cdot}{Br}}:$$

Count the number of electron domains around the central atom. There are two electron domains around Hg.

Since there are two electron domains around Hg, the electron-domain geometry is **linear**.

In addition, since there are no lone pairs around the central atom, the geometry is also **linear** ($AB_2$).

You could establish the geometry of $HgBr_2$ by measuring its dipole moment. If mercury(II) bromide were bent, it would have a measurable dipole moment. Experimentally, it has no dipole moment and therefore must be linear.

9.71 Geometry: bent; $sp^3$ hybridization.

9.73 a. Looking at the electronic configuration for $N_2$ shown in Figure 9.18 of the text, we write the electronic configuration for $P_2$.

$$[Ne_2](\sigma_{3s})^2(\sigma_{3s}^{\star})^2(\pi_{3p_y})^2(\pi_{3p_z})^2(\sigma_{3p_x})^2$$

b. Past the $Ne_2$ core configuration, there are 8 bonding electrons and 2 antibonding electrons. The bond order is:

$$\textbf{bond order} = \tfrac{1}{2}(8 - 2) = \mathbf{3}$$

c. All the electrons in the electronic configuration are paired. $P_2$ is **diamagnetic**.

9.75 To predict the bond angles for the molecules, you would have to draw the Lewis structure and determine the geometry using the VSEPR model. From the geometry, you can predict the bond angles.

a. $BeCl_2$: $AB_2$, 180° (linear).

b. $BCl_3$: $AB_3$, 120° (trigonal planar).

c. $CCl_4$: $AB_4$, 109.5° (tetrahedral).

d. $CH_3Cl$: $AB_4$, 109.5° (tetrahedral with a possible slight distortion resulting from the different sizes of the chlorine and hydrogen atoms).

e. $Hg_2Cl_2$: Each mercury atom is of the $AB_2$ type. The entire molecule is linear, 180° bond angles.

f. $SnCl_2$: $AB_2$ w/one lone pair, roughly 120° (bent).

g. $H_2O_2$: The atom arrangement is HOOH. Each oxygen atom is of the $AB_2$ w/two lone pairs and the H–O–O angles will be roughly 109.5°.

h. $SnH_4$: $AB_4$, 109.5° (tetrahedral).

9.77 a. The Lewis structure is (lone pairs on F atoms not shown):

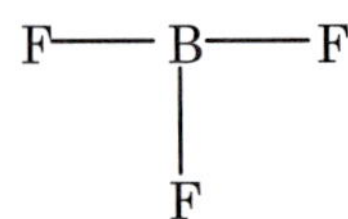


The shape will be trigonal ***planar*** ($AB_3$)

b. The Lewis structure is (lone pairs on O atoms not shown):

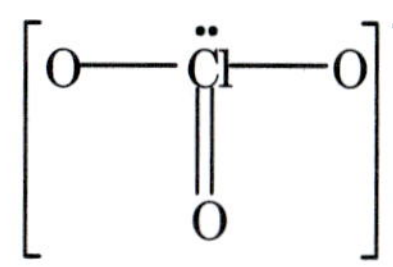


The molecule will be a trigonal pyramid (***nonplanar***).

c. The Lewis structure is:

H—C≡N:

The molecule is ***polar***.

d. The Lewis structure is (lone pairs on F atoms not shown):

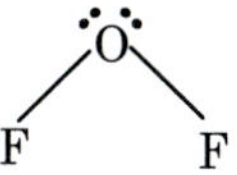


The molecule is bent and therefore ***polar***.

e. The Lewis structure is:

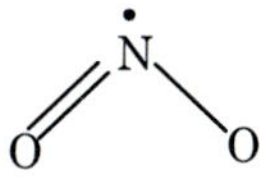


The nitrogen atom is an $AB_2$ w/one lone, unshared electron rather than the usual *pair*. As a result, the repulsion will not be as great and the O–N–O angle will be ***greater than 120°*** expected for $AB_2$ (w/one lone pair) geometry. Experiment shows the angle to be around 135°.

9.79 **Strategy:** For each molecule, draw the Lewis structure, use the VSEPR model to determine its molecular geometry, and then determine whether the individual bond dipoles cancel.

**Setup:** a. The Lewis structure is:

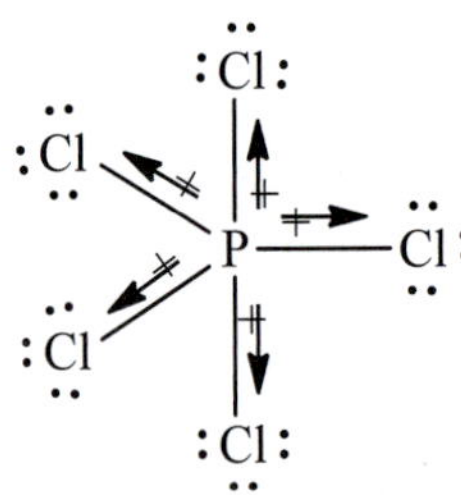


With five identical electron domains around the central atom, the electron-domain and molecular geometries are trigonal bipyramidal. The equatorial bond dipoles will cancel one another and the axial bond dipoles will also cancel each other.

b. The Lewis structure is:

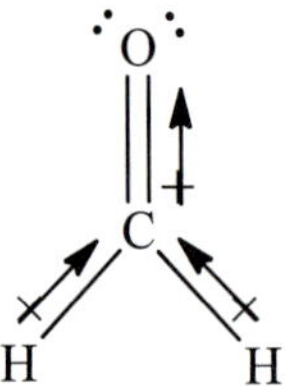


The bond dipoles, although symmetrically distributed around the C atom, are not identical and therefore will not sum to zero.

**Solution:** a. Nonpolar

b. Polar

9.81 Only $ICl_2^-$ and $CdBr_2$ will be linear. The rest are bent.

9.83 a. **Strategy:** The steps for determining the hybridization of the central atom in a molecule are:

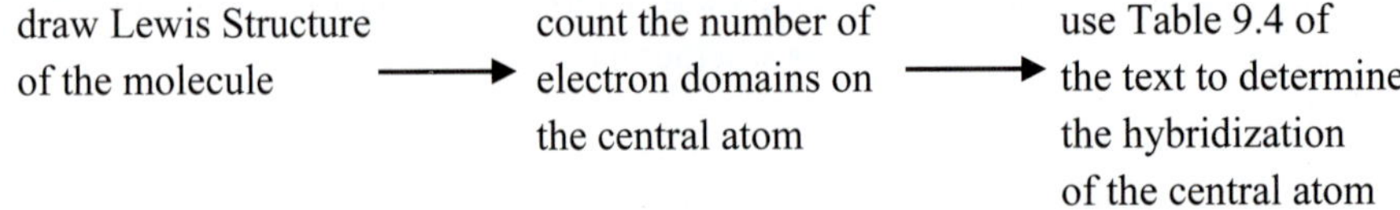


**Solution:** The geometry around each nitrogen is identical. To complete an octet of electrons around N, you must add a lone pair of electrons. Count the number of electron domains around N.

There are three electron domains around each N.

Since there are three electron domains around N, and none of them is a lone pair, the electron-domain geometry is trigonal planar and we conclude that each N is ***sp*$^2$ hybridized**.

b. **Strategy:** Keep in mind that the polarity of a molecule depends on both the difference in electronegativities of the elements present and its geometry. A molecule can have polar bonds (if the bonded atoms have different electronegativities), but it may not be polar if it has a highly symmetrical geometry.

**Solution:** An N–F bond is polar because F is more electronegative than N. The structure on the right is polar because the two N–F bond moments do not cancel each other out and so the molecule has a net dipole moment. On the other hand, the two N–F bond moments in the left-hand structure cancel. The sum of bond dipoles will be *zero*. Therefore, the molecule on the left will be nonpolar.

9.85 a. The Lewis structure is:

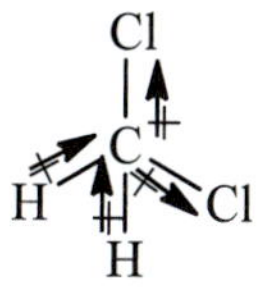

Cl
C
H Cl
H

The dipoles are not equivalent, so they do not sum to zero. The molecule is polar.

b.

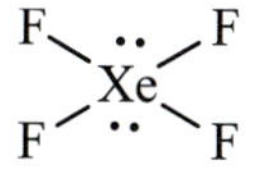

F F
Xe
F F

The dipoles are arranged symmetrically and are equivalent, so they sum to zero. The molecule is nonpolar.

9.87 a. Trigonal bipyramidal: The axial atoms are the same type and the equatorial atoms are all the other type.

Square planar: The opposite atoms are the same type.

Octahedral: Each pair of opposite atoms is the same type.

b. Octahedral: Each pair of opposite atoms is the same type.

9.89 Assume that the skeletal structure of the molecule is O C C C O. Drawing the Lewis dot structure of the molecule results in the following structure.

O=C=C=C=O

The molecule is linear and symmetric about the molecular axis, so we do not expect the molecule to possess a dipole moment.

9.91 C has no *d* orbitals but Si does (3*d*). Thus, $H_2O$ molecules can add to Si in hydrolysis (valence-shell expansion).

9.93 The carbons are all $sp^2$ hybridized. The nitrogen double bonded to the carbon in the ring is $sp^2$ hybridized. The other nitrogens are $sp^3$ hybridized.

9.95 $F_2$ has 8 electrons in bonding orbitals and 6 electrons in antibonding orbitals, giving it a bond order of 1. $F_2^-$ has 8 electrons in bonding orbitals and 7 electrons in antibonding orbitals, giving it a bond order of 1/2.

9.97 One of the bending modes of $CO_2$ is illustrated below (see problem 9.99). As the molecule vibrates, it deviates from its equilibrium linear geometry, producing a transient dipole moment.

The $CO_2$ molecular can also vibrate by an asymmetric shift in the positions of the atoms along the molecular axis, as illustrated below (see problem 9.99 also). This causes an asymmetric distribution of charge that also creates a transient dipole moment.

In both cases, the transient dipole moments disappear as the molecule relaxes to its equilibrium geometry.

9.99 Referring to Figure 9.18 of the text, we see that $C_2$ and $N_2$ have a similar energy order for their molecular orbitals. Therefore, we expect $CN^-$ to have the same order.

$$[He_2](\sigma_{2s})^2(\sigma_{2s}^{\star})^2(\pi_{2p_y})^2(\pi_{2p_z})^2(\sigma_{2p_x})^2$$

**CO** is isoelectronic with $CN^-$ (both contain 14 electrons).

9.101 The Lewis structure of $O_2$ shows 4 pairs of electrons on the two oxygen atoms. From Figure 9.18 of the text, we see that these 8 valence electrons are placed in the $\sigma_{2p_x}$, $\pi_{2p_y}$, $\pi_{2p_z}$, $\pi^{\star}_{2p_y}$, and $\pi^{\star}_{2p_z}$ orbitals. For all the electrons to be paired, energy is needed to flip the spin in one of the antibonding molecular orbitals ($\pi^{\star}_{2p_y}$ or $\pi^{\star}_{2p_z}$). According to Hund's rule, this arrangement is less stable than the ground-state configuration shown in Figure 9.18, and hence the Lewis structure shown actually corresponds to an excited state of the oxygen molecule.

9.103 There are eight valence electrons (four from tin and one each from four hydrogens), and the dot structure indicates four electron domains, all bonding. The electron-domain and molecular geometries are both tetrahedral.

9.105 a. $C_6H_8O_6$

b. Five central O atoms $sp^3$; hybridization of the sixth (double bonded) peripheral O atom is $sp^2$. Three C atoms $sp^2$; three C atoms $sp^3$.

c. The five central O atoms bent. The three $sp^2$ C atoms are trigonal planar. The three $sp^3$ C atoms are tetrahedral.

9.107 a. Although the O atoms are $sp^3$ hybridized, they are locked in a planar structure by the benzene rings. The molecule is symmetrical and therefore is not polar.

b. 20 $\sigma$ bonds and 6 $\pi$ bonds.

9.109 a.

$$:\overset{-}{C}\equiv\overset{+}{O}:$$

This is the only reasonable Lewis structure for CO. The electronegativity difference between O and C suggests that electron density should concentrate on the O atom, but assigning formal charges places a negative charge on the C atom. Therefore, we expect CO to have a small dipole moment.

b. CO is isoelectronic with $N_2$, bond order 3. This agrees with the triple bond in the Lewis structure.

c. Since C has a negative formal charge, it is more likely to form bonds with $Fe^{2+}$. (OC—$Fe^{2+}$ rather than CO—$Fe^{2+}$). More elaborate analysis of the orbitals involved shows that this is indeed the case.

9.111 The S—S bond is a normal 2-electron shared pair covalent bond. Each S is $sp^3$ hybridized, so the X—S—S angle is about 109°.

9.113 In 1,2-dichloroethane, the two C atoms are joined by a sigma bond. Rotation about a sigma bond does not destroy the bond, and the bond is therefore free (or relatively free) to rotate. Thus, all angles are permitted and the molecule is nonpolar because the C–Cl bond moments cancel each other because of the averaging effect brought about by rotation. In *cis*-dichloroethylene the two C–Cl bonds are locked in position. The π bond between the C atoms prevents rotation (in order to rotate, the π bond must be broken, using an energy source such as light or heat). Therefore, there is no rotation about the C=C in *cis*-dichloroethylene, and the molecule is polar.

9.115 The balanced chemical equation is:

$$S_8(s) + 16SO_3(g) \rightarrow 24SO_2(g)$$

The oxidation states for the sulfur in each species are:

$S_8$: 0 $SO_3$: +6 $SO_2$: +4

To determine the hybridization of the sulfur in each species, draw the Lewis structures.

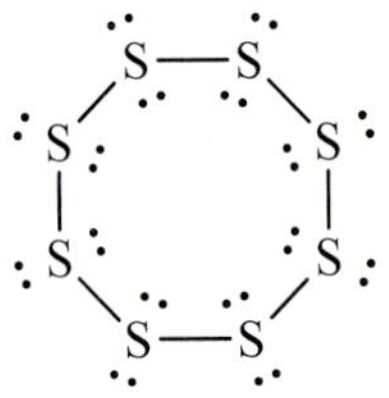

Each S is $sp^3$ hybridized.

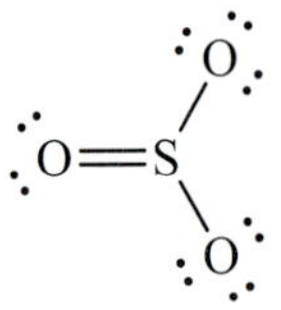

The S is $sp^2$ hybridized.

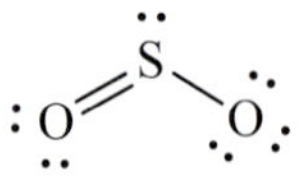

The S is $sp^2$ hybridized.

Determine the mass of sulfur trioxide required for the reaction:

$$(1.00 \text{ kg } S_8)\left(\frac{1000 \text{ g } S_8}{1 \text{ kg } S_8}\right)\left(\frac{1 \text{ mol } S_8}{256.56 \text{ g } S_8}\right)\left(\frac{16 \text{ mol } SO_3}{1 \text{ mol } S_8}\right)\left(\frac{80.07 \text{ g } SO_3}{1 \text{ mol } SO_3}\right)\left(\frac{1 \text{ kg } SO_3}{1000 \text{ g } SO_3}\right) = 4.99 \text{ kg } SO_3$$

Determine the mass of sulfur dioxide produced in the reaction:

$$(1.00 \text{ kg } S_8)\left(\frac{1000 \text{ g } S_8}{1 \text{ kg } S_8}\right)\left(\frac{1 \text{ mol } S_8}{256.56 \text{ g } S_8}\right)\left(\frac{24 \text{ mol } SO_2}{1 \text{ mol } S_8}\right)\left(\frac{64.07 \text{ g } SO_2}{1 \text{ mol } SO_2}\right)\left(\frac{1 \text{ kg } SO_2}{1000 \text{ g } SO_2}\right) = 5.99 \text{ kg } SO_2$$

# Chapter 10

# Gases

10.13 **Strategy:** We use the conversion factors provided in Table 10.2 in the text to convert a pressure in mmHg to atm, bar, torr, and Pa.

**Solution:** Converting to atm:

$$\textbf{?atm} = 375 \text{ mmHg} \times \frac{133.322 \text{ Pa}}{1 \text{ mmHg}} \times \frac{1 \text{ atm}}{101{,}325 \text{ Pa}} = \textbf{0.493 atm}$$

We could also have solved this by remembering that 760 mmHg = 1 atm.

$$\textbf{?atm} = 375 \text{ mmHg} \times \frac{1 \text{ atm}}{760 \text{ mmHg}} = \textbf{0.493 atm}$$

Converting to bar:

$$\textbf{?bar} = 375 \text{ mmHg} \times \frac{133.322 \text{ Pa}}{1 \text{ mmHg}} \times \frac{1 \text{ bar}}{1 \times 10^5 \text{ Pa}} = \textbf{0.500 bar}$$

Converting to torr:

$$\textbf{?torr} = 375 \text{ mmHg} \times \frac{133.322 \text{ Pa}}{1 \text{ mmHg}} \times \frac{1 \text{ torr}}{133.322 \text{ Pa}} = \textbf{375 torr}$$

Note that because 1 mmHg and 1 torr are both equal to 133.322 Pa, this could be simplified by recognizing that 1 torr = 1 mmHg.

$$375 \text{ mmHg} \times \frac{1 \text{ torr}}{1 \text{ mmHg}} = \textbf{375 torr}$$

Converting to Pa:

$$\textbf{?Pa} = 375 \text{ mmHg} \times \frac{1 \text{ atm}}{760 \text{ mmHg}} \times \frac{101{,}325 \text{ Pa}}{1 \text{ atm}} = \textbf{5.00} \times \textbf{10}^{\textbf{4}} \textbf{ Pa}$$

10.15 **Strategy:** This problem is similar to Sample Problem 10.1. We can use the equation derived in the sample problem to solve for the height of the column of methanol. The equation ispressure = height ×

density × gravitational constant

The gravitational constant is 9.80665 $m/s^2$.

**Solution:** Solving the equation for height of the column gives

$$\text{height} = \frac{\text{pressure}}{\text{density} \times \text{gravitational constant}}$$

Recall that for units to cancel properly, pressure must be expressed in Pa (1 Pa = 1 $kg/m \cdot s^2$) and density must be expressed in $kg/m^3$. Converting the information given in the problem to the appropriate units,

$$?\text{Pa} = 1\ \text{atm} \times \frac{101{,}325\ \text{Pa}}{1\ \text{atm}} = 101{,}325\ \text{Pa}$$

Note that we consider atmospheric pressure to be exactly 1 atm.

$$?\text{kg/m}^3 = \frac{0.787\ \text{g}}{1\ \text{cm}^3} \times \frac{1\ \text{kg}}{1000\ \text{g}} \times \left(\frac{100\ \text{cm}}{1\ \text{m}}\right)^3 = 787\ \text{kg/m}^3$$

Remember that when a unit is raised to a power, any conversion factor used must be raised to the same power. Substituting into the equation to solve for height gives

$$\textbf{height} = \frac{101{,}325\ \text{Pa}}{\left(787\ \text{kg/m}^3\right)\left(9.80665\ \text{m/s}^2\right)} = \frac{101{,}325\ \text{kg/m} \cdot \text{s}^2}{\left(787\ \text{kg/m}^3\right)\left(9.80665\ \text{m/s}^2\right)} = \mathbf{13.1\ m}$$

**Think About It:** It's easier to see how the units cancel if we express the pressure in pascals using base SI units.

10.17 **Strategy:** This problem is very similar to Sample Problem 10.1. We can use the equation derived in the sample problem to solve for the pressure exerted by the column of toluene. The equation is

pressure = height × density × gravitational constant

The gravitational constant is 9.80665 $m/s^2$.

**Solution:** Recall that for units to cancel properly, density must be expressed in $kg/m^3$.

$$?\text{kg/m}^3 = \frac{0.867\ \text{g}}{1\ \text{cm}^3} \times \frac{1\ \text{kg}}{1000\ \text{g}} \times \left(\frac{100\ \text{cm}}{1\ \text{m}}\right)^3 = 867\ \text{kg/m}^3$$

Remember that when a unit is raised to a power, any conversion factor used must be raised to the same power. Substituting into the equation to solve for pressure gives

$$\textbf{pressure} = 87\text{ m} \times \frac{867\text{ kg}}{\text{m}^3} \times \frac{9.80665\text{ m}}{\text{s}^2} = 7.40 \times 10^5\text{ kg/m}\cdot\text{s}^2 = 7.40 \times 10^5\text{ Pa} \times \frac{1\text{ atm}}{101{,}325\text{ Pa}} = \textbf{7.3 atm}$$

**Think About It:** It's easier to see how the units cancel if we express the pressure in pascals using base SI units.

10.21 **Strategy:** This is a Boyle's law problem. Temperature and the amount of gas are both constant. Therefore, we can use Equation 10.3 to solve for the final volume.

**Solution:**

| Initial Conditions | Final Conditions |
|---|---|
| $P_1 = 0.970$ atm | $P_2 = 0.541$ atm |
| $V_1 = 25.6$ mL | $\mathbf{V_2 = ?}$ |

$$\mathbf{V_2} = \frac{P_1V_1}{P_2} = \frac{(0.970\text{ atm})(25.6\text{ mL})}{0.541\text{ atm}} = \textbf{45.9 mL}$$

**Think About It:** Make sure that Boyle's law is obeyed. If the pressure decreases at constant temperature, the volume must increase.

10.23 **Strategy:** The amount of gas and the temperature remain constant in this problem. We can use Equation 10.3 (Boyle's Law) to solve for the unknown pressure.

**Solution:**

| Initial Conditions | Final Conditions |
|---|---|
| $P_1 = 1.00$ atm $= 760$ mmHg | $\mathbf{P_2 = ?}$ |
| $V_1 = 7.15$ L | $V_2 = 9.25$ L |

$$P_1V_1 = P_2V_2$$

$$\mathbf{P_2} = \frac{P_1V_1}{V_2} = \frac{(760\text{ mmHg})(7.15\text{ L})}{9.25\text{ L}} = \textbf{587 mmHg}$$

10.25 **Strategy:** Pressure is held constant in this problem. Only volume and temperature change. This is a Charles's law problem. We use Equation 10.5 to solve for the unknown volume.

$$\frac{V_1}{T_1} = \frac{V_2}{T_2}$$

**Solution:** Initial Conditions — Final Conditions

$T_1 = 35°C + 273 = 308\text{ K}$ — $T_2 = 72° + 273 = 345\text{ K}$

$V_1 = 28.4\text{ L}$ — $\boldsymbol{V_2 = ?}$

$$\boldsymbol{V_2} = \frac{V_1T_2}{T_1} = \frac{(28.4\text{ L})(345\text{ K})}{308\text{ K}} = \mathbf{31.8\ L}$$

**Think About It:** Make sure you express temperatures in kelvins and that Charles's law is obeyed. At constant pressure, when temperature increases, volume should also increase.

10.27 The balanced equation is:

$$4NH_3(g) + 5O_2(g) \longrightarrow 4NO(g) + 6H_2O(g)$$

Recall that Avogadro's Law states that the volume of a gas is directly proportional to the number of moles of gas at constant temperature and pressure. The ammonia and nitric oxide coefficients in the balanced equation are the same, so **one volume** of nitric oxide must be obtained from **one volume** of ammonia.

Could you have reached the same conclusion if you had noticed that nitric oxide is the only nitrogen-containing product and that ammonia is the only nitrogen-containing reactant?

10.29 a. If the final temperature of the sample is above the boiling point, it would still be in the gas phase. The diagram that best represents this is choice **(d)**.

b. If the final temperature of the sample is below its boiling point, it will condense to a liquid. The liquid will have a vapor pressure, so some of the sample will remain in the gas phase. The diagram that best represents this is choice **(b)**.

10.35 **Strategy:** This problem gives the amount, volume, and temperature of CO gas. Is the gas undergoing a change in any of its properties?

**Solution:** Because no changes in gas properties occur, we can use the ideal gas equation to calculate the pressure. Rearranging Equation 10.9 of the text, we write: $P = \dfrac{nRT}{V}$

$$\boldsymbol{P} = \frac{(6.9\ \text{mol})\left(0.0821\dfrac{\text{L}\cdot\text{atm}}{\text{mol}\cdot\text{K}}\right)(355\ \text{K})}{(30.4\ \text{L})} = \mathbf{6.6\ atm}$$

**Think About It:** Don't forget to convert temperatures from °C to K!

10.37 In this problem, the moles of gas and the volume the gas occupies are constant. Temperature and pressure change. We use the equation derived in Problem 10.53:

$$\frac{P_1}{T_1} = \frac{P_2}{T_2}$$

| Initial Conditions | Final Conditions |
|---|---|
| $P_1 = 1.00$ atm | $\boldsymbol{P_2} = \textbf{?}$ |
| $T_1 = 273$ K | $T_2 = 210°\text{C} + 273 = 483$ K |

Solving for the final pressure gives

$$\boldsymbol{P_2} = \frac{P_1T_2}{T_1} = \frac{(1.00\ \text{atm})(483\ \text{K})}{273\ \text{K}} = \mathbf{1.8\ atm}$$

10.39 **Strategy:** In this problem, the moles of gas and the pressure on the gas are constant. Temperature and volume both change. This is a Charles's law problem and we use Equation 10.5 to solve it. Note that the way the problem is stated, it is $V_1$ that is unknown.

$$\frac{V_1}{T_1} = \frac{V_2}{T_2}$$

| Initial Conditions | Final Conditions |
|---|---|
| $\boldsymbol{V_1} = \textbf{?}$ | $V_2 = 0.67$ L |
| $T_1 = 36.5 + 273.15 = 309.7$ K | $T_2 = 22.5°\text{C} + 273 = 295.7$ K |

**Solution:** Solving Equation 10.5 for the original volume of the gas gives

$$V_1 = \frac{V_2T_1}{T_2} = \frac{(0.67\text{ L})(309.7\text{ K})}{295.7\text{ K}} = \mathbf{0.70\text{ L}}$$

10.41 In the problem, temperature and pressure are given. If we can determine the moles of $CO_2$, we can calculate the volume it occupies using the ideal gas equation.

$$?\text{ mol CO}_2 = 124.3\text{ g CO}_2 \times \frac{1\text{ mol CO}_2}{44.01\text{ g CO}_2} = 2.8244\text{ mol CO}_2$$

We now substitute into the ideal gas equation to calculate volume of $CO_2$.

$$V_{CO_2} = \frac{nRT}{P} = \frac{(2.8244\text{ mol})\left(0.08206\frac{\text{L}\cdot\text{atm}}{\text{mol}\cdot\text{K}}\right)(273.15\text{ K})}{1\text{ atm}} = \mathbf{63.31\text{ L}}$$

Note that because there are four significant figures in the mass, we use more significant figures than we usually do for $R$ and for the temperature. We also carried an extra digit in the calculated number of moles to avoid rounding error in the final result.

Alternatively, we could use the fact that 1 mole of an ideal gas occupies a volume of 22.41 L at STP. After calculating the moles of $CO_2$, we can use this fact as a conversion factor to convert to volume of $CO_2$.

$$?\text{ L CO}_2 = 2.8244\text{ mol} \times \frac{22.41\text{ L}}{1\text{ mol}} = \mathbf{63.29\text{ L}}$$

The slight difference in the results of our two calculations is due to rounding the volume occupied by 1 mole of an ideal gas to 22.41 L.

10.43 The molar mass of $CO_2$ = 44.01 g/mol. Since $PV = nRT$, we write:

$$P = \frac{nRT}{V}$$

$$\mathbf{P} = \frac{\left(0.050\text{ g} \times \frac{1\text{ mol}}{44.01\text{ g}}\right)\left(0.0821\frac{\text{L}\cdot\text{atm}}{\text{mol}\cdot\text{K}}\right)(30 + 273)\text{K}}{4.6\text{ L}} = \mathbf{6.1 \times 10^{-3}\text{ atm}}$$

10.45 **Strategy:** We can calculate the molar mass of a gas if we know its density, temperature, and pressure. What temperature and pressure units should we use?

**Solution:** We need to use Equation 10.11 of the text to calculate the molar mass of the gas.

$$\mathcal{M} = \frac{dRT}{P}$$

Before substituting into the above equation, we need to calculate the density and check that the other known quantities ($P$ and $T$) have the appropriate units.

$$d = \frac{7.10 \text{ g}}{5.40 \text{ L}} = 1.31 \text{ g/L}$$

$$T = 44°\text{C} + 273 = 317 \text{ K}$$

$$P = 741 \text{ torr} \times \frac{1 \text{ atm}}{760 \text{ torr}} = 0.975 \text{ atm}$$

Calculate the molar mass by substituting in the known quantities.

$$\mathcal{M} = \frac{\left(1.31 \frac{\text{g}}{\text{L}}\right)\left(0.0821 \frac{\text{L} \cdot \text{atm}}{\text{mol} \cdot \text{K}}\right)(317 \text{ K})}{0.975 \text{ atm}} = \mathbf{35.0 \text{ g/mol}}$$

Alternatively, we can solve for the molar mass by writing:

$$\text{molar mass of compound} = \frac{\text{mass of compound}}{\text{moles of compound}}$$

Mass of compound is given in the problem (7.10 g), so we need to solve for moles of compound in order to calculate the molar mass.

$$n = \frac{PV}{RT}$$

$$n = \frac{(0.975 \text{ atm})(5.40 \text{ L})}{\left(0.0821 \frac{\text{L} \cdot \text{atm}}{\text{mol} \cdot \text{K}}\right)(317 \text{ K})} = 0.202 \text{ mol}$$

Now, we can calculate the molar mass of the gas.

$$\textbf{molar mass of compound} = \frac{\text{mass of compound}}{\text{moles of compound}} = \frac{7.10 \text{ g}}{0.202 \text{ mol}} = \mathbf{35.1 \text{ g/mol}}$$

10.47 The number of particles in 1 L of gas at STP is:

$$\text{Number of particles} = 1.0 \text{ L} \times \frac{1 \text{ mol}}{22.41 \text{ L}} \times \frac{6.022 \times 10^{23} \text{ particles}}{1 \text{ mol}} = 2.69 \times 10^{22} \text{ particles}$$

**Number of $N_2$ molecules** = $0.78 \times 2.69 \times 10^{22}$ particles = $\mathbf{2.1 \times 10^{22}}$ **$N_2$ molecules**

**Number of $O_2$ molecules** = $0.21 \times 2.69 \times 10^{22}$ particles = $\mathbf{5.6 \times 10^{22}\ O_2}$ **molecules**

**Number of Ar atoms** = $0.01 \times 2.69 \times 10^{22}$ particles = $\mathbf{2.7 \times 10^{20}\ O_2}$ **molecules**

10.49 The density can be calculated from the ideal gas equation.

$$d = \frac{P\mathcal{M}}{RT}$$

$$\mathcal{M} = 1.008 \text{ g/mol} + 79.90 \text{ g/mol} + 80.91 \text{ g/mol}$$

$$T = 46°\text{C} + 273 = 319 \text{ K}$$

$$P = 733 \text{ mmHg} \times \frac{1 \text{ atm}}{760 \text{ mmHg}} = 0.964 \text{ atm}$$

$$\boldsymbol{d} = \frac{(0.964 \text{ atm})\left(\dfrac{80.91 \text{ g}}{1 \text{ mol}}\right)}{319 \text{ K}} \times \frac{\text{mol} \cdot \text{K}}{0.0821 \text{ L} \cdot \text{atm}} = \mathbf{2.98 \text{ g/L}}$$

Alternatively, we can solve for the density by writing:

$$\text{density} = \frac{\text{mass}}{\text{volume}}$$

Assuming that we have 1 mole of HBr, the mass is 80.91 g. The volume of the gas can be calculated using the ideal gas equation.

$$V = \frac{nRT}{P}$$

$$V = \frac{(1 \text{ mol})\left(0.0821 \dfrac{\text{L} \cdot \text{atm}}{\text{mol} \cdot \text{K}}\right)(319 \text{ K})}{0.964 \text{ atm}} = 27.2 \text{ L}$$

Now, we can calculate the density of HBr gas.

$$\textbf{density} = \frac{\text{mass}}{\text{volume}} = \frac{80.91 \text{ g}}{27.2 \text{ L}} = \mathbf{2.97 \text{ g/L}}$$

10.51 This is an extension of an ideal gas law calculation involving molar mass. If you determine the molar mass of the gas, you will be able to determine the molecular formula from the empirical formula.

$$\mathcal{M} = \frac{dRT}{P}$$

Calculate the density, then substitute its value into the equation above.

$$d = \frac{0.100\ \text{g}}{22.1\ \text{mL}} \times \frac{1000\ \text{mL}}{1\ \text{L}} = 4.52\ \text{g/L}$$

$$T(\text{K}) = 20°\text{C} + 273 = 293\ \text{K}$$

$$\mathcal{M} = \frac{\left(4.52\ \frac{\text{g}}{\text{L}}\right)\left(0.0821\frac{\text{L}\cdot\text{atm}}{\text{mol}\cdot\text{K}}\right)(293\ \text{K})}{1.02\ \text{atm}} = 107\ \text{g/mol}$$

Compare the empirical mass to the molar mass.

$$\text{empirical mass} = 32.07\ \text{g/mol} + 4(19.00\ \text{g/mol}) = 108.07\ \text{g/mol}$$

Remember, the molar mass will be a whole number multiple of the empirical mass. In this case, the $\frac{\text{molar mass}}{\text{empirical mass}} \approx 1$. Therefore, the molecular formula is the same as the empirical formula, $\mathbf{SF_4}$.

10.53 **Strategy:** In this problem, the moles of gas are constant. Use the combined gas law (Equation 10.8b).

$$\frac{P_1V_1}{T_1} = \frac{P_2V_2}{T_2}$$

Because $V$ is constant, the above equation reduces to

$$\frac{P_1}{T_1} = \frac{P_2}{T_2}$$

**Solution:** The final temperature is given by:

$$T_2 = \frac{T_1P_2}{P_1}$$

$$\boldsymbol{T_2} = \frac{(298\ \text{K})(5.00\ \text{atm})}{0.800\ \text{L}} = 1.86 \times 10^3\ \text{K} = \mathbf{1590°C}$$

**Think About It:** At a temperature of 1590°C, the pressure in the vessel would be 5.00 atm. Any temperature higher than 1590°C would result in a pressure above 5.00 atm and the vessel would burst.

10.55 **Strategy:** From the moles of $CH_4$ that reacts, we can calculate the moles of $CO_2$ produced. From the balanced equation, we see that 1 mol $CH_4$ is stoichiometrically equivalent to 1 mol $CO_2$. Once moles of $CO_2$ are determined, we can use the ideal gas equation to calculate the volume of $CO_2$.

**Solution:** First let's calculate moles of $CO_2$ produced.

$$? \text{ mol } CO_2 = 15.0 \text{ mol } CH_4 \times \frac{1 \text{ mol } CO_2}{1 \text{ mol } CH_4} = 15.0 \text{ mol } CO_2$$

Now, we can substitute moles, temperature, and pressure into the ideal gas equation to solve for volume of $CO_2$.

$$V = \frac{nRT}{P}$$

$$V_{CO_2} = \frac{(15.0 \text{ mol})\left(0.0821 \frac{\text{L} \cdot \text{atm}}{\text{mol} \cdot \text{K}}\right)(23 + 273)\text{K}}{0.985 \text{ atm}} = \mathbf{3.70 \times 10^2\ L}$$

10.57 From the amount of glucose that reacts (5.97 g), we can calculate the theoretical yield of $CO_2$. We can then compare the theoretical yield to the actual yield given in the problem (1.44 L) to determine the percent yield.

First, let's determine the moles of $CO_2$ that can be produced theoretically. Then, we can use the ideal gas equation to determine the volume of $CO_2$.

$$? \text{ mol } CO_2 = 5.97 \text{ g glucose} \times \frac{1 \text{ mol glucose}}{180.2 \text{ g glucose}} \times \frac{2 \text{ mol } CO_2}{1 \text{ mol glucose}} = 0.0663 \text{ mol } CO_2$$

Now, substitute moles, pressure, and temperature into the ideal gas equation to calculate the volume of $CO_2$.

$$V = \frac{nRT}{P}$$

$$V_{CO_2} = \frac{(0.0663 \text{ mol})\left(0.0821 \frac{\text{L} \cdot \text{atm}}{\text{mol} \cdot \text{K}}\right)(293 \text{ K})}{0.984 \text{ atm}} = 1.62 \text{ L}$$

This is the theoretical yield of $CO_2$. The actual yield, which is given in the problem, is 1.44 L. We can now calculate the percent yield.

$$\text{percent yield} = \frac{\text{actual yield}}{\text{theoretical yield}} \times 100\%$$

$$\textbf{percent yield} = \frac{1.44 \text{ L}}{1.62 \text{ L}} \times 100\% = \mathbf{88.9\%}$$

10.59 **Strategy:** We can calculate the moles of M consumed, and the moles of $H_2$ gas produced. By comparing the number of moles of M consumed to the number of moles $H_2$ produced, we can determine the mole ratio in the balanced equation.

**Solution:** First let's calculate the moles of the metal (M) consumed.

$$\text{mol M} = 0.225 \text{ g M} \times \frac{1 \text{ mol M}}{27.0 \text{ g M}} = 8.33 \times 10^{-3} \text{ mol M}$$

Solve the ideal gas equation algebraically for $n_{H_2}$. Then, calculate the moles of $H_2$ by substituting the known quantities into the equation.

$$P = 741 \text{ mmHg} \times \frac{1 \text{ atm}}{760 \text{ mmHg}} = 0.975 \text{ atm}$$

$$T = 17°\text{C} + 273 = 290 \text{ K}$$

$$n_{H_2} = \frac{PV_{H_2}}{RT}$$

$$n_{H_2} = \frac{(0.975 \text{ atm})(0.303 \text{ L})}{\left(0.0821 \frac{\text{L}\cdot\text{atm}}{\text{mol}\cdot\text{K}}\right)(290 \text{ K})} = 1.24 \times 10^{-2} \text{ mol } H_2$$

Compare the number moles of $H_2$ produced to the number of moles of M consumed.

$$\frac{1.24 \times 10^{-2} \text{ mol } H_2}{8.33 \times 10^{-3} \text{ mol M}} \approx 1.5$$

This means that the mole ratio of $H_2$ to M is 1.5 : 1.

We can now write the balanced equation since we know the mole ratio between $H_2$ and M.

The unbalanced equation is:

$$\text{M}(s) + \text{HCl}(aq) \longrightarrow 1.5\text{H}_2(g) + \text{M}_x\text{Cl}_y(aq)$$

We have 3 atoms of H on the products side of the reaction, so a 3 must be placed in front of HCl. The ratio of M to Cl on the reactants side is now 1 : 3. Therefore the formula of the metal chloride must be $MCl_3$.

The balanced equation is:

$$\mathbf{M}(s) + \mathbf{3HCl}(aq) \longrightarrow \mathbf{1.5H_2}(g) + \mathbf{MCl_3}(aq)$$

From the formula of the metal chloride, we determine that the charge of the metal is +3.

Therefore, the formula of the metal oxide and the metal sulfate are $\mathbf{M_2O_3}$ and $\mathbf{M_2(SO_4)_3}$, respectively.

10.61 From the moles of $CO_2$ produced, we can calculate the amount of calcium carbonate that must have reacted. We can then determine the percent by mass of $CaCO_3$ in the 3.00 g sample.

The balanced equation is:

$$CaCO_3(s) + 2HCl(aq) \longrightarrow CO_2(g) + CaCl_2(aq) + H_2O(l)$$

The moles of $CO_2$ produced can be calculated using the ideal gas equation.

$$n_{CO_2} = \frac{PV_{CO_2}}{RT}$$

$$n_{CO_2} = \frac{\left(792\text{ mmHg} \times \frac{1\text{ atm}}{760\text{ mmHg}}\right)(0.656\text{ L})}{\left(0.0821\frac{\text{L}\cdot\text{atm}}{\text{mol}\cdot\text{K}}\right)(20 + 273\text{ K})} = \mathbf{2.84 \times 10^{-2}\ mol\ CO_2}$$

The balanced equation shows a 1:1 mole ratio between $CO_2$ and $CaCO_3$. Therefore, $2.84 \times 10^{-2}$ mole of $CaCO_3$ must have reacted.

$$?\text{ g CaCO}_3\text{ reacted} = (2.84 \times 10^{-2}\text{ mol CaCO}_3) \times \frac{100.1\text{ g CaCO}_3}{1\text{ mol CaCO}_3} = 2.84\text{ g CaCO}_3$$

The percent by mass of the $CaCO_3$ sample is:

$$\mathbf{\%\ CaCO_3} = \frac{2.84\text{ g}}{3.00\text{ g}} \times 100\% = \mathbf{94.7\%}$$

**Assumption:** The impurity (or impurities) must not react with HCl to produce $CO_2$ gas.

10.63 The balanced equation is:

$$\mathbf{C_2H_5OH(\mathit{l}) + 3O_2(\mathit{g}) \longrightarrow 2CO_2(\mathit{g}) + 3H_2O(\mathit{l})}$$

The moles of $O_2$ needed to react with 185 g ethanol are:

$$185\text{ g C}_2\text{H}_5\text{OH} \times \frac{1\text{ mol C}_2\text{H}_5\text{OH}}{46.07\text{ g C}_2\text{H}_5\text{OH}} \times \frac{3\text{ mol O}_2}{1\text{ mol C}_2\text{H}_5\text{OH}} = 12.05\text{ mol O}_2$$

12.05 moles of $O_2$ correspond to a volume of:

$$V_{O_2} = \frac{n_{O_2}RT}{P} = \frac{(12.05 \text{ mol } O_2)\left(0.0821\frac{\text{L}\cdot\text{atm}}{\text{mol}\cdot\text{K}}\right)(318 \text{ K})}{\left(793 \text{ mmHg}\times\frac{1 \text{ atm}}{760 \text{ mmHg}}\right)} = 302 \text{ L } O_2$$

Since air is 21.0 percent $O_2$ by volume, we can write:

$$\boldsymbol{V_{air}} = V_{O_2}\left(\frac{100\% \text{ air}}{21\% \ O_2}\right) = (302 \text{ L } O_2)\left(\frac{100\% \text{ air}}{21\% \ O_2}\right) = \mathbf{1.44 \times 10^3 \ L \ air}$$

10.67 Dalton's law states that the total pressure of the mixture is the sum of the partial pressures.

a.
$$\boldsymbol{P_{total}} = 0.32 \text{ atm} + 0.15 \text{ atm} + 0.42 \text{ atm} = \mathbf{0.89 \ atm}$$

b. We know:

| Initial Conditions | Final Conditions |
|---|---|
| $P_1 = 0.15 \text{ atm} + 0.42 \text{ atm} = 0.57 \text{ atm}$ | $P_2 = 1.0 \text{ atm}$ |
| $T_1 = 15°\text{C} + 273 = 288 \text{ K}$ | $T_2 = 273 \text{ K}$ |
| $V_1 = 2.5 \text{ L}$ | $\boldsymbol{V_2 = ?}$ |

$$\frac{P_1V_1}{n_1T_1} = \frac{P_2V_2}{n_2T_2}$$

Because $n_1 = n_2$, we can write:

$$V_2 = \frac{P_1V_1T_2}{P_2T_1}$$

$$\boldsymbol{V_2} = \frac{(0.57 \text{ atm})(2.5 \text{ L})(273 \text{ K})}{(1.0 \text{ atm})(288 \text{ K})} = \mathbf{1.4 \ L \ at \ STP}$$

10.69
$$P_{Total} = P_1 + P_2 + P_3 + \ldots + P_n$$

In this case,

$$P_{Total} = P_{Ne} + P_{He} + P_{H_2O}$$

$$P_{Ne} = P_{Total} - P_{He} - P_{H_2O}$$

$$\boldsymbol{P_{Ne}} = 745 \text{ mm Hg} - 368 \text{ mmHg} - 28.3 \text{ mmHg} = \mathbf{349 \ mmHg}$$

10.71 **Strategy:** To solve for moles of $H_2$ generated, we must first calculate the partial pressure of $H_2$ in the mixture. What gas law do we need? How do we convert from moles of $H_2$ to amount of Zn reacted?

**Solution:** Dalton's law of partial pressure states that

$$P_{\text{Total}} = P_1 + P_2 + P_3 + \ldots + P_n$$

In this case,

$$P_{\text{Total}} = P_{H_2} + P_{H_2O}$$

$$P_{H_2} = P_{\text{Total}} - P_{H_2O}$$

$$P_{H_2} = 0.980 \text{ atm} - (23.8 \text{ mmHg})\left(\frac{1 \text{ atm}}{760 \text{ mmHg}}\right) = 0.949 \text{ atm}$$

Now that we know the pressure of $H_2$ gas, we can calculate the moles of $H_2$. Then, using the mole ratio from the balanced equation, we can calculate moles of Zn.

$$n_{H_2} = \frac{P_{H_2}V}{RT}$$

$$n_{H_2} = \frac{(0.949 \text{ atm})(7.80 \text{ L})}{(25 + 273)\text{K}} \times \frac{\text{mol} \cdot \text{K}}{0.0821 \text{ L} \cdot \text{atm}} = 0.303 \text{ mol } H_2$$

Using the mole ratio from the balanced equation and the molar mass of zinc, we can now calculate the grams of zinc consumed in the reaction.

$$\textbf{? g Zn} = 0.303 \text{ mol } H_2 \times \frac{1 \text{ mol Zn}}{1 \text{ mol } H_2} \times \frac{65.39 \text{ g Zn}}{1 \text{ mol Zn}} = \textbf{19.8 g Zn}$$

10.73

$$P_i = X_i P_T$$

We need to determine the mole fractions of each component in order to determine their partial pressures. To calculate mole fraction, write the balanced chemical equation to determine the correct mole ratio.

$$2NH_3(g) \longrightarrow N_2(g) + 3H_2(g)$$

The mole fractions of $N_2$ and $H_2$ are:

$$X_{N_2} = \frac{1 \text{ mol}}{3 \text{ mol} + 1 \text{ mol}} = 0.250$$

$$X_{H_2} = \frac{3 \text{ mol}}{3 \text{ mol} + 1 \text{ mol}} = 0.750$$

The partial pressures of $N_2$ and $H_2$ are:

$$P_{N_2} = X_{N_2}P_T = (0.250)(866 \text{ mmHg}) = \mathbf{217\ mmHg}$$

$$P_{H_2} = X_{H_2}P_T = (0.750)(866 \text{ mmHg}) = \mathbf{650\ mmHg}$$

10.75 a. Box 2 b. Box 2

10.83 **Strategy:** To calculate the root-mean-square speed, we use Equation 10.17 of the text. What units should we use for $R$ and $\mathcal{M}$ so the $u_{rms}$ will be expressed in units of m/s?

**Solution:** To calculate $u_{rms}$, the units of $R$ should be 8.314 J/mol·K, and because 1 J = 1 kg·m$^2$/s$^2$, the units of molar mass must be kg/mol.

First, let's calculate the molar masses ($\mathcal{M}$) of $N_2$, $O_2$, and $O_3$. Remember, $\mathcal{M}$ must be in units of kg/mol.

$$\mathcal{M}_{N_2} = 2(14.01 \text{ g/mol}) = 28.02\frac{\text{g}}{\text{mol}} \times \frac{1 \text{ kg}}{1000 \text{ g}} = 0.02802 \text{ kg/mol}$$

$$\mathcal{M}_{O_2} = 2(16.00 \text{ g/mol}) = 32.00\frac{\text{g}}{\text{mol}} \times \frac{1 \text{ kg}}{1000 \text{ g}} = 0.03200 \text{ kg/mol}$$

$$\mathcal{M}_{O_3} = 3(16.00 \text{ g/mol}) = 48.00\frac{\text{g}}{\text{mol}} \times \frac{1 \text{ kg}}{1000 \text{ g}} = 0.04800 \text{ kg/mol}$$

Now, we can substitute into Equation 10.17 of the text.

$$u_{rms} = \sqrt{\frac{3RT}{\mathcal{M}}}$$

$$u_{rms}(N_2) = \sqrt{\frac{(3)\left(8.314\frac{\text{J}}{\text{mol}\cdot\text{K}}\right)(-23 + 273)\text{K}}{\left(0.02802\frac{\text{kg}}{\text{mol}}\right)}}$$

$$\boldsymbol{u_{rms}(N_2) = 472\ \text{m/s}}$$

Similarly, $\boldsymbol{u_{rms}(O_2) = 441\ \text{m/s}}$, $\boldsymbol{u_{rms}(O_3) = 360\ \text{m/s}}$

**Think About It:** Since the molar masses of the gases increase in the order: $N_2 < O_2 < O_3$, we expect the lightest gas ($N_2$) to move the fastest on average and the heaviest gas ($O_3$) to move the slowest on average. This is confirmed in the above calculation.

10.85

$$\textbf{RMS speed} = \sqrt{\frac{(2.0^2 + 2.2^2 + 2.6^2 + 2.7^2 + 3.3^2 + 3.5^2)(\text{m/s})^2}{6}} = \textbf{2.8 m/s}$$

$$\textbf{Average speed} = \frac{(2.0 + 2.2 + 2.6 + 2.7 + 3.3 + 3.5)\text{m/s}}{6} = \textbf{2.7 m/s}$$

The root-mean-square value is always greater than the average value, because squaring favors the larger values compared to just taking the average value.

10.87 The rate of effusion is the number of molecules passing through a porous barrier in a given time. The longer it takes, the slower the rate of effusion. Therefore, Equation 10.18 of the text can be written as

$$\frac{r_1}{r_2} = \frac{t_2}{t_1} = \sqrt{\frac{\mathcal{M}_2}{\mathcal{M}_1}}$$

where $t_1$ and $t_2$ are the times of effusion for gases 1 and 2, respectively.

The molar mass of $N_2$ is 28.02 g/mol. We write

$$\frac{15.0\text{ min}}{12.0\text{ min}} = \sqrt{\frac{\mathcal{M}}{28.02\text{ g/mol}}}$$

where $\mathcal{M}$ is the molar mass of the unknown gas. Solving for $\mathcal{M}$, we obtain

$$\mathcal{M} = \left(\frac{15.0\text{ min}}{12.0\text{ min}}\right)^2 \times 28.02\text{ g/mol} = \textbf{43.8 g/mol}$$

The gas is **carbon dioxide, $CO_2$** (molar mass = 44.01 g/mol). During the fermentation of glucose, ethanol and carbon dioxide are produced.

10.89 Graham's Law states that the rate of effusion of a gas is inversely proportional to the square root of its molar mass. Thus, the gas with the greater molar mass will effuse more slowly.

a. Since more of the yellow gas was able to escape the container in the same amount of time, the rate of effusion for the yellow gas is greater. Therefore, the molar mass of the yellow has is lower.

b. More of the red gas escaped, so the molar mass of the blue gas is greater.

10.95 We convert the temperature to units of kelvins, then substitute the given quantities into the ideal gas equation.

$$T(\text{K}) = 27°\text{C} + 273 = 300 \text{ K}$$

$$P = \frac{nRT}{V} = \frac{(10.0 \text{ mol})\left(0.0821 \frac{\text{L} \cdot \text{atm}}{\text{mol} \cdot \text{K}}\right)(300 \text{ K})}{1.50 \text{ L}} = 164 \text{ atm}$$

Now, we can compare the ideal pressure to the actual pressure by calculating the percent error.

$$\% \text{ error} = \frac{164 \text{ atm} - 130 \text{ atm}}{130 \text{ atm}} \times 100\% = 26.2\%$$

Based on the large percent error, we conclude that under this condition of high pressure, the gas behaves in a **non-ideal** manner.

10.97 When $a$ and $b$ are zero, the van der Waals equation simply becomes the ideal gas equation. In other words, an ideal gas has zero for the $a$ and $b$ values of the van der Waals equation. It therefore stands to reason that the gas with the smallest values of $a$ and $b$ will behave most like an ideal gas under a specific set of pressure and temperature conditions. Of the choices given in the problem, the gas with the smallest $a$ and $b$ values is **Ne** (see Table 10.6).

10.99 We need to determine the molar mass of the gas. Comparing the molar mass to the empirical mass will allow us to determine the molecular formula.

$$n = \frac{PV}{RT} = \frac{(0.74 \text{ atm})\left(97.2 \text{ mL} \times \frac{0.001 \text{ L}}{1 \text{ mL}}\right)}{\left(0.0821 \frac{\text{L} \cdot \text{atm}}{\text{mol} \cdot \text{K}}\right)(200 + 273)\text{K}} = 1.85 \times 10^{-3} \text{ mol}$$

$$\text{molar mass} = \frac{0.145 \text{ g}}{1.85 \times 10^{-3} \text{ mol}} = 78.4 \text{ g/mol}$$

The empirical mass of CH = 13.02 g/mol

Since $\frac{78.4 \text{ g/mol}}{13.02 \text{ g/mol}} = 6.02 \approx 6$, the molecular formula is $(CH)_6$ or $\mathbf{C_6H_6}$.

10.101 a. The total pressure in (i) is 2.0 atm. The gas is represented by a total of 9 spheres in a volume of 2.0 L. In (ii), the volume is only 1.0 L but the gas is represented by a total of 9 spheres. The same amount of gas in half the volume will exert twice the pressure. Therefore, $\boldsymbol{P_{ii}}$ **= 4.0 atm**. In (iii), the gas is represented by 12 spheres in a 2.0 L volume. It contains 1/3 more spheres than (i) in the same volume, so its pressure will be 1/3 greater. $\boldsymbol{P_{iii}}$ **= 2.67 atm**.

b. When the valves are opened, the gases will distribute themselves among the flasks and the pressure will be the same throughout. The total number of spheres is 9 + 9 + 12 = 30 spheres. The spheres will be distributed among the three flasks in a total volume of 2.0 L + 1.0 L + 2.0 L = 5.0 L. Therefore, there will be 30 ÷ 5 = 6 spheres per liter. 6 spheres per liter corresponds to a pressure of 2.67 atm. (Flask iii originally contains 6 spheres per liter.) Because there are equal numbers of red spheres and blue spheres (15 each), their partial pressure will be equal. $\boldsymbol{P_A}$ **= 1.33 atm,** $\boldsymbol{P_B}$ **= 1.33 atm.**

10.103 a.

$$\mathbf{2KClO_3(\mathit{s}) \rightarrow 2KCl(\mathit{s}) + 3O_2(\mathit{g})}$$

b. First, calculate the moles of $KClO_3$ in 20.4 g. The molar mass of $KClO_3$ is 122.55 g/mol.

$$20.4 \text{ g } KClO_3 \times \frac{1 \text{ mol } KClO_3}{122.55 \text{ g } KClO_3} = 0.1665 \text{ mol } KClO_3$$

According to the balanced equation, 3 moles of $O_2$ form for every 2 moles of $KClO_3$ that decompose. Therefore,

$$0.1665 \text{ mol } KClO_3 \times \frac{3 \text{ mol } O_2}{2 \text{ mol } KClO_3} = 0.2498 \text{ mol } O_2$$

Finally, we use the ideal gas equation to determine the volume of $O_2$.

$$V_{O_2} = \frac{nRT}{P} = \frac{(0.2498 \text{ mol})\left(0.08206 \frac{\text{L}\cdot\text{atm}}{\text{mol}\cdot\text{K}}\right)(291.5 \text{ K})}{0.962 \text{ atm}} = \mathbf{6.21 \text{ L}}$$

10.105 a.

$$C_3H_8(g) + 5O_2(g) \longrightarrow 3CO_2(g) + 4H_2O(g)$$

b. From the balanced equation, we see that there is a 1:3 mole ratio between $C_3H_8$ and $CO_2$.

$$\mathbf{? \text{ L } CO_2} = 7.45 \text{ g } C_3H_8 \times \frac{1 \text{ mol } C_3H_8}{44.09 \text{ g } C_3H_8} \times \frac{3 \text{ mol } CO_2}{1 \text{ mol } C_3H_8} \times \frac{22.414 \text{ L } CO_2}{1 \text{ mol } CO_2} = \mathbf{11.4 \text{ L } CO_2}$$

10.107 Calculate the initial number of moles of NO and $O_2$ using the ideal gas equation.

$$n_{NO} = \frac{P_{NO}V}{RT} = \frac{(0.500 \text{ atm})(4.00 \text{ L})}{\left(0.0821 \frac{\text{L}\cdot\text{atm}}{\text{mol}\cdot\text{K}}\right)(298 \text{ K})} = 0.0817 \text{ mol NO}$$

$$n_{O_2} = \frac{P_{O_2}V}{RT} = \frac{(1.00\text{ atm})(2.00\text{ L})}{\left(0.0821\frac{\text{L}\cdot\text{atm}}{\text{mol}\cdot\text{K}}\right)(298\text{ K})} = 0.0817\text{ mol } O_2$$

Determine which reactant is the limiting reactant. The number of moles of NO and $O_2$ calculated above are equal; however, the balanced equation shows that twice as many moles of NO are needed compared to $O_2$. Thus, NO is the limiting reactant.

Determine the molar amounts of NO, $O_2$, and $NO_2$ after complete reaction.

**mol NO = 0 mol** (All NO is consumed during reaction)

$$\textbf{mol NO}_2 = 0.0817\text{ mol NO} \times \frac{2\text{ mol NO}_2}{2\text{ mol NO}} = \textbf{0.0817 mol NO}_2$$

$$\text{mol O}_2\text{ consumed} = 0.0817\text{ mol NO} \times \frac{1\text{ mol O}_2}{2\text{ mol NO}} = 0.0409\text{ mol O}_2\text{ consumed}$$

$$\textbf{mol O}_2\textbf{ remaining} = 0.0817\text{ mol O}_2\text{ initial} - 0.0409\text{ mol O}_2\text{ consumed} = \textbf{0.0408 mol O}_2$$

Calculate the partial pressures of $O_2$ and $NO_2$ using the ideal gas equation.

Volume of entire apparatus = 2.00 L + 4.00 L = 6.00 L

$$T(\text{K}) = 25°\text{C} + 273 = 298\text{ K}$$

$$P_{O_2} = \frac{n_{O_2}RT}{V} = \frac{(0.0408\text{ mol})\left(0.0821\frac{\text{L}\cdot\text{atm}}{\text{mol}\cdot\text{K}}\right)(298\text{ K})}{(6.00\text{ L})} = \textbf{0.166 atm}$$

$$P_{NO_2} = \frac{n_{NO_2}RT}{V} = \frac{(0.0817\text{ mol})\left(0.0821\frac{\text{L}\cdot\text{atm}}{\text{mol}\cdot\text{K}}\right)(298\text{ K})}{(6.00\text{ L})} = \textbf{0.333 atm}$$

10.109 a. First, the total pressure ($P_{Total}$) of the mixture of carbon dioxide and hydrogen must be determined at a given temperature in a container of known volume. Next, the carbon dioxide can be removed by reaction with sodium hydroxide.

$$CO_2(g) + 2NaOH(aq) \longrightarrow Na_2CO_3(aq) + H_2O(l)$$

The pressure of the hydrogen gas that remains can now be measured under the same conditions of temperature and volume. Finally, the partial pressure of $CO_2$ can be calculated.

$$P_{CO_2} = P_{Total} - P_{H_2}$$

b. The most direct way to measure the partial pressures would be to use a mass spectrometer to measure the mole fractions of the gases. The partial pressures could then be calculated from the mole fractions and the total pressure. Another way to measure the partial pressures would be to realize that helium has a much lower boiling point than nitrogen. Therefore, nitrogen gas can be removed by lowering the temperature until nitrogen liquefies. Helium will remain as a gas. As in part (a), the total pressure is measured first. Then, the pressure of helium can be measured after the nitrogen is removed. Finally, the pressure of nitrogen is simply the difference between the total pressure and the pressure of helium.

10.111 The reactions are:

$$Na_2CO_3(s) + 2HCl(aq) \longrightarrow 2NaCl(aq) + H_2O(l) + CO_2(g)$$

$$MgCO_3(s) + 2HCl(aq) \longrightarrow MgCl_2(aq) + H_2O(l) + CO_2(g)$$

First, let's calculate the moles of $CO_2$ produced using the ideal gas equation. We carry an extra significant figure throughout this calculation to limit rounding errors.

$$n_{CO_2} = \frac{PV}{RT} = \frac{(1.24\text{ atm})(1.67\text{ L})}{\left(0.0821\dfrac{\text{L}\cdot\text{atm}}{\text{mol}\cdot\text{K}}\right)(26 + 273)\text{K}} = 0.08436\text{ mol CO}_2$$

Since there is a 1:1 mole ratio between $CO_2$ and both reactants ($Na_2CO_3$ and $MgCO_3$), then 0.08436 mole of the mixture must have reacted.

$$\text{mol Na}_2\text{CO}_3 + \text{mol MgCO}_3 = 0.08436\text{ mol}$$

Let $x$ be the mass of $Na_2CO_3$ in the mixture, then $(7.63 - x)$ is the mass of $MgCO_3$ in the mixture.

$$\left[x\text{ g Na}_2\text{CO}_3 \times \frac{1\text{ mol Na}_2\text{CO}_3}{106.0\text{ g Na}_2\text{CO}_3}\right] + \left[(7.63 - x)\text{g MgCO}_3 \times \frac{1\text{ mol MgCO}_3}{84.32\text{ g MgCO}_3}\right] = 0.08436\text{ mol}$$

$$0.009434x - 0.01186x + 0.09049 = 0.08436$$

$$x = 2.527\text{ g} = \text{mass of Na}_2\text{CO}_3\text{ in the mixture}$$

The percent composition by mass of $Na_2CO_3$ in the mixture is:

$$\textbf{mass \% Na}_2\textbf{CO}_3 = \frac{\text{mass Na}_2\text{CO}_3}{\text{mass of mixture}} \times 100\% = \frac{2.527\text{ g}}{7.63\text{ g}} \times 100\% = \textbf{33.1\% Na}_2\textbf{CO}_3$$

10.113 Using the ideal gas equation, we can calculate the moles of water that would be vaporized. We can then convert to mass of water vaporized.

$$n = \frac{PV}{RT} = \frac{\left(187.5 \text{ mmHg} \times \dfrac{1 \text{ atm}}{760 \text{ mmHg}}\right)(2.500 \text{ L})}{\left(0.0821 \dfrac{\text{L} \cdot \text{atm}}{\text{mol} \cdot \text{K}}\right)(65 + 273)\text{K}} = 0.0222 \text{ mol } H_2O$$

$$\textbf{? g } \mathbf{H_2O} \textbf{ vaporized} = 0.0222 \text{ mol } H_2O \times \frac{18.02 \text{ g } H_2O}{1 \text{ mol } H_2O} = \mathbf{0.400 \text{ g } H_2O}$$

10.115 The value of $a$ indicates how strongly molecules of a given type of gas attract one anther. $\mathbf{C_6H_6}$ has the greatest intermolecular attractions due to its larger size compared to the other choices. Therefore, it has the largest $a$ value.

10.117 a. Use the formula for volume of a sphere to calculate the volume of a sphere of radius $2r$. This is the excluded volume defined by two molecules.

$$\text{excluded volume defined by two molecules} = \frac{4}{3}\pi(2r)^3 = 8\left(\frac{4}{3}\pi r^3\right)$$

b. Multiply the excluded volume per molecule by Avogadro's number to determine excluded volume per mole.

$$\text{excluded volume per mole} = \frac{8\left(\dfrac{4}{3}\pi r^3\right)}{2 \text{ molecules}} \times \frac{N_A \text{ molecules}}{1 \text{ mole}} = \frac{4N_A\left(\dfrac{4}{3}\pi r^3\right)}{1 \text{ mole}}$$

The volume actually occupied by a mole of molecules with radius $r$ is $N_A\left(\frac{4}{3}\pi r^3\right)$.

10.119

From problem 10.117, the excluded volume is $\dfrac{4N_A\left(\frac{4}{3}\pi r^3\right)}{1 \text{ mole}}$.

So,

$$b = \frac{4N_A\left(\dfrac{4}{3}\pi r^3\right)}{1 \text{ mole}} = 0.0315 \text{ L/mol}$$

Rearranging to solve for $r^3$, we get:

$$r^3 = \frac{0.0315 \text{ L/mol}}{4N_A\left(\frac{4}{3}\pi\right)/\text{mol}} = \frac{0.0315 \text{ L}}{(4)\left(6.022\times10^{23}\right)\left(\frac{4}{3}\pi\right)}$$
$$= 3.122\times10^{27} \text{ L}$$

By definintion, 1 L = 1 $(dm)^3$, so

$$r^3 = 3.122\times10^{27} \text{ L} = 3.122\times10^{27} \ (dm)^3$$
$$r = 1.462\times10^{-9} \ dm$$
$$= 146 \ pm$$

10.121 The partial pressure of carbon dioxide is higher in the winter because carbon dioxide is utilized less by photosynthesis in plants.

10.123 First, let's calculate the root-mean-square speed of $N_2$ at 25°C.

$$u_{\text{rms}}(N_2) = \sqrt{\frac{3RT}{\mathcal{M}}} = \sqrt{\frac{3(8.314 \text{ J/mol}\cdot\text{K})(298 \text{ K})}{0.02802 \text{ kg/mol}}} = 515 \text{ m/s}$$

Now, we can calculate the temperature at which He atoms will have this same root-mean-square speed.

$$u_{\text{rms}}(\text{He}) = \sqrt{\frac{3RT}{\mathcal{M}}}$$

$$515 \text{ m/s} = \sqrt{\frac{3(8.314 \text{ J/mol}\cdot\text{K})T}{0.004003 \text{ kg/mol}}}$$

$$\mathbf{T = 42.6 \ K}$$

10.125 **Radon**, because it is radioactive so that its mass is constantly changing (decreasing). The number of radon atoms is not constant.

10.127 The reaction between Zn and HCl is: $Zn(s) + 2HCl(aq) \rightarrow H_2(g) + ZnCl_2(aq)$

From the amount of $H_2(g)$ produced, we can determine the amount of Zn consumed. Then, using the original mass of the sample, we can calculate the mass % of Zn in the sample.

$$n_{H_2} = \frac{PV_{H_2}}{RT}$$

$$n_{H_2} = \frac{\left(728 \text{ mmHg} \times \frac{1 \text{ atm}}{760 \text{ mmHg}}\right)(1.26 \text{ L})}{\left(0.0821 \frac{\text{L} \cdot \text{atm}}{\text{mol} \cdot \text{K}}\right)(22 + 273)\text{K}} = 0.0498 \text{ mol } H_2$$

Since the mole ratio between $H_2$ and Zn is 1:1, the amount of Zn consumed is also 0.0498 mole. Converting to grams of Zn, we find:

$$0.0498 \text{ mol Zn} \times \frac{65.39 \text{ g Zn}}{1 \text{ mol Zn}} = 3.26 \text{ g Zn}$$

The mass percent of Zn in the 6.11 g sample is:

$$\textbf{mass \% Zn} = \frac{\text{mass Zn}}{\text{mass sample}} \times 100\% = \frac{3.26 \text{ g}}{6.11 \text{ g}} \times 100\% = \textbf{53.4\%}$$

10.129 Assuming a volume of 1.00 L, we can determine the average molar mass of the mixture.

$$n_{\text{mixture}} = \frac{PV}{RT} = \frac{(0.98 \text{ atm})(1.00 \text{ L})}{\left(0.0821 \frac{\text{L} \cdot \text{atm}}{\text{mol} \cdot \text{K}}\right)(25 + 273)\text{K}} = 0.0401 \text{ mol}$$

The average molar mass of the mixture is:

$$\overline{\mathcal{M}} = \frac{2.7 \text{ g}}{0.0401 \text{ mol}} = \textbf{67 g/mol}$$

Now, we can calculate the mole fraction of each component of the mixture. Once we determine the mole fractions, we can calculate the partial pressure of each gas.

$$X_{NO_2}\mathcal{M}_{NO_2} + X_{N_2O_4}\mathcal{M}_{N_2O_4} = 67 \text{ g/mol}$$

$$X_{NO_2}(46.01 \text{ g/mol}) + (1 - X_{NO_2})(92.02 \text{ g/mol}) = 67 \text{ g/mol}$$

$$46.01X_{NO_2} - 92.02X_{NO_2} + 92.02 = 67$$

$$X_{NO_2} = 0.54 \quad \text{and} \quad X_{N_2O_4} = 1 - 0.54 = 0.46$$

Finally, the partial pressures are:

$$P_{NO_2} = X_{NO_2}P_T \qquad P_{N_2O_4} = X_{N_2O_4}P_T$$

$$\boldsymbol{P_{NO_2}} = (0.54)(0.98 \text{ atm}) = \textbf{0.53 atm}$$

$$\boldsymbol{P_{N_2O_4}} = (0.46)(0.98 \text{ atm}) = \textbf{0.45 atm}$$

10.131 **Strategy:** Find the number of moles of each substance. Since the temperature and pressure are the same for each sample, the sample with the largest number of moles of particles will have the greatest volume.

**Solution:** a. 0.82 mol He (given)

b.
$$(24 \text{ g N}_2)\left(\frac{1 \text{ mol N}_2}{28 \text{ g N}_2}\right) = 0.86 \text{ mol N}_2$$

c.
$$\left(5.0\times10^{23} \text{ Cl}_2\right)\left(\frac{1 \text{ mol Cl}_2}{6.022\times10^{23} \text{ Cl}_2}\right) = 0.83 \text{ mol Cl}_2$$

10.133 The law of conservation of energy (or the first law of thermodynamics) states that energy cannot be created or destroyed. While individual gas particles constantly exchange energy with other particles and with the container walls, the overall energy remains constant. There is not violation of energy conservation.

10.135 a. Of the substances listed in Figure 10.23, $Cl_2$ and $NH_3$ have normal boiling points that are significantly higher than the others, so we may conclude that the intermolecular attractions in these two substances are relatively large. These *strong attractions* lead to a measured molar volume that is less than the molar volume of an ideal gas.

b. For Ar, $H_2$, He, $N_2$, Ne, and $O_2$, the normal boiling points are relatively low, indicating relatively weak intermolecular attractions. Since the attractions are weak, it is the *excluded volume* that dominates the deviation from non-ideal behavior, and excluded volume causes the actual volume to be higher than expected ideally.

10.137 We start with Graham's Law as this problem relates to effusion of gases. Using Graham's Law, we can calculate the effective molar mass of the mixture of CO and $CO_2$. Once the effective molar mass of the mixture is known, we can determine the mole fraction of each component. Because $n \propto V$ at constant $T$ and $P$, the volume fraction = mole fraction.

$$\frac{r_{\text{He}}}{r_{\text{mix}}} = \sqrt{\frac{\mathcal{M}_{\text{mix}}}{\mathcal{M}_{\text{He}}}}$$

$$\mathcal{M}_{\text{mix}} = \left(\frac{r_{\text{He}}}{r_{\text{mix}}}\right)^2 \mathcal{M}_{\text{He}}$$

$$\mathcal{M}_{mix} = \left(\frac{\dfrac{29.7\text{ mL}}{2.00\text{ min}}}{\dfrac{10.0\text{ mL}}{2.00\text{ min}}}\right)^2 (4.003\text{ g/mol}) = 35.31\text{ g/mol}$$

Now that we know the molar mass of the mixture, we can calculate the mole fraction of each component.

$$X_{CO} + X_{CO_2} = 1$$

and

$$X_{CO_2} = 1 - X_{CO}$$

The mole fraction of each component multiplied by its molar mass will give the contribution of that component to the effective molar mass.

$$X_{CO}\mathcal{M}_{CO} + X_{CO_2}\mathcal{M}_{CO_2} = \mathcal{M}_{mix}$$

$$X_{CO}\mathcal{M}_{CO} + (1 - X_{CO})\mathcal{M}_{CO_2} = \mathcal{M}_{mix}$$

$$X_{CO}(28.01\text{ g/mol}) + (1 - X_{CO})(44.01\text{ g/mol}) = 35.31\text{ g/mol}$$

$$28.01X_{CO} + 44.01 - 44.01X_{CO} = 35.31$$

$$16.00X_{CO} = 8.70$$

$$X_{CO} = 0.544$$

At constant $P$ and $T$, $n \propto V$. Therefore, volume fraction = mole fraction. As a result,

$$\%\text{ of CO by volume} = \mathbf{54.4\%}$$

$$\%\text{ of } CO_2 \text{ by volume} = 1 - \%\text{ of CO by volume} = \mathbf{45.6\%}$$

10.139 Warm air rises because of its buoyancy. This buoyancy is a direct result of the decreased density of the warm air relative to the surrounding air. On a molecular level, the decreased density of the warm air can be accounted for by considering that the molecules in the warm air move with more speed and thus more kinetic energy at higher temperatures. These more-energetic molecules are able to open up a larger "bubble" of volume within the surrounding air than they would otherwise, thus making the density less than the surrounding air.

10.141 a.

$$\frac{188\text{ g CO}}{1\text{ hr}} \times \frac{1\text{ mol CO}}{28.01\text{ g CO}} \times \frac{1\text{ hr}}{60\text{ min}} = \mathbf{0.112\text{ mol CO/min}}$$

b. 1000 ppm means that there are 1000 particles of gas per 1,000,000 particles of air. The pressure of a gas is directly proportional to the number of particles of gas. We can calculate the partial pressure of CO in atmospheres, assuming that atmospheric pressure is 1 atm.

$$\frac{1000 \text{ particles}}{1{,}000{,}000 \text{ particles}} \times 1 \text{ atm} = 1.0 \times 10^{-3} \text{ atm}$$

A partial pressure of $1.0 \times 10^{-3}$ atm CO is lethal.

The volume of the garage (in L) is:

$$(6.0 \text{ m} \times 4.0 \text{ m} \times 2.2 \text{ m}) \times \left(\frac{1 \text{ cm}}{0.01 \text{ m}}\right)^3 \times \frac{1 \text{ L}}{1000 \text{ cm}^3} = 5.3 \times 10^4 \text{ L}$$

From part (a), we know the rate of CO production per minute. In one minute the partial pressure of CO will be:

$$P_{CO} = \frac{nRT}{V} = \frac{(0.112 \text{ mol})\left(0.0821 \frac{\text{L} \cdot \text{atm}}{\text{mol} \cdot \text{K}}\right)(20 + 273)\text{K}}{5.3 \times 10^4 \text{ L}} = 5.1 \times 10^{-5} \text{ atm CO/min}$$

How many minutes will it take for the partial pressure of CO to reach the lethal level, $1.0 \times 10^{-3}$ atm?

$$\textbf{? min} = (1.0 \times 10^{-3} \text{ atm CO}) \times \frac{1 \text{ min}}{5.1 \times 10^{-5} \text{ atm CO}} = \mathbf{2.0 \times 10^1 \text{ min}}$$

10.143 The balanced equation is: $CO_2(g) + 2NH_3(g) \longrightarrow (NH_2)_2CO(s) + H_2O(g)$

First, we can calculate the moles of $NH_3$ needed to produce 1.0 ton of urea. Then, we can use the ideal gas equation to calculate the volume of $NH_3$.

$$? \text{ mol } NH_3 = 1.0 \text{ ton urea} \times \frac{2000 \text{ lb}}{1 \text{ ton}} \times \frac{453.6 \text{ g}}{1 \text{ lb}} \times \frac{1 \text{ mol urea}}{60.06 \text{ g urea}} \times \frac{2 \text{ mol } NH_3}{1 \text{ mol urea}} = 3.0 \times 10^4 \text{ mol } NH_3$$

$$V_{NH_3} = \frac{nRT}{P} = \frac{(3.0 \times 10^4 \text{ mol})\left(0.0821 \frac{\text{L} \cdot \text{atm}}{\text{mol} \cdot \text{K}}\right)(200 + 273)\text{K}}{150 \text{ atm}} = \mathbf{7.8 \times 10^3 \text{ L } NH_3}$$

10.145 a. We can calculate the moles of mercury vapor using the ideal gas equation, but first we need to know the volume of the room in liters.

$$V_{\text{room}} = (15.2\text{ m})(6.6\text{ m})(2.4\text{ m}) \times \left(\frac{1\text{ cm}}{0.01\text{ m}}\right)^3 \times \frac{1\text{ L}}{1000\text{ cm}^3} = 2.4 \times 10^5\text{ L}$$

$$n_{\text{Hg}} = \frac{PV}{RT} = \frac{(1.7 \times 10^{-6}\text{ atm})(2.4 \times 10^5\text{ L})}{\left(0.0821\frac{\text{L}\cdot\text{atm}}{\text{mol}\cdot\text{K}}\right)(20 + 273)\text{K}} = 0.017\text{ mol Hg}$$

Converting to mass:

$$\textbf{? g Hg} = 0.017\text{ mol Hg} \times \frac{200.6\text{ g Hg}}{1\text{ mol Hg}} = \mathbf{3.4\text{ g Hg}}$$

b. The concentration of Hg vapor in the room is:

$$\frac{3.4\text{ g Hg}}{(15.2\text{ m})(6.6\text{ m})(2.4\text{ m})} = 0.014\text{ g Hg/m}^3 = \mathbf{14\text{ mg Hg/m}^3}$$

**Yes**, this far exceeds the air quality regulation of 0.050 mg $\text{Hg/m}^3$ of air.

c. Physical: The sulfur powder covers the Hg surface, thus retarding the rate of evaporation.

Chemical: Sulfur reacts slowly with Hg to form HgS. HgS has no measurable vapor pressure.

10.147 The volume of the bulb can be calculated using the ideal gas equation. Pressure and temperature are given in the problem. Moles of air must be calculated before the volume can be determined.

$$\text{Mass of air} = 91.6843\text{ g} - 91.4715\text{ g} = 0.2128\text{ g air}$$

$$\text{Molar mass of air} = (0.78 \times 28.02\text{ g/mol}) + (0.21 \times 32.00\text{ g/mol}) + (0.01 \times 39.95\text{ g/mol}) = 29\text{ g/mol}$$

$$\text{moles air} = 0.2128\text{ g air} \times \frac{1\text{ mol air}}{29\text{ g air}} = 7.3 \times 10^{-3}\text{ mol air}$$

Now, we can calculate the volume of the bulb.

$$\mathbf{V_{bulb}} = \frac{nRT}{P} = \frac{(7.3 \times 10^{-3}\text{ mol})\left(0.0821\frac{\text{L}\cdot\text{atm}}{\text{mol}\cdot\text{K}}\right)(23 + 273)\text{K}}{\left(744\text{ mmHg} \times \frac{1\text{ atm}}{760\text{ mmHg}}\right)} = 0.18\text{ L} = \mathbf{1.8 \times 10^2\text{ mL}}$$

10.149 Let's calculate the pressure of ammonia using both the ideal gas equation and van der Waal's equation.

We can then calculate the percent error in using the ideal gas equation.

Ideal gas: $$P = \frac{nRT}{V} = \frac{(5.00\ \text{mol})(0.0821\ \text{L}\cdot\text{atm}/\text{mol}\cdot\text{K})(300\ \text{K})}{1.92\ \text{L}} = 64.1\ \text{atm}$$

van der Waals: $$P = \frac{nRT}{(V - nb)} - \frac{an^2}{V^2}$$

$$P = \frac{(5.00\ \text{mol})(0.0821\ \text{L}\cdot\text{atm/mol}\cdot\text{K})(300\ \text{K})}{1.92\ \text{L} - (5.00\ \text{mol})(0.0371\ \text{L/mol})} - \frac{(4.17\ \text{atm}\cdot\text{L}^2/\text{mol}^2)(5.00\ \text{mol})^2}{(1.92\ \text{L})^2}$$

$$P = 71.0\ \text{atm} - 28.3\ \text{atm} = 42.7\ \text{atm}$$

The $a$ and $b$ values used are from Table 10.6 of the text.

The percent error in using the ideal gas equation is:

$$\%\ \text{error} = \frac{|P_{\text{vdw}} - P_{\text{ideal}}|}{P_{\text{vdw}}} \times 100\%$$

$$\%\ \text{error} = \frac{|42.7\ \text{atm} - 64.1\ \text{atm}|}{42.7\ \text{atm}} \times 100\% = \mathbf{50.1\%}$$

10.151 First, calculate the moles of $H_2$ formed.

$$?\ \text{mol H}_2 = 3.12\ \text{g Al} \times \frac{1\ \text{mol Al}}{26.98\ \text{g}} \times \frac{3\ \text{mol H}_2}{2\ \text{mol Al}} = 0.173\ \text{mol}$$

Next, calculate the volume of $H_2$ produced using the ideal gas equation.

$$V_{\text{H}_2} = \frac{n_{\text{H}_2}RT}{P} = \frac{(0.173\ \text{mol})\left(0.0821\ \frac{\text{L}\cdot\text{atm}}{\text{mol}\cdot\text{K}}\right)(296\ \text{K})}{(1.00\ \text{atm})} = \mathbf{4.20\ L}$$

10.153 The moles of $O_2$ can be calculated from the ideal gas equation. The mass of $O_2$ can then be calculated using the molar mass as a conversion factor.

$$n_{\text{O}_2} = \frac{PV}{RT} = \frac{(132\ \text{atm})(120\ \text{L})}{\left(0.0821\frac{\text{L}\cdot\text{atm}}{\text{mol}\cdot\text{K}}\right)(22 + 273)\text{K}} = 654\ \text{mol O}_2$$

$$\mathbf{?\ g\ O_2} = 654\ \text{mol O}_2 \times \frac{32.00\ \text{g O}_2}{1\ \text{mol O}_2} = \mathbf{2.09 \times 10^4\ g\ O_2 = 20.9\ kg\ O_2}$$

The volume of $O_2$ gas under conditions of 1.00 atm pressure and a temperature of 22°C can be calculated using the ideal gas equation. The moles of $O_2$ = 654 moles.

$$V_{O_2} = \frac{n_{O_2}RT}{P} = \frac{(654\ \text{mol})\left(0.0821\dfrac{\text{L}\cdot\text{atm}}{\text{mol}\cdot\text{K}}\right)(22+273)\text{K}}{1.00\ \text{atm}} = \mathbf{1.58 \times 10^4\ L\ O_2}$$

10.155 a. The equation to calculate the root-mean-square speed is Equation 10.17 of the text. Let's calculate $u_{mp}$ and $u_{rms}$ at 25°C (298 K). Recall that the molar mass of $N_2$ must be in units of kg/mol, because the units of $R$ are J/mol·K and 1 J = 1 kg·m$^2$/s$^2$.

$$u_{mp} = \sqrt{\frac{2RT}{\mathcal{M}}} \qquad u_{rms} = \sqrt{\frac{3RT}{\mathcal{M}}}$$

$$u_{mp} = \sqrt{\frac{2(8.314\ \text{J/mol}\cdot\text{K})(298\ \text{K})}{0.02802\ \text{kg/mol}}} \qquad u_{rms} = \sqrt{\frac{3(8.314\ \text{J/mol}\cdot\text{K})(298\ \text{K})}{0.02802\ \text{kg/mol}}}$$

$$\mathbf{u_{mp} = 421\ m/s} \qquad \mathbf{u_{rms} = 515\ m/s}$$

The most probable speed ($u_{mp}$) will always be slower than the root-mean-square speed. We can derive a general relation between the two speeds.

$$\frac{u_{mp}}{u_{rms}} = \frac{\sqrt{\dfrac{2RT}{\mathcal{M}}}}{\sqrt{\dfrac{3RT}{\mathcal{M}}}} = \sqrt{\frac{\dfrac{2RT}{\mathcal{M}}}{\dfrac{3RT}{\mathcal{M}}}}$$

$$\frac{u_{mp}}{u_{rms}} = \sqrt{\frac{2}{3}} = 0.816$$

This relation indicates that the most probable speed ($u_{mp}$) will be 81.6% of the root-mean-square speed ($u_{rms}$) at a given temperature.

b. We can derive a relationship between the most probable speeds at $T_1$ and $T_2$.

$$\frac{u_{mp}(1)}{u_{mp}(2)} = \frac{\sqrt{\dfrac{2RT_1}{\mathcal{M}}}}{\sqrt{\dfrac{2RT_2}{\mathcal{M}}}} = \sqrt{\frac{\dfrac{2RT_1}{\mathcal{M}}}{\dfrac{2RT_2}{\mathcal{M}}}}$$

$$\frac{u_{mp}(1)}{u_{mp}(2)} = \sqrt{\frac{T_1}{T_2}}$$

Looking at the diagram, let's assume that the most probable speed at $T_1$ = 300 K is 500 m/s, and the most probable speed at $T_2$ is 1000 m/s. Substitute into the above equation to solve for $T_2$.

$$\frac{500 \text{ m/s}}{1000 \text{ m/s}} = \sqrt{\frac{300 \text{ K}}{T_2}}$$

$$(0.5)^2 = \frac{300}{T_2}$$

$$\boldsymbol{T_2 = 1200 \text{ K}}$$

10.157 The air inside the egg expands with increasing temperature. The increased pressure can cause the egg to crack.

10.159 a. This is a Boyle's Law problem, pressure and volume vary. Assume that the pressure at the water surface is 1 atm. The pressure that the diver experiences 36 ft below water is:

$$1 \text{ atm} + \left(36 \text{ ft} \times \frac{1 \text{ atm}}{33 \text{ ft}}\right) = 2.1 \text{ atm}$$

$$P_1V_1 = P_2V_2 \qquad \text{or} \qquad \frac{V_1}{V_2} = \frac{P_2}{P_1}$$

$$\frac{V_1}{V_2} = \frac{2.1 \text{ atm}}{1 \text{ atm}} = \mathbf{2.1}$$

The diver's lungs would increase in volume **2.1 times** by the time he reaches the surface.

b.

$$P_{O_2} = X_{O_2}P_T$$

$$P_{O_2} = \frac{n_{O_2}}{n_{O_2} + n_{N_2}}P_T$$

At constant temperature and pressure, the volume of a gas is directly proportional to the number of moles of gas. We can write:

$$P_{O_2} = \frac{V_{O_2}}{V_{O_2} + V_{N_2}}P_T$$

We know the partial pressure of $O_2$ in air, and we know the total pressure exerted on the diver. Plugging these values into the above equation gives:

$$0.20\text{ atm} = \frac{V_{O_2}}{V_{O_2} + V_{N_2}}(4.00\text{ atm})$$

$$\frac{V_{O_2}}{V_{O_2} + V_{N_2}} = \frac{0.20\text{ atm}}{4.00\text{ atm}} = \mathbf{0.050}$$

In other words, the air that the diver breathes must have an oxygen content of **5% by volume**.

10.161 The reaction is:

$$HCO_3^-(aq) + H^+(aq) \longrightarrow H_2O(l) + CO_2(g)$$

The mass of $HCO_3^-$ reacted is:

$$3.29\text{ g tablet} \times \frac{32.5\%\ HCO_3^-}{100\%\text{ tablet}} = 1.07\text{ g } HCO_3^-$$

$$\text{mol } CO_2 \text{ produced} = 1.07\text{ g } HCO_3^- \times \frac{1\text{ mol } HCO_3^-}{61.02\text{ g } HCO_3^-} \times \frac{1\text{ mol } CO_2}{1\text{ mol } HCO_3^-} = 0.0175\text{ mol } CO_2$$

$$\mathbf{V_{CO_2}} = \frac{n_{CO_2}RT}{P} = \frac{(0.0175\text{ mol } CO_2)\left(0.0821\frac{\text{L}\cdot\text{atm}}{\text{mol}\cdot\text{K}}\right)(37 + 273)\text{K}}{(1.00\text{ atm})} = 0.445\text{ L} = \mathbf{445\text{ mL}}$$

10.163 a. Write a balanced chemical equation.

$$2NaHCO_3(s) \longrightarrow Na_2CO_3(s) + CO_2(g) + H_2O(g)$$

First, calculate the moles of $CO_2$ produced.

$$?\text{ mol } CO_2 = 5.0\text{ g } NaHCO_3 \times \frac{1\text{ mol } NaHCO_3}{84.01\text{ g } NaHCO_3} \times \frac{1\text{ mol } CO_2}{2\text{ mol } NaHCO_3} = 0.030\text{ mol}$$

Next, calculate the volume of $CO_2$ produced using the ideal gas equation.

$$T(\text{K}) = 180°\text{C} + 273 = 453\text{ K}$$

$$V_{CO_2} = \frac{n_{CO_2}RT}{P}$$

$$V_{CO_2} = \frac{(0.030\text{ mol})\left(0.0821\dfrac{\text{L}\cdot\text{atm}}{\text{mol}\cdot\text{K}}\right)(453\text{ K})}{(1.3\text{ atm})} = \mathbf{0.86\ L}$$

b. The balanced chemical equation for the decomposition of $NH_4HCO_3$ is

$$NH_4HCO_3(s) \longrightarrow NH_3(g) + CO_2(g) + H_2O(g)$$

The advantage in using the ammonium salt is that more gas is produced per gram of reactant. The disadvantage is that one of the gases is ammonia. The strong odor of ammonia would ***not*** make the ammonium salt a good choice for baking.

10.165 a. First, convert density to units of g/L.

$$\frac{0.426\text{ kg}}{1\text{ m}^3} \times \frac{1000\text{ g}}{1\text{ kg}} \times \left(\frac{0.01\text{ m}}{1\text{ cm}}\right)^3 \times \frac{1000\text{ cm}^3}{1\text{ L}} = 0.426\text{ g/L}$$

Let's assume a volume of 1.00 L of air. This air sample will have a mass of 0.426 g. Converting to moles of air:

$$0.426\text{ g air} \times \frac{1\text{ mol air}}{29.0\text{ g air}} = 0.0147\text{ mol air}$$

Now, we can substitute into the ideal gas equation to calculate the air temperature.

$$\boldsymbol{T} = \frac{PV}{nR} = \frac{\left(210\text{ mmHg} \times \dfrac{1\text{ atm}}{760\text{ mmHg}}\right)(1.00\text{ L})}{(0.0147\text{ mol})\left(0.0821\dfrac{\text{L}\cdot\text{atm}}{\text{mol}\cdot\text{K}}\right)} = \mathbf{229\ K = -44^\circ C}$$

b. To determine the percent decrease in oxygen gas, let's compare moles of $O_2$ at the top of Mt. Everest to the moles of $O_2$ at sea level.

$$\frac{n_{O_2}(\text{Mt. Everest})}{n_{O_2}(\text{sea level})} = \frac{\dfrac{P_{O_2}(\text{Mt. Everest})V}{RT}}{\dfrac{P_{O_2}(\text{sea level})V}{RT}}$$

$$\frac{n_{O_2}(\text{Mt. Everest})}{n_{O_2}(\text{sea level})} = \frac{P_{O_2}(\text{Mt. Everest})}{P_{O_2}(\text{sea level})} = \frac{210\text{ mmHg}}{760\text{ mmHg}} = 0.276$$

This calculation indicates that there is only 27.6% as much oxygen at the top of Mt. Everest compared to sea level. Therefore, the percent decrease in oxygen gas from sea level to the top of Mt. Everest is **72.4%**.

10.167 The molar volume is the volume of 1 mole of gas under the specified conditions.

$$V = \frac{nRT}{P} = \frac{(1\ \text{mol})\left(0.0821\dfrac{\text{L}\cdot\text{atm}}{\text{mol}\cdot\text{K}}\right)(220\ \text{K})}{\left(6.0\ \text{mmHg} \times \dfrac{1\ \text{atm}}{760\ \text{mmHg}}\right)} = \mathbf{2.3 \times 10^3\ L}$$

10.169 a.

$$CaO(s) + CO_2(g) \longrightarrow CaCO_3(s)$$

$$BaO(s) + CO_2(g) \longrightarrow BaCO_3(s)$$

b. First, we need to find the number of moles of $CO_2$ consumed in the reaction. We can do this by calculating the initial moles of $CO_2$ in the flask and then comparing it to the $CO_2$ remaining after the reaction.

<u>Initially</u>:

$$n_{CO_2} = \frac{PV}{RT} = \frac{\left(746\ \text{mmHg} \times \dfrac{1\ \text{atm}}{760\ \text{mmHg}}\right)(1.46\ \text{L})}{\left(0.0821\dfrac{\text{L}\cdot\text{atm}}{\text{mol}\cdot\text{K}}\right)(35 + 273)\text{K}} = 0.0567\ \text{mol}\ CO_2$$

<u>Remaining</u>:

$$n_{CO_2} = \frac{PV}{RT} = \frac{\left(252\ \text{mmHg} \times \dfrac{1\ \text{atm}}{760\ \text{mmHg}}\right)(1.46\ \text{L})}{\left(0.0821\dfrac{\text{L}\cdot\text{atm}}{\text{mol}\cdot\text{K}}\right)(35 + 273)\text{K}} = 0.0191\ \text{mol}\ CO_2$$

Thus, the amount of $CO_2$ consumed in the reaction is: (0.0567 mol – 0.0191 mol) = 0.0376 mol $CO_2$.

Since the mole ratio between $CO_2$ and both reactants (CaO and BaO) is 1:1, 0.0376 mole of the mixture must have reacted. We can write:

$$\text{mol CaO} + \text{mol BaO} = 0.0376\ \text{mol}$$

Let $x$ = mass of CaO in the mixture, then $(4.88 - x)$ = mass of BaO in the mixture. We can write:

$$\left[x\ \text{g CaO} \times \frac{1\ \text{mol CaO}}{56.08\ \text{g CaO}}\right] + \left[(4.88 - x)\text{g BaO} \times \frac{1\ \text{mol BaO}}{153.3\ \text{g BaO}}\right] = 0.0376\ \text{mol}$$

$$0.01783x - 0.006523x + 0.0318 = 0.0376$$

$$x = 0.513\ \text{g} = \text{mass of CaO in the mixture}$$

$$\text{mass of BaO in the mixture} = 4.88 - x = 4.37 \text{ g}$$

The percent compositions by mass in the mixture are:

$$\textbf{CaO}: \frac{0.513 \text{ g}}{4.88 \text{ g}} \times 100\% = \mathbf{10.5\%}$$

$$\textbf{BaO}: \frac{4.37 \text{ g}}{4.88 \text{ g}} \times 100\% = \mathbf{89.5\%}$$

10.171 First, write the balanced equation.

$$\text{Zn}(s) + 2\text{HCl}(aq) \longrightarrow \text{Zn}^{2+}(aq) + \text{H}_2(g) + 2\text{Cl}^-(aq)$$

Determine the limiting reagent and calculate the moles of hydrogen gas produced.

$$(2.675 \text{ g Zn})\left(\frac{1 \text{ mol Zn}}{65.41 \text{ g Zn}}\right)\left(\frac{1 \text{ mol H}_2}{1 \text{ mol Zn}}\right) = 0.0409 \text{ mol H}_2$$

$$(0.1000 \text{ L HCl solution})\left(\frac{1.75 \text{ mol HCl}}{\text{L solution}}\right)\left(\frac{1 \text{ mol H}_2}{1 \text{ mol HCl}}\right) = 0.0875 \text{ mol H}_2$$

Zn is the limiting reagent, so 0.0409 mol $H_2$ are produced.

Use the Ideal Gas Law to calculate the volume of $H_2$ produced.

$$PV = nRT \quad \text{and} \quad w = -P\Delta V, \text{ so}$$

$$\begin{aligned} w &= -nRT \\ &= -(0.0409 \text{ mol H}_2)\left(8.314 \frac{\text{J}}{\text{mol} \cdot \text{K}}\right)(296.95 \text{ K}) \\ &= -101.0 \text{ J} \end{aligned}$$

Therefore, 101.0 J of work is performed by the system.

# Chapter 11

# Intermolecular Forces and the Physical Properties of Liquids and Solids

11.9 **Strategy:** Classify the species into three categories: ionic, polar (possessing a dipole moment), and nonpolar. Keep in mind that dispersion forces exist between *all* species.

**Solution:** The three molecules are essentially nonpolar. There is little difference in electronegativity between carbon and hydrogen. Thus, the only type of intermolecular attraction in these molecules is dispersion forces. Other factors being equal, the molecule with the greater number of electrons will exert greater intermolecular attractions. By looking at the molecular formulas you can predict that the order of increasing boiling points will be $CH_4 < C_3H_8 < C_4H_{10}$.

Butane would be a liquid in winter (boiling point −44.5°C), and on the coldest days even propane would become a liquid (boiling point −0.5°C). Only **methane** would remain gaseous (boiling point −161.6°C).

11.11 a. Benzene ($C_6H_6$) molecules are nonpolar. Only **dispersion** forces will be present.

b. Chloroform ($CH_3Cl$) molecules are polar (why?). **Dispersion and dipole-dipole** forces will be present.

c. Phosphorus trifluoride ($PF_3$) molecules are polar. **Dispersion and dipole-dipole** forces will be present.

d. Sodium chloride (NaCl) is an ionic compound. **Dispersion and ionic** forces will be present.

e. Carbon disulfide ($CS_2$) molecules are nonpolar. Only **dispersion** forces will be present.

11.13 In this problem you must identify the species capable of hydrogen bonding among themselves, not with water. In order for a molecule to be capable of hydrogen bonding with another molecule like itself, it must have at least one hydrogen atom bonded to N, O, or F. Of the choices, only **(e) $CH_3COOH$** (acetic acid) shows this structural feature. The others cannot form hydrogen bonds among themselves.

11.15 **Strategy:** The molecule with the stronger intermolecular forces will have the higher boiling point. If a molecule contains an N–H, O–H, or F–H bond it can form intermolecular hydrogen bonds. A hydrogen bond is a particularly strong dipole-dipole intermolecular attraction.

**Solution:** 1-butanol has greater intermolecular forces because it can form hydrogen bonds. (It contains an O–H bond.) Therefore, it has the higher boiling point. Diethyl ether molecules do contain both oxygen atoms and hydrogen atoms. However, all the hydrogen atoms are bonded to carbon, not oxygen. There is no hydrogen bonding in diethyl ether, because carbon is not electronegative enough.

11.17 a. Xe: it is larger and therefore stronger dispersion forces.

b. $CS_2$: it is larger (both molecules nonpolar) and therefore stronger dispersion forces.

c. $Cl_2$: it is larger (both molecules nonpolar) and therefore stronger dispersion forces.

d. LiF: it is an ionic compound, and the ion-ion attractions are much stronger than the dispersion forces between $F_2$ molecules.

e. $NH_3$: it can form hydrogen bonds and $PH_3$ cannot.

11.19 **Strategy:** Classify the species into three categories: ionic, polar (possessing a dipole moment), and nonpolar. Also look for molecules that contain an N–H, O–H, or F–H bond, which are capable of forming intermolecular hydrogen bonds. Keep in mind that dispersion forces exist between *all* species.

**Solution:** a. Water has O–H bonds. Therefore, water molecules can form hydrogen bonds. The attractive forces that must be overcome are **dispersion and dipole-dipole, including hydrogen bonding**.

b. Bromine ($Br_2$) molecules are nonpolar. The forces that must be overcome are **dispersion only**.

c. Iodine ($I_2$) molecules are nonpolar. The forces that must be overcome are **dispersion only**.

d. In this case, the F–F bond must be broken. This is an *intra*molecular force between two F atoms, not an *inter*molecular force between $F_2$ molecules. The attractive forces that must be overcome are **covalent bonds**.

11.21 The compound with $-NO_2$ and $-OH$ groups on adjacent carbons can form hydrogen bonds with itself (*intra*molecular hydrogen bonds). Such bonds do not contribute to *inter*molecular attraction and do not help raise the melting point of the compound. The other compound, with the $-NO_2$ and $-OH$ groups on opposite sides of the ring, can form only *inter*molecular hydrogen bonds; therefore it will take a higher temperature to escape into the gas phase.

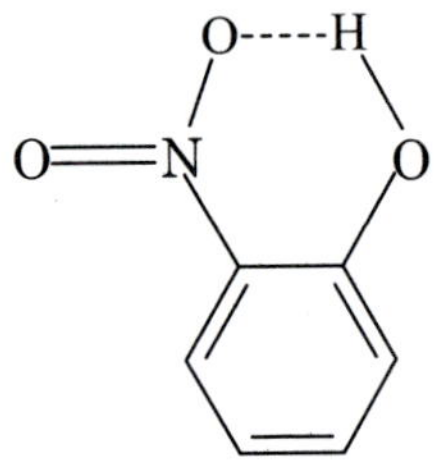

O-----H

O═N O

11.33 Using Equation 11.4 of the text:

$$\ln\frac{P_1}{P_2} = \frac{\Delta H_{vap}}{R}\left(\frac{1}{T_2} - \frac{1}{T_1}\right)$$

$$\ln\left(\frac{1}{2}\right) = \left(\frac{\Delta H_{vap}}{8.314\ \text{J/K}\cdot\text{mol}}\right)\left(\frac{1}{373\ \text{K}} - \frac{1}{348\ \text{K}}\right) = \Delta H_{vap}\left(\frac{-1.93\times 10^{-4}}{8.314\ \text{J/mol}}\right)$$

$$\boldsymbol{\Delta H_{vap}} = 2.99\times 10^{4}\ \text{J/mol} = \mathbf{29.9\ kJ/mol}$$

11.35 Ethylene glycol has two –OH groups, allowing it to exert strong intermolecular forces through hydrogen bonding. Its viscosity should fall between ethanol (1 OH group) and glycerol (3 OH groups).

11.37 Using the Clausius-Clapeyron equation (Equation 11.1), we see that a plot of ln $P$ vs. $1/T$ is a straight line whose slope is proportional to $\Delta H_{vap}$. The relative positions of the points ($1/T$, ln($P$)) for each liquid is shown below. (Note that higher temperature corresponds to smaller values of $1/T$.)

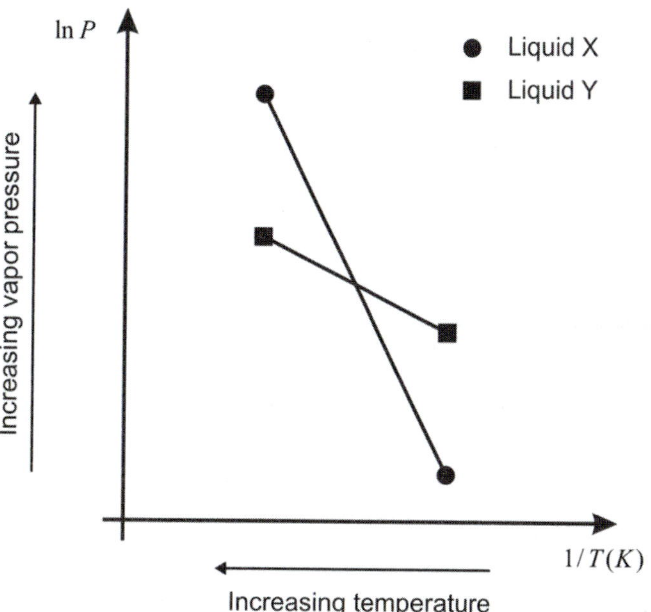


According to the graph, liquid X has a more negative slope. So, we can conclude that liquid X has a larger $\Delta H_{\text{vap}}$ than does liquid Y.

11.45 A corner sphere is shared equally among eight unit cells, so only one-eighth of each corner sphere "belongs" to any one unit cell. A face-centered sphere is divided equally between the two unit cells sharing the face. A body-centered sphere belongs entirely to its own unit cell.

In a *simple cubic cell* there are eight corner spheres. One-eighth of each belongs to the individual cell giving a total of **one sphere** per cell. In a *body-centered cubic cell*, there are eight corner spheres and one body-center sphere giving a total of **two spheres** per unit cell (one from the corners and one from the body-center). In a *face-center cubic cell*, there are eight corner spheres and six face-centered spheres (six faces). The total number would be **four spheres**: one from the corners and three from the faces.

11.47 **Strategy:** First, we need to calculate the volume (in cm$^3$) occupied by 1 mole of Ba atoms. Next, we calculate the volume that a Ba atom occupies. Once we have these two pieces of information, we can multiply them together to end up with the number of Ba atoms per mole of Ba.

$$\frac{\text{number of Ba atoms}}{\text{cm}^3} \times \frac{\text{cm}^3}{1 \text{ mol Ba}} = \frac{\text{number of Ba atoms}}{1 \text{ mol Ba}}$$

**Solution:** The volume that contains one mole of barium atoms can be calculated from the density using the following strategy:

$$\frac{\text{volume}}{\text{mass of Ba}} \rightarrow \frac{\text{volume}}{\text{mol Ba}}$$

$$\frac{1\text{ cm}^3}{3.50\text{ g Ba}} \times \frac{137.3\text{ g Ba}}{1\text{ mol Ba}} = \frac{39.23\text{ cm}^3}{1\text{ mol Ba}}$$

We carry an extra significant figure in this calculation to limit rounding errors. Next, the volume that contains two barium atoms is the volume of the body-centered cubic unit cell. Some of this volume is empty space because packing is only 68.0 percent efficient. But, this will not affect our calculation.

$$V = a^3$$

Let's also convert to $\text{cm}^3$.

$$V = (502\text{ pm})^3 \times \left(\frac{1\times 10^{-12}\text{ m}}{1\text{ pm}}\right)^3 \times \left(\frac{1\text{ cm}}{0.01\text{ m}}\right)^3 = \frac{1.265\times 10^{-22}\text{ cm}^3}{2\text{ Ba atoms}}$$

We can now calculate the number of barium atoms in one mole using the strategy presented above.

$$\frac{\text{number of Ba atoms}}{\text{cm}^3} \times \frac{\text{cm}^3}{1\text{ mol Ba}} = \frac{\text{number of Ba atoms}}{1\text{ mol Ba}}$$

$$\frac{2\text{ Ba atoms}}{1.265\times 10^{-22}\text{ cm}^3} \times \frac{39.23\text{ cm}^3}{1\text{ mol Ba}} = \mathbf{6.20\times 10^{23}\text{ atoms/mol}}$$

This is close to Avogadro's number, $6.022 \times 10^{23}$ particles/mol.

11.49 The mass of the unit cell is the mass in grams of two europium atoms.

$$m = \frac{2\text{ Eu atoms}}{1\text{ unit cell}} \times \frac{1\text{ mol Eu}}{6.022\times 10^{23}\text{ Eu atoms}} \times \frac{152.0\text{ g Eu}}{1\text{ mol Eu}} = 5.048\times 10^{-22}\text{ g Eu/unit cell}$$

$$V = \frac{5.048\times 10^{-22}\text{ g}}{1\text{ unit cell}} \times \frac{1\text{ cm}^3}{5.26\text{ g}} = 9.60\times 10^{-23}\text{ cm}^3\text{/unit cell}$$

The edge length ($a$) is:

$$\boldsymbol{a} = V^{1/3} = (9.60\times 10^{-23}\text{ cm}^3)^{1/3} = 4.58\times 10^{-8}\text{ cm} = \mathbf{458\text{ pm}}$$

11.51 **Strategy:** Recall that a corner atom is shared with 8 unit cells and therefore only 1/8 of corner atom is within a given unit cell. Also recall that a face atom is shared with 2 unit cells and therefore 1/2 of a face atom is within a given unit cell. See Figure 11.18 of the text.

**Solution:** In a face-centered cubic unit cell, there are atoms at each of the eight corners, and there is one atom in each of the six faces. Only one-half of each face-centered atom and one-eighth of each corner atom belongs to the unit cell.

$$\text{X atoms/unit cell} = (8 \text{ corner atoms})(1/8 \text{ atom per corner}) = 1 \text{ X atom/unit cell}$$

$$\text{Y atoms/unit cell} = (6 \text{ face-centered atoms})(1/2 \text{ atom per face}) = 3 \text{ Y atoms/unit cell}$$

The unit cell is the smallest repeating unit in the crystal; therefore, the empirical formula is $\mathbf{XY_3}$.

11.53 Rearranging the Equation 11.5, we have:

$$\lambda = \frac{2d\sin\theta}{n} = \frac{2(282 \text{ pm})(\sin 23.0^\circ)}{1} = 220 \text{ pm} = \mathbf{0.220 \text{ nm}}$$

11.55 The cell is face-centered cubic, determined by the positions of $O^{2-}$ ions, so there are four $O^{2-}$ ions. There are also four $Zn^{2+}$ ions. Therefore, the formula of zinc oxide is **ZnO**.

11.59 See Table 11.4 of the text. The properties listed are those of a **molecular solid**.

11.61 In a molecular crystal the lattice points are occupied by molecules. Of the solids listed, the ones that are composed of molecules are $\mathbf{Se_8}$, **HBr**, $\mathbf{CO_2}$, $\mathbf{P_4O_6}$, and $\mathbf{SiH_4}$. In covalent crystals, atoms are held together in an extensive three-dimensional network entirely by covalent bonds. Of the solids listed, the ones that are composed of atoms held together by covalent bonds are **Si** and **C**.

11.63 Diamond: each carbon atom is covalently bonded to four other carbon atoms. Because these bonds are strong and uniform, diamond is a very hard substance. Graphite: the carbon atoms in each layer are linked by strong bonds, but the layers are bound by weak dispersion forces. As a result, graphite may be cleaved easily between layers and is not hard.

In graphite, all atoms are $sp^2$ hybridized; each atom is covalently bonded to three other atoms. The remaining unhybridized $2p$ orbital is used in pi bonding forming a delocalized molecular orbital. The electrons are free to move around in this extensively delocalized molecular orbital making graphite a good conductor of electricity in directions along the planes of carbon atoms.

11.85 *Step 1:* Warming ice to the melting point.

$$q_1 = ms\Delta T = (866 \text{ g } H_2O)(2.03 \text{ J/g°C})[0 - (-15)°C] = 26.4 \text{ kJ}$$

*Step 2:* Converting ice at the melting point to liquid water at 0°C. (See Table 11.8 of the text for the heat of fusion of water.)

$$q_2 = 866 \text{ g } H_2O \times \frac{1 \text{ mol}}{18.02 \text{ g } H_2O} \times \frac{6.01 \text{ kJ}}{1 \text{ mol}} = 289 \text{ kJ}$$

*Step 3:* Heating water from 0°C to 100°C.

$$q_3 = ms\Delta T = (866 \text{ g } H_2O)(4.184 \text{ J/g°C})[(100 - 0)\text{°C}] = 362 \text{ kJ}$$

*Step 4:* Converting water at 100°C to steam at 100°C. (See Table 11.6 of the text for the heat of vaporization of water.)

$$q_4 = 866 \text{ g } H_2O \times \frac{1 \text{ mol}}{18.02 \text{ g } H_2O} \times \frac{40.79 \text{ kJ}}{1 \text{ mol}} = 1.96 \times 10^3 \text{ kJ}$$

*Step 5:* Heating steam from 100°C to 146°C.

$$q_5 = ms\Delta T = (866 \text{ g } H_2O)(1.99 \text{ J/g°C})[(146 - 100)\text{°C}] = 79.3 \text{ kJ}$$

$$q_{\text{total}} = q_1 + q_2 + q_3 + q_4 + q_5 = \mathbf{2.72 \times 10^3 \text{ kJ}}$$

**Think About It:** How would you set up and work this problem if you were computing the heat lost in cooling steam from 126°C to ice at −10°C?

11.87 a. Other factors being equal, liquids evaporate faster at higher temperatures.

b. The greater the surface area, the greater the rate of evaporation.

c. Weak intermolecular forces imply a high vapor pressure and rapid evaporation.

11.89 Two phase changes occur in this process. First, the liquid is turned to solid (freezing), then the solid ice is turned to gas (sublimation).

11.91 When steam condenses to liquid water at 100°C, it releases a large amount of heat equal to the enthalpy of vaporization. Thus steam at 100°C exposes one to more heat than an equal amount of water at 100°C.

11.95 Initially, the ice melts because of the increase in pressure. As the wire sinks into the ice, the water above the wire refreezes. Eventually the wire actually moves completely through the ice block without cutting it in half.

11.97 Region labels: The region containing point A is the solid region. The region containing point B is the liquid region. The region containing point C is the gas region.

a. Ice would melt. (If heating continues, the liquid water would eventually boil and become a vapor.)

b. Liquid water would vaporize.

c. Water vapor would solidify without becoming a liquid.

11.99 a. A low surface tension means the attraction between molecules making up the surface is weak. Water has a high surface tension; water bugs could not "walk" on the surface of a liquid with a low surface tension.

b. A low critical temperature means a gas is very difficult to liquefy by cooling. This is the result of weak intermolecular attractions. Helium has the lowest known critical temperature (5.3 K).

c. A low boiling point means weak intermolecular attractions. It takes little energy to separate the particles. All ionic compounds have extremely high boiling points.

d. A low vapor pressure means it is difficult to remove molecules from the liquid phase because of high intermolecular attractions. Substances with low vapor pressures have high boiling points (why?).

Thus, only choice **(d)** indicates strong intermolecular forces in a liquid. The other choices indicate weak intermolecular forces in a liquid.

11.101 The properties of hardness, high melting point, poor conductivity, and so on, could place boron in either the ionic or covalent categories. However, boron atoms will not alternately form positive and negative ions to achieve an ionic crystal. The structure is **covalent** because the units are single boron atoms.

11.103 **$CCl_4$**. Generally, the larger the molecule greater its polarizability. Recall that polarizability is the ease with which the electron distribution in an atom or molecule can be distorted.

11.105 **Strategy:** Use the Bragg equation to solve for the angle of diffraction.

**Setup:** The Bragg equation is:

$$2d\sin\theta = n\lambda$$

Convert the distance between the layers to nanometers:

$$0.154 \text{ nm} = 154 \text{ pm}$$

**Solution:** Rearrange the Bragg equation to solve for the angle.

$$2d\sin\theta = n\lambda$$

$$\sin\theta = \frac{n\lambda}{2d}$$

$$\theta = \sin^{-1}\left(\frac{n\lambda}{2d}\right)$$

$$= \sin^{-1}\left(\frac{(1)(154\text{ pm})}{(2)(188\text{ pm})}\right)$$

$$= 24.2°$$

11.107 The vapor pressure of mercury (as well as all other substances) is 760 mmHg at its normal boiling point.

11.109 It has reached the critical point; the point of critical temperature ($T_c$) and critical pressure ($P_c$).

11.111 **Crystalline $SiO_2$**. Its regular structure results in a more efficient packing.

11.113 a. **False**. Permanent dipoles are usually much stronger than temporary dipoles.

b. **False**. The hydrogen atom must be bonded to N, O, or F.

c. **True**.

11.115 **Strategy:** Use the Bragg equation to solve for *d*.

**Setup:** $\Theta = 19.3°$, $n = 1$, and $\lambda = 154$ pm.

**Solution:**

$$d = \frac{n\lambda}{2\sin\theta} = \frac{154\text{ pm}}{2\sin 19.3°} = 233\text{ pm}$$

11.117 a. **$K_2S$**: Ionic forces are much stronger than the dipole-dipole forces in $(CH_3)_3N$.

b. **$Br_2$**: Both molecules are nonpolar; but $Br_2$ has a larger mass. (The boiling point of $Br_2$ is 50°C and that of $C_4H_{10}$ is –0.5°C.)

11.119 $CH_4$ is a tetrahedral, nonpolar molecule that can only exert weak dispersion type attractive forces. $SO_2$ is bent (why?) and possesses a dipole moment, which gives rise to stronger dipole-dipole attractions. **$SO_2$** will behave less ideally because it is polar and has greater intermolecular forces.

11.121 The standard enthalpy change for the formation of gaseous iodine from solid iodine is simply the difference between the standard enthalpies of formation of the products and the reactants in the equation:

$$I_2(s) \rightarrow I_2(g)$$

$$\Delta H_{\text{vap}} = \Delta H_f^\circ[I_2(g)] - \Delta H_f^\circ[I_2(s)] = 62.4 \text{ kJ/mol} - 0 \text{ kJ/mol} = \mathbf{62.4\ kJ/mol}$$

11.123 Smaller ions can approach water molecules more closely, resulting in larger ion-dipole interactions. The greater the ion-dipole interaction, the larger is the heat of hydration.

11.125 a. Using data from Appendix 2, for the process: $Br_2(l) \rightarrow Br_2(g)$

$$\boldsymbol{\Delta H^\circ} = \Delta H_f^\circ[Br_2(g)] - \Delta H_f^\circ[Br_2(l)] = (1)(30.7 \text{ kJ/mol}) - 0 = \mathbf{30.7\ kJ/mol}$$

b. Using data for the process: $Br_2(g) \rightarrow 2Br(g)$

$$\Delta H^\circ = \mathbf{192.5\ kJ/mol}$$

As expected, the bond enthalpy represented in part (b) is much greater than the energy of vaporization represented in part (a). It requires more energy to break the bond than to vaporize the molecule.

11.127 a. Decreases b. No change c. No change

11.129

$$CaCO_3(s) \rightarrow CaO(s) + CO_2(g)$$

Initial state: one solid phase, final state: two solid phase components and one gas phase component. $CaCO_3$ and CaO constitute two separate solid phases because they are separated by well-defined boundaries.

11.131 a. Pumping allows Ar atoms to escape, thus removing heat from the liquid phase. Eventually the liquid freezes.

b. The slope of the solid-liquid line of cyclohexane is positive. Therefore, its melting point increases with pressure.

c. These droplets are super-cooled liquids.

d. When the dry ice is added to water, it sublimes. The cold $CO_2$ gas generated causes nearby water vapor to condense, hence the appearance of fog.

11.133 The time required to cook food depends on the boiling point of the water in which it is cooked. The boiling point of water increases when the pressure inside the cooker increases.

11.135 a. Extra heat produced when steam condenses at 100°C.

b. Avoids extraction of ingredients by boiling in water.

11.137 The fuel source for the Bunsen burner is most likely methane gas. When methane burns in air, carbon dioxide and water are produced.

$$CH_4(g) + 2O_2(g) \rightarrow CO_2(g) + 2H_2O(g)$$

The water vapor produced during the combustion condenses to liquid water when it comes in contact with the outside of the cold beaker.

11.139 First, we need to calculate the volume (in $cm^3$) occupied by 1 mole of Fe atoms. Next, we calculate the volume that a Fe atom occupies. Once we have these two pieces of information, we can multiply them together to end up with the number of Fe atoms per mole of Fe.

$$\frac{\text{number of Fe atoms}}{\text{cm}^3} \times \frac{\text{cm}^3}{\text{1 mol Fe}} = \frac{\text{number of Fe atoms}}{\text{1 mol Fe}}$$

The volume that contains one mole of iron atoms can be calculated from the density using the following strategy:

$$\frac{\text{volume}}{\text{mass of Fe}} \rightarrow \frac{\text{volume}}{\text{mol Fe}}$$

$$\frac{1\text{ cm}^3}{7.874\text{ g Fe}} \times \frac{55.85\text{ g Fe}}{1\text{ mol Fe}} = \frac{7.093\text{ cm}^3}{1\text{ mol Fe}}$$

Next, the volume that contains two iron atoms is the volume of the body-centered cubic unit cell. Some of this volume is empty space because packing is only 68.0 percent efficient. But, this will not affect our calculation.

$$V = a^3$$

Let's also convert to $cm^3$.

$$V = (286.7\ \text{pm})^3 \times \left(\frac{1 \times 10^{-12}\ \text{m}}{1\ \text{pm}}\right)^3 \times \left(\frac{1\ \text{cm}}{0.01\ \text{m}}\right)^3 = \frac{2.357 \times 10^{-23}\ \text{cm}^3}{2\ \text{Fe atoms}}$$

We can now calculate the number of iron atoms in one mole using the strategy presented above.

$$\frac{\text{number of Fe atoms}}{\text{cm}^3} \times \frac{\text{cm}^3}{1\ \text{mol Fe}} = \frac{\text{number of Fe atoms}}{1\ \text{mol Fe}}$$

$$\frac{2\ \text{Fe atoms}}{2.357 \times 10^{-23}\ \text{cm}^3} \times \frac{7.093\ \text{cm}^3}{1\ \text{mol Ba}} = \mathbf{6.019 \times 10^{23}\ Fe\ atoms/mol}$$

The small difference between the above number and $6.022 \times 10^{23}$ is the result of rounding off and using rounded values for density and other constants.

11.141 If we know the values of $\Delta H_{vap}$ and $P$ of a liquid at one temperature, we can use the Clausius-Clapeyron equation, Equation 11.4 of the text, to calculate the vapor pressure at a different temperature. At 65.0°C, we can calculate $\Delta H_{vap}$ of methanol. Because this is the boiling point, the vapor pressure will be 1 atm (760 mmHg).

First, we calculate $\Delta H_{vap}$. From Appendix 2 of the text, $\Delta H_f^\circ[CH_3OH(l)] = -238.7$ kJ/mol

$$CH_3OH(l) \rightarrow CH_3OH(g)$$

$$\Delta H_{vap} = \Delta H_f^\circ[CH_3OH(g)] - \Delta H_f^\circ[CH_3OH(l)]$$

$$\Delta H_{vap} = -201.2\ \text{kJ/mol} - (-238.7\ \text{kJ/mol}) = 37.5\ \text{kJ/mol}$$

Next, we substitute into Equation 11.4 of the text to solve for the vapor pressure of methanol at 25°C.

$$\ln\frac{P_1}{P_2} = \frac{\Delta H_{vap}}{R}\left(\frac{1}{T_2} - \frac{1}{T_1}\right)$$

$$\ln\frac{P_1}{760} = \frac{37.5 \times 10^3\ \text{J/mol}}{8.314\ \text{J/mol}\cdot\text{K}}\left(\frac{1}{338\ \text{K}} - \frac{1}{298\ \text{K}}\right)$$

$$\ln\frac{P_1}{760} = -1.79$$

Taking the antiln of both sides gives: $\mathbf{P_1 = 127\ mmHg}$

11.143 If half the water remains in the liquid phase, there is 1.0 g of water vapor. We can derive a relationship between vapor pressure and temperature using the ideal gas equation.

$$P = \frac{nRT}{V} = \frac{\left(1.0\text{ g H}_2\text{O} \times \dfrac{1\text{ mol H}_2\text{O}}{18.02\text{ g H}_2\text{O}}\right)\left(0.0821\dfrac{\text{L}\cdot\text{atm}}{\text{mol}\cdot\text{K}}\right)T}{9.6\text{ L}} = (4.75\times 10^{-4})T\text{ atm}$$

Converting to units of mmHg:

$$(4.75\times 10^{-4})T\text{ atm} \times \frac{760\text{ mmHg}}{1\text{ atm}} = 0.36T\text{ mmHg}$$

To determine the temperature at which only half the water remains, we set up the following table and refer to Table 11.5 of the text. The calculated value of vapor pressure that most closely matches the vapor pressure in Table 11.5 would indicate the approximate value of the temperature.

| T($K$) | $P_{H_2O}$ mmHg(from Table11.5) | (0.36 $T$) mmHg |
|---|---|---|
| 313 | 55.3 | 112.7 |
| 318 | 71.9 | 114.5 |
| 323 | 92.5 | 116.3 |
| 328 | 118.0 | 118.1 (closest match) |
| 333 | 149.4 | 119.9 |
| 338 | 187.5 | 121.7 |

Therefore, the temperature is about 328 K = **55°C** at which half the water has vaporized.

11.145 Use equation 10.10 to compute the density of the ideal gas:

$$d = \frac{P\mathcal{M}}{RT} = \frac{(101{,}325\text{ Pa})(20.008\times10^{-3}\text{ kg/mol})}{(8.314\text{ J/mol}\cdot\text{K})(273.15+19.5)(\text{K})} = 0.833\text{ kg/m}^3 = 0.833\text{ g/L}$$

This density is nearly 4 times smaller than the experimental value of 3.10 g/L. **The hydrogen-bonding interactions in HF are relatively strong, and since the ideal gas law ignores intermolecular forces, it underestimates significantly the density of HF gas near its boiling point.**

11.147 Of the three compounds, only fluoromethane has a permanent dipole moment. Assuming that the dispersion forces are about the same in all three substances, we predict that **fluoromethane** will have the highest boiling point.

11.149 a. Two triple points: Diamond/graphite/liquid and graphite/liquid/vapor.

b. Diamond.

c. Apply high pressure at high temperature.

11.151 a. ~2.3 K.

b. ~10 atm.

c. ~5 K.

d. No. There is no solid-vapor phase boundary.

11.153 Ethanol mixes well with water. The mixture has a lower surface tension and readily flows out of the ear channel.

11.155 Ratio of separate strands to hydrogen-bonded double helix = $e^{-\Delta E/RT}$

$$\Delta E = 100 \text{ base pairs} \times 10 \text{ kJ/mol}\cdot\text{base pair} = 1000 \text{ kJ/mol}$$

$$-(1000 \text{ kJ/mol})/[(8.314\times 10^{-3} \text{ kJ/mol}\cdot\text{K})(300 \text{ K})] = e$$

$$\textbf{Ratio} = e^{-401} \approx 0$$

There are essentially no separate strands in solution at 300 K.

11.157 The molecules are all polar. The F atoms can form H-bonds with water and other –OH and –NH groups in the membrane, so water solubility plus easy attachment to the membrane would allow these molecules to pass the blood-brain barrier.

11.159 The two main reasons for spraying the trees with water are:

1) When water freezes it releases heat, helping keep the fruit warm enough not to freeze.

$$H_2O(l) \rightarrow H_2O(s) \qquad -\Delta H_{\text{fus}} = -6.01 \text{ kJ/mol}$$

2) A layer of ice is a thermal insulator.

# Chapter 12

# Modern Materials

12.3 The monomer must have a triple bond.

12.5 There are two possible isomers, depending on whether the two H atoms in the double bond are cis- or trans-.

```
 [ H    C2H5  H       ]          [ H    C2H5             ]
   |    |     |                    |    |
 --C----C-----C==C--               --C----C------C==C--
   |    |        |                 |    |      |   |
   CH3  CH3      H    ]n           CH3  CH3    H   H     ]n
```

12.9 (1) Produce the alkoxide:

$Sc(s) + 2C_2H_5OH(l) \rightarrow Sc(OC_2H_5)(alc) + 2H^+(alc)$ ("alc" indicates a solution in alcohol)

(2) Hydrolyze to produce hydroxide pellets:

$$Sc(OC_2H_5)(alc) + 2H_2O(l) \rightarrow Sc(OH)_2(s) + 2C_2H_5OH(alc)$$

(3) Sinter pellets to produce ceramic: $Sc(OH)_2(s) \rightarrow ScO(s) + 2H_2O(g)$

12.11 Bakelite is best described as a themosetting composite polymer.

12.15 No. These polymers are too flexible, and liquid crystals require long, relatively rigid molecules.

12.19 As shown, it is an alternating condensation copolymer of the polyester class.

12.21 Metal amalgams expand with age; composite fillings tend to shrink.

12.25 Each carbon atom is attached to three other carbon atoms in a (near) plane. Thus, $sp^2$ hybrids are used.

12.27 Each carbon atom is attached to three other carbon atoms in a (near) plane. Thus, $sp^2$ hybrids are used.

12.31 a. 4 + 5: n-type (one more $e^-$)

b. 4 + 3: p-type (one less $e^-$)

12.35 BSCCO-2212 is $Bi_2Sr_2CaCu_2O_8$, composed of the ions $Bi^{3+}$, $Sr^{2+}$, $Ca^{2+}$, $Cu^{2+}$ and $O^{2-}$

BSCCO-2201 would therefore be $Bi_2Sr_2Ca_0Cu_1O_6$ or $\mathbf{Bi_2Sr_2CuO_6}$

12.37 In a plastic (organic) polymer, there are covalent bonds, disulfide (covalent) bonds, H-bonds and dispersion forces. In ceramics, there are mostly ionic and network covalent bonds.

12.39 Two are +2 ([Ar] $3d^9$), one is +3 ([Ar] $3d^8$). The +3 oxidation state is unusual for copper.

12.41 It is amphoteric, since it reacts with both acid and base (OR = alkoxide)

$$M(OR)_n + nH^+(aq) \rightarrow M^{n+}(aq) + nHOR$$

$$M(OR)_n + nOH^-(aq) \rightarrow M(OH)_n + nOR^-$$

12.43 The green light has a shorter wavelength (higher energy) than red, so the LED in the exit sign has the greater band gap.

12.45 Fluoroapatite is less soluble than hydroxyapatite, particularly in acidic solutions. Dental fillings must also be insoluble.

# Chapter 13:

# Physical Properties of Solutions

13.9 In predicting solubility, remember the saying: "**Like dissolves like.**" A nonpolar solute will dissolve in a nonpolar solvent; ionic compounds will generally dissolve in polar solvents due to favorable ion-dipole interactions; solutes that can form hydrogen bonds with a solvent will also have high solubility in the solvent. **Naphthalene and benzene are nonpolar, whereas CsF is ionic.**

13.11 The order of increasing solubility is: $\mathbf{O_2 < Br_2 < LiCl < CH_3OH}$. Methanol is miscible with water because of strong hydrogen bonding. LiCl is an ionic solid and is very soluble because of the high polarity of the water molecules. Both oxygen and bromine are nonpolar and exert only weak dispersion forces. Bromine is a larger molecule and is therefore more polarizable and susceptible to dipole–induced dipole attractions.

13.15 **Strategy:** We are given the mass of the solute and either the mass of the solution, or the mass of the solvent. Percent by mass can be determined using Equation 13.2.

**Solution:** a. $\dfrac{5.75\text{ g NaBr}}{67.9\text{ g soln}} \times 100\% = \mathbf{8.47\%}$

b. $\dfrac{24.6\text{ g KCl}}{(24.6+114)\text{ g soln}} \times 100\% = \mathbf{17.7\%}$

c. $\dfrac{4.8\text{ g toluene}}{(4.8+39)\text{ g soln}} \times 100\% = \mathbf{11\%}$

13.17 a. The molality is the number of moles of sucrose (molar mass 342.3 g/mol) divided by the mass of the solvent (water) in kg.

$$\text{mol sucrose} = 14.3\text{ g sucrose} \times \frac{1\text{ mol}}{342.3\text{ g sucrose}} = 0.0418\text{ mol}$$

$$\textbf{molality} = \frac{0.0418\text{ mol sucrose}}{0.685\text{ kg H}_2\text{O}} = \mathbf{0.0610}\ \boldsymbol{m}$$

b.
$$\textbf{molality} = \frac{7.15 \text{ mol ethylene glycol}}{3.505 \text{ kg } H_2O} = \mathbf{2.04}\ \boldsymbol{m}$$

13.19 In each case we consider one liter of solution.

$$\text{mass of solution} = \text{volume} \times \text{density}$$

a.
$$\text{mass of sugar} = 1.22 \text{ mol sugar} \times \frac{342.3 \text{ g sugar}}{1 \text{ mol sugar}} = 418 \text{ g sugar} \times \frac{1 \text{ kg}}{1000 \text{ g}} = 0.418 \text{ kg sugar}$$

$$\text{mass of soln} = 1000 \text{ mL} \times \frac{1.12 \text{ g}}{1 \text{ mL}} = 1120 \text{ g} \times \frac{1 \text{ kg}}{1000 \text{ g}} = 1.120 \text{ kg}$$

$$\textbf{molality} = \frac{1.22 \text{ mol sugar}}{(1.120 - 0.418) \text{ kg } H_2O} = \mathbf{1.7}\ \boldsymbol{m}$$

b.
$$\text{mass of NaOH} = 0.87 \text{ mol NaOH} \times \frac{40.00 \text{ g NaOH}}{1 \text{ mol NaOH}} = 35 \text{ g NaOH}$$

$$\text{mass solvent } (H_2O) = 1040 \text{ g} - 35 \text{ g} = 1005 \text{ g} = 1.005 \text{ kg}$$

$$\textbf{molality} = \frac{0.87 \text{ mol NaOH}}{1.005 \text{ kg } H_2O} = \mathbf{0.87}\ \boldsymbol{m}$$

c.
$$\text{mass of } NaHCO_3 = 5.24 \text{ mol } NaHCO_3 \times \frac{84.01 \text{ g } NaHCO_3}{1 \text{ mol } NaHCO_3} = 440 \text{ g } NaHCO_3$$

$$\text{mass solvent } (H_2O) = 1190 \text{ g} - 440 \text{ g} = 750 \text{ g} = 0.75 \text{ kg}$$

$$\textbf{molality} = \frac{5.24 \text{ mol } NaHCO_3}{0.75 \text{ kg } H_2O} = \mathbf{7.0}\ \boldsymbol{m}$$

13.21 We find the volume of ethanol in 1.00 L of 75 proof gin. Note that 75 proof means $\left(\frac{75}{2}\right)\%$.

$$\text{Volume} = 1.00 \text{ L} \times \left(\frac{75}{2}\right)\% = 0.38 \text{ L} = 3.8 \times 10^2 \text{ mL}$$

$$\textbf{Ethanol mass} = (3.8 \times 10^2 \text{ mL}) \times \frac{0.798 \text{ g}}{1 \text{ mL}} = \mathbf{3.0 \times 10^2\ g}$$

13.23 **Strategy:** In this problem, the masses of both solute and solvent are given, and the density is given.

**Solution:** In order to determine molarity, we need to know the volume of the solution. We add the masses of solute and solvent to get the total mass, and use the density to determine the volume.

$$35.0\text{ g }NH_3 + 75.0\text{ g }H_2O = 110.0\text{ g soln}$$

$$\text{Volume of soln} = 110.0\text{ g soln} \times \frac{1\text{ mL soln}}{0.982\text{ g soln}} = 112\text{ mL soln}$$

We then convert the mass of $NH_3$ to moles, and divide by the volume in liters to get molarity.

$$35.0\text{ g }NH_3 \times \frac{1\text{ mol }NH_3}{17.03\text{ g }NH_3} = 2.055\text{ mol }NH_3$$

$$\textbf{Molarity} = \frac{2.055\text{ mol }NH_3}{0.112\text{ L soln}} = \mathbf{18.3}\ \boldsymbol{M}$$

To calculate molality, we use Equation 13.1.

$$\textbf{Molality} = \frac{2.055\text{ mol }NH_3}{0.0750\text{ kg }H_2O} = \mathbf{27.4}\ \boldsymbol{m}$$

**Think About It:** Remember in these calculations to express volume of solution in L and mass of solvent in kg.

13.25 Assuming that $N_2$ and $O_2$ are the only dissolved gases, the mole fractions in the total dissolved gas can be determined as follows:

We multiply each partial pressure by the corresponding Henry's law constant (solubility).

For $O_2$, we have

$$0.20\text{ atm} \times 1.3 \times 10^{-3}\text{ mol/L}\cdot\text{atm} = 2.6 \times 10^{-4}\text{ mol/L}$$

For $N_2$, we have

$$0.80\text{ atm} \times 6.8 \times 10^{-4}\text{ mol/L}\cdot\text{atm} = 5.44 \times 10^{-4}\text{ mol/L}$$

In a liter of water, then, there will be $2.6 \times 10^{-4}$ mol $O_2$ and $5.44 \times 10^{-4}$ mol $N_2$. The mole fractions are

$$\chi(N_2) = \frac{5.44\times10^{-4}\text{ mol }N_2}{(2.6\times10^{-4} + 5.44\times10^{-4})\text{mol}} = \mathbf{0.677}$$

$$\chi(\mathbf{O_2}) = \frac{2.6\times10^{-4}\ \text{mol}\ O_2}{(2.6\times10^{-4} + 5.44\times10^{-4})\text{mol}} = \mathbf{0.323}$$

Due to the greater solubility of oxygen, it has a larger mole fraction in solution than it does in the air.

13.33 At 75°C, 155 g of $KNO_3$ dissolves in 100 g of water to form 255 g of solution. When cooled to 25°C, only 38.0 g of $KNO_3$ remain dissolved. This means that (155 – 38.0) g = 117 g of $KNO_3$ will crystallize. The amount of $KNO_3$ formed when 100 g of saturated solution at 75°C is cooled to 25°C can be found by a simple unit conversion.

$$100\ \text{g saturated soln} \times \frac{117\ \text{g}\ KNO_3\ \text{crystallized}}{255\ \text{g saturated soln}} = \mathbf{45.9\ g\ KNO_3}$$

13.35 **Strategy:** The given solubility allows us to calculate Henry's law constant (*k*), which can then be used to determine the concentration of $CO_2$ at 0.0003 atm.

**Solution:** First, calculate the Henry's law constant, *k*, using the concentration of $CO_2$ in water at 1 atm.

$$k = \frac{c}{P} = \frac{0.034\ \text{mol/L}}{1\ \text{atm}} = 0.034\ \text{mol/L}\cdot\text{atm}$$

For atmospheric conditions we write:

$$c = kP = (0.034\ \text{mol/L}\cdot\text{atm})(0.00030\ \text{atm}) = \mathbf{1.0\times10^{-5}\ mol/L}$$

13.37 According to Henry's law, the solubility of a gas in a liquid increases as the pressure increases ($c = kP$). The soft drink tastes flat at the bottom of the mine because the carbon dioxide pressure is greater and the dissolved gas is not released from the solution. As the miner goes up in the elevator, the atmospheric carbon dioxide pressure decreases and dissolved gas is released from his stomach.

13.39 **Strategy:** The mass of $CO_2$ liberated from the container is given by the difference of the two mass measurements. Assume that, at the time of the second mass measurement, all the $CO_2$ has been released and that the amount of $CO_2$ in the vapor phase of the unopened soda is negligible. Then, we can use the following conversions to calculate the pressure of $CO_2$ in the unopened soda.

$$\text{mass of}\ CO_2\ \text{lost} \rightarrow \text{moles of}\ CO_2\ \text{in solution} \rightarrow \text{pressure of}\ CO_2\ \text{above solution}$$

**Solution:** Mass of $CO_2$ lost = 853.5 g – 851.3 g = 2.2 g $CO_2$

$$\text{moles } CO_2 = (2.2 \text{ g } CO_2)\left(\frac{1 \text{ mol } CO_2}{44.01 \text{ g } CO_2}\right) = 0.050 \text{ mol } CO_2$$

Using Henry's law for the final conversion:

$$c = kP$$

$$\frac{0.050 \text{ mol } CO_2}{0.4524 \text{ L}} = (3.4\times10^{-2} \text{ mol/L}\cdot\text{atm})P$$

$$\boldsymbol{P = 3.3 \text{ atm}}$$

**This pressure is only an estimate since we ignored the amount of $CO_2$ that was present in the unopened container in the gas phase.**

13.41 The dissolution of the red solute is exothermic, because increasing the temperature causes it to precipitate.

The dissolution of the green solute is endothermic, because increasing the temperature causes more solute to dissolve.

The numerical value of $\Delta H_{soln}$ is greater for the red solute, since changing the temperature produces a greater difference in solubility.

13.57 The first step is to find the number of moles of sucrose and of water.

$$\text{Moles sucrose} = 396 \text{ g} \times \frac{1 \text{ mol}}{342.3 \text{ g}} = 1.16 \text{ mol sucrose}$$

$$\text{Moles water} = 624 \text{ g} \times \frac{1 \text{ mol}}{18.02 \text{ g}} = 34.6 \text{ mol water}$$

The mole fraction of water is:

$$X_{H_2O} = \frac{34.6 \text{ mol}}{34.6 \text{ mol} + 1.16 \text{ mol}} = 0.968$$

The vapor pressure of the solution is found as follows:

$$\boldsymbol{P_{solution}} = X_{H_2O} \times P^\circ_{H_2O} = (0.968)(31.8 \text{ mmHg}) = \mathbf{30.8 \text{ mmHg}}$$

13.59 Let us call benzene component 1 and camphor component 2.

$$P_1 = X_1 P_1^\circ = \left(\frac{n_1}{n_1 + n_2}\right) P_1^\circ$$

$$n_1 = 98.5 \text{ g benzene} \times \frac{1 \text{ mol}}{78.11 \text{ g}} = 1.26 \text{ mol benzene}$$

$$n_2 = 24.6 \text{ g camphor} \times \frac{1 \text{ mol}}{152.2 \text{ g}} = 0.162 \text{ mol camphor}$$

$$\boldsymbol{P_1} = \frac{1.26 \text{ mol}}{(1.26 + 0.162) \text{ mol}} \times 100.0 \text{ mmHg} = \mathbf{88.6\ mmHg}$$

13.61

$$\Delta P = X_{\text{urea}} P^{\circ}_{\text{water}}$$

$$2.50 \text{ mmHg} = X_{\text{urea}}(31.8 \text{ mmHg})$$

$$X_{\text{urea}} = 0.0786$$

The number of moles of water is:

$$n_{\text{water}} = 658 \text{ g } H_2O \times \frac{1 \text{ mol } H_2O}{18.02 \text{ g } H_2O} = 36.51 \text{ mol } H_2O$$

$$X_{\text{urea}} = \frac{n_{\text{urea}}}{n_{\text{water}} + n_{\text{urea}}}$$

$$0.0786 = \frac{n_{\text{urea}}}{36.51 + n_{\text{urea}}}$$

$$n_{\text{urea}} = 3.11 \text{ mol}$$

$$\textbf{mass of urea} = 3.11 \text{ mol urea} \times \frac{60.06 \text{ g urea}}{1 \text{ mol urea}} = \mathbf{187\ g\ urea}$$

13.63

$$\boldsymbol{m} = \frac{\Delta T_f}{K_f} = \frac{1.1°\text{C}}{1.86°\text{C}/m} = \mathbf{0.59}\ \boldsymbol{m}$$

13.65 We first find the number of moles of gas using the ideal gas equation.

$$n = \frac{PV}{RT} = \frac{\left(748 \text{ mmHg} \times \frac{1 \text{ atm}}{760 \text{ mmHg}}\right)(4.00 \text{ L})}{(27 + 273) \text{ K}} \times \frac{\text{mol} \cdot \text{K}}{0.0821 \text{ L} \cdot \text{atm}} = 0.160 \text{ mol}$$

$$\text{molality} = \frac{0.160 \text{ mol}}{0.0750 \text{ kg benzene}} = 2.13\ m$$

$$\Delta T_f = K_f m = (5.12°\text{C}/m)(2.13\ m) = 10.9°\text{C}$$

$$\textbf{freezing point} = 5.5°C - 10.9°C = \mathbf{-5.4°C}$$

13.67 a. NaCl is a strong electrolyte. The concentration of particles (ions) is double the concentration of NaCl. Note that because the density of water is 1 g/mL, 135 mL of water has a mass of 135 g.

The number of moles of NaCl is:

$$21.2\text{ g NaCl} \times \frac{1\text{ mol}}{58.44\text{ g}} = 0.363\text{ mol NaCl}$$

Next, we can find the changes in boiling and freezing points ($i = 2$)

$$m = \frac{0.363\text{ mol}}{0.135\text{ kg}} = 2.70\ m$$

$$\Delta T_b = iK_bm = 2(0.52°C/m)(2.70\ m) = 2.8°C$$

$$\Delta T_f = iK_fm = 2(1.86°C/m)(2.70\ m) = 10.0°C$$

The *boiling point* is **102.8°C**; the *freezing point* is **–10.0°C**.

b. Urea is a nonelectrolyte. The particle concentration is just equal to the urea concentration.

The molality of the urea solution is:

$$\text{moles urea} = 15.4\text{ g urea} \times \frac{1\text{ mol urea}}{60.06\text{ g urea}} = 0.256\text{ mol urea}$$

$$m = \frac{0.256\text{ mol urea}}{0.0667\text{ kg } H_2O} = 3.84\ m$$

$$\Delta T_b = iK_bm = 1(0.52°C/m)(3.84\ m) = 2.0°C$$

$$\Delta T_f = iK_fm = 1(1.86°C/m)(3.84\ m) = 7.14°C$$

The *boiling point* is **102.0°C**; the *freezing point* is **–7.14°C**.

13.69 Both NaCl and $CaCl_2$ are strong electrolytes. Urea and sucrose are nonelectrolytes. The NaCl or $CaCl_2$ will yield more particles per mole of the solid dissolved, resulting in greater freezing point depression. Also, sucrose and urea would make a mess when the ice melts.

13.71 The temperature and molarity of the two solutions are the same. If we divide Equation 13.11 of the text for one solution by the same equation for the other, we can find the ratio of the van't Hoff factors in terms of the osmotic pressures ($i = 1$ for urea).

$$\frac{\pi_{CaCl_2}}{\pi_{urea}} = \frac{iMRT}{MRT} = i = \frac{0.605 \text{ atm}}{0.245 \text{ atm}} = \mathbf{2.47}$$

13.73 For this problem, we first need to calculate the pressure necessary to support a column of water 105 m high. We use the equation $P = hdg$, where $h$, $d$, and $g$ are the column height in m, the density of water in kg/m$^3$, and the gravitational constant, respectively.

$$P = 105 \text{ m} \times 1.00 \times 10^3 \frac{\text{kg}}{\text{m}^3} \times 9.81 \frac{\text{m}}{\text{s}^2} = 1.03 \times 10^6 \text{ kg/m·s}^2 = 1.03 \times 10^6 \text{ Pa} \times \frac{1 \text{ atm}}{101{,}325 \text{ Pa}} = 10.16 \text{ atm}$$

Because atmospheric pressure contributes 1 atm, the osmotic pressure required is only **9.16 atm**.

13.75 $CaCl_2$ is an ionic compound and is therefore an electrolyte in water. Assuming that $CaCl_2$ is a strong electrolyte and completely dissociates (no ion pairs, van't Hoff factor $i = 3$), the total ion concentration will be $3 \times 0.35 = 1.05$ *m*, which is larger than the urea (nonelectrolyte) concentration of 0.90 *m*.

a. The **$CaCl_2$** solution will show a larger boiling point elevation.

b. The freezing point of the **urea** solution will be higher because the $CaCl_2$ solution will have a larger freezing point depression.

c. The **$CaCl_2$** solution will have a larger vapor pressure lowering.

13.77 Assume that all the salts are completely dissociated. Calculate the molality of the ions in the solutions.

| | |
|---|---|
| 0.10 *m* $Na_3PO_4$: | 0.10 *m* × 4 ions/unit = 0.40 *m* |
| 0.35 *m* NaCl: | 0.35 *m* × 2 ions/unit = 0.70 *m* |
| 0.20 *m* $MgCl_2$: | 0.20 *m* × 3 ions/unit = 0.60 *m* |
| 0.15 *m* $C_6H_{12}O_6$: | nonelectrolyte, 0.15 *m* |
| 0.15 *m* $CH_3COOH$: | weak electrolyte, slightly greater than 0.15 *m* |

The solution with the lowest molality will have the highest freezing point (smallest freezing point depression):
**0.15 *m* $C_6H_{12}O_6$ > 0.15 *m* $CH_3COOH$ > 0.10 *m* $Na_3PO_4$ > 0.20 *m* $MgCl_2$ > 0.35 *m* NaCl.**

13.79 a. $Na_2SO_4$ b. $MgSO_4$ c. KBr

13.83 First, from the freezing point depression we can calculate the molality of the solution. See Table 13.2 of the text for the normal freezing point and $K_f$ value for benzene.

$$\Delta T_f = (5.5 - 4.3)°C = 1.2°C$$

$$m = \frac{\Delta T_f}{K_f} = \frac{1.2°C}{5.12°C/m} = 0.23\ m$$

Multiplying the molality by the mass of solvent (in kg) gives moles of unknown solute. Then, dividing the mass of solute (in g) by the moles of solute, gives the molar mass of the unknown solute.

$$?\text{ mol of unknown solute} = \frac{0.23\text{ mol solute}}{1\text{ kg benzene}} \times 0.0250\text{ kg benzene}$$

$$= 0.0058\text{ mol solute}$$

$$\textbf{molar mass of unknown} = \frac{2.50\text{ g}}{0.0058\text{ mol}} = \mathbf{4.3 \times 10^2\ g/mol}$$

The empirical molar mass of $C_6H_5P$ is 108.1 g/mol. Therefore, the molecular formula is $(C_6H_5P)_4$ or $\mathbf{C_{24}H_{20}P_4}$.

13.85 **Strategy:** We are asked to calculate the molar mass of the polymer. Grams of the polymer are given in the problem, so we need to solve for moles of polymer.

From the osmotic pressure of the solution, we can calculate the molarity of the solution. Then, from the molarity, we can determine the number of moles in 0.8330 g of the polymer. What units should we use for $\pi$ and temperature?

**Setup:**

$$\text{molar mass of polymer} = \frac{\text{grams of polymer}}{\text{moles of polymer}}$$

**Solution:** First, we calculate the molarity using Equation 13.8 of the text.

$$\pi = MRT$$

$$M = \frac{\pi}{RT} = \frac{\left(5.20\text{ mmHg} \times \frac{1\text{ atm}}{760\text{ mmHg}}\right)}{298\text{ K}} \times \frac{\text{mol}\cdot\text{K}}{0.0821\text{ L}\cdot\text{atm}} = 2.80 \times 10^{-4}\ M$$

Multiplying the molarity by the volume of solution (in L) gives moles of solute (polymer).

$$?\text{ mol of polymer} = (2.80 \times 10^{-4}\text{ mol/L})(0.170\text{ L}) = 4.76 \times 10^{-5}\text{ mol polymer}$$

Lastly, dividing the mass of polymer (in g) by the moles of polymer, gives the molar mass of the polymer.

$$\textbf{molar mass of polymer} = \frac{0.8330 \text{ g polymer}}{4.76 \times 10^{-5} \text{ mol polymer}} = \mathbf{1.75 \times 10^4 \text{ g/mol}}$$

13.87 We use the osmotic pressure data to determine the molarity.

$$M = \frac{\pi}{RT} = \frac{4.61 \text{ atm}}{(20 + 273)\text{K}} \times \frac{\text{mol} \cdot \text{K}}{0.0821 \text{ L} \cdot \text{atm}} = 0.192 \text{ mol/L}$$

Next we use the density and the solution mass to find the volume of the solution.

$$\text{mass of soln} = 6.85 \text{ g} + 100.0 \text{ g} = 106.9 \text{ g soln}$$

$$\text{volume of soln} = 106.9 \text{ g soln} \times \frac{1 \text{ mL}}{1.024 \text{ g}} = 104.4 \text{ mL} = 0.1044 \text{ L}$$

Multiplying the molarity by the volume (in L) gives moles of solute (carbohydrate).

$$\text{mol of solute} = M \times \text{L} = (0.192 \text{ mol/L})(0.1044 \text{ L}) = 0.02004 \text{ mol solute}$$

Finally, dividing mass of carbohydrate by moles of carbohydrate gives the molar mass of the carbohydrate.

$$\textbf{molar mass} = \frac{6.85 \text{ g carbohydrate}}{0.02004 \text{ mol carbohydrate}} = \mathbf{342 \text{ g/mol}}$$

13.89 Use the osmotic pressure to calculate the molar concentration of dissolved particles in the HB solution.

$$M = \frac{\pi}{RT} = \frac{2.83 \text{ atm}}{\left(0.0821 \frac{\text{L} \cdot \text{atm}}{\text{mol} \cdot \text{K}}\right)(298 \text{ K})} = 0.1157\ M$$

When an HB molecule ionizes it produces one $H_2B^+$ ion and one $OH^-$ ion.

$$\text{HB}(aq) + \text{H}_2\text{O}(l) \rightleftarrows \text{H}_2\text{B}^+(aq) + \text{OH}^-(aq)$$

The concentration of HB is originally 0.100 *M*. It decreases by $x$ and the concentrations of $H_2B^+$ and $OH^-$ each increase by $x$. Thus, the concentration of dissolved particles is $0.100 - x + 2x = 0.100 + x$. We solve for $x$ and determine what percentage $x$ is of the original concentration, 0.100 *M*.

$$\textbf{\% ionization} = \frac{(0.1157 - 0.100)}{0.100} \times 100\% = \mathbf{15.7\%}$$

13.93 a. Vitamin D is fat soluble, because it contains mainly large hydrocarbon groups that are predominantly nonpolar.

b. Vitamin $B_2$ is water soluble, because it contains many polar OH groups.

13.95 First, we can calculate the molality of the solution from the freezing point depression.

$$\Delta T_f = (5.12)m$$

$$(5.5 - 3.5) = (5.12)m$$

$$m = 0.39$$

Next, from the definition of molality, we can calculate the moles of solute.

$$m = \frac{\text{mol solute}}{\text{kg solvent}}$$

$$0.39\ m = \frac{\text{mol solute}}{80 \times 10^{-3}\ \text{kg benzene}}$$

$$\text{mol solute} = 0.031\ \text{mol}$$

The molar mass ($\mathcal{M}$) of the solute is:

$$\frac{3.8\ \text{g}}{0.031\ \text{mol}} = \mathbf{1.2 \times 10^2\ g/mol}$$

The molar mass of $CH_3COOH$ is 60.05 g/mol. Since the molar mass of the solute calculated from the freezing point depression is twice this value, the structure of the solute most likely is a dimer that is held together by hydrogen bonds.

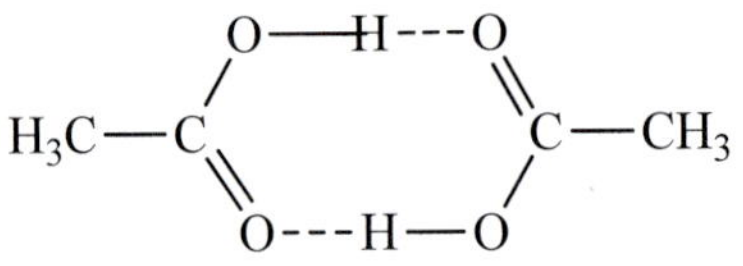

```
            O—H---O
           /       \\
    H3C—C          C—CH3
           \\       /
            O---H—O
```

13.97 As the water freezes, dissolved minerals in the water precipitate from solution. The minerals refract light and create an opaque appearance.

13.99

$$\Delta T_f = iK_fm$$

$$\boldsymbol{i} = \frac{\Delta T_f}{K_f m} = \frac{2.6}{(1.86)(0.40)} = \mathbf{3.5}$$

13.101 Vitamin $B_6$ is water soluble, because of the polar OH groups.

13.103 Vitamin A is fat soluble. Although it contains one OH group, most of the molecule is composed of nonpolar hydrocarbon groups.

13.105 Reverse osmosis involves no phase changes and is usually cheaper than distillation or freezing.

To reverse the osmotic migration of water across a semipermeable membrane, an external pressure exceeding the osmotic pressure must be applied. To find the osmotic pressure of 0.70 $M$ NaCl solution, we must use the van't Hoff factor because NaCl is a strong electrolyte and the total ion concentration becomes 2(0.70 $M$) = 1.4 $M$.

The osmotic pressure of sea water is:

$$\pi = iMRT = 2(0.70 \text{ mol/L})(0.0821 \text{ L·atm/mol·K})(298 \text{ K}) = \mathbf{34 \text{ atm}}$$

To cause reverse osmosis a pressure in excess of 34 atm must be applied.

13.107 a. Solubility decreases with increasing lattice energy.

b. Ionic compounds are more soluble in a polar solvent.

c. Solubility increases with enthalpy of hydration of the cation and anion.

13.109 Let's assume we have 100 g of solution. The 100 g of solution will contain 70.0 g of $HNO_3$ and 30.0 g of $H_2O$.

$$\text{mol solute } (HNO_3) = 70.0 \text{ g } HNO_3 \times \frac{1 \text{ mol } HNO_3}{63.02 \text{ g } HNO_3} = 1.11 \text{ mol } HNO_3$$

$$\text{kg solvent } (H_2O) = 30.0 \text{ g } H_2O \times \frac{1 \text{ kg}}{1000 \text{ g}} = 0.0300 \text{ kg } H_2O$$

$$\textbf{molality} = \frac{1.11 \text{ mol } HNO_3}{0.0300 \text{ kg } H_2O} = \mathbf{37.0}\ \boldsymbol{m}$$

To calculate the density, let's again assume we have 100 g of solution. Since,

$$d = \frac{\text{mass}}{\text{volume}}$$

we know the mass (100 g) and therefore need to calculate the volume of the solution. We know from the molarity that 15.9 mol of $HNO_3$ are dissolved in a solution volume of 1000 mL. In 100 g of solution, there are 1.11 moles $HNO_3$ (calculated above). What volume will 1.11 moles of $HNO_3$ occupy?

$$1.11 \text{ mol HNO}_3 \times \frac{1000 \text{ mL soln}}{15.9 \text{ mol HNO}_3} = 69.8 \text{ mL soln}$$

Dividing the mass by the volume gives the density.

$$d = \frac{100 \text{ g}}{69.8 \text{ mL}} = \mathbf{1.43\ g/mL}$$

13.111 $NH_3$ can form hydrogen bonds with water; $NCl_3$ cannot. (Like dissolves like.)

13.113 We can calculate the molality of the solution from the freezing point depression.

$$\Delta T_f = K_f m$$

$$0.203 = 1.86\, m$$

$$m = \frac{0.203}{1.86} = 0.109\, m$$

The molality of the original solution was 0.106 $m$. Some of the solution has ionized to $H^+$ and $CH_3COO^-$.

| | $CH_3COOH$ $\rightleftarrows$ | $CH_3COO^-$ + | $H^+$ |
|---|---|---|---|
| Initial | 0.106 $m$ | 0 | 0 |
| Change | $-x$ | $+x$ | $+x$ |
| Equil. | 0.106 $m - x$ | $x$ | $x$ |

At equilibrium, the total concentration of species in solution is 0.109 $m$.

$$(0.106 - x) + 2x = 0.109\, m$$

$$x = 0.003\, m$$

The percentage of acid that has undergone ionization is:

$$\frac{0.003\, m}{0.106\, m} \times 100\% = \mathbf{3\%}$$

13.115 First, we calculate the number of moles of HCl in 100 g of solution.

$$n_{\text{HCl}} = 100 \text{ g soln} \times \frac{37.7 \text{ g HCl}}{100 \text{ g soln}} \times \frac{1 \text{ mol HCl}}{36.46 \text{ g HCl}} = 1.03 \text{ mol HCl}$$

Next, we calculate the volume of 100 g of solution.

$$V = 100\text{ g} \times \frac{1\text{ mL}}{1.19\text{ g}} \times \frac{1\text{ L}}{1000\text{ mL}} = 0.0840\text{ L}$$

Finally, the molarity of the solution is:

$$\frac{1.03\text{ mol}}{0.0840\text{ L}} = \mathbf{12.3}\ \boldsymbol{M}$$

13.117 Let the mass of NaCl be $x$ g. Then, the mass of sucrose is $(10.2 - x)$g.

We know that the equation representing the osmotic pressure is:

$$\pi = MRT$$

$\pi$, $R$, and $T$ are given. Using this equation and the definition of molarity, we can calculate the percentage of NaCl in the mixture.

$$\text{molarity} = \frac{\text{mol solute}}{\text{L soln}}$$

Remember that NaCl dissociates into two ions in solution; therefore, we multiply the moles of NaCl by two.

$$\text{mol solute} = 2\left(x\text{ g NaCl} \times \frac{1\text{ mol NaCl}}{58.44\text{ g NaCl}}\right) + \left((10.2 - x)\text{g sucrose} \times \frac{1\text{ mol sucrose}}{342.3\text{ g sucrose}}\right)$$

$$\text{mol solute} = 0.03422x + 0.02980 - 0.002921x$$

$$\text{mol solute} = 0.03130x + 0.02980$$

$$\text{Molarity of solution} = \frac{\text{mol solute}}{\text{L soln}} = \frac{(0.03130x + 0.02980)\text{mol}}{0.250\text{ L}}$$

Substitute molarity into the equation for osmotic pressure to solve for $x$.

$$\pi = MRT$$

$$7.32\text{ atm} = \left(\frac{(0.03130x + 0.02980)\text{mol}}{0.250\text{ L}}\right)\left(0.0821\frac{\text{L}\cdot\text{atm}}{\text{mol}\cdot\text{K}}\right)(296\text{ K})$$

$$.0753 = 0.03130x + 0.02980$$

$$x = 1.45\text{ g} = \text{mass of NaCl}$$

$$\textbf{Mass \% NaCl} = \frac{1.45\text{ g}}{10.2\text{ g}} \times 100\% = \mathbf{14.2\%}$$

13.119 To solve for the molality of the solution, we need the moles of solute (urea) and the kilograms of solvent (water). If we assume that we have 1 mole of water, we know the mass of water. Using the change in vapor pressure, we can solve for the mole fraction of urea and then the moles of urea.

Using Equation 13.5 of the text, we solve for the mole fraction of urea.

$$\Delta P = 23.76\text{ mmHg} - 22.98\text{ mmHg} = 0.78\text{ mmHg}$$

$$\Delta P = X_2 P_1^\circ = X_{\text{urea}} P_{\text{water}}^\circ$$

$$X_{\text{urea}} = \frac{\Delta P}{P_{\text{water}}^\circ} = \frac{0.78\text{ mmHg}}{23.76\text{ mmHg}} = 0.033$$

Assuming that we have 1 mole of water, we can now solve for moles of urea.

$$X_{\text{urea}} = \frac{\text{mol urea}}{\text{mol urea} + \text{mol water}}$$

$$0.033 = \frac{n_{\text{urea}}}{n_{\text{urea}} + 1}$$

$$0.033 n_{\text{urea}} + 0.033 = n_{\text{urea}}$$

$$0.033 = 0.967 n_{\text{urea}}$$

$$n_{\text{urea}} = 0.034\text{ mol}$$

1 mole of water has a mass of 18.02 g or 0.01802 kg. We now know the moles of solute (urea) and the kilograms of solvent (water), so we can solve for the molality of the solution.

$$\boldsymbol{m} = \frac{\text{mol solute}}{\text{kg solvent}} = \frac{0.034\text{ mol}}{0.01802\text{ kg}} = \mathbf{1.9}\ \boldsymbol{m}$$

13.121 $2H_2O_2 \rightarrow 2H_2O + O_2$

$$10\text{ mL} \times \frac{3.0\text{ g } H_2O_2}{100\text{ mL}} \times \frac{1\text{ mol } H_2O_2}{34.02\text{ g } H_2O_2} \times \frac{1\text{ mol } O_2}{2\text{ mol } H_2O_2} = 4.4 \times 10^{-3}\text{ mol } O_2$$

a. Using the ideal gas law:

$$V = \frac{nRT}{P} = \frac{(4.4 \times 10^{-3}\text{ mol } O_2)(0.0821\text{ L}\cdot\text{atm/mol}\cdot\text{K})(273\text{ K})}{1.0\text{ atm}} = \mathbf{99\ mL} = \mathbf{0.099\ L}$$

b. The ratio of the volumes: $\dfrac{99\text{ mL}}{10\text{ mL}} = \mathbf{9.9}$

13.123 **Strategy:** From the figure, note that the normal boiling point of the solution is elevated by about 1°C. Use the boiling point elevation equation (Equation 13.6) with $K_b = 2.53$ °C/m (Table 13.2).

**Solution:**

$$\Delta T_b = K_b m$$

$$1°C = (2.53\ °C/m)m$$

$$m = \frac{1°C}{2.53\ °C/m} \approx 0.4m$$

The solution concentration is **about 0.4 molal.**

13.125 Starting with $n = kP$ and substituting into the ideal gas equation ($PV = nRT$), we find:

$$PV = (kP)RT$$

$$V = kRT$$

This equation shows that the volume of a gas that dissolves in a given amount of solvent is dependent on the *temperature*, not the pressure of the gas.

13.127 At equilibrium, the vapor pressure of benzene over each beaker must be the same. Assuming ideal solutions, this means that the mole fraction of benzene in each beaker must be identical at equilibrium. Consequently, the mole fraction of solute is also the same in each beaker, even though the solutes are different in the two solutions. Assuming the solute to be non-volatile, equilibrium is reached by the transfer of benzene, via the vapor phase, from beaker A to beaker B.

The mole fraction of naphthalene in beaker A at equilibrium can be determined from the data given. The number of moles of naphthalene is given, and the moles of benzene can be calculated using its molar mass and knowing that 100 g − 7.0 g = 93.0 g of benzene remain in the beaker.

$$X_{C_{10}H_8} = \frac{0.15\text{ mol}}{0.15\text{ mol} + \left(93.0\text{ g benzene} \times \frac{1\text{ mol benzene}}{78.11\text{ g benzene}}\right)} = 0.112$$

Now, let the number of moles of unknown compound be $n$. Assuming all the benzene lost from beaker A is transferred to beaker B, there are 100 g + 7.0 g = 107 g of benzene in the beaker. Also, recall that the mole fraction of solute in beaker B is equal to that in beaker A at equilibrium (0.112). The mole fraction of the unknown compound is:

$$X_{\text{unknown}} = \frac{n}{n + \left(107\text{ g benzene} \times \frac{1\text{ mol benzene}}{78.11\text{ g benzene}}\right)}$$

$$0.112 = \frac{n}{n + 1.370}$$

$$n = 0.1728 \text{ mol}$$

There are 31 grams of the unknown compound dissolved in benzene. The molar mass of the unknown is:

$$\frac{31 \text{ g}}{0.173 \text{ mol}} = \mathbf{1.8 \times 10^2 \text{ g/mol}}$$

Temperature is assumed constant and ideal behavior is also assumed. Both solutes are assumed to be nonvolatile.

13.129 a. At reduced pressure, the solution is supersaturated with $CO_2$.

b. As the escaping $CO_2$ expands it cools, condensing water vapor in the air to form fog.

13.131 At equilibrium, the concentrations in the 2 beakers are equal. Let $x$ L be the change in volume.

$$\frac{n_A}{(0.050 - x)(\text{L})} = \frac{n_B}{(50 + x)(\text{L})}$$

$$\frac{(0.050 \text{ L})(1.0\ M)}{(0.050 - x)(\text{L})} = \frac{(0.050 \text{ L})(2.0\ M)}{(0.050 + x)(\text{L})}$$

$$(0.050 + x)\cdot(1.0) = (0.050 - x)\cdot(2.0) \text{ (all units cancel)}$$

$$3x = 0.050$$

$$x = .0167 \text{ L} = 16.7 \text{ mL} \quad (x \text{ has a unit of liters})$$

The final volumes are:

$$(50 - 16.7) \text{ mL} = \mathbf{33 \text{ mL}}$$

$$(50 + 16.7) \text{ mL} = \mathbf{67 \text{ mL}}$$

13.133 Egg yolk contains lecithins which solubilize oil in water (See Figure 13.18 of the text). The nonpolar oil becomes soluble in water because the nonpolar tails of lecithin dissolve in the oil, and the polar heads of the lecithin molecules dissolve in polar water (like dissolves like).

13.135 For this problem we must find the solution mole fractions, the molality, and the molarity. For molarity, we can assume the solution to be so dilute that its density is 1.00 g/mL. We first find the number of moles of lysozyme and of water.

$$n_{\text{lysozyme}} = 0.100 \text{ g} \times \frac{1 \text{ mol}}{13930 \text{ g}} = 7.18 \times 10^{-6} \text{ mol}$$

$$n_{\text{water}} = 150\text{ g} \times \frac{1\text{ mol}}{18.02\text{ g}} = 8.32\text{ mol}$$

Vapor pressure lowering: $\Delta P = X_{\text{lysozyme}} P^{\circ}_{\text{water}} = \dfrac{n_{\text{lysozyme}}}{n_{\text{lysozyme}} + n_{\text{water}}}(23.76\text{ mmHg})$

$$\boldsymbol{\Delta P} = \frac{7.18\times 10^{-6}\text{ mol}}{[(7.18\times 10^{-6}) + 8.32]\text{mol}}(23.76\text{ mmHg}) = \mathbf{2.05\times 10^{-5}\ mmHg}$$

Freezing point depression:

$$\boldsymbol{\Delta T_f} = K_f m = (1.86°\text{C}/m)\left(\frac{7.18\times 10^{-6}\text{ mol}}{0.150\text{ kg}}\right) = \mathbf{8.90\times 10^{-5}\ °C}$$

Boiling point elevation:

$$\boldsymbol{\Delta T_b} = K_b m = (0.52°\text{C}/m)\left(\frac{7.18\times 10^{-6}\text{ mol}}{0.150\text{ kg}}\right) = \mathbf{2.5\times 10^{-5}\ °C}$$

Osmotic pressure:

As stated above, we assume the density of the solution is 1.00 g/mL. The volume of the solution will be 150 mL.

$$\boldsymbol{\pi} = MRT = \left(\frac{7.18\times 10^{-6}\text{ mol}}{0.150\text{ L}}\right)(0.0821\text{ L}\cdot\text{atm/mol}\cdot\text{K})(298\text{ K}) = \mathbf{1.17\times 10^{-3}\ atm} = \mathbf{0.889\ mmHg}$$

With the exception of osmotic pressure, these changes in colligative properties are essentially negligible.

13.137

$$\Delta T_f = K_f m$$

$$m = \frac{\Delta T_f}{K_f} = \frac{60^{\circ}\text{C}}{1.86^{\circ}\text{C}/m} = 32\ m$$

This is an extremely high molal concentration.

13.139 The pill is in a hypotonic solution. Consequently, by osmosis, water moves across the semipermeable membrane into the pill. The increase in pressure pushes the elastic membrane to the right, causing the drug to exit through the small holes at a constant rate.

13.141 a. Runoff of the salt solution into the soil increases the salinity of the soil. If the soil becomes hypertonic relative to the tree cells, osmosis would reverse, and the tree would lose water to the soil and eventually die of dehydration.

b. Assuming the collecting duct acts as a semipermeable membrane, water would flow from the urine into the hypertonic fluid, thus returning water to the body.

13.143 Using Equation 13.8 of the text, we find the molarity of the solution.

$$M = \frac{\pi}{RT} = \frac{7.5\text{ atm}}{\left(0.0821\frac{\text{L}\cdot\text{atm}}{\text{mol}\cdot\text{K}}\right)(310\text{ K})} = \mathbf{0.295}\ \boldsymbol{M}$$

This is the combined concentrations of all the ions. For dilute solutions, $M \approx m$. Therefore, the freezing point of blood can be calculated using Equation 13.7.

$$\Delta T_f = K_f m = (1.86°\text{C}/m)(0.295\ m) = 0.55°\text{C}$$

Thus, the freezing point of blood should be about **–0.55°C**.

13.145 a. Using Equation 13.8 of the text, we find the molarity of the solution.

$$M = \frac{\pi}{RT} = \frac{0.257\text{ atm}}{(0.0821\text{ L}\cdot\text{atm/mol}\cdot\text{K})(298\text{ K})} = 0.0105\text{ mol/L}$$

This is the combined concentrations of all the ions. The amount dissolved in 10.0 mL (0.01000 L) is

$$?\text{ moles} = \frac{0.0105\text{ mol}}{1\text{ L}} \times 0.0100\text{ L} = 1.05 \times 10^{-4}\text{ mol}$$

Since the mass of this amount of protein is 0.225 g, the apparent molar mass is

$$\frac{0.225\text{ g}}{1.05 \times 10^{-4}\text{ mol}} = \mathbf{2.14 \times 10^3\ g/mol}$$

b. We need to use a van't Hoff factor to take into account the fact that the protein is a strong electrolyte. The van't Hoff factor will be $i = 21$ (why?).

$$M = \frac{\pi}{iRT} = \frac{0.257\text{ atm}}{(21)(0.0821\text{ L}\cdot\text{atm/mol}\cdot\text{K})(298\text{ K})} = 5.00 \times 10^{-4}\text{ mol/L}$$

This is the actual concentration of the protein. The amount in 10.0 mL (0.0100 L) is

$$\frac{5.00 \times 10^{-4} \text{ mol}}{1 \text{ L}} \times 0.0100 \text{ L} = 5.00 \times 10^{-6} \text{ mol}$$

Therefore the actual molar mass is:

$$\frac{0.225 \text{ g}}{5.00 \times 10^{-6} \text{ mol}} = \mathbf{4.50 \times 10^4 \text{ g/mol}}$$

13.147 **Strategy:** First, we can determine the empirical formula from mass percent data. Then, we can determine the molar mass from the freezing-point depression. Finally, from the empirical formula and the molar mass, we can find the molecular formula.

**Solution:** **METHOD 1:** If we assume that we have 100 g of the compound, then each percentage can be converted directly to grams.

$$\text{Moles C} = 80.78 \text{ g} \times \frac{1 \text{ mol}}{12.01 \text{ g}} = 6.726 \text{ mol C}$$

$$\text{Moles H} = 13.56 \text{ g} \times \frac{1 \text{ mol}}{1.008 \text{ g}} = 13.45 \text{ mol H}$$

$$\text{Moles O} = 5.66 \text{ g} \times \frac{1 \text{ mol}}{16.00 \text{ g}} = 0.354 \text{ mol O}$$

This gives the formula: $C_{6.726}H_{13.45}O_{0.354}$. Dividing through by the smallest subscript (0.354) gives the smallest whole numbers possible.

$$\text{C: } \frac{6.726}{0.354} = 19.00 \qquad \text{H: } \frac{13.45}{0.354} = 37.99 \qquad \text{O: } \frac{0.354}{0.354} = 1.00$$

Thus, the empirical formula is $C_{19}H_{38}O$.

The freezing point depression is $\Delta T_f = 5.5°C - 3.37°C = 2.16°C$. This implies a solution molality of:

$$m = \frac{\Delta T_f}{K_f} = \frac{2.16^\circ\text{C}}{5.12^\circ\text{C}/m} = 0.416\ m$$

Since the solvent mass is 8.50 g or 0.00850 kg, the amount of solute is:

$$\frac{0.416 \text{ mol}}{1 \text{ kg benzene}} \times 0.00850 \text{ kg benzene} = 3.54 \times 10^{-3} \text{ mol}$$

Since 1.00 g of the sample represents $3.54 \times 10^{-3}$ mol, the molar mass is:

$$\textbf{molar mass} = \frac{1.00 \text{ g}}{3.54 \times 10^{-3} \text{ mol}} = \textbf{282.5 g/mol}$$

The mass of the empirical formula is 282.5 g/mol, so the molecular formula is the same as the empirical formula, $\mathbf{C_{19}H_{38}O}$.

**METHOD 2:** As in Method 1, we determine the molar mass of the unknown from the freezing point data. Once the molar mass is known, we can multiply the mass % of each element (converted to a decimal) by the molar mass to convert to grams of each element. From the grams of each element, the moles of each element can be determined and hence the mole ratio in which the elements combine.

Use the freezing point data as above to determine the molar mass.

$$\textbf{molar mass} = \textbf{282.5 g/mol}$$

Multiply the mass % (converted to a decimal) of each element by the molar mass to convert to grams of each element. Then, use the molar mass to convert to moles of each element.

$$n_C = (0.8078) \times (282.5 \text{ g}) \times \frac{1 \text{ mol C}}{12.01 \text{ g C}} = \textbf{19.00 mol C}$$

$$n_H = (0.1356) \times (282.5 \text{ g}) \times \frac{1 \text{ mol H}}{1.008 \text{ g H}} = \textbf{38.00 mol H}$$

$$n_O = (0.0566) \times (282.5 \text{ g}) \times \frac{1 \text{ mol O}}{16.00 \text{ g O}} = \textbf{1.00 mol O}$$

Since we used the molar mass to calculate the moles of each element present in the compound, this method directly gives the molecular formula. The formula is $\mathbf{C_{19}H_{38}O}$.

13.149 The desired process is for (fresh) water to move from a more concentrated solution (seawater) to pure solvent. This is an example of reverse osmosis, and external pressure must be provided to overcome the osmotic pressure of the seawater. The source of the pressure here is the water pressure, which increases with increasing depth. The osmotic pressure of the seawater is:

$$\pi = MRT$$

$$\pi = (0.70\ M)(0.0821 \text{ L·atm/mol·K})(293 \text{ K})$$

$$\pi = 16.8 \text{ atm}$$

The water pressure at the membrane depends on the height of the sea above it, *i.e.* the depth. $P = hdg$ where $h$, $d$, and $g$ are the height of the sea above the membrane, the density of the solution, and the gravitational constant, respectively; and fresh water will begin to pass through the membrane when $P = \pi$. Substituting $\pi$

$= P$ into the equation gives:

$$\pi = hdg$$

and

$$h = \frac{\pi}{dg}$$

Before substituting into the equation to solve for $h$, we need to convert atm to pascals, and the density to units of $kg/m^3$. These conversions will give a height in units of meters.

$$16.8 \text{ atm} \times \frac{1.01325 \times 10^5 \text{ Pa}}{1 \text{ atm}} = 1.70 \times 10^6 \text{ Pa}$$

1 Pa = 1 $N/m^2$ and 1 N = 1 $kg{\cdot}m/s^2$. Therefore, we can write $1.70 \times 10^6$ Pa as $1.70 \times 10^6\ kg/m{\cdot}s^2$

$$\frac{1.03 \text{ g}}{1 \text{ cm}^3} \times \frac{1 \text{ kg}}{1000 \text{ g}} \times \left(\frac{100 \text{ cm}}{1 \text{ m}}\right)^3 = 1.03 \times 10^3 \text{ kg/m}^3$$

$$\boldsymbol{h} = \frac{\pi}{dg} = \frac{1.70\times10^6 \dfrac{\text{kg}}{\text{m}\cdot\text{s}^2}}{\left(9.81\dfrac{\text{m}}{\text{s}^2}\right)\left(1.03\times10^3 \dfrac{\text{kg}}{\text{m}^3}\right)} = \mathbf{168\ m}$$

# Chapter 14

# Chemical Kinetics

14.5 In general for a reaction $aA + bB \rightarrow cC + dD$

$$\text{rate} = -\frac{1}{a}\frac{\Delta[A]}{\Delta t} = -\frac{1}{b}\frac{\Delta[B]}{\Delta t} = \frac{1}{c}\frac{\Delta[C]}{\Delta t} = \frac{1}{d}\frac{\Delta[D]}{\Delta t}$$

a.

$$\text{rate} = -\frac{\Delta[H_2]}{\Delta t} = -\frac{\Delta[I_2]}{\Delta t} = \frac{1}{2}\frac{\Delta[HI]}{\Delta t}$$

b.

$$\text{rate} = -\frac{1}{5}\frac{\Delta[Br^-]}{\Delta t} = -\frac{\Delta[BrO_3^-]}{\Delta t} = -\frac{1}{6}\frac{\Delta[H^+]}{\Delta t} = \frac{1}{3}\frac{\Delta[Br_2]}{\Delta t}$$

Note that because the reaction is carried out in the aqueous phase, we do not monitor the concentration of water.

14.7 **Strategy:** he rate is defined as the change in concentration of a reactant or product with time. Each change-in-concentration term is divided by the corresponding stoichiometric coefficient. Terms involving *reactants* are preceded by a minus sign because reactant concentrations decrease as a reaction progresses—and *reaction* rates are always expressed as positive quantities.

$$\text{Rate} = -\frac{1}{2}\frac{\Delta[NO]}{\Delta t} \qquad \frac{\Delta[NO]}{\Delta t} = -0.066\ M/\text{s}$$

**Solution:** a. If the concentration of NO is changing at the rate of –0.066 $M$/s, the rate at which $NO_2$ is being formed is

$$-\frac{1}{2}\frac{\Delta[NO]}{\Delta t} = \frac{1}{2}\frac{\Delta[NO_2]}{\Delta t}$$

$$\frac{\Delta[NO_2]}{\Delta t} = \mathbf{0.066\ \mathit{M}/s}$$

The rate at which $NO_2$ is forming is **0.066 *M*/s**.

b.

$$-\frac{1}{2}\frac{\Delta[NO]}{\Delta t} = -\frac{\Delta[O_2]}{\Delta t}$$

$$\frac{\Delta[O_2]}{\Delta t} = \frac{-0.066\ M/s}{2} = \mathbf{-0.033\ \textit{M}/s}$$

The negative sign indicates that the concentration of molecular oxygen is decreasing as the reaction progresses. The rate at which molecular oxygen is reacting is **0.033 *M*/s**.

14.15

$$\textbf{rate} = k[NH_4^+][NO_2^-] = (3.0 \times 10^{-4}\ /M\cdot s)(0.36\ M)(0.075\ M) = \mathbf{8.1 \times 10^{-6}\ \textit{M}/s}$$

14.17 **Strategy:** We are given a set of concentrations and rate data and asked to determine the order of the reaction and the value of the rate constant. To determine the order of the reaction, we need to find the rate law for the reaction. We assume that the rate law takes the form

$$\text{rate} = k[A]^x[B]^y$$

How do we use the data to determine $x$ and $y$? Once the orders of the reactants are known, we can calculate $k$ using the set of rate and concentrations from any one of the experiments.

**Solution:** By comparing the first and second sets of data, we see that changing [B] does not affect the rate of the reaction. Therefore, the reaction is zero order in B. By comparing the first and third sets of data, we see that doubling [A] doubles the rate of the reaction.

$$\frac{\text{rate}_3}{\text{rate}_1} = \frac{6.40 \times 10^{-1}\ M/s}{3.20 \times 10^{-1}\ M/s} = 2 = \frac{k(3.00)^x(1.50)^y}{k(1.50)^x(1.50)^y}$$

Therefore,

$$\frac{(3.00)^x}{(1.50)^x} = 2^x = 2$$

and $x = 1$. This shows that **the reaction is first order in A and first order overall.**

$$\text{rate} = k[A]$$

From the set of data for the first experiment:

$$3.20 \times 10^{-1}\ M/s = k(1.50\ M)$$

$$\mathbf{\textit{k} = 0.213\ s^{-1}}$$

**Think About It:** What would be the value of $k$ if you had used the second or third set of data? Should $k$ be constant?

14.19 a. 2 b. 0 c. 2.5 d. 3

14.21 The graph below is a plot of ln $P$ vs. time. Since the plot is linear, the reaction is 1st order.

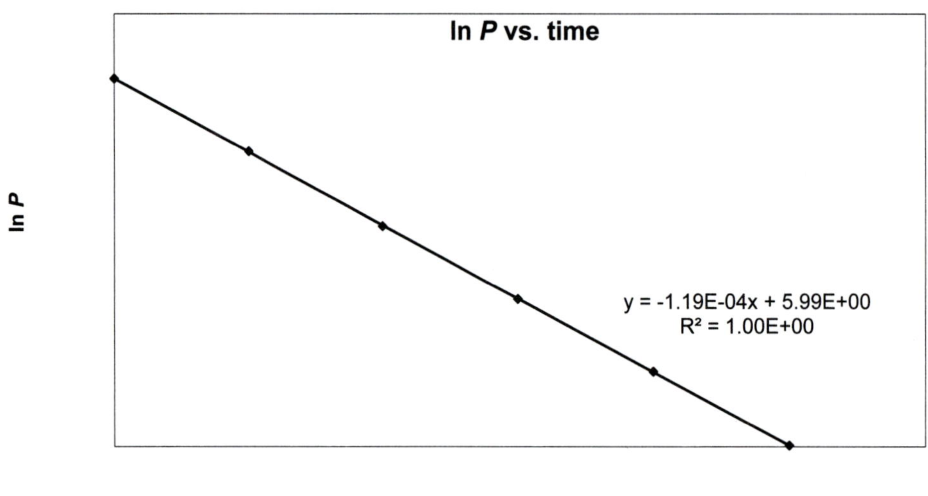


$$\text{Slope} = -k$$

$$\mathbf{k = 1.19 \times 10^{-4}\ s^{-1}}$$

14.27 We know that half of the substance decomposes in a time equal to the half-life, $t_{1/2}$. This leaves half of the compound. Half of what is left decomposes in a time equal to another half-life, so that only one quarter of the original compound remains. We see that 75% of the original compound has decomposed after two half–lives. Thus two half-lives equal one hour, or the half-life of the decay is 30 min.

$$100\%\ \text{starting compound} \xrightarrow{t_{1/2}} 50\%\ \text{starting compound} \xrightarrow{t_{1/2}} 25\%\ \text{starting compound}$$

Using first order kinetics, we can solve for $k$ using Equation 14.3 of the text, with $[A]_0 = 100$ and $[A] = 25$,

$$\ln\frac{[A]_t}{[A]_0} = -kt$$

$$\ln\frac{25}{100} = -k(60\ \text{min})$$

$$k = -\frac{\ln(0.25)}{60\text{ min}} = 0.023\text{ min}^{-1}$$

Then, substituting $k$ into Equation 14.5 of the text, you arrive at the same answer for $t_{1/2}$.

$$t_{\frac{1}{2}} = \frac{0.693}{k} = \frac{0.693}{0.023\text{ min}^{-1}} = \mathbf{30\text{ min}}$$

14.29 a. Since the reaction is known to be second-order, the relationship between reactant concentration and time is given by Equation 14.6 of the text. The problem supplies the rate constant and the initial (time = 0) concentration of NOBr. The concentration after 22s can be found easily.

$$\frac{1}{[\text{NOBr}]_t} = kt + \frac{1}{[\text{NOBr}]_0}$$

$$\frac{1}{[\text{NOBr}]_t} = (0.80/M\cdot\text{s})(22\text{s}) + \frac{1}{0.086\ M}$$

$$\frac{1}{[\text{NOBr}]_t} = 29\ M^{-1}$$

$$\mathbf{[NOBr] = 0.034}\ \boldsymbol{M}$$

**Think About It:** If the reaction were first order with the same $k$ and initial concentration, could you calculate the concentration after 22 s? If the reaction were first order and you were given the $t_{1/2}$, could you calculate the concentration after 22 s?

b. The half-life for a second-order reaction *is* dependent on the initial concentration. The half-lives can be calculated using Equation 14.7 of the text.

$$t_{\frac{1}{2}} = \frac{1}{k[\text{A}]_0}$$

$$t_{\frac{1}{2}} = \frac{1}{(0.80/M\cdot\text{s})(0.072\ M)}$$

$$t_{\frac{1}{2}} = \mathbf{17\ s}$$

For an initial concentration of 0.054 *M*, you should find $t_{\frac{1}{2}} = \mathbf{23\ s}$. Note that the half-life of a second-order reaction is inversely proportional to the initial reactant concentration.

14.31 We are told that the reaction is second-order. Therefore, the rate law is rate = $k\,[\text{P}]^2$.

Given the rate constant and the concentration of the protein (P), we can calculate the rate:

$$\textbf{rate} = 6.2 \times 10^{-3}/M{\cdot}\text{s}\ (2.7 \times 10^{-4}\ M)^2 = \mathbf{4.5 \times 10^{-10}\ \mathit{M}/s}$$

Using Equation 14.6, we can calculate the amount of time necessary for the concentration of P to drop to $2.7 \times 10^{-5}\ M$.

$$\frac{1}{[\mathrm{P}]_t} = \left(6.2\times10^{-3}\ /\ M\cdot s\right)(t) + \frac{1}{[\mathrm{P}]_0}$$

$$\frac{1}{2.7\times10^{-5}\,M} = 6.2\times10^{-3}\ /\ M\cdot s(t) + \frac{1}{2.7\times10^{-4}\,M}$$

$$t = \mathbf{5.4 \times 10^6\ s}$$

14.33 a. The relative rates containers i, ii, and iii are **4:3:6**.

b. The relative rates would be unaffected, each absolute rate would decrease by 50%.

c. Because half-life of a first-order reaction does not depend on reactant concentration, the relative half-lives are **1:1:1**.

14.41 **Strategy:** Equation 14.8 relates the rate constant to the frequency factor, the activation energy, and the temperature. Remember to make sure the units of $R$ and $E_a$ are consistent.

**Solution:** The appropriate value of $R$ is 8.314 J/K mol, not 0.0821 L·atm/mol·K. You must also use the activation energy value of 63,000 J/mol (why?). Once the temperature has been converted to kelvins, the rate constant is:

$$k = Ae^{-E_a/RT} = (8.7\times10^{12}\ \mathrm{s^{-1}})e^{-\left[\frac{63000\ \mathrm{J/mol}}{(8.314\ \mathrm{J/mol{\cdot}K})(348\ \mathrm{K})}\right]} = (8.7\times10^{12}\ \mathrm{s^{-1}})(3.5\times10^{-10})$$

$$\mathbf{\mathit{k} = 3.0 \times 10^3\ s^{-1}}$$

**Think About It:** Can you tell from the units of $k$ what the order of the reaction is?

14.43 Let $k_1$ be the rate constant at 295 K and $2k_1$ the rate constant at 305 K. Using Equation 14.11, we write:

$$\ln\frac{k_1}{2k_1} = \frac{E_a}{R}\left(\frac{1}{T_2} - \frac{1}{T_1}\right)$$

$$-0.693 = \frac{E_a}{8.314 \text{ J/K}\cdot\text{mol}}\left(\frac{1}{305 \text{ K}} - \frac{1}{295 \text{ K}}\right)$$

$$\boldsymbol{E_a = 5.18\times10^4 \textbf{ J/mol} = 51.8 \textbf{ kJ/mol}}$$

14.45 Using Equation 14.11,

$$\ln\frac{1}{2} = \frac{E_a}{R}\left(\frac{1}{T_2} - \frac{1}{T_1}\right)$$

Remember to convert temperatures to the Kelvin scale.

$$-0.693 = \frac{E_a}{8.314\times10^{-3} \text{ kJ/mol}\cdot\text{K}}\left(\frac{1}{275.4 \text{ K}} - \frac{1}{272.1 \text{ K}}\right)$$

$$\boldsymbol{E_a} = \frac{(-0.693)(8.314\times10^{-3} \text{ kJ/mol}\cdot\text{K})}{-4.4\times10^{-5}\text{K}^{-1}} = \boldsymbol{1.3\times10^2 \textbf{ kJ/mol}}$$

For maximum freshness, fish should be frozen immediately after capture and kept frozen until cooked.

14.47 One form of the Arrhenius equation (Equation 14.11 of the text) relates the rate constant at one temperature to the rate constant at another temperature. The data are: $T_1 = 250°\text{C} = 523$ K, $T_2 = 150°\text{C} = 423$ K, and $k_1/k_2 = 1.50\times10^3$. Substituting into Equation 14.11,

$$\ln\frac{k_1}{k_2} = \frac{E_a}{R}\left(\frac{1}{T_2} - \frac{1}{T_1}\right)$$

$$\ln(1.50\times10^3) = \frac{E_a}{8.314 \text{ J/mol}\cdot\text{K}}\left(\frac{1}{423 \text{ K}} - \frac{1}{523 \text{ K}}\right)$$

$$7.31 = \frac{E_a}{8.314\frac{\text{J}}{\text{mol}\cdot\text{K}}}\left(4.52\times10^{-4}\frac{1}{\text{K}}\right)$$

$$\boldsymbol{E_a = 1.3\times10^5 \textbf{ J/mol} = 1.3\times10^2 \textbf{ kJ/mol}}$$

14.49 **Diagram (a).** At a higher temperature, the reaction would proceed at a higher rate, resulting in the formation of more product.

14.59 a. The order of the reaction is simply the sum of the exponents in the rate law (Section 14.2 of the text). The reaction is **second-order**.

b. The rate law reveals the identity of the substances participating in the slow or rate-determining step of a reaction mechanism. This rate law implies that the slow step involves the reaction of a molecule of NO with a molecule of $Cl_2$. If this is the case, then **the first step is the slower (rate-determining) step**.

14.61 The experimentally determined rate law is first order in $H_2$ and second order in NO. In Mechanism I the slow step is bimolecular and the rate law would be:

$$\text{rate} = k[H_2][NO]$$

**Mechanism I** can be discarded.

The rate-determining step in Mechanism II involves the simultaneous collision of two NO molecules with one $H_2$ molecule. The rate law would be:

$$\text{rate} = k[H_2][NO]^2$$

**Mechanism II is a possibility**.

In Mechanism III we assume the forward and reverse reactions in the first fast step are in dynamic equilibrium, so their rates are equal:

$$k_f[NO]^2 = k_r[N_2O_2]$$

The slow step is bimolecular and involves collision of a hydrogen molecule with a molecule of $N_2O_2$. The rate would be:

$$\text{rate} = k_2[H_2][N_2O_2]$$

If we solve the dynamic equilibrium equation of the first step for $[N_2O_2]$ and substitute into the above equation, we have the rate law:

$$\text{rate} = \frac{k_2 k_f}{k_r}[H_2][NO]^2 = k[H_2][NO]^2$$

**Mechanism III is also a possibility**.

**Think About It:** Can you suggest an experiment that might help to decide between the two mechanisms?

14.71 Catalysts lower the activation energy but do not change the enthalpies of the reactants and products (see Figure 14.15 in the text). Look for a pair of figures that differ only in the height of the activation barrier. Figures (i) and (iv) show reactants at the same energy and products at the same energy, but show two different barrier heights. So, **(i) and (iv)** represent the same reaction catalyzed and un-catalyzed.

14.73 Temperature, energy of activation, concentration of reactants, and a catalyst.

14.75 Strictly, the **temperature must be specified** whenever the rate or rate constant of a reaction is quoted.

14.77 Using the diagrams, we see that the half-life of the reaction is 20 s. We rearrange Equation 14.5 to solve for rate constant:

$$k = \frac{0.693}{20\,\text{s}} = \mathbf{0.035\ s^{-1}}$$

14.79 **Strategy:** Since the reaction is first-order, use a plot of the given data and Equation 14.4 to determine $k$. Then use the value of $k$ in Equation 14.5 to find $t_{1/2}$. Use a spreadsheet to fit a straight line to the plot of ln($P$) vs. $t$, as shown below.

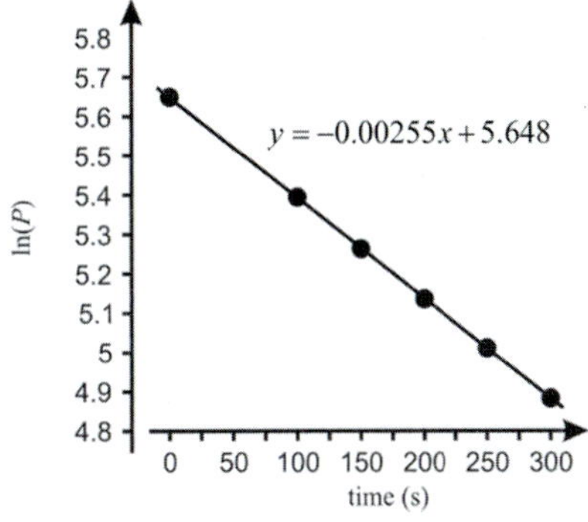


**Solution:** According to equation 14.4, the slope of ln($P$) vs. $t$ is $-k$ so $k = 0.00255\ \text{s}^{-1}$. Using Equation 14.5,

$$t_{1/2} = \frac{0.693}{0.00255\ \text{s}^{-1}} = \mathbf{272\ s}$$

14.81 Since the methanol contains no oxygen–18, the oxygen atom must come from the phosphate group and not the water. The mechanism must involve a bond–breaking process like:

$$CH_3—O \wr P(=O)(—O—H)(—O—H)$$

14.83 Most transition metals have several stable oxidation states. This allows the metal atoms to act as either a source or a receptor of electrons in a broad range of reactions.

14.85 a. To determine the rate law, we must determine the exponents in the equation

$$\text{rate} = k[\text{CH}_3\text{COCH}_3]^x[\text{Br}_2]^y[\text{H}^+]^z$$

To determine the order of the reaction with respect to $CH_3COCH_3$, find two experiments in which the $[Br_2]$ and $[H^+]$ are held constant. Compare the data from experiments (1) and (5). When the concentration of $CH_3COCH_3$ is increased by a factor of 1.33, the reaction rate increases by a factor of 1.33. Thus, the reaction is first-order in $CH_3COCH_3$.

To determine the order with respect to $Br_2$, compare experiments (1) and (2). When the $Br_2$ concentration is doubled, the reaction rate does not change. Thus, the reaction is zero-order in $Br_2$.

To determine the order with respect to $H^+$, compare experiments (1) and (3). When the $H^+$ concentration is doubled, the reaction rate doubles. Thus, the reaction is first-order in $H^+$.

The rate law is:

$$\textbf{rate} = \boldsymbol{k}[\textbf{CH}_3\textbf{COCH}_3][\textbf{H}^+]$$

b. Rearrange the rate law from part (a), solving for $k$.

$$k = \frac{\text{rate}}{[\text{CH}_3\text{COCH}_3][\text{H}^+]}$$

Substitute the data from any one of the experiments to calculate $k$. Using the data from Experiment (1),

$$\boldsymbol{k} = \frac{5.7 \times 10^{-5}\ M/\text{s}}{(0.30\ M)(0.050\ M)} = \mathbf{3.8 \times 10^{-3}\ /\boldsymbol{M} \cdot s}$$

(The units $/M{\cdot}s$ can also be expressed as $M^{-1}s^{-1}$.)

c. Let $k_2$ be the rate constant for the slow step:

$$\text{rate} = k_2[\text{CH}_3\text{–}\overset{\overset{+}{\text{OH}}}{\overset{\|}{\text{C}}}\text{—CH}_3][\text{H}_2\text{O}] \qquad (1)$$

Let $k_1$ and $k_{-1}$ be the rate constants for the forward and reverse steps in the fast equilibrium.

$$k_1[\text{CH}_3\text{COCH}_3][\text{H}_3\text{O}^+] = k_{-1}[\text{CH}_3\text{-}\overset{\overset{+}{\text{OH}}}{\overset{\|}{\text{C}}}\text{—CH}_3][\text{H}_2\text{O}] \qquad (2)$$

Therefore, Equation (1) becomes

$$\text{rate} = \frac{k_1 k_2}{k_{-1}}[CH_3COCH_3][H_3O^+]$$

which is the same as (a), where $\boldsymbol{k = k_1k_2/k_{-1}}$.

14.87 $Fe^{3+}$ undergoes a redox cycle: $Fe^{3+} \rightarrow Fe^{2+} \rightarrow Fe^{3+}$

| | |
|---|---|
| $Fe^{3+}$ oxidizes $I^-$: | $2Fe^{3+} + 2I^- \rightarrow 2Fe^{2+} + I_2$ |
| $Fe^{2+}$ reduces $S_2O_8^{2-}$: | $2Fe^{2+} + S_2O_8^{2-} \rightarrow 2Fe^{3+} + 2SO_4^{2-}$ |
| Overall Reaction: | $2I^- + S_2O_8^{2-} \rightarrow I_2 + 2SO_4^{2-}$ |

The uncatalyzed reaction is slow because both $I^-$ and $S_2O_8^{2-}$ are negatively charged which makes their mutual approach unfavorable.

14.89 For a rate law, *zero order* means that the exponent is zero. In other words, the reaction rate is just equal to a constant; it doesn't change as time passes.

a. **(i)** The rate law would be:

$$\text{rate} = k[A]^0 = k$$

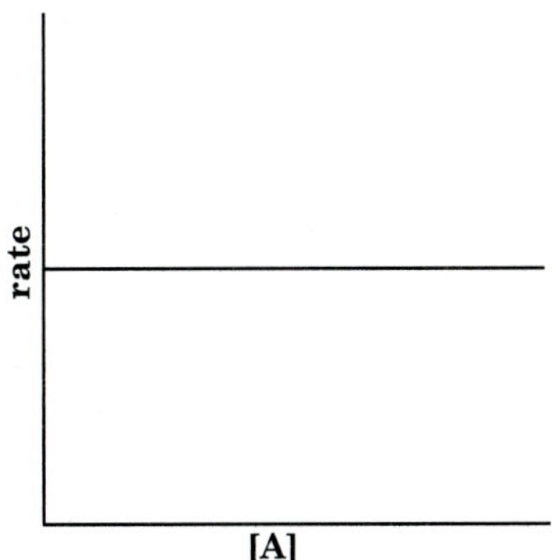


**(ii)** The integrated zero-order rate law is: $[A] = -kt + [A]_0$. Therefore, a plot of [A] versus time should be a straight line with a slope equal to $-k$.

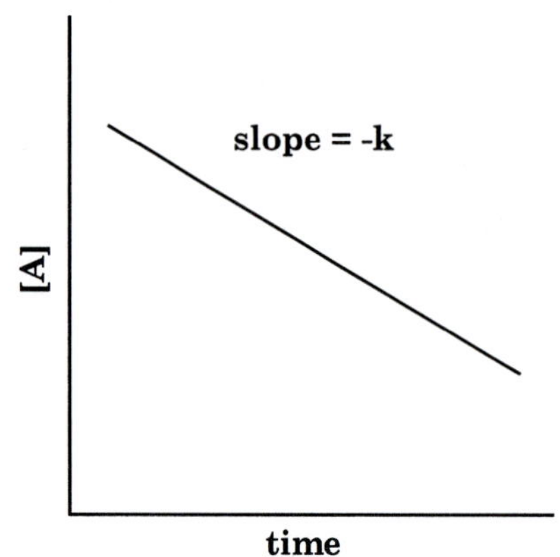


b. $$[A] = [A]_0 - kt$$

At $t_{\frac{1}{2}}$, $[A] = \dfrac{[A]_0}{2}$. Substituting into the above equation:

$$\frac{[A]_0}{2} = [A]_0 - kt_{\frac{1}{2}}$$

$$\boldsymbol{t_{\frac{1}{2}} = \frac{[A]_0}{2k}}$$

$$k = \frac{[A]_0}{2t_{\frac{1}{2}}}$$

c. When $[A] = 0$,

$$[A]_0 = kt$$

$$t = \frac{[A]_0}{k}$$

Substituting for $k$,

$$t = \frac{[A]_0}{\dfrac{[A]_0}{2t_{\frac{1}{2}}}}$$

$$\boldsymbol{t = 2t_{\frac{1}{2}}}$$

This indicates that the integrated rate law is no longer valid after ***two*** half-lives.

14.91 There are three gases present and we can measure only the total pressure of the gases. To measure the partial pressure of azomethane at a particular time, we must withdraw a sample of the mixture, analyze and determine the mole fractions. Then,

$$P_{\text{azomethane}} = P_T X_{\text{azomethane}}$$

This is a rather tedious process if many measurements are required. A mass spectrometer will help.

14.93

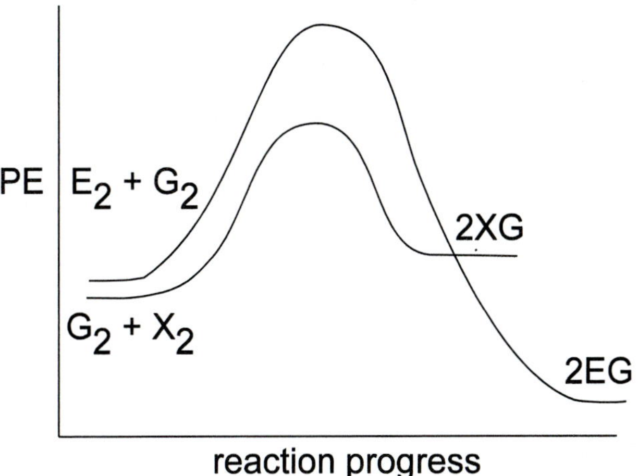


14.95 a. A catalyst works by changing the reaction mechanism, thus lowering the activation energy.

b. A catalyst changes the reaction mechanism.

c. A catalyst does not change the enthalpy of reaction.

d. A catalyst increases the forward rate of reaction.

e. A catalyst increases the reverse rate of reaction.

14.97 At very high $[H_2]$,

$$k_2[H_2] >> 1$$

$$\textbf{rate} = \frac{k_1[NO]^2[H_2]}{k_2[H_2]} = \frac{k_1}{k_2}[NO]^2$$

At very low $[H_2]$,

$$k_2[H_2] \ll 1$$

$$\textbf{rate} = \frac{k_1[NO]^2[H_2]}{1} = \boldsymbol{k_1[NO]^2[H_2]}$$

The result from Problem 14.80 agrees with the rate law determined for low $[H_2]$.

14.99 First we plot the data for the reaction: $2N_2O_5 \rightarrow 4NO_2 + O_2$

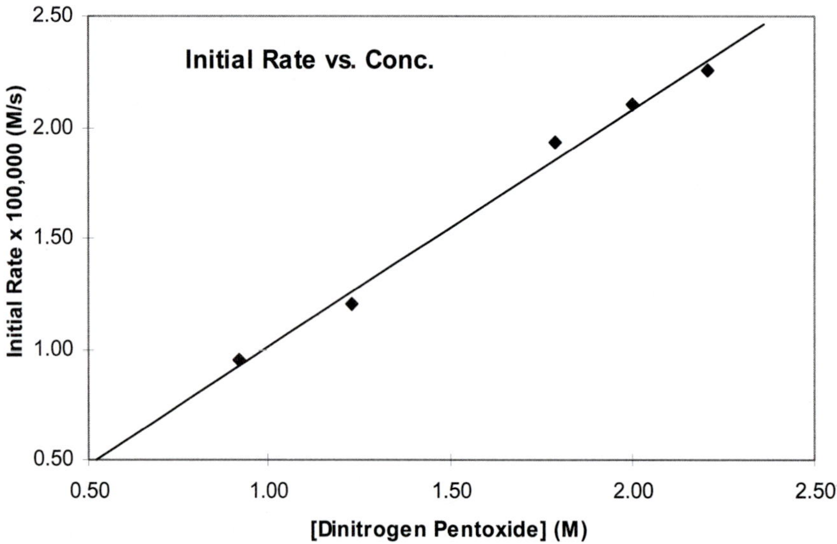


The data is linear, which means that the initial rate is directly proportional to the concentration of $N_2O_5$.

Thus, the rate law is:

$$\textbf{Rate} = \boldsymbol{k[N_2O_5]}$$

The rate constant $k$ can be determined from the slope of the graph $\left(\frac{\Delta(\text{Initial Rate})}{\Delta[N_2O_5]}\right)$ or by using any set of data.

$$\boldsymbol{k = 1.0 \times 10^{-5}\ s^{-1}}$$

Note that the rate law is ***not*** Rate $= k[N_2O_5]^2$, as we might expect from the balanced equation. In general, the order of a reaction must be determined by experiment; it cannot be deduced from the coefficients in the balanced equation.

14.101 The red bromine vapor absorbs photons of blue light and dissociates to form bromine atoms.

$$Br_2 \rightarrow 2Br\cdot$$

The bromine atoms collide with methane molecules and abstract hydrogen atoms.

$$Br\cdot + CH_4 \rightarrow HBr + \cdot CH_3$$

The methyl radical then reacts with $Br_2$, giving the observed product and regenerating a bromine atom to start the process over again:

$$\cdot CH_3 + Br_2 \rightarrow CH_3Br + Br\cdot$$

$$Br\cdot + CH_4 \rightarrow HBr + \cdot CH_3$$

and so on...

14.103 a. The units of the rate constant show the reaction to be second-order, meaning the rate law is most likely:

$$\text{Rate} = k[H_2][I_2]$$

We can use the ideal gas equation to solve for the concentrations of $H_2$ and $I_2$. We can then solve for the initial rate in terms of $H_2$ and $I_2$ and then convert to the initial rate of formation of HI. We carry an extra significant figure throughout this calculation to minimize rounding errors.

$$n = \frac{PV}{RT}$$

$$\frac{n}{V} = M = \frac{P}{RT}$$

Since the total pressure is 1658 mmHg and there are equimolar amounts of $H_2$ and $I_2$ in the vessel, the partial pressure of each gas is 829 mmHg.

$$[H_2] = [I_2] = \frac{\left(829\ \text{mmHg} \times \dfrac{1\ \text{atm}}{760\ \text{mmHg}}\right)}{\left(0.0821\dfrac{\text{L}\cdot\text{atm}}{\text{mol}\cdot\text{K}}\right)(400 + 273)\,\text{K}} = 0.01974\ M$$

Let's convert the units of the rate constant to $/M{\cdot}\text{min}$, and then we can substitute into the rate law to solve for rate.

$$k = 2.42\times 10^{-2}\,\frac{1}{M\cdot\text{s}} \times \frac{60\ \text{s}}{1\ \text{min}} = 1.452\frac{1}{M\cdot\text{min}}$$

$$\text{Rate} = k[H_2][I_2]$$

$$\text{Rate} = \left(1.452\frac{1}{M\cdot\text{min}}\right)(0.01974\ M)(0.01974\ M) = 5.658\times 10^{-4}\ M/\text{min}$$

We know that,

$$\text{Rate} = \frac{1}{2}\frac{\Delta[\text{HI}]}{\Delta t}$$

or

$$\frac{\Delta[\text{HI}]}{\Delta t} = 2 \times \text{Rate} = (2)(5.658 \times 10^{-4}\ M/\text{min}) = \mathbf{1.13 \times 10^{-3}}\ \boldsymbol{M}\mathbf{/min}$$

b. We can use the second-order integrated rate law to calculate the concentration of $H_2$ after 10.0 minutes. We can then substitute this concentration back into the rate law to solve for rate.

$$\frac{1}{[\text{H}_2]_t} = kt + \frac{1}{[\text{H}_2]_0}$$

$$\frac{1}{[\text{H}_2]_t} = \left(1.452\frac{1}{M\cdot\text{min}}\right)(10.0\ \text{min}) + \frac{1}{0.01974\ M}$$

$$[\text{H}_2]_t = 0.01534\ M$$

We can now substitute this concentration back into the rate law to solve for rate. The concentration of $I_2$ after 10.0 minutes will also equal 0.01534 *M*.

$$\text{Rate} = k[\text{H}_2][\text{I}_2]$$

$$\text{Rate} = \left(1.452\frac{1}{M\cdot\text{min}}\right)(0.01534\ M)(0.01534\ M) = 3.417 \times 10^{-4}\ M/\text{min}$$

We know that,

$$\text{Rate} = \frac{1}{2}\frac{\Delta[\text{HI}]}{\Delta t}$$

or

$$\frac{\Delta[\text{HI}]}{\Delta t} = 2 \times \text{Rate} = (2)(3.417 \times 10^{-4}\ M/\text{min}) = \mathbf{6.83 \times 10^{-4}}\ \boldsymbol{M}\mathbf{/min}$$

The concentration of HI after 10.0 minutes is:

$$[\text{HI}]_t = ([\text{H}_2]_0 - [\text{H}_2]_t) \times 2$$

$$\mathbf{[HI]}_t = (0.01974\ M - 0.01534\ M) \times 2 = \mathbf{8.8 \times 10^{-3}}\ \boldsymbol{M}$$

14.105 a. We can write the rate law for an elementary step directly from the stoichiometry of the balanced reaction. In this rate-determining elementary step three molecules must collide simultaneously (one X and two Y's). This makes the reaction termolecular, and consequently the rate law must be third order: first order in X and second order in Y.

The rate law is:

$$\textbf{rate} = \boldsymbol{k}\textbf{[X][Y]}^2$$

b. The value of the rate constant can be found by solving algebraically for $k$.

$$\boldsymbol{k} = \frac{\text{rate}}{[\text{X}][\text{Y}]^2} = \frac{3.8 \times 10^{-3}\ M/\text{s}}{(0.26\ M)(0.88\ M)^2} = \mathbf{1.9 \times 10^{-2}}\ \boldsymbol{M}^{\mathbf{-2}}\mathbf{s}^{\mathbf{-1}} \text{ or } \mathbf{0.019}\ \boldsymbol{M}^{\mathbf{-2}}\mathbf{s}^{\mathbf{-1}}$$

14.107 a. According to the given information, the ratio $k_1/k_2$ is larger for reaction A than for reaction B:

$$\frac{k_1(A)}{k_2(A)} > \frac{k_1(B)}{k_2(B)}.$$

Since the natural log function ln($x$) is always increasing with increasing $x$, taking the natural log of both sides of the above inequality does not affect the direction of the inequality symbol:

$$\ln\left(\frac{k_1(A)}{k_2(A)}\right) > \ln\left(\frac{k_1(B)}{k_2(B)}\right).$$

Now, use Equation 14.11 to substitute for each side of the inequality:

$$\frac{E_a(A)}{R}\left(\frac{1}{T_2} - \frac{1}{T_1}\right) > \frac{E_a(B)}{R}\left(\frac{1}{T_2} - \frac{1}{T_1}\right).$$

Multiply both sides by $\dfrac{R}{\dfrac{1}{T_2} - \dfrac{1}{T_1}}$ (a *negative* quantity) to get:

$E_a(A) < E_a(B)$ (note the change in the direction of the inequality symbol).

So, we can conclude that **the activation energy of reaction B is larger than that of reaction A.**

b. $E_a \approx 0$. Orientation factor is not important.

14.109 a.

$$\frac{\Delta[\text{B}]}{\Delta t} = k_1[\text{A}] - k_2[\text{B}]$$

b. If $\frac{\Delta[B]}{\Delta t} = 0$, then, from part (a) of this problem:

$$k_1[A] = k_2[B]$$

$$\mathbf{[B] = \frac{k_1}{k_2}[A]}$$

14.111 a. There are three elementary steps: A → B, B → C, and C → D.

b. There are two intermediates: B and C.

c. The third step, C → D, is rate determining because it has the largest activation energy.

d. The overall reaction is exothermic.

14.113 Initially, the number of moles of gas in terms of the volume is:

$$n = \frac{PV}{RT} = \frac{(0.350 \text{ atm})V}{\left(0.0821\frac{\text{L}\cdot\text{atm}}{\text{mol}\cdot\text{K}}\right)(450 + 273)\text{K}} = 5.90 \times 10^{-3}\, V$$

We can calculate the concentration of dimethyl ether from the following equation.

$$\ln\frac{[(CH_3)_2O]_t}{[(CH_3)_2O]_0} = -kt$$

$$\frac{[(CH_3)_2O]_t}{[(CH_3)_2O]_0} = e^{-kt}$$

Since, the volume is held constant, it will cancel out of the equation. The concentration of dimethyl ether after 8.0 minutes (480 s) is:

$$[(CH_3)_2O]_t = \left(\frac{5.90 \times 10^{-3}\, V}{V}\right) e^{-\left(3.2 \times 10^{-4}\frac{1}{\text{s}}\right)(480\text{ s})}$$

$$[(CH_3)_2O]_t = 5.06 \times 10^{-3}\, M$$

After 8.0 min, the concentration of $(CH_3)_2O$ has decreased by $(5.90 \times 10^{-3} - 5.06 \times 10^{-3})M$ or $8.4 \times 10^{-4}\, M$. Since three moles of product form for each mole of dimethyl ether that reacts, the concentrations of the products are $(3)(8.4 \times 10^{-4}\, M) = 2.5 \times 10^{-3}\, M$.

The pressure of the system after 8.0 minutes is:

$$P = \frac{nRT}{V} = \left(\frac{n}{V}\right)RT = MRT$$

$$P = [(5.06 \times 10^{-3}) + (2.5 \times 10^{-3})]M \times (0.0821\ \text{L·atm/mol·K})(723\ \text{K})$$

$$\mathbf{P = 0.45\ atm}$$

14.115 a. Catalyst: $Mn^{2+}$; intermediates: $Mn^{3+}$, $Mn^{4+}$

First step is rate-determining.

b. Without the catalyst, the reaction would be a termolecular one involving 3 cations! ($Tl^+$ and two $Ce^{4+}$). The reaction would be slow.

c. The catalyst is a homogeneous catalyst because it has the same phase (aqueous) as the reactants.

14.117 **Strategy:** According to the potential energy diagram shown in Figure 14.12 of the text, the reverse activation $E_{a,r}$ is equal to the forward activation energy $E_{a,f}$ plus the change in potential energy from reactants to products (see below).

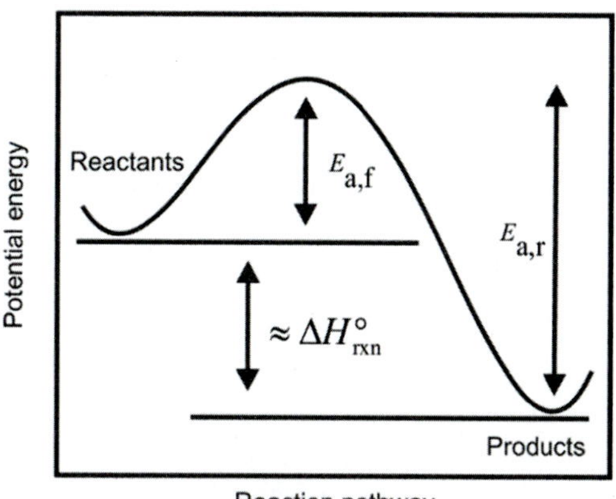


Using $\Delta H^\circ_{rxn}$ as an approximation for the change in the potential energies, we can write

$E_{a,r} \approx E_{a,f} + \Delta H^\circ_{rxn}$.

14.119

$$t_{\frac{1}{2}} \propto \frac{1}{[A]_0^{n-1}}$$

$$t_{\frac{1}{2}} = C\frac{1}{[A]_0^{n-1}}$$, where $C$ is a proportionality constant.

Substituting in for zero, first, and second-order reactions gives:

$$n = 0 \qquad t_{\frac{1}{2}} = C\frac{1}{[A]_0^{-1}} = C[A]_0$$

$$n = 1 \qquad t_{\frac{1}{2}} = C\frac{1}{[A]_0^{0}} = C$$

$$n = 2 \qquad t_{\frac{1}{2}} = C\frac{1}{[A]_0}$$

**Think About It:** Compare these results with those in Table 14.5 of the text. What is $C$ in each case?

14.121 a. The first-order rate constant can be determined from the half-life.

$$t_{\frac{1}{2}} = \frac{0.693}{k}$$

$$k = \frac{0.693}{t_{\frac{1}{2}}} = \frac{0.693}{28.1\ \text{yr}} = \mathbf{0.0247\ yr^{-1}}$$

b. See Problem 14.94. Mathematically, the amount left after ten half–lives is:

$$\left(\frac{1}{2}\right)^{10} = \mathbf{9.8 \times 10^{-4}}$$

c. If 99.0% has disappeared, then 1.0% remains. The ratio of $[A]_t/[A]_0$ is 1.0%/100% or 0.010/1.00. Substitute into the first-order integrated rate law, Equation 14.3 of the text, to determine the time.

$$\ln\frac{[A]_t}{[A]_0} = -kt$$

$$\ln\frac{0.010}{1.0} = -(0.0247\ \text{yr}^{-1})t$$

$$-4.61 = -(0.0247\ \text{yr}^{-1})t$$

$$t = \mathbf{187\ yr}$$

14.123 First, solve for the rate constant, $k$, from the half-life of the decay.

$$t_{\frac{1}{2}} = 2.44 \times 10^5\ \text{yr} = \frac{0.693}{k}$$

$$k = \frac{0.693}{2.44 \times 10^5\ \text{yr}} = 2.84 \times 10^{-6}\ \text{yr}^{-1}$$

Now, we can calculate the time for the plutonium to decay from $5.0 \times 10^2$ g to $1.0 \times 10^2$ g using the equation for a first-order reaction relating concentration and time.

$$\ln\frac{[\text{A}]_t}{[\text{A}]_0} = -kt$$

$$\ln\frac{1.0 \times 10^2}{5.0 \times 10^2} = -(2.84 \times 10^{-6}\ \text{yr}^{-1})t$$

$$-1.61 = -(2.84 \times 10^{-6}\ \text{yr}^{-1})t$$

$$t = \mathbf{5.7 \times 10^5\ yr}$$

14.125 The half-life is related to the initial concentration of A by

$$t_{\frac{1}{2}} \propto \frac{1}{[\text{A}]_0^{n-1}}$$

According to the data given, the half-life doubled when $[\text{A}]_0$ was halved. This is only possible if the half-life is inversely proportional to $[\text{A}]_0$. Substituting $n = 2$ into the above equation gives:

$$t_{\frac{1}{2}} \propto \frac{1}{[\text{A}]_0}$$

Looking at this equation, it is clear that if $[\text{A}]_0$ is halved, the half-life would double. The reaction is **second-order**.

We use Equation 14.7 of the text to calculate the rate constant.

$$t_{\frac{1}{2}} = \frac{1}{k[\text{A}]_0}$$

$$\boldsymbol{k} = \frac{1}{[\text{A}]_0 t_{\frac{1}{2}}} = \frac{1}{(1.20\ M)(2.0\ \text{min})} = \mathbf{0.42\ /\mathit{M}\cdot min}$$

14.127 a. The rate law for the reaction is:

$$\text{rate} = k[\text{Hb}][\text{O}_2]$$

We are given the rate constant and the concentration of Hb and $O_2$, so we can substitute in these quantities to solve for rate.

$$\text{rate} = (2.1 \times 10^6\ /M\cdot\text{s})(8.0 \times 10^{-6}\ M)(1.5 \times 10^{-6}\ M)$$

$$\textbf{rate} = \mathbf{2.5 \times 10^{-5}\ \textit{M}/s}$$

b. If $HbO_2$ is being formed at the rate of $2.5 \times 10^{-5}$ *M*/s, then $O_2$ is being consumed at the same rate, **$2.5 \times 10^{-5}$ *M*/s**. Note the 1:1 mole ratio between $O_2$ and $HbO_2$.

c. The rate of formation of $HbO_2$ increases, but the concentration of Hb remains the same. Assuming that temperature is constant, we can use the same rate constant as in part (a). We substitute rate, [Hb], and the rate constant into the rate law to solve for $O_2$ concentration.

$$\text{rate} = k[\text{Hb}][\text{O}_2]$$

$$1.4 \times 10^{-4}\ M/\text{s} = (2.1 \times 10^6\ /M\cdot\text{s})(8.0 \times 10^{-6}\ M)[\text{O}_2]$$

$$\mathbf{[O_2] = 8.3 \times 10^{-6}\ \textit{M}}$$

14.129 Lowering the temperature would slow all chemical reactions, which would be especially important for those that might damage the brain.

14.131 First, we write an overall balanced equation.

$$\text{P} \rightarrow \text{P}^*$$

$$\text{P}^* \rightarrow \text{P}_2$$

---

$$\text{P} \rightarrow \tfrac{1}{2}\text{P}_2$$

The average molar mass is given by:

$$\overline{\mathcal{M}} = \frac{([\text{P}]_t\mathcal{M} + 2[\text{P}_2]\mathcal{M})\dfrac{\text{mol}}{\text{L}} \times \dfrac{\text{g}}{\text{mol}}}{([\text{P}]_t + [\text{P}_2])\dfrac{\text{mol}}{\text{L}}} \qquad (1)$$

where $\mathcal{M}$ is the molar mass of P and $[\text{P}]_t$ is the concentration of P at a later time in the reaction. Note that in the numerator $[\text{P}_2]$ is multiplied by 2 because the molar mass of $P_2$ is double that of P. Also note that the units work out to give units of molar mass, g/mol.

Based on the stoichiometry of the reaction, the concentration of $[P_2]$ is:

$$[P_2] = \frac{[P]_0 - [P]_t}{2}$$

Substituting back into Equation (1) gives:

$$\overline{\mathcal{M}} = \frac{\left([P]_t\mathcal{M} + 2\frac{[P]_0 - [P]_t}{2}\mathcal{M}\right)}{\left([P]_t + \frac{[P]_0 - [P]_t}{2}\right)} = \frac{([P]_t\mathcal{M} + [P]_0\mathcal{M} - [P]_t\mathcal{M})}{\left([P]_t + \frac{1}{2}[P]_0 - \frac{1}{2}[P]_t\right)} = \frac{2\mathcal{M}[P]_0}{[P]_0 + [P]_t} \qquad (2)$$

In the proposed mechanism, the denaturation step is rate-determining. Thus,

$$\text{Rate} = k[P]$$

Because we are looking at change in concentration over time, we need the first-order integrated rate law, Equation 14.3 of the text.

$$\ln\frac{[P]_t}{[P]_0} = -kt$$

$$\frac{[P]_t}{[P]_0} = e^{-kt}$$

$$[P]_t = [P]_0e^{-kt}$$

Substituting into Equation (2) gives:

$$\overline{\mathcal{M}} = \frac{2\mathcal{M}[P]_0}{[P]_0 + [P]_0e^{-kt}} = \frac{2\mathcal{M}}{1 + e^{-kt}}$$

or

$$\frac{2\mathcal{M} - \overline{\mathcal{M}}}{\overline{\mathcal{M}}} = e^{-kt}$$

$$\ln\left(\frac{2\mathcal{M} - \overline{\mathcal{M}}}{\overline{\mathcal{M}}}\right) = -kt$$

The rate constant, $k$, can be determined by plotting $\ln\left(\frac{2\mathcal{M} - \overline{\mathcal{M}}}{\overline{\mathcal{M}}}\right)$ versus $t$. The plot will give a straight line with a slope of $-k$.

14.133

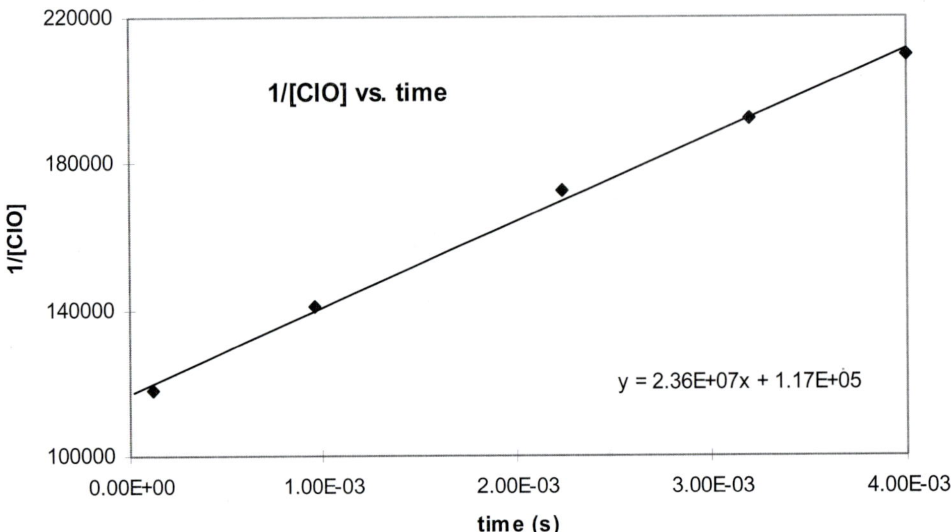


Reaction is **second-order** because a plot of 1/[ClO] vs. time is a straight line. The slope of the line equals the rate constant, $k$.

$$\boldsymbol{k} = \text{Slope} = \mathbf{2.4 \times 10^7} \; \boldsymbol{/M}\mathbf{\cdot s} \text{ or } \mathbf{2.4 \times 10^7} \; \boldsymbol{M}^{\mathbf{-1}}\mathbf{s^{-1}}$$

14.135 a. The relationship between half-life and rate constant is given in Equation 14.5 of the text. Rearranging to solve for rate constant gives

$$k = \frac{0.693}{t_{\frac{1}{2}}}$$

$$k = \frac{0.693}{19.8 \text{ min}}$$

$$\boldsymbol{k} = \mathbf{0.0350 \; min^{-1}}$$

b. Following the same procedure as in part (a), we find the rate constant at 70°C to be $1.58 \times 10^{-3}$ min$^{-1}$. We now have two values of rate constants ($k_1$ and $k_2$) at two temperatures ($T_1$ and $T_2$). This information allows us to calculate the activation energy, $E_a$, using Equation 14.11 of the text.

$$\ln\frac{k_1}{k_2} = \frac{E_a}{R}\left(\frac{1}{T_2} - \frac{1}{T_1}\right)$$

$$\ln\left(\frac{0.0350 \text{ min}^{-1}}{1.58 \times 10^{-3} \text{ min}^{-1}}\right) = \frac{E_a}{(8.314 \text{ J/mol}\cdot\text{K})}\left(\frac{1}{343 \text{ K}} - \frac{1}{373 \text{ K}}\right)$$

$$E_a = 1.1 \times 10^5 \text{ J/mol} = \mathbf{110\ kJ/mol}$$

c. Since all the steps are elementary steps, we can deduce the rate law simply from the equations representing the steps. The rate laws are:

Initiation:

$$\text{rate} = k_i[R_2]$$

Propagation:

$$\text{rate} = k_p[M][M_1]$$

Termination:

$$\text{rate} = k_t[M'][M'']$$

The reactant molecules are the ethylene monomers, and the product is polyethylene. Recalling that intermediates are species that are formed in an early elementary step and consumed in a later step, we see that they are the radicals M′·, M″·, and so on. (The R· species also qualifies as an intermediate.)

d. The growth of long polymers would be favored by a high rate of propagations and a low rate of termination. Since the rate law of propagation depends on the concentration of monomer, an increase in the concentration of ethylene would increase the propagation (growth) rate. From the rate law for termination we see that a low concentration of the radical fragment M′· or M″· would lead to a slower rate of termination. This can be accomplished by using a low concentration of the initiator, $R_2$.

# Chapter 15

# Chemical Equilibrium

15.9 a. $$\frac{[N_2][H_2O]^2}{[NO]^2[H_2]^2}$$

b. $$K = \frac{(0.082)(4.64)^2}{(0.31)^2(0.16)^2} = 7.2 \times 10^2$$

15.11 The problem states that the system is at equilibrium, so we simply substitute the equilibrium concentrations into the equilibrium expression to calculate $K_c$.

*Step 1:* Calculate the concentrations of the components in units of mol/L. The molarities can be calculated by simply dividing the number of moles by the volume of the flask.

$$[H_2] = \frac{2.50 \text{ mol}}{12.0 \text{ L}} = 0.208\ M$$

$$[S_2] = \frac{1.35 \times 10^{-5} \text{ mol}}{12.0 \text{ L}} = 1.13 \times 10^{-6}\ M$$

$$[H_2S] = \frac{8.70 \text{ mol}}{12.0 \text{ L}} = 0.725\ M$$

*Step 2:* Once the molarities are known, $K_c$ can be found by substituting the molarities into the equilibrium expression.

$$K_c = \frac{[H_2S]^2}{[H_2]^2[S_2]} = \frac{(0.725)^2}{(0.208)^2(1.13 \times 10^{-6})} = \mathbf{1.08 \times 10^7}$$

**Think About It:** If you forget to convert moles to moles/liter, will you get a different answer? Under what circumstances will the two answers be the same?

15.13 The equilibrium expression for this reaction is

$$K_c = \frac{[B]^2}{[A]^2}$$

According to the first diagram, there are 6 white spheres (A) and 4 red spheres (B). The equilibrium expression becomes

$$K_c = \frac{[B]^2}{[A]^2} = \frac{(4)^2}{(6)^2} = \frac{16}{36}$$

In the second diagram, there are 9 white spheres (A). For this reaction to also be at equilibrium,

$$\frac{16}{36} = \frac{x^2}{(9)^2} \text{ or } \frac{x^2}{81}$$

$$x^2 = 36$$

$$x = 6$$

there must be **6** red spheres.

15.21 When the equation for a reversible reaction is written in the opposite direction, the equilibrium constant becomes the reciprocal of the original equilibrium constant.

$$\mathbf{K'} = \frac{1}{K} = \frac{1}{4.17 \times 10^{-34}} = \mathbf{2.40 \times 10^{33}}$$

15.23 **Strategy:** The relationship between $K_c$ and $K_P$ is given by Equation 15.4 of the text. What is the change in the number of moles of gases from reactant to product? Recall that

$$\Delta n = \text{moles of gaseous products} - \text{moles of gaseous reactants}$$

What unit of temperature should we use?

**Solution:** The relationship between $K_c$ and $K_P$ is given by Equation 15.4 of the text.

$$K_P = K_c(0.0821\,T)^{\Delta n}$$

Rearrange the equation relating $K_P$ and $K_c$, solving for $K_c$.

$$K_c = \frac{K_P}{(0.0821T)^{\Delta n}}$$

Because $T = 623$ K and $\Delta n = 3 - 2 = 1$, we have:

$$\boldsymbol{K_c} = \frac{K_P}{(0.0821T)^{\Delta n}} = \frac{1.8 \times 10^{-5}}{(0.0821)(623\ \text{K})} = \mathbf{3.5 \times 10^{-7}}$$

15.25 The equilibrium expressions are:

a.

$$K_c = \frac{[NH_3]^2}{[N_2][H_2]^3}$$

Substituting the given equilibrium concentration gives:

$$\boldsymbol{K_c} = \frac{(0.25)^2}{(0.11)(1.91)^3} = \mathbf{0.082} \text{ or } \mathbf{8.2 \times 10^{-2}}$$

b.

$$K_c = \frac{[NH_3]}{[N_2]^{\frac{1}{2}}[H_2]^{\frac{3}{2}}}$$

Substituting the given equilibrium concentration gives:

$$\boldsymbol{K_c} = \frac{(0.25)}{(0.11)^{\frac{1}{2}}(1.91)^{\frac{3}{2}}} = \mathbf{0.29}$$

**Think About It:** Is there a relationship between the $K_c$ values from parts (a) and (b)?

15.27 Because pure solids do not enter into an equilibrium expression, we can calculate $K_P$ directly from the pressure that is due solely to $CO_2(g)$.

$$\boldsymbol{K_P} = P_{CO_2} = \mathbf{0.105}$$

Now, we can convert $K_P$ to $K_c$ using the following equation.

$$K_P = K_c(0.0821\,T)^{\Delta n}$$

$$K_c = \frac{K_P}{(0.0821T)^{\Delta n}}$$

$$\boldsymbol{K_c} = \frac{0.105}{(0.0821 \times 623)^{(1-0)}} = \mathbf{2.05 \times 10^{-3}}$$

15.29 **Strategy:** Because they are constant quantities, the concentrations of solids and liquids do not appear in the equilibrium expressions for heterogeneous systems. The total pressure at equilibrium that is given is due to both $NH_3$ and $CO_2$. Note that for every 1 atm of $CO_2$ produced, 2 atm of $NH_3$ will be produced due to the stoichiometry of the balanced equation. Using this ratio, we can calculate the partial pressures of $NH_3$ and $CO_2$ at equilibrium.

**Solution:** Expression for the reaction is

$$K_P = P_{NH_3}^2 P_{CO_2}$$

The total pressure in the flask (0.363 atm) is a sum of the partial pressures of $NH_3$ and $CO_2$.

$$P_T = P_{NH_3} + P_{CO_2} = 0.363 \text{ atm}$$

Let the partial pressure of $CO_2 = x$. From the stoichiometry of the balanced equation, you should find that $P_{NH_3} = 2P_{CO_2}$. Therefore, the partial pressure of $NH_3 = 2x$. Substituting into the equation for total pressure gives:

$$P_T = P_{NH_3} + P_{CO_2} = 2x + x = 3x$$

$$3x = 0.363 \text{ atm}$$

$$x = P_{CO_2} = 0.121 \text{ atm}$$

$$P_{NH_3} = 2x = 0.242 \text{ atm}$$

Substitute the equilibrium pressures into the equilibrium expression to solve for $K_P$.

$$\boldsymbol{K_P} = P_{NH_3}^2 P_{CO_2} = (0.242)^2(0.121) = \mathbf{7.09 \times 10^{-3}}$$

15.31 To obtain $2SO_2$ as a reactant in the final equation, we must reverse the first equation and multiply by two. For the equilibrium, $2SO_2(g) \rightleftarrows 2S(s) + 2O_2(g)$

$$K_c''' = \left(\frac{1}{K_c'}\right)^2 = \left(\frac{1}{4.2 \times 10^{52}}\right)^2 = 5.7 \times 10^{-106}$$

Now we can add the above equation to the second equation to obtain the final equation. Since we add the two equations, the equilibrium constant is the product of the equilibrium constants for the two reactions.

$$2SO_2(g) \rightleftarrows 2S(s) + 2O_2(g) \qquad K_c''' = 5.7 \times 10^{-106}$$

$$2S(s) + 3O_2(g) \rightleftarrows 2SO_3(g) \qquad K_c'' = 9.8 \times 10^{128}$$

$$2SO_2(g) + O_2(g) \rightleftarrows 2SO_3(g) \qquad \mathbf{K_c} = K_c''' \times K_c'' = \mathbf{5.6 \times 10^{23}}$$

15.33 Let $x$ be the initial pressure of NOBr. Using the balanced equation, we can write expressions for the partial pressures at equilibrium.

$$P_{NOBr} = (1 - 0.34)x = 0.66x$$

$$P_{NO} = 0.34x$$

$$P_{Br_2} = 0.17x$$

The sum of these is the total pressure.

$$0.66x + 0.34x + 0.17x = 1.17x = 0.25 \text{ atm}$$

$$x = 0.214 \text{ atm}$$

The equilibrium pressures are then

$$P_{NOBr} = 0.66(0.214) = 0.141 \text{ atm}$$

$$P_{NO} = 0.34(0.214) = 0.073 \text{ atm}$$

$$P_{Br_2} = 0.17(0.214) = 0.036 \text{ atm}$$

We find $K_P$ by substitution.

$$\mathbf{K_P} = \frac{(P_{NO})^2 P_{Br_2}}{(P_{NOBr})^2} = \frac{(0.073)^2(0.036)}{(0.141)^2} = \mathbf{9.6 \times 10^{-3}}$$

The relationship between $K_P$ and $K_c$ is given by

$$K_P = K_c(RT)^{\Delta n}$$

We find $K_c$ (for this system $\Delta n = +1$)

$$\mathbf{K_c} = \frac{K_P}{(RT)^{\Delta n}} = \frac{K_P}{RT} = \frac{9.6 \times 10^{-3}}{(0.0821 \times 298)} = \mathbf{3.9 \times 10^{-4}}$$

15.35

$$K = K'K''$$

$$K = (6.5 \times 10^{-2})(6.1 \times 10^{-5})$$

$$\mathbf{K = 4.0 \times 10^{-6}}$$

15.37 Note that we are comparing similar reactions at equilibrium – two reactants producing one product, all with coefficients of one in the balanced equation.

a. **The reaction, A + C ⇄ AC has the largest equilibrium constant**. Of the three diagrams, there is the most product present at equilibrium.

b. **The reaction, A + D ⇄ AD has the smallest equilibrium constant**. Of the three diagrams, there is the least amount of product present at equilibrium.

15.39 Given:

$$K_P = \frac{P_{SO_3}^2}{P_{SO_2}^2 P_{O_2}} = 5.60 \times 10^4$$

Initially, the total pressure is (0.350 + 0.762) atm or 1.112 atm. As the reaction progresses from left to right toward equilibrium there will be a decrease in the number of moles of molecules present. (Note that 2 moles of $SO_2$ react with 1 mole of $O_2$ to produce 2 moles of $SO_3$, or, at constant volume, three atmospheres of reactants forms two atmospheres of products.) Since pressure is directly proportional to the number of molecules present, at equilibrium the total pressure will be less than 1.112 atm. **The equilibrium pressure is less than the original pressure.**

15.41 **Strategy:** The equilibrium constant $K_c$ is given, and we start with a mixture of CO and $H_2O$. From the stoichiometry of the reaction, we can determine the concentration of $H_2$ at equilibrium. Knowing the partial pressure of $H_2$ and the volume of the container, we can calculate the number of moles of $H_2$.

**Solution:** The balanced equation shows that one mole of water will combine with one mole of carbon monoxide to form one mole of hydrogen and one mole of carbon dioxide. (This is the *reverse* reaction.) Let $x$ be the depletion in the concentration of either CO or $H_2O$ at equilibrium (why can $x$ serve to represent either quantity?). The equilibrium concentration of hydrogen must then also be equal to $x$. The changes are summarized as shown in the table.

| | $H_2$ | + $CO_2$ | ⇄ $H_2O$ | + CO |
|---|---|---|---|---|
| Initial ($M$): | 0 | 0 | 0.0300 | 0.0300 |
| Change ($M$): | $+x$ | $+x$ | $-x$ | $-x$ |
| Equilibrium ($M$): | $x$ | $x$ | $(0.0300 - x)$ | $(0.0300 - x)$ |

The equilibrium constant is:

$$K_c = \frac{[H_2O][CO]}{[H_2][CO_2]} = 0.534$$

Substituting,

$$\frac{(0.0300 - x)^2}{x^2} = 0.534$$

Taking the square root of both sides, we obtain:

$$\frac{(0.0300 - x)}{x} = \sqrt{0.534}$$

$$x = 0.0173\ M$$

The number of moles of $H_2$ formed is:

$$0.0173\ \text{mol/L} \times 10.0\ \text{L} = \mathbf{0.173\ mol\ H_2}$$

15.43 Notice that the balanced equation requires that for every two moles of HBr consumed, one mole of $H_2$ and one mole of $Br_2$ must be formed. Let $2x$ be the depletion in the concentration of HBr at equilibrium. The equilibrium concentrations of $H_2$ and $Br_2$ must therefore each be $x$. The changes are shown in the table.

| | $H_2$ | + $Br_2$ | $\rightleftarrows$ 2HBr |
|---|---|---|---|
| Initial ($M$): | 0 | 0 | 0.267 |
| Change ($M$): | $+x$ | $+x$ | $-2x$ |
| Equilibrium ($M$): | $x$ | $x$ | $(0.267 - 2x)$ |

The equilibrium constant relationship is given by:

$$K_c = \frac{[HBr]^2}{[H_2][Br_2]}$$

Substitution of the equilibrium concentration expressions gives

$$K_c = \frac{(0.267 - 2x)^2}{x^2} = 2.18 \times 10^6$$

Taking the square root of both sides we obtain:

$$\frac{0.267 - 2x}{x} = 1.48 \times 10^3$$

$$x = 1.80 \times 10^{-4}$$

The equilibrium concentrations are:

$$[\mathbf{H_2}] = [\mathbf{Br_2}] = \mathbf{1.80 \times 10^{-4}}\ \boldsymbol{M}$$

$$[\mathbf{HBr}] = 0.267 - 2(1.80 \times 10^{-4}) = \mathbf{0.267}\ \boldsymbol{M}$$

**Think About It:** If the depletion in the concentration of HBr at equilibrium were defined as $x$, rather than $2x$, what would be the appropriate expressions for the equilibrium concentrations of $H_2$ and $Br_2$? Should the final answers be different in this case?

15.45 pressures are desired, we calculate $K_P$.

$$K_P = K_c(0.0821\,T)^{\Delta n} = (4.63 \times 10^{-3})(0.0821 \times 800)^1 = 0.304$$

| | $COCl_2(g)$ | $\rightleftarrows$ | $CO(g)$ | + | $Cl_2(g)$ |
|---|---|---|---|---|---|
| Initial (atm): | 0.760 | | 0.000 | | 0.000 |
| Change (atm): | $-x$ | | $+x$ | | $+x$ |
| Equilibrium (atm): | $(0.760 - x)$ | | $x$ | | $x$ |

$$\frac{x^2}{(0.760 - x)} = 0.304$$

$$x^2 + 0.304x - 0.231 = 0$$

$$x = 0.352 \text{ atm}$$

At equilibrium:

$$P_{COCl_2} = (0.760 - 0.352)\text{atm} = \mathbf{0.408\ atm}$$

$$P_{CO} = P_{Cl_2} = \mathbf{0.352\ atm}$$

15.47 xpression for the system is:

$$K_P = \frac{(P_{CO})^2}{P_{CO_2}}$$

The total pressure can be expressed as:

$$P_{\text{total}} = P_{CO_2} + P_{CO}$$

If we let the partial pressure of CO be $x$, then the partial pressure of $CO_2$ is:

$$P_{CO_2} = P_{total} - x = (4.50 - x)\text{atm}$$

Substitution gives the equation:

$$K_P = \frac{(P_{CO})^2}{P_{CO_2}} = \frac{x^2}{(4.50 - x)} = 1.52$$

This can be rearranged to the quadratic:

$$x^2 + 1.52x - 6.84 = 0$$

The solutions are $x = 1.96$ and $x = -3.48$; only the positive result has physical significance (why?). The equilibrium pressures are

$$\boldsymbol{P_{CO}} = x = \mathbf{1.96\ atm}$$

$$\boldsymbol{P_{CO_2}} = (4.50 - 1.96) = \mathbf{2.54\ atm}$$

15.49 **Strategy:** We are given the concentrations of all species in a biochemical reaction and must determine whether the reaction will proceed in the forward direction or the reverse direction in order to establish equilibrium. This requires calculating the value of $Q$ and comparing it to the value of $K$, which is given in the problem.

**Solution:** The reaction quotient is

$$Q = \frac{[\alpha\text{-ketoglutarate}][\text{L-alanine}]}{[\text{L-glutamate}][\text{pyruvate}]} = \frac{(1.6 \times 10^{-2})(6.25 \times 10^{-3})}{(3.0 \times 10^{-5})(3.3 \times 10^{-4})} = 1.01 \times 10^4$$

Because $K$ is 1.11, $Q$ is greater than $K$. Thus, the reaction will proceed to the left (the reverse reaction will occur) in order to establish equilibrium. **The forward reaction will not occur.**

15.55 a. Removal of $CO_2(g)$ from the system: **The equilibrium would shift to the right.**

b. Addition of more solid $Na_2CO_3$: **The equilibrium would be unaffected**. $[Na_2CO_3]$ does not appear in the equilibrium expression.

c. Removal of some of the solid $NaHCO_3$: **The equilibrium would be unaffected**. Same reason as (b).

15.57 **Strategy:** A change in pressure can affect only the volume of a gas, but not that of a solid or liquid because solids and liquids are much less compressible. The stress applied is an increase in pressure. According to Le Châtelier's principle, the system will adjust to partially offset this stress. In other words, the system will adjust to decrease the pressure. This can be achieved by shifting to the side of the equation that has fewer moles of gas. Recall that pressure is directly proportional to moles of gas: $PV = nRT$ so $P \propto n$.

**Solution:** a. Changes in pressure ordinarily do not affect the concentrations of reacting species in condensed phases because liquids and solids are virtually incompressible. Pressure change should have **no effect** on this system.

b. **No effect**. Same situation as (a).

c. Only the product is in the gas phase. An increase in pressure should favor the reaction that decreases the total number of moles of gas. The equilibrium should **shift to the left**, that is, the amount of B should decrease and that of A should increase.

d. In this equation there are equal moles of gaseous reactants and products. A shift in either direction will have no effect on the total number of moles of gas present. There will be **no effect** on the position of the equilibrium when the pressure is increased.

e. A shift in the direction of the reverse reaction (**shift to the left**) will have the result of decreasing the total number of moles of gas present.

15.59 **Strategy:** (a) What does the sign of $\Delta H°$ indicate about the heat change (endothermic or exothermic) for the forward reaction? (b) The stress is the addition of $Cl_2$ gas. How will the system adjust to partially offset the stress? (c) The stress is the removal of $PCl_3$ gas. How will the system adjust to partially offset the stress? (d) The stress is an increase in pressure. The system will adjust to decrease the pressure. Remember, pressure is directly proportional to moles of gas. (e) What is the function of a catalyst? How does it affect a reacting system not at equilibrium? at equilibrium?

**Solution:** a. The stress applied is the heat added to the system. Note that the reaction is endothermic ($\Delta H° > 0$). Endothermic reactions absorb heat from the surroundings; therefore, we can think of heat as a reactant.

$$\text{heat} + PCl_5(g) \rightleftarrows PCl_3(g) + Cl_2(g)$$

The system will adjust to remove some of the added heat by undergoing a decomposition reaction (**Shift to the right**)

b. The stress is the addition of $Cl_2$ gas. The system will shift in the direction to remove some of the added $Cl_2$. The system **shifts to the left** until equilibrium is reestablished.

c. The stress is the removal of $PCl_3$ gas. The system will shift to replace some of the $PCl_3$ that was removed. The system **shifts to the right** until equilibrium is reestablished.

d. The stress applied is an increase in pressure. The system will adjust to remove the stressby decreasing the pressure. Recall that pressure is directly proportional to the number of moles of gas. In the balanced equation we see 1 mole of gas on the reactants side and 2 moles of gas on the products side. The pressure can be decreased by shifting to the side with the fewer moles of gas. The system will **shift to the left** to reestablish equilibrium.

e. The function of a catalyst is to increase the rate of a reaction. If a catalyst is added to the reacting system not at equilibrium, the system will reach equilibrium faster than if left undisturbed. If a system is already at equilibrium, as in this case, the addition of a catalyst will not affect either the concentrations of reactant and product, or the equilibrium constant. **A catalyst has no effect on equilibrium position.**

15.61 **No change**. A catalyst has no effect on the position of the equilibrium.

15.63 For this system,

$$K_P = P_{CO_2}$$

This means that to remain at equilibrium, the pressure of carbon dioxide must stay at a fixed value as long as the temperature remains the same.

a. If the volume is increased, the pressure of $CO_2$ will drop (Boyle's law, pressure and volume are inversely proportional). Some $CaCO_3$ will break down to form more $CO_2$ and CaO. (**Shift to right**)

b. Assuming that the amount of added solid CaO is not so large that the volume of the system is altered significantly, there should be **no effect**. If a huge amount of CaO were added, this would have the effect of reducing the volume of the container. What would happen then?

c. Assuming that the amount of $CaCO_3$ removed doesn't alter the container volume significantly, there should be **no effect**. Removing a huge amount of $CaCO_3$ will have the effect of increasing the container volume. The result in that case will be the same as in part (a).

d. The pressure of $CO_2$ will be greater and will exceed the value of $K_P$. Some $CO_2$ will combine with CaO to form more $CaCO_3$. **(Shift to left)**

e. Carbon dioxide combines with aqueous NaOH according to the equation

$$CO_2(g) + NaOH(aq) \rightarrow NaHCO_3(aq)$$

This will have the effect of reducing the $CO_2$ pressure and causing more $CaCO_3$ to break down to $CO_2$ and CaO. (**Shift to the right**)

f. Carbon dioxide does not react with hydrochloric acid, but $CaCO_3$ does.

$$CaCO_3(s) + 2HCl(aq) \rightarrow CaCl_2(aq) + CO_2(g) + H_2O(l)$$

The $CO_2$ produced by the action of the acid will combine with CaO as discussed in (d) above. (**Shift to the left**)

g. This is a decomposition reaction. Decomposition reactions are endothermic. Increasing the temperature will favor this reaction and produce more $CO_2$ and CaO. (**Shift to the right**)

15.65 a. $\mathbf{2O_3(g) \rightleftarrows 3O_2(g) \quad \Delta H^\circ_{rxn} = -284.4\ kJ/mol}$

b. **Equilibrium would shift to the left**. The number of $O_3$ molecules would increase and the number of $O_2$ molecules would decrease.

15.67 a. Since the total pressure is 1.00 atm, the sum of the partial pressures of NO and $Cl_2$ is

$$1.00\text{ atm} - \text{partial pressure of NOCl} = 1.00\text{ atm} - 0.64\text{ atm} = 0.36\text{ atm}$$

The stoichiometry of the reaction requires that the partial pressure of NO be twice that of $Cl_2$. Hence, the partial pressure of NO is **0.24 atm** and the partial pressure of $Cl_2$ is **0.12 atm.**

b. The equilibrium constant $K_P$ is found by substituting the partial pressures calculated in part (a) into the equilibrium expression.

$$K_P = \frac{P_{NO}^2 P_{Cl_2}}{P_{NOCl}^2} = \frac{(0.24)^2(0.12)}{(0.64)^2} = \mathbf{0.017}$$

15.69 The equilibrium expression for this system is given by:

$$K_P = P_{CO_2} P_{H_2O}$$

a. In a closed vessel the decomposition will stop when the product of the partial pressures of $CO_2$ and $H_2O$ equals $K_P$. Adding more sodium bicarbonate will have **no effect**.

b. In an open vessel, $CO_2(g)$ and $H_2O(g)$ will escape from the vessel, and the partial pressures of $CO_2$ and $H_2O$ will never become large enough for their product to equal $K_P$. Therefore, equilibrium will never be established. Adding more sodium bicarbonate will result in the production of **more $CO_2$ and $H_2O$**.

15.71 a. relates $K_P$ and $K_c$ is:

$$K_P = K_c(0.0821\,T)^{\Delta n}$$

For this reaction, $\Delta n = 3 - 2 = 1$

$$\mathbf{K_c} = \frac{K_P}{(0.0821\text{T})} = \frac{2 \times 10^{-42}}{(0.0821 \times 298)} = \mathbf{8 \times 10^{-44}}$$

b. **A mixture of $H_2$ and $O_2$ can be kept at room temperature because of a very large activation energy.** The reaction of hydrogen with oxygen is infinitely slow without a catalyst or an initiator. The action of a single spark on a mixture of these gases results in the explosive formation of water.

15.73 a. Calculate the value of $K_P$ by substituting the equilibrium partial pressures into the equilibrium expression.

$$\mathbf{K_P} = \frac{P_B}{P_A^2} = \frac{(0.60)}{(0.60)^2} = \mathbf{1.7}$$

b. The total pressure is the sum of the partial pressures for the two gaseous components, A and B. We can write:

$$P_A + P_B = 1.5 \text{ atm}$$

and

$$P_B = 1.5 - P_A$$

Substituting into the expression for $K_P$ gives:

$$K_P = \frac{(1.5 - P_A)}{P_A^2} = 1.7$$

$$1.7P_A^2 + P_A - 1.5 = 0$$

Solving the quadratic equation, we obtain:

$$\mathbf{P_A = 0.69\ atm}$$

and by difference,

$$\mathbf{P_B = 0.81\ atm}$$

Check that substituting these equilibrium concentrations into the equilibrium expression gives the equilibrium constant calculated in part (a).

$$K_P = \frac{P_B}{P_A^2} = \frac{0.81}{(0.69)^2} = 1.7$$

15.75 We start with a table.

| | $A_2$ | + | $B_2$ | $\rightleftarrows$ | $2AB$ |
|---|---|---|---|---|---|
| Initial (mol): | 1 | | 3 | | 0 |
| Change (mol): | $-\frac{x}{2}$ | | $-\frac{x}{2}$ | | $+x$ |
| Equilibrium (mol): | $1 - \frac{x}{2}$ | | $3 - \frac{x}{2}$ | | $x$ |

After the addition of 2 moles of A,

| | $A_2$ | + | $B_2$ | $\rightleftarrows$ | $2AB$ |
|---|---|---|---|---|---|
| Initial (mol): | $3 - \frac{x}{2}$ | | $3 - \frac{x}{2}$ | | $x$ |
| Change (mol): | $-\frac{x}{2}$ | | $-\frac{x}{2}$ | | $+x$ |
| Equilibrium (mol): | $3 - x$ | | $3 - x$ | | $2x$ |

We write two different equilibrium constants expressions for the two tables.

$$K = \frac{[AB]^2}{[A_2][B_2]}$$

$$K = \frac{x^2}{\left(1 - \frac{x}{2}\right)\left(3 - \frac{x}{2}\right)} \quad \text{and} \quad K = \frac{(2x)^2}{(3 - x)(3 - x)}$$

We equate the equilibrium expressions and solve for $x$.

$$\frac{x^2}{\left(1 - \frac{x}{2}\right)\left(3 - \frac{x}{2}\right)} = \frac{(2x)^2}{(3 - x)(3 - x)}$$

$$\frac{1}{\frac{1}{4}(x^2 - 8x + 12)} = \frac{4}{x^2 - 6x + 9}$$

$$-6x + 9 = -8x + 12$$

$$x = 1.5$$

We substitute $x$ back into one of the equilibrium expressions to solve for $K$.

$$\boldsymbol{K} = \frac{(2x)^2}{(3 - x)(3 - x)} = \frac{(3)^2}{(1.5)(1.5)} = \mathbf{4.0}$$

Substitute $x$ into the other equilibrium expression to see if you obtain the same value for $K$. Note that we used moles rather than molarity for the concentrations, because the volume, $V$, cancels in the equilibrium expressions.

15.77 Set up a table that contains the initial concentrations, the change in concentrations, and the equilibrium concentration. Assume that the vessel has a volume of 1 L.

| | $H_2$ | $+ Cl_2$ | $\rightleftarrows$ | $2HCl$ |
|---|---|---|---|---|
| Initial ($M$): | 0.47 | 0 | | 3.59 |
| Change ($M$): | $+x$ | $+x$ | | $-2x$ |
| Equilibrium ($M$): | $(0.47 + x)$ | $x$ | | $(3.59 - 2x)$ |

Substitute the equilibrium concentrations into the equilibrium expression, then solve for $x$. Since $\Delta n = 0$, $K_c = K_P$.

$$K_c = \frac{[HCl]^2}{[H_2][Cl_2]} = \frac{(3.59 - 2x)^2}{(0.47 + x)x} = 193$$

Solving the quadratic equation,

$$x = 0.103$$

Having solved for $x$, calculate the equilibrium concentrations of all species.

$$[H_2] = 0.573\ M \qquad [Cl_2] = 0.103\ M \qquad [HCl] = 3.384\ M$$

Since we assumed that the vessel had a volume of 1 L, the above molarities also correspond to the number of moles of each component.

From the mole fraction of each component and the total pressure, we can calculate the partial pressure of each component.

$$\text{Total number of moles} = 0.573 + 0.103 + 3.384 = 4.06\ \text{mol}$$

$$P_{H_2} = \frac{0.573}{4.06} \times 2.00 = \mathbf{0.28\ atm}$$

$$P_{Cl_2} = \frac{0.103}{4.06} \times 2.00 = \mathbf{0.051\ atm}$$

$$P_{HCl} = \frac{3.384}{4.06} \times 2.00 = \mathbf{1.67\ atm}$$

15.79 This is a difficult problem. Express the equilibrium number of moles in terms of the initial moles and the change in number of moles ($x$). Next, calculate the mole fraction of each component. Using the mole fraction, you should come up with a relationship between partial pressure and total pressure for each component. Substitute the partial pressures into the equilibrium expression to solve for the total pressure, $P_T$.

The reaction is:

| | $N_2$ | + | $3\ H_2$ | $\rightleftarrows$ | $2\ NH_3$ |
|---|---|---|---|---|---|
| Initial (mol): | 1 | | 3 | | 0 |
| Change (mol): | $-x$ | | $-3x$ | | $2x$ |
| Equilibrium (mol): | $(1 - x)$ | | $(3 - 3x)$ | | $2x$ |

$$\text{Mole fraction of } NH_3 = \frac{\text{mol of } NH_3}{\text{total number of moles}}$$

$$X_{NH_3} = \frac{2x}{(1 - x) + (3 - 3x) + 2x} = \frac{2x}{4 - 2x}$$

$$0.21 = \frac{2x}{4 - 2x}$$

$$x = 0.35 \text{ mol}$$

Substituting $x$ into the following mole fraction equations, the mole fractions of $N_2$ and $H_2$ can be calculated.

$$X_{N_2} = \frac{1 - x}{4 - 2x} = \frac{1 - 0.35}{4 - 2(0.35)} = 0.20$$

$$X_{H_2} = \frac{3 - 3x}{4 - 2x} = \frac{3 - 3(0.35)}{4 - 2(0.35)} = 0.59$$

The partial pressures of each component are equal to the mole fraction multiplied by the total pressure.

$P_{NH_3} = 0.21P_T$ $\quad$ $P_{N_2} = 0.20P_T$ $\quad$ $P_{H_2} = 0.59P_T$

e the partial pressures above (in terms of $P_T$) into the equilibrium expression, and solve for $P_T$.

$$K_P = \frac{P_{NH_3}^2}{P_{H_2}^3 P_{N_2}}$$

$$4.31 \times 10^{-4} = \frac{(0.21)^2 P_T^2}{(0.59P_T)^3(0.20P_T)}$$

$$4.31 \times 10^{-4} = \frac{1.07}{P_T^2}$$

$$\mathbf{P_T = 5.0 \times 10^1 \text{ atm}}$$

15.81 **Strategy:** The concentrations of solids and pure liquids do not appear in $K_C$. So, even though the given equilibrium is heterogeneous, Equation 15.4 still applies.

**Solution:** There are 3 moles of gas molecules on the product side and 6 moles on the reactant side. So, $\boldsymbol{\Delta n} = 3 - 6 = \mathbf{-3}$.

15.83 Of the original 1.05 moles of $Br_2$, 1.20% has dissociated. The amount of $Br_2$ dissociated in molar concentration is:

$$[Br_2] = 0.0120 \times \frac{1.05\text{ mol}}{0.980\text{ L}} = 0.0129\ M$$

Setting up a table:

| | $Br_2(g)$ | $\rightleftarrows$ | $2Br(g)$ |
|---|---|---|---|
| Initial ($M$): | $\frac{1.05\text{ mol}}{0.980\text{ L}} = 1.07\ M$ | | 0 |
| Change ($M$): | −0.0129 | | +2(0.0129) |
| Equilibrium ($M$): | 1.06 | | 0.0258 |

$$K_c = \frac{[Br]^2}{[Br_2]} = \frac{(0.0258)^2}{1.06} = \mathbf{6.28 \times 10^{-4}}$$

15.85 a.

| | fructose | $\rightleftarrows$ | glucose |
|---|---|---|---|
| Initial ($M$): | 0.244 | | 0 |
| Change ($M$): | −0.131 | | +0.131 |
| Equilibrium ($M$): | 0.113 | | 0.131 |

Calculating the equilibrium constant,

$$K_c = \frac{[\text{glucose}]}{[\text{fructose}]} = \frac{0.131}{0.113} = \mathbf{1.16}$$

b.

$$\textbf{Percent converted} = \frac{\text{amount of fructose converted}}{\text{original amount of fructose}} \times 100\%$$

$$= \frac{0.131}{0.244} \times 100\% = \mathbf{53.7\%}$$

15.87 **There is a temporary dynamic equilibrium between the melting ice cubes and the freezing of water between the ice cubes.**

15.89 We first must find the initial concentrations of all the species in the system.

$$[H_2]_0 = \frac{0.714 \text{ mol}}{2.40 \text{ L}} = 0.298\ M$$

$$[I_2]_0 = \frac{0.984 \text{ mol}}{2.40 \text{ L}} = 0.410\ M$$

$$[HI]_0 = \frac{0.886 \text{ mol}}{2.40 \text{ L}} = 0.369\ M$$

Calculate the reaction quotient by substituting the initial concentrations into the appropriate equation.

$$Q_c = \frac{[HI]_0^2}{[H_2]_0[I_2]_0} = \frac{(0.369)^2}{(0.298)(0.410)} = 1.11$$

We find that $Q_c$ is less than $K_c$. The equilibrium will shift to the right, decreasing the concentrations of $H_2$ and $I_2$ and increasing the concentration of HI.

We set up the usual table. Let $x$ be the decrease in concentration of $H_2$ and $I_2$.

| | $H_2$ | + | $I_2$ | $\rightleftarrows$ | 2 HI |
|---|---|---|---|---|---|
| Initial ($M$): | 0.298 | | 0.410 | | 0.369 |
| Change ($M$): | $-x$ | | $-x$ | | $+2x$ |
| Equilibrium ($M$): | $(0.298 - x)$ | | $(0.410 - x)$ | | $(0.369 + 2x)$ |

The equilibrium expression is:

$$K_c = \frac{[HI]^2}{[H_2][I_2]} = \frac{(0.369 + 2x)^2}{(0.298 - x)(0.410 - x)} = 54.3$$

This becomes the quadratic equation

$$50.3x^2 - 39.9x + 6.49 = 0$$

The smaller root is $x = 0.228\ M$. (The larger root is physically impossible.)

Having solved for $x$, calculate the equilibrium concentrations.

$$\mathbf{[H_2]} = (0.298 - 0.228)\ M = \mathbf{0.07\ M}$$

$$\mathbf{[I_2]} = (0.410 - 0.228)\ M = \mathbf{0.18\ M}$$

$$\mathbf{[HI]} = [0.36 + 2(0.228)]\ M = \mathbf{0.83\ M}$$

15.91 The gas cannot be (a) because the color became lighter with heating. Heating (a) to 150°C would produce some HBr, which is colorless and would lighten rather than darken the gas.

The gas cannot be (b) because $Br_2$ doesn't dissociate into Br atoms at 150°C, so the color shouldn't change.

The gas must be **(c). $N_2O_4$(colorless) → $2NO_2$(brown) is consistent with the observations. The reaction is endothermic so heating darkens the color. Above 150°C, the $NO_2$ breaks up into colorless NO and $O_2$:**

$$2NO_2(g) \rightarrow 2NO(g) + O_2(g)$$

**An increase in pressure shifts the equilibrium back to the left, restoring the color by producing $NO_2$.**

15.93 a. **Color deepens.** b. **Increases.** c. **Decreases.** d. **Increases.** e. **Unchanged.**

15.95 In this problem, you are asked to calculate $K_c$.

*Step 1:* Calculate the initial concentration of NOCl. We carry an extra significant figure throughout this calculation to minimize rounding errors.

$$[NOCl]_0 = \frac{2.50 \text{ mol}}{1.50 \text{ L}} = 1.667\ M$$

*Step 2:* Let's represent the change in concentration of NOCl as $-2x$. Setting up a table:

| | $2NOCl(g)$ | $\rightleftarrows$ | $2NO(g)$ | + | $Cl_2(g)$ |
|---|---|---|---|---|---|
| Initial ($M$): | 1.667 | | 0 | | 0 |
| Change ($M$): | $-2x$ | | $+2x$ | | $+x$ |
| Equilibrium ($M$): | $1.667 - 2x$ | | $2x$ | | $x$ |

If 28.0 percent of the NOCl has dissociated at equilibrium, the amount consumed is:

$$(0.280)(1.667\ M) = 0.4668\ M$$

In the table above, we have represented the amount of NOCl that reacts as $2x$. Therefore,

$$2x = 0.4668\ M$$

$$x = 0.2334\ M$$

The equilibrium concentrations of NOCl, NO, and $Cl_2$ are:

$$[NOCl] = (1.667 - 2x)M = (1.667 - 0.4668)M = 1.200\ M$$

$$[NO] = 2x = 0.4668\ M$$

$$[Cl_2] = x = 0.2334\ M$$

*Step 3:* The equilibrium constant $K_c$ can be calculated by substituting the above concentrations into the equilibrium expression.

$$\boldsymbol{K_c} = \frac{[NO]^2[Cl_2]}{[NOCl]^2} = \frac{(0.4668)^2(0.2334)}{(1.200)^2} = \mathbf{0.035} \text{ or } \mathbf{3.5 \times 10^{-2}}$$

15.97 a. Assuming the self-ionization of water occurs by a single elementary step mechanism, the equilibrium constant is just the ratio of the forward and reverse rate constants.

$$\boldsymbol{K} = \frac{k_f}{k_r} = \frac{k_1}{k_{-1}} = \frac{2.4 \times 10^{-5}}{1.3 \times 10^{11}} = \mathbf{1.8 \times 10^{-16}}$$

b. The product can be written as:

$$[H^+][OH^-] = K[H_2O]$$

What is $[H_2O]$? It is the concentration of pure water. One liter of water has a mass of 1000 g (density = 1.00 g/mL). The number of moles of $H_2O$ is:

$$1000 \text{ g} \times \frac{1 \text{ mol}}{18.0 \text{ g}} = 55.5 \text{ mol}$$

The concentration of water is 55.5 mol/1.00 L or 55.5 *M*. The product is:

$$\mathbf{[H^+][OH^-]} = (1.8 \times 10^{-16})(55.5) = \mathbf{1.0 \times 10^{-14}}$$

We assume the concentration of hydrogen ion and hydroxide ion are equal.

$$\mathbf{[H^+][OH^-]} = (1.0 \times 10^{-14})^{1/2} = \mathbf{1.0 \times 10^{-7}}\ \boldsymbol{M}$$

15.99 First, let's calculate the initial concentration of ammonia.

$$[NH_3] = \frac{14.6 \text{ g} \times \frac{1 \text{ mol } NH_3}{17.03 \text{ g } NH_3}}{4.00 \text{ L}} = 0.214\ M$$

Let's set up a table to represent the equilibrium concentrations. We represent the amount of $NH_3$ that reacts as $2x$.

| | $2NH_3(g)$ | $\rightleftarrows$ | $N_2(g)$ | + | $3H_2(g)$ |
|---|---|---|---|---|---|
| Initial (*M*): | 0.214 | | 0 | | 0 |
| Change (*M*): | $-2x$ | | $+x$ | | $+3x$ |
| Equilibrium (*M*): | $0.214 - 2x$ | | $x$ | | $3x$ |

Substitute into the equilibrium expression to solve for $x$.

$$K_c = \frac{[N_2][H_2]^3}{[NH_3]^2}$$

$$0.83 = \frac{(x)(3x)^3}{(0.214 - 2x)^2} = \frac{27x^4}{(0.214 - 2x)^2}$$

Taking the square root of both sides of the equation gives:

$$0.91 = \frac{5.20x^2}{0.214 - 2x}$$

Rearranging,

$$5.20x^2 + 1.82x - 0.195 = 0$$

Solving the quadratic equation gives the solutions:

$$x = 0.086\ M \text{ and } x = -0.44\ M$$

The positive root is the correct answer. The equilibrium concentrations are:

$$\mathbf{[NH_3]} = 0.214 - 2(0.086) = \mathbf{0.042\ \mathit{M}}$$

$$\mathbf{[N_2]} = \mathbf{0.086\ \mathit{M}}$$

$$\mathbf{[H_2]} = 3(0.086) = \mathbf{0.26\ \mathit{M}}$$

15.101 a. From the balanced equation

| | $N_2O_4$ | $\rightleftarrows$ | $2NO_2$ |
|---|---|---|---|
| Initial (mol): | 1 | | 0 |
| (mol): | $-x$ | | $+2x$ |
| Equilibrium (mol): | $(1 - x)$ | | $2x$ |

The total moles in the system = (moles $N_2O_4$ + moles $NO_2$) = $[(1 - x) + 2x] = 1 + x$. If the total pressure in the system is $P$, then:

$$P_{N_2O_4} = \frac{1-x}{1+x}P \text{ and } P_{NO_2} = \frac{2x}{1+x}P$$

$$K_P = \frac{P^2_{NO_2}}{P_{N_2O_4}} = \frac{\left(\frac{2x}{1+x}\right)^2 P^2}{\left(\frac{1-x}{1+x}\right)P}$$

$$\boldsymbol{K_P = \frac{\left(\frac{4x^2}{1+x}\right)P}{1-x} = \frac{4x^2}{1-x^2}P}$$

b. Rearranging the $K_P$ expression:

$$4x^2P = K_P - x^2K_P$$

$$x^2(4P + K_P) = K_P$$

$$x^2 = \frac{K_P}{4P + K_P}$$

$$x = \sqrt{\frac{K_P}{4P+K_P}}$$

$K_P$ is a constant (at constant temperature). **If $P$ increases, the fraction $\frac{4x^2}{1-x^2}$ (and therefore $x$) must decrease. Equilibrium shifts to the left to produce less $NO_2$ and more $N_2O_4$ as predicted.**

15.103 **Strategy:** According to Equation 15.4, $K_P$ and $K_C$ are the same when $\Delta n = 0$:

$$K_P = K_c(0.0821T)^{\Delta n} = K_c(0.0821T)^0 = K_c \cdot 1 = K_c$$

So, look for reactions that show no change in the number of gas molecules as the reaction occurs. Also, note that Equation 15.4 does not apply to *heterogeneous* equilibria if liquid phases are *solutions*.

**Solution:** a. $K_P \neq K_C$ ($\Delta n \neq 0$)

b. $K_P \neq K_C$ ($\Delta n \neq 0$)

c. **Equation 15.4 is not applicable** (some species are aqueous).

d. $\mathbf{K_P = K_C}$ **($\Delta n = 0$)**

e. $K_P \neq K_C$ ($\Delta n \neq 0$)

f. **Equation 15.4 is not applicable** (some species are aqueous).

g. $\mathbf{K_P = K_C}$ **($\Delta n = 0$)**

h. $K_P \neq K_C$ ($\Delta n \neq 0$)

15.105 a. There is only one gas phase component, $O_2$. The equilibrium constant is simply

$$\boldsymbol{K_P} = P_{O_2} = \mathbf{0.49\ atm}$$

b. From the ideal gas equation, we can calculate the moles of $O_2$ produced by the decomposition of CuO.

$$n_{O_2} = \frac{PV}{RT} = \frac{(0.49\text{ atm})(2.0\text{ L})}{(0.0821\text{ L}\cdot\text{atm/K}\cdot\text{mol})(1297\text{ K})} = 9.2 \times 10^{-3}\text{ mol } O_2$$

From the balanced equation,

$$(9.2 \times 10^{-3}\text{ mol } O_2) \times \frac{4\text{ mol CuO}}{1\text{ mol } O_2} = 3.7 \times 10^{-2}\text{ mol CuO decomposed}$$

$$\textbf{Fraction of CuO decomposed} = \frac{\text{amount of CuO lost}}{\text{original amount of CuO}}$$

$$= \frac{3.7 \times 10^{-2}\text{ mol}}{0.16\text{ mol}} = \mathbf{0.23\ (23\%)}$$

c. If a 1.0 mol sample were used, the pressure of oxygen would still be the same (0.49 atm) and it would be due to the same quantity of $O_2$. Remember, a pure solid does not affect the equilibrium position. The moles of CuO lost would still be $3.7 \times 10^{-2}$ mol. Thus the fraction decomposed would be:

$$\frac{0.037}{1.0} = \mathbf{0.037\ (3.7\%)}$$

d. If the number of moles of CuO were less than 0.037 mol, the equilibrium could not be established because the pressure of $O_2$ would be less than 0.49 atm. Therefore, the smallest number of moles of CuO needed to establish equilibrium must be slightly **greater than 0.037 mol.**

15.107 **Potassium is more volatile than sodium. Therefore, its removal shifts the equilibrium from left to right.**

15.109 Initially, the pressure of $SO_2Cl_2$ is 9.00 atm. The pressure is held constant, so after the reaction reaches equilibrium, $P_{SO_2Cl_2} + P_{SO_2} + P_{Cl_2} = 9.00$ atm . The amount (pressure) of $SO_2Cl_2$ reacted must equal the pressure of $SO_2$ and $Cl_2$ produced for the pressure to remain constant. If we let $P_{SO_2} + P_{Cl_2} = x$ , then the pressure of $SO_2Cl_2$ reacted must be $2x$. We set up a table showing the initial pressures, the change in pressures, and the equilibrium pressures.

| | $SO_2Cl_2(g)$ | $\rightleftarrows$ | $SO_2(g)$ | + | $Cl_2(g)$ |
|---|---|---|---|---|---|
| Initial (atm): | 9.00 | | 0 | | 0 |
| Change (atm): | $-2x$ | | $+x$ | | $+x$ |
| Equilibrium (atm): | $9.00 - 2x$ | | $x$ | | $x$ |

Again, note that the change in pressure for $SO_2Cl_2$ ($-2x$) does not match the stoichiometry of the reaction, because we are expressing changes in pressure. The total pressure is kept at 9.00 atm throughout.

$$K_P = \frac{P_{SO_2}P_{Cl_2}}{P_{SO_2Cl_2}}$$

$$2.05 = \frac{(x)(x)}{9.00 - 2x}$$

$$x^2 + 4.10x - 18.45 = 0$$

Solving the quadratic equation, $x = 2.71$ atm. At equilibrium,

$$P_{SO_2Cl_2} = 9.00 - 2(2.71) = \mathbf{3.58\ atm}$$

$$P_{SO_2} = P_{Cl_2} = x = \mathbf{2.71\ atm}$$

15.111 We carry an additional significant figure throughout this calculation to minimize rounding errors. The initial molarity of $SO_2Cl_2$ is:

$$[SO_2Cl_2] = \frac{6.75 \text{ g } SO_2Cl_2 \times \dfrac{1 \text{ mol } SO_2Cl_2}{135.0 \text{ g } SO_2Cl_2}}{2.00 \text{ L}} = 0.02500\ M$$

The concentration of $SO_2$ at equilibrium is:

$$[SO_2] = \frac{0.0345 \text{ mol}}{2.00 \text{ L}} = 0.01725\ M$$

Since there is a 1:1 mole ratio between $SO_2$ and $SO_2Cl_2$, the concentration of $SO_2$ at equilibrium (0.01725 *M*) equals the concentration of $SO_2Cl_2$ reacted. The concentrations of $SO_2Cl_2$ and $Cl_2$ at equilibrium are:

| | $SO_2Cl_2(g)$ | $\rightleftarrows$ | $SO_2(g)$ | + | $Cl_2(g)$ |
|---|---|---|---|---|---|
| Initial (*M*): | 0.02500 | | 0 | | 0 |
| Change (*M*): | −0.01725 | | +0.01725 | | +0.01725 |
| Equilibrium (*M*): | 0.00775 | | 0.01725 | | 0.01725 |

Substitute the equilibrium concentrations into the equilibrium expression to calculate $K_c$.

$$\boldsymbol{K_c} = \frac{[SO_2][Cl_2]}{[SO_2Cl_2]} = \frac{(0.01725)(0.01725)}{(0.00775)} = \mathbf{3.8 \times 10^{-2}} \text{ or } \mathbf{0.038}$$

15.113 a.

$$K_P = \frac{P_{NO}^2}{P_{N_2}P_{O_2}} = \frac{P_{NO}^2}{(3.0)(0.012)} = 2.9 \times 10^{-11}$$

$$P_{NO} = \mathbf{1.0 \times 10^{-6}\ atm}$$

b.

$$4.0 \times 10^{-31} = \frac{P_{NO}^2}{(0.78)(0.21)}$$

$$P_{NO} = \mathbf{2.6 \times 10^{-16}\ atm}$$

c. Since $K_P$ increases with temperature, it is **endothermic**.

d. **Lightening. The electrical energy promotes the endothermic reaction.**

15.115 For a 100% yield, 2.00 moles of $SO_3$ would be formed (why?). An 80% yield means 2.00 moles × (0.80) = 1.60 moles $SO_3$ is formed.

The amount of $SO_2$ remaining at equilibrium = (2.00 – 1.60) mol = 0.40 mol

The amount of $O_2$ reacted = $\frac{1}{2}$ × (amount of $SO_2$ reacted) = ($\frac{1}{2}$ × 1.60) mol = 0.80 mol

The amount of $O_2$ remaining at equilibrium = (2.00 – 0.80) mol = 1.20 mol

Total moles at equilibrium = moles $SO_2$ + moles $O_2$ + moles $SO_3$ = (0.40 + 1.20 + 1.60) mol = 3.20 moles

$$P_{SO_2} = \frac{0.40}{3.20}P_{\text{total}} = 0.125\,P_{\text{total}}$$

$$P_{O_2} = \frac{1.20}{3.20}P_{\text{total}} = 0.375\,P_{\text{total}}$$

$$P_{SO_3} = \frac{1.60}{3.20}P_{\text{total}} = 0.500\,P_{\text{total}}$$

$$K_P = \frac{P_{SO_3}{}^2}{P_{SO_2}{}^2 P_{O_2}}$$

$$0.13 = \frac{(0.500\,P_{\text{total}})^2}{(0.125\,P_{\text{total}})^2(0.375\,P_{\text{total}})}$$

$$\boldsymbol{P_{\text{total}} = 3.3\times10^2\ \text{atm}}$$

15.117 a. $\boldsymbol{K_P} = P_{Hg} = 0.0020\ \text{mmHg} = 2.6\times10^{-6}\ \text{atm} = \boldsymbol{2.6\times10^{-6}}$ (equil. constants are expressed without units)

$$\boldsymbol{K_c} = \frac{K_P}{(0.0821T)^{\Delta n}} = \frac{2.6\times10^{-6}}{(0.0821\times299)^1} = \boldsymbol{1.1\times10^{-7}}$$

b.

$$\text{Volume of lab} = (6.1\ \text{m})(5.3\ \text{m})(3.1\ \text{m}) = 100\ \text{m}^3$$

$$[\text{Hg}] = K_c$$

$$\textbf{Total mass of Hg vapor} = \frac{1.1\times10^{-7}\ \text{mol}}{1\ \text{L}}\times\frac{200.6\ \text{g}}{1\ \text{mol}}\times\frac{1\ \text{L}}{1000\ \text{cm}^3}\times\left(\frac{1\ \text{cm}}{0.01\ \text{m}}\right)^3\times100\ \text{m}^3 = \mathbf{2.2\ g}$$

The concentration of mercury vapor in the room is:

$$\frac{2.2 \text{ g}}{100 \text{ m}^3} = 0.022 \text{ g/m}^3 = \mathbf{22 \text{ mg/m}^3}$$

**Yes**. This concentration exceeds the safety limit of 0.05 mg/m$^3$.

15.119 a. **Shifts to right.**

b. **Shifts to right.**

c. **No change.**

d. **No change.**

e. **No change.**

f. **Shifts to left.**

15.121 **Panting decreases the concentration of $CO_2$ because $CO_2$ is exhaled during respiration. This decreases the concentration of carbonate ions, shifting the equilibrium to the left. Less $CaCO_3$ is produced. Two possible solutions would be either to cool the chickens' environment or to feed them carbonated water.**

15.123 a. **A catalyst speeds up the rates of the forward and reverse reactions to the same extent.**

b. **A catalyst would not change the energies of the reactant and product.**

c. **The first reaction is exothermic. Raising the temperature would favor the reverse reaction, increasing the amount of reactant and decreasing the amount of product at equilibrium. The equilibrium constant, *K*, would decrease. The second reaction is endothermic. Raising the temperature would favor the forward reaction, increasing the amount of product and decreasing the amount of reactant at equilibrium. The equilibrium constant, *K*, would increase.**

d. **A catalyst lowers the activation energy for the forward and reverse reactions to the same extent. Adding a catalyst to a reaction mixture will simply cause the mixture to reach equilibrium sooner. The same equilibrium mixture could be obtained without the catalyst, but we might have to wait longer for equilibrium to be reached. If the same equilibrium position is reached, with or without a**

**catalyst, then the equilibrium constant is the same.**

15.125 a. To determine $\Delta H°$, we need to plot $\ln K_P$ versus $1/T$ ($y$ vs. $x$).

| $\ln K_P$ | $1/T$ |
|---|---|
| 4.93 | 0.00167 |
| 1.63 | 0.00143 |
| –0.83 | 0.00125 |
| –2.77 | 0.00111 |
| –4.34 | 0.00100 |

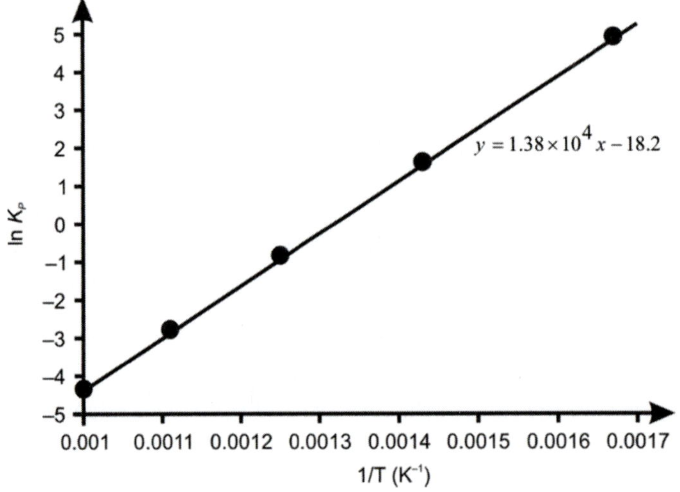


The slope of the plot equals $-\Delta H°/R$.

$$1.38 \times 10^4\ \text{K} = -\frac{\Delta H°}{8.314\ \text{J/mol}\cdot\text{K}}$$

$$\Delta H° = \mathbf{-1.15 \times 10^5\ J/mol} = \mathbf{-115\ kJ/mol}$$

b. **We start by writing the van't Hoff equation at two different temperatures.**

$$\mathbf{ln}K_1 = \frac{-\Delta H°}{RT_1} + C$$

$$\mathbf{ln}K_2 = \frac{-\Delta H°}{RT_2} + C$$

$$\mathbf{ln}K_1 - \mathbf{ln}K_2 = \frac{-\Delta H°}{RT_1} - \frac{-\Delta H°}{RT_2}$$

$$\mathbf{ln}\frac{K_1}{K_2} = \frac{\Delta H°}{R}\left(\frac{1}{T_2} - \frac{1}{T_1}\right)$$

**Assuming an endothermic reaction, $\Delta H° > 0$ and $T_2 > T_1$. Then, $\frac{\Delta H°}{R}\left(\frac{1}{T_2} - \frac{1}{T_1}\right) < 0$, meaning that $\ln\frac{K_1}{K_2} < 0$ or $K_1 < K_2$. A larger $K_2$ indicates that there are more products at equilibrium as the temperature is raised. This agrees with LeChatelier's principle that an increase in temperature favors the forward endothermic reaction. The opposite of the above discussion holds for an exothermic reaction.**

c. Treating

$$H_2O(l) \rightleftarrows H_2O(g) \qquad \Delta H_{vap} = ?$$

as a heterogeneous equilibrium, $K_P = P_{H_2O}$.

We substitute into the equation derived in part (a) to solve for $\Delta H_{vap}$.

$$\ln\frac{K_1}{K_2} = \frac{\Delta H°}{R}\left(\frac{1}{T_2} - \frac{1}{T_1}\right)$$

$$\ln\frac{31.82 \text{ mmHg}}{92.51 \text{ mmHg}} = \frac{\Delta H°}{8.314 \text{ J/mol}\cdot\text{K}}\left(\frac{1}{323 \text{ K}} - \frac{1}{303 \text{ K}}\right)$$

$$-1.067 = \Delta H°(-2.466 \times 10^{-5})$$

$$\Delta H° = \mathbf{4.34 \times 10^4\ J/mol} = \mathbf{434\ kJ/mol}$$

# Chapter 16

# Acids and Bases

16.3 Tables 16.1 and 16.2 of the text list important Brønsted acids and bases and their respective conjugates.

a. **both** (why?)

b. **base**

c. **acid**

d. **base**

e. **acid**

f. **base**

g. **base**

h. **base**

i. **acid**

j. **acid**

16.5 Recall that the conjugate base of a Brønsted acid is the species that remains when *one* proton has been removed from the acid.

a. nitrite ion: $\mathbf{NO_2^-}$

b. hydrogen sulfate ion (also called bisulfate ion): $\mathbf{HSO_4^-}$

c. hydrogen sulfide ion (also called bisulfide ion): $\mathbf{HS^-}$

d. cyanide ion: $\mathbf{CN^-}$

e. formate ion: $\mathbf{HCOO^-}$

16.7 The conjugate base of any acid is simply the acid minus one proton.

a. $\mathbf{CH_2ClCOO^-}$

b. $\mathbf{IO_4^-}$

c. $\mathbf{H_2PO_4^-}$

d. $\mathbf{HPO_4^{2-}}$

e. $\mathbf{PO_4^{3-}}$

f. $\mathbf{HSO_4^-}$

g. $\mathbf{SO_4^{2-}}$

h. $\mathbf{IO_3^-}$

i. $\mathbf{SO_3^{2-}}$

j. $\mathbf{NH_3}$

k. $\mathbf{HS^-}$

l. $\mathbf{S^{2-}}$

m. $\mathbf{OCl^-}$

16.15 **Strategy:** The equilibrium concentrations of $H^+$ and $OH^-$ must satisfy Equation 16.1, $K_W = [H^+][OH^-] = 1.0 \times 10^{-14}$ (25°C). Substitute the given concentrations and solve for $[H^+]$:

$$[H^+] = \frac{1.0 \times 10^{-14}}{[OH^-]}$$

**Solution:** a.

$$[H^+] = \frac{1.0 \times 10^{-14}}{2.50 \times 10^{-2}} = \mathbf{4.0 \times 10^{-13}}\ \boldsymbol{M}$$

b.

$$[H^+] = \frac{1.0 \times 10^{-14}}{1.67 \times 10^{-5}} = \mathbf{6.0 \times 10^{-10}}\ \boldsymbol{M}$$

c.

$$[H^+] = \frac{1.0 \times 10^{-14}}{8.62 \times 10^{-3}} = \mathbf{1.2 \times 10^{-12}}\ \boldsymbol{M}$$

d.

$$[H^+] = \frac{1.0 \times 10^{-14}}{1.75 \times 10^{-12}} = \mathbf{5.7 \times 10^{-3}}\ \boldsymbol{M}$$

16.17 **Strategy:** Use Equation 16.1 to calculate the $[H_3O^+]$ concentrations.

$$K_w = [H_3O^+][OH^-]$$

**Setup:** $K_w = 5.13 \times 10^{-13}$ at 100°C. Rearranging Equation 16.1 to solve for $[H_3O^+]$ gives:

$$[H_3O^+] = \frac{5.13 \times 10^{-13}}{[OH^-]}$$

**Solution:** a.

$$[H_3O^+] = \frac{5.13 \times 10^{-13}}{2.50 \times 10^{-2}} = \mathbf{2.05 \times 10^{-11}}\ \boldsymbol{M}$$

b.

$$[H_3O^+] = \frac{5.13 \times 10^{-13}}{1.67 \times 10^{-5}} = \mathbf{3.07 \times 10^{-8}}\ \boldsymbol{M}$$

c.

$$[H_3O^+] = \frac{5.13 \times 10^{-13}}{8.62 \times 10^{-3}} = \mathbf{5.95 \times 10^{-11}}\ \boldsymbol{M}$$

d.

$$[H_3O^+] = \frac{5.13 \times 10^{-13}}{1.75 \times 10^{-12}} = \mathbf{2.93 \times 10^{-1}}\ \boldsymbol{M}$$

16.21

$$[H^+] = 1.4 \times 10^{-3}\ M$$

$$\mathbf{[OH^-]} = \frac{K_w}{[H^+]} = \frac{1.0 \times 10^{-14}}{1.4 \times 10^{-3}} = \mathbf{7.1 \times 10^{-12}}\ \boldsymbol{M}$$

16.23 a. HCl is a strong acid, so the concentration of hydrogen ion is also 0.0010 *M*. (What is the concentration of chloride ion?) We use the definition of pH.

$$\mathbf{pH} = -\log[H^+] = -\log(0.0010) = \mathbf{3.00}$$

b. KOH is an ionic compound and completely dissociates into ions. We first find the concentration of hydrogen ion.

$$[H^+] = \frac{K_w}{[OH^-]} = \frac{1.0 \times 10^{-14}}{0.76} = 1.3 \times 10^{-14}\ M$$

The pH is then found from its defining equation

$$\mathbf{pH} = -\log[H^+] = -\log[1.3 \times 10^{-14}] = \mathbf{13.89}$$

16.25 **Strategy:** Here we are given the pH of a solution and asked to calculate $[H^+]$. Because pH is defined as $pH = -\log[H^+]$, we can solve for $[H^+]$ by taking the antilog of the pH; that is, $[H^+] = 10^{-pH}$.

**Solution:** a. From Equation 16.2 of the text:

$$pH = -\log[H^+] = 2.42$$

$$\log[H^+] = -2.42$$

To calculate $[H^+]$, we need to take the antilog of −2.42.

$$\mathbf{[H^+]} = 10^{-2.42} = \mathbf{3.8 \times 10^{-3}}\ \boldsymbol{M}$$

**Check**: Because the pH is between 2 and 3, we can expect $[H^+]$ to be between $1 \times 10^{-2}$ *M* and $1 \times 10^{-3}$ *M*. Therefore, the answer is reasonable.

b.

$$pH = -\log[H^+] = 11.24$$

$$\log[H^+] = -11.21$$

To calculate $[H^+]$, we need to take the antilog of $-11.21$.

$$[H^+] = 10^{-11.21} = \mathbf{6.2 \times 10^{-12}} \; \boldsymbol{M}$$

**Check**: Because the pH is between 11 and 12, we can expect $[H^+]$ to be between $1 \times 10^{-11}$ *M* and $1 \times 10^{-12}$ *M*. Therefore, the answer is reasonable.

c.

$$pH = -\log[H^+] = 6.96$$

$$\log[H^+] = -6.96$$

To calculate $[H^+]$, we need to take the antilog of $-6.96$.

$$[H^+] = 10^{-6.96} = \mathbf{1.1 \times 10^{-7}} \; \boldsymbol{M}$$

d.

$$pH = -\log[H^+] = 15.00$$

$$\log[H^+] = -15.00$$

To calculate $[H^+]$, we need to take the antilog of $-15.00$.

$$[H^+] = 10^{-15.00} = \mathbf{1.0 \times 10^{-15}} \; \boldsymbol{M}$$

16.27 The pH can be found by using Equation 16.6 of the text.

$$pH = 14.00 - pOH = 14.00 - 9.40 = 4.60$$

The hydrogen ion concentration can be found as follows.

$$4.60 = -\log[H^+]$$

Taking the antilog of both sides:

$$[H^+] = \mathbf{2.5 \times 10^{-5}} \; \boldsymbol{M}$$

16.29 We can calculate the $OH^-$ concentration from the pOH.

$$pOH = 14.00 - pH = 14.00 - 10.00 = 4.00$$

$$[OH^-] = 10^{-pOH} = 1.0 \times 10^{-4}\ M$$

Since NaOH is a strong base, it ionizes completely. The $OH^-$ concentration equals the initial concentration of NaOH.

$$[NaOH] = 1.0 \times 10^{-4}\ mol/L$$

So, we need to prepare 546 mL of $1.0 \times 10^{-4}\ M$ NaOH.

This is a dimensional analysis problem. We need to perform the following unit conversions.

$$mol/L \rightarrow mol\ NaOH \rightarrow grams\ NaOH$$

$$546\ mL = 0.546\ L$$

$$\textbf{? g NaOH} = 546\ mL \times \frac{1.0 \times 10^{-4}\ mol\ NaOH}{1000\ mL\ soln} \times \frac{40.00\ g\ NaOH}{1\ mol\ NaOH} = \mathbf{2.2 \times 10^{-3}\ g\ NaOH}$$

16.31

| pH | $[H^+]$ | Solution is: |
|---|---|---|
| < 7 | $> 1.0 \times 10^{-7}\ M$ | acid |
| > 7 | $< 1.0 \times 10^{-7}\ M$ | basic |
| = 7 | $= 1.0 \times 10^{-7}\ M$ | neutral |

16.37 a. **–0.009** b. **1.46** c. **5.82**

16.39 **Strategy:** $HNO_3$ is a strong acid. Therefore, the concentration of $HNO_3$ is equal to the concentration of hydronium ion. We use Equation 16.3 to determine $[H_3O^+]$:

**Solution:** a.

$$[HNO_3] = [H_3O^+] = 10^{-4.21} = \mathbf{6.2 \times 10^{-5}\ M}$$

b.

$$[HNO_3] = [H_3O^+] = 10^{-3.55} = \mathbf{2.8 \times 10^{-4}\ M}$$

c.

$$[HNO_3] = [H_3O^+] = 10^{-0.98} = \mathbf{0.10\ M}$$

16.41 **Strategy:** We use Equation 16.4 to determine pOH and Equation 16.6 to determine pH. The hydroxide ion concentration of a monobasic strong base, such as LiOH and NaOH, is equal to the base concentration. In the case of a dibasic base such as $Ba(OH)_2$, the hydroxide concentration is twice the base concentration.

**Solution:** a.

$$[OH^-] = [LiOH] = 1.24\ M.$$

$$\textbf{pOH} = -\log [OH^-] = -\log (1.24) = \mathbf{-0.093}$$

$$\textbf{pH} = 14.00 - pOH = 14.00 - (-0.093) = \mathbf{14.09}$$

b.

$$[OH^-] = \mathbf{2}[Ba(OH)_2] = 2(0.22\ M) = 0.44\ M.$$

$$\textbf{pOH} = -\log [OH^-] = -\log (0.44) = \mathbf{0.36}$$

$$\textbf{pH} = 14.00 - pOH = 14.00 - 0.36 = \mathbf{13.64}$$

c.

$$[OH^-] = [NaOH] = 0.085\ M.$$

$$\textbf{pOH} = -\log [OH^-] = -\log (0.085) = \mathbf{1.07}$$

$$\textbf{pH} = 14.00 - pOH = 14.00 - 1.07 = \mathbf{12.93}$$

16.43 **Strategy:** We use Equation 16.6 to determine pOH and Equation 16.5 to determine hydroxide ion concentration. The concentration of a monobasic strong base such as LiOH is equal to the hydroxide concentration. In the case of a dibasic base such as $Ba(OH)_2$, the base concentration is *half* the hydroxide concentration.

**Solution:** a.

$$pOH = 14.00 - pH = 14.00 - 11.04 = 2.96$$

$$\mathbf{[KOH]} = [OH^-] = 10^{-pOH} = 10^{-2.96} = \mathbf{1.1 \times 10^{-3}\ \textit{M}}$$

b.

$$[OH^-] = 10^{-pOH} = 10^{-2.96} = 1.1 \times 10^{-3}\ M$$

$$\mathbf{[Ba(OH)_2]} = [OH^-] \times \frac{1\ \text{mol Ba(OH)}_2}{2\ \text{mol OH}^-} = \frac{1.1 \times 10^{-3}\ M}{2} = \mathbf{5.5 \times 10^{-4}\ \textit{M}}$$

16.51 **Strategy:** Recall that a weak acid only partially ionizes in water. We are given the initial quantity of a weak acid ($CH_3COOH$) and asked to calculate the pH. For this we will need to calculate $[H^+]$, so we follow the procedure outlined in Section 16.5 of the text.

**Solution:** We set up a table for the dissociation.

| | $C_6H_5COOH(aq)$ | $\rightleftarrows$ | $H^+(aq)$ | + | $C_6H_5COO^-(aq)$ |
|---|---|---|---|---|---|
| Initial ($M$): | 0.10 | | 0.00 | | 0.00 |

| | | | |
|---|---|---|---|
| Change ($M$): | $-x$ | $+x$ | $+x$ |
| Equilibrium ($M$): | $(0.10 - x)$ | $x$ | $x$ |

$$K_a = \frac{[H^+][C_6H_5COO^-]}{[C_6H_5COOH]}$$

$$6.5 \times 10^{-5} = \frac{x^2}{(0.10 - x)}$$

Assuming that $x$ is small compared to 0.10, we neglect it in the denominator:

$$6.5 \times 10^{-5} = \frac{x^2}{0.10}$$

$$x = 2.55 \times 10^{-3}\ M = [H^+]$$

$$\mathbf{pH} = -\log(2.55 \times 10^{-3}) = \mathbf{2.59}$$

This problem could also have been solved using the quadratic equation. The difference in pH obtained using the approximation $x \approx 0$ is very small. (Using the quadratic equation would have given pH = 2.60, a difference of less than 0.4%.)

16.53 We set up a table for the dissociation.

| | $HCN(aq)$ | $\rightleftarrows$ | $H^+(aq)$ | + | $CN^-(aq)$ |
|---|---|---|---|---|---|
| Initial ($M$): | 0.095 | | 0.00 | | 0.00 |
| Change ($M$): | $-x$ | | $+x$ | | $+x$ |
| Equilibrium ($M$): | $(0.095 - x)$ | | $x$ | | $x$ |

$$K_a = \frac{[H^+][CN^-]}{[HCN]}$$

$$4.9 \times 10^{-10} = \frac{x^2}{(0.095 - x)}$$

Assuming that $x$ is small compared to 0.095, we neglect it in the denominator:

$$4.9 \times 10^{-10} = \frac{x^2}{0.095}$$

$$x = 6.82 \times 10^{-6}\ M = [H^+]$$

$$\mathbf{pH} = -\log(6.82 \times 10^{-6}) = \mathbf{5.17}$$

16.55 **Strategy:** Formic acid is a weak acid, and some of it will ionize in solution according to the equilibrium

$$HF(aq) + H_2O(l) \rightleftarrows F^-(aq) + H_3O^+(aq)$$

$$K_a = 1.7 \times 10^{-4} = \frac{[H^+][F^-]}{[HF]}$$

The percent ionization of a weak acid is given by Equation 16.7:

$$\%\text{ ionization} = \frac{[H^+]}{[HA]_0} \times 100\%,$$

where $[HA]_0$ is the initial concentration of the acid (that is, the concentration assuming there is *no* ionization). Using the $K_a$ expression above, set up a concentration table and solve for $[H^+]$. Then, use this value to compute the percent ionization.

**Solution:** a.

| | $HF(aq)$ + | $H_2O(l)$ | $\rightleftarrows$ | $F^-(aq)$ | + $H_3O^+(aq)$ |
|---|---|---|---|---|---|
| Initial ($M$): | 0.016 | | | 0 | 0 |
| Change ($M$): | $-x$ | | | $+x$ | $+x$ |
| Equilibrium ($M$): | $0.016 - x$ | | | $x$ | $x$ |

$$1.7 \times 10^{-4} = \frac{(x)(x)}{(0.016 - x)} = \frac{x^2}{(0.016 - x)} \quad (1)$$

As a first approximation, assume that $x$ is small compared to 0.016 so that $0.016 - x \approx 0.016$.

$$1.7 \times 10^{-4} \approx \frac{x^2}{0.016}$$

$$x \approx \sqrt{(1.7 \times 10^{-4})(0.016)} = 0.0016$$

Note that this approximate value of $x$ is 10% of 0.016, so it is probably not appropriate to have assumed that $0.016 - x \approx 0.016$. Instead, attempt to solve for $x$ without making any approximations. Start by clearing the denominator of equation (1) and then expanding the result to get:

$$x^2 + 1.7 \times 10^{-4}x - 2.72 \times 10^{-6} = 0$$

This equation is a *quadratic equation* of the form $ax^2 + bx + c = 0$, the solutions of which are given by (see Appendix 1):

$$x = \frac{-b \pm \sqrt{b^2 - 4ac}}{2a}.$$

For this case, $a = 1$, $b = 1.7\times10^{-4}$, and $c = -2.72\times10^{-6}$. Substituting these values into the quadratic formula gives:

$$x = \frac{-1.7\times10^{-4} \pm \sqrt{\left(1.7 \times 10^{-4}\right)^2 - 4(1)\left(-2.72 \times 10^{-6}\right)}}{(2)(1)}$$

$$= \frac{-1.7 \times 10^{-4} \pm 3.30 \times 10^{-3}}{2}$$

$$x = 1.6 \times 10^{-3}\ M \quad \text{or} \quad x = -1.7 \times 10^{-3}\ M$$

The negative solution is not physically reasonable since there is no such thing as a negative concentration. The solution is $x = 1.6\times10^{-3}\ M$.

Finally,

$$\textbf{\% ionization} = \frac{[\mathrm{H}^+]}{[\mathrm{HF}]_0} \times 100\% = \frac{x}{0.016} \times 100\% = \frac{.0016}{0.016} \times 100\% = \mathbf{10\%}$$

**Think About It:** Did the result obtained using the quadratic formula differ significantly from the approximate solution? What if the given $K_a$ value had three or more significant digits instead of two?

b. Proceed in the same manner as part (a). Note though that the initial concentration is even less than it was in part (a), making it even more necessary to avoid the approximate solution.

| | $HF(aq)$ + | $H_2O(l)$ | $\rightleftarrows$ | $F^-(aq)$ | + $H_3O^+(aq)$ |
|---|---|---|---|---|---|
| Initial ($M$): | $5.7\times10^{-4}$ | | | 0 | 0 |
| Change ($M$): | $-x$ | | | $+x$ | $+x$ |
| Equilibrium ($M$): | $5.7\times10^{-4} - x$ | | | $x$ | $x$ |

$$1.7 \times 10^{-4} = \frac{x^2}{\left(5.7 \times 10^{-4} - x\right)}$$

$$x^2 + 1.7 \times 10^{-4}x - 9.69 \times 10^{-8} = 0$$

$$x = \frac{-1.7 \times 10^{-4} \pm \sqrt{\left(1.7 \times 10^{-4}\right)^2 - 4(1)\left(-9.69 \times 10^{-8}\right)}}{(2)(1)}$$

$$= \frac{-1.7 \times 10^{-4} \pm 6.45 \times 10^{-4}}{2}$$

$$x = 4.8 \times 10^{-4}\ M \quad \text{or} \quad x = -8.2 \times 10^{-4}\ M$$

The negative solution is rejected, giving $x = 6.3{\times}10^{3}$ M. Thus,

$$\textbf{\% ionization} = \frac{[\mathrm{H^+}]}{[\mathrm{HF}]_0} \times 100\% = \frac{x}{5.7 \times 10^{4}} \times 100\% = \frac{4.8 \times 10^{-4}}{5.7 \times 10^{-4}} \times 100\% = \mathbf{84\%}$$

c. Proceed in the same manner as part (a). Note though that the initial concentration is very large, and we may safely seek an approximate solution.

| | $HF(aq)$ + | $H_2O(l)$ | ⇄ | $F^-(aq)$ | + $H_3O^+(aq)$ |
|---|---|---|---|---|---|
| Initial ($M$): | 1.75 | | | 0 | 0 |
| Change ($M$): | $-x$ | | | $+x$ | $+x$ |
| Equilibrium ($M$): | $1.75 - x$ | | | $x$ | $x$ |

$$1.7 \times 10^{-4} = \frac{x^2}{(1.75 - \mathrm{x})} \approx \frac{x^2}{1.75}$$

$$x \approx \sqrt{(1.7 \times 10^{-4})(1.75)} = 0.017\ M$$

$$\textbf{\% ionization} = \frac{[\mathrm{H^+}]}{[\mathrm{HF}]_0} \times 100\% = \frac{x}{1.75} \times 100\% = \frac{0.017}{1.75} \times 100\% = \mathbf{0.97\%}$$

16.57 **Strategy:** We are asked to compute $K_a$ for the equilibrium

$$HA(aq) + H_2O(l) \rightleftarrows A^-(aq) + H_3O^+(aq)$$

$$K_a = \frac{[\mathrm{H^+}][\mathrm{A^-}]}{[\mathrm{HA}]} \quad (1)$$

We are given the percent ionization of the acid, which is related to $[H^+]$ and $[HA]_0$ by Equation 16.7:

$$\%\ \text{ionization} = \frac{[\mathrm{H^+}]}{[\mathrm{HA}]_0} \times 100\% \quad \text{or} \quad [\mathrm{H^+}] = \left(\frac{\%\ \text{ionization}}{100\%}\right)[\mathrm{HA}]_0 \quad (2),$$

where $[HA]_0$ is the initial concentration of the acid (that is, the concentration assuming there is *no* ionization). Since the percent ionization (0.92%) is very small, it is safe to assume that $[HA] \approx [HA]_0$. Using equation (2) above for $[H^+]$, along with the fact that $[H^+] = [A^-]$, substitute into equation (1) and evaluate $K_a$.

**Solution:**

$$K_a = \frac{[H^+][A^-]}{[HA]}$$

$$\approx \frac{\left(\left(\frac{\%\text{ ionization}}{100\%}\right)[HA]_0\right)^2}{[HA]_0}$$

$$\approx \left(\frac{\%\text{ ionization}}{100\%}\right)^2 [HA]_0$$

$$\approx (0.92/100)^2 (0.015)$$

$$\boldsymbol{K_a \approx 1.3 \times 10^{-6}}$$

16.59 **Strategy:** We use the pH to determine $[H^+]_{eq}$, and set up a table for the dissociation of the weak acid, filling in what we know. The initial concentration of weak acid is 0.19 *M*. The initial concentrations of $H^+$ and the anion are both zero; and because they are both products of the ionization of the weak acid, their equilibrium concentrations are equal: $[H^+]_{eq} = [A^-]_{eq}$.

**Solution:**

$$[H^+] = 10^{-4.52} = 3.02 \times 10^{-5}\ M$$

| | $HA(aq)$ | $\rightleftarrows$ | $H^+(aq)$ | $+\ A^-(aq)$ |
|---|---|---|---|---|
| Initial (*M*): | 0.19 | | 0 | 0 |
| Change (*M*): | $-3.02 \times 10^{-5}$ | | $+3.02 \times 10^{-5}$ | $+3.02 \times 10^{-5}$ |
| Equilibrium (*M*): | $(0.19 - 3.02 \times 10^{-5})$ | | $3.02 \times 10^{-5}$ | $3.02 \times 10^{-5}$ |

The amount of HA ionized is very small compared to the initial concentration:

$$0.19 - 3.02 \times 10^{-5} \approx 0.19$$

$$\text{Therefore, } K_a = \frac{[H^+]_{eq}[A^-]_{eq}}{[HA]_{eq}} = \frac{(3.02 \times 10^{-5})^2}{0.19} = \mathbf{4.8 \times 10^{-9}}$$

16.61 A pH of 3.26 corresponds to a $[H^+]$ of $5.5 \times 10^{-4}$ *M*. Let the original concentration of formic acid be *x*. If the concentration of $[H^+]$ is $5.5 \times 10^{-4}$ *M*, that means that $5.5 \times 10^{-4}$ *M* of HCOOH ionized because of the 1:1 mole ratio between HCOOH and $H^+$.

| | $HCOOH(aq)$ | $\rightleftarrows$ | $H^+(aq)$ | + $HCOO^-(aq)$ |
|---|---|---|---|---|
| Initial ($M$): | $x$ | | 0 | 0 |
| Change ($M$): | $-5.5 \times 10^{-4}$ | | $+5.5 \times 10^{-4}$ | $+5.5 \times 10^{-4}$ |
| Equilibrium ($M$): | $x - (5.5 \times 10^{-4})$ | | $5.5 \times 10^{-4}$ | $5.5 \times 10^{-4}$ |

Substitute $K_a$ and the equilibrium concentrations into the ionization constant expression to solve for $x$.

$$K_a = \frac{[H^+][HCOO^-]}{[HCOOH]}$$

$$1.7 \times 10^{-4} = \frac{(5.5 \times 10^{-4})^2}{x - (5.5 \times 10^{-4})}$$

$$\mathbf{x = [HCOOH] = 2.3 \times 10^{-3}\ M}$$

16.63 The maximum possible concentration of hydrogen ion in a 0.10 $M$ solution of HA is 0.10 $M$. This is the case if HA is a strong acid. If HA is a weak acid, the hydrogen ion concentration is less than 0.10 $M$. The pH corresponding to 0.10 $M$ $[H^+]$ is 1.00. (Why three digits?) For a smaller $[H^+]$ the pH is larger than 1.00 (why?).

a. **false**, the pH is greater than 1.00

b. **false**, they are equal

c. **true**

d. **false**

16.65 a. **strong base** b. **weak base** c. **weak base** d. **weak base** e. **strong base**

16.69 **Strategy:** Weak bases only partially ionize in water.

$$B(aq) + H_2O(l) \rightleftarrows BH^+(aq) + OH^-(aq)$$

Note that the concentration of the weak base given refers to the initial concentration before ionization has started. The pH of the solution, on the other hand, refers to the situation at equilibrium. To calculate $K_b$, we need to know the concentrations of all three species, [B], $[BH^+]$, and $[OH^-]$ at equilibrium. We ignore the ionization of water as a source of $OH^-$ ions.

**Solution:** We proceed as follows.

*Step 1:* The major species in solution are B, $OH^-$, and the conjugate acid $BH^+$.

*Step 2:* First, we need to calculate the hydroxide ion concentration from the pH value. Calculate the pOH from the pH. Then, calculate the $OH^-$ concentration from the pOH.

$$\text{pOH} = 14.00 - \text{pH} = 14.00 - 10.66 = 3.34$$

$$\text{pOH} = -\log[OH^-]$$

$$-\text{pOH} = \log[OH^-]$$

Taking the antilog of both sides of the equation,

$$10^{-\text{pOH}} = [OH^-]$$

$$[OH^-] = 10^{-3.34} = 4.57 \times 10^{-4}\ M$$

*Step 3:* If the concentration of $OH^-$ is $4.57 \times 10^{-4}\ M$ at equilibrium, that must mean that $4.57 \times 10^{-4}\ M$ of the base ionized. We summarize the changes.

| | $B(aq)$ | + | $H_2O(l)$ | $\rightleftarrows$ | $BH^+(aq)$ | + $OH^-(aq)$ |
|---|---|---|---|---|---|---|
| Initial ($M$): | 0.30 | | | | 0 | 0 |
| Change ($M$): | $-4.57 \times 10^{-4}$ | | | | $+4.57 \times 10^{-4}$ | $+4.57 \times 10^{-4}$ |
| Equilibrium ($M$): | $0.30 - (4.57 \times 10^{-4})$ | | | | $4.57 \times 10^{-4}$ | $4.57 \times 10^{-4}$ |

*Step 4:* Substitute the equilibrium concentrations into the ionization constant expression to solve for $K_b$.

$$K_b = \frac{[BH^+][OH^-]}{[B]}$$

$$\boldsymbol{K_b} = \frac{\left(4.57 \times 10^{-4}\right)^2}{\left(0.30 - 4.57 \times 10^{-4}\right)} = \mathbf{6.97 \times 10^{-7}}$$

16.71 We set up an equilibrium table to solve for $[OH^-]$, and use pOH to get pH. We use B and $BH^+$ to represent the weak base and its protonated form, respectively.

| | $B(aq)$ | + | $H_2O(l)$ | $\rightleftarrows$ | $BH^+(aq)$ | + $OH^-(aq)$ |
|---|---|---|---|---|---|---|
| Initial ($M$): | 0.61 | | | | 0.00 | 0.00 |
| Change ($M$): | $-x$ | | | | $+x$ | $+x$ |
| Equilibrium ($M$): | $0.61 - x$ | | | | $x$ | $x$ |

$$[OH^-] = \frac{1.5 \times 10^{-4}[B]}{[BH^+]}$$

$$K_b = \frac{[BH^+][OH^-]}{[B]}$$

$$[OH^-] = \frac{1.5 \times 10^{-4}[B]}{[BH^+]}$$

$$x = \frac{1.5 \times 10^{-4}(0.61 - x)}{x}$$

Assuming $x$ is small relative to 0.61, then

$$x = \frac{(1.5 \times 10^{-4})(0.61)}{x}$$

$$\mathbf{x = 0.0096\ M = [OH^-]}$$

$$\text{pOH} = -\log(0.0096) = 2.02$$

$$\textbf{pH} = 14.00 - 2.02 = \mathbf{11.98}$$

16.75 **Strategy:** We calculate the $K_b$ value for a conjugate base using Equation 16.7.

**Solution:**

$$K_b(CN^-) = \frac{K_w}{K_a(HCN)} = \frac{1.0 \times 10^{-14}}{4.9 \times 10^{-10}} = \mathbf{2.0 \times 10^{-5}}$$

$$K_b(F^-) = \frac{K_w}{K_a(HF)} = \frac{1.0 \times 10^{-14}}{7.1 \times 10^{-4}} = \mathbf{1.4 \times 10^{-11}}$$

$$K_b(CH_3COO^-) = \frac{K_w}{K_a(CH_3COOH)} = \frac{1.0 \times 10^{-14}}{1.8 \times 10^{-5}} = \mathbf{5.6 \times 10^{-10}}$$

$$K_b(HCO_3^-) = \frac{K_w}{K_a(H_2CO_3)} = \frac{1.0 \times 10^{-14}}{4.2 \times 10^{-7}} = \mathbf{2.4 \times 10^{-8}}$$

16.77 a. HA has the largest $K_a$ value. Therefore, **$A^-$ has the smallest $K_b$ value.**

b. HB has the smallest $K_a$ value. Therefore, **$B^-$ is the strongest base.**

16.81 The pH of a 0.040 *M* HCl solution (strong acid) is: pH = –log(0.040) = **1.40**. Follow the procedure for calculating the pH of a diprotic acid to calculate the pH of the sulfuric acid solution.

**Strategy:** Determining the pH of a diprotic acid in aqueous solution is more involved than for a monoprotic acid. The first stage of ionization for $H_2SO_4$ goes to completion. We follow the procedure for determining the pH of a strong acid for this stage. The conjugate base produced in the first ionization ($HSO_4^-$) is a weak acid. We follow the procedure for determining the pH of a weak acid for this stage.

**Solution:** We proceed according to the following steps.

*Step 1:* $H_2SO_4$ is a strong acid. The first ionization stage goes to completion. The ionization of $H_2SO_4$ is

$$H_2SO_4(aq) \rightarrow H^+(aq) + HSO_4^-(aq)$$

The concentrations of all the species ($H_2SO_4$, $H^+$, and $HSO_4^-$) before and after ionization can be represented as follows.

| | $H_2SO_4(aq)$ | $\rightarrow$ | $H^+(aq)$ | $+\ HSO_4^-(aq)$ |
|---|---|---|---|---|
| Initial (*M*): | 0.040 | | 0 | 0 |
| Change (*M*): | –0.040 | | +0.040 | +0.040 |
| Equilibrium (*M*): | 0 | | 0.040 | 0.040 |

*Step 2:* Now, consider the second stage of ionization. $HSO_4^-$ is a weak acid. Set up a table showing the concentrations for the second ionization stage. Let *x* be the change in concentration. Note that the initial concentration of $H^+$ is 0.040 *M* from the first ionization.

| | $HSO_4^-(aq)$ | $\rightleftarrows$ | $H^+(aq)$ | $+\ SO_4^{2-}(aq)$ |
|---|---|---|---|---|
| Initial (*M*): | 0.040 | | 0.040 | 0 |
| Change (*M*): | $-x$ | | $+x$ | $+x$ |
| Equilibrium (*M*): | $0.040 - x$ | | $0.040 + x$ | $x$ |

Write the ionization constant expression for $K_a$. Then, solve for *x*. You can find the $K_a$ value in Table 16.8 of the text.

$$K_a = \frac{[H^+][SO_4^{2-}]}{[HSO_4^-]}$$

$$1.3 \times 10^{-2} = \frac{(0.040 + x)(x)}{(0.040 - x)}$$

Since $K_a$ is quite large, we cannot make the assumptions that

$$0.040 - x \approx 0.040 \quad \text{and} \quad 0.040 + x \approx 0.040$$

Therefore, we must solve a quadratic equation.

$$x^2 + 0.053x - (5.2 \times 10^{-4}) = 0$$

$$x = \frac{-0.053 \pm \sqrt{(0.053)^2 - 4(1)(-5.2 \times 10^{-4})}}{2(1)}$$

$$x = \frac{-0.053 \pm 0.070}{2}$$

$$x = 8.5 \times 10^{-3}\ M \quad \text{or} \quad x = -0.062\ M$$

The second solution is physically impossible because you cannot have a negative concentration. The first solution is the correct answer.

*Step 3:* Having solved for $x$, we can calculate the $H^+$ concentration at equilibrium. We can then calculate the pH from the $H^+$ concentration.

$$[H^+] = 0.040\ M + x = [0.040 + (8.5 \times 10^{-3})]M = 0.049\ M$$

$$\textbf{pH} = -\log(0.049) = \textbf{1.31}$$

**Think About It:** Without doing any calculations, could you have known that the pH of the sulfuric acid would be lower (more acidic) than that of the hydrochloric acid?

16.83 For the first stage of ionization:

$$H_2CO_3(aq) \rightleftharpoons H_3O^+(aq) + HCO_3^-(aq)$$

| | $H_2CO_3(aq)$ | $H_3O^+(aq)$ | $HCO_3^-(aq)$ |
|---|---|---|---|
| Initial ($M$): | 0.025 | 0.00 | 0.00 |
| Change ($M$): | $-x$ | $+x$ | $+x$ |
| Equilibrium ($M$): | $(0.025 - x)$ | $+x$ | $+x$ |

$$K_{a_1} = \frac{[H_3O^+][HCO_3^-]}{[H_2CO_3]}$$

$$4.2 \times 10^{-7} = \frac{x^2}{(0.025 - x)} \approx \frac{x^2}{0.025}$$

$$x = 1.0 \times 10^{-4}\ M$$

For the second ionization,

| | $HCO_3^-(aq)$ | $\rightleftarrows$ | $H_3O^+(aq)$ | $+\ CO_3^{2-}(aq)$ |
|---|---|---|---|---|
| Initial (*M*): | $1.0 \times 10^{-4}$ | | $1.0 \times 10^{-4}$ | 0.00 |
| Change (*M*): | $-y$ | | $+y$ | $+y$ |
| Equilibrium (*M*): | $(1.0 \times 10^{-4}) - y$ | | $(1.0 \times 10^{-4}) + y$ | $y$ |

$$K_{a_2} = \frac{[H_3O^+][CO_3^{2-}]}{[HCO_3^-]}$$

$$4.8 \times 10^{-11} = \frac{[(1.0 \times 10^{-4}) + y](y)}{1.0\times10^{-4} - y} \approx \frac{(1.0 \times 10^{-4})(y)}{(1.0 \times 10^{-4})}$$

$$y = 4.8 \times 10^{-11}\ M$$

Since $HCO_3^-$ is a very weak acid, there is little ionization at this stage. Therefore we have:

$$\mathbf{[H_3O^+] = [HCO_3^-] = 1.0 \times 10^{-4}\ \mathit{M}} \text{ and } \mathbf{[CO_3^{2-}]} = y = \mathbf{4.8 \times 10^{-11}\ \mathit{M}}$$

16.85 Oxalic acid is diprotic. Because its first and second ionization constants differ only by a factor of about 1000, we consider both the first and second ionizations to determine $[H^+]$. We set up an equilibrium table:

| | $H_2C_2O_4(aq)$ | $\rightleftarrows$ | $H^+(aq)$ | $+\ HC_2O_4^-(aq)$ |
|---|---|---|---|---|
| Initial (*M*): | 0.25 | | 0.00 | 0.00 |
| Change (*M*): | $-x$ | | $+x$ | $+x$ |
| Equilibrium (*M*): | $(0.25 - x)$ | | $+x$ | $+x$ |

$$6.5 \times 10^{-2} = \frac{x^2}{0.25 - x}$$

Using the quadratic equation to solve for $x$ we find

$$x = 0.099$$

For the second ionization:

| | $HC_2O_4^-(aq)$ | $\rightleftarrows$ | $H^+(aq)$ | $+\ C_2O_4^{2-}(aq)$ |
|---|---|---|---|---|
| Initial (*M*): | 0.099 | | 0.099 | 0.00 |
| Change (*M*): | $-y$ | | $+y$ | $+y$ |
| Equilibrium (*M*): | $(0.099 - y)$ | | $0.099+y$ | $y$ |

$$6.1 \times 10^{-5} = \frac{(0.099 + y)y}{(0.099 - y)}$$

We neglect $y$ with respect to 0.099 and find

$$y = 6.1 \times 10^{-5}$$

The concentration of hydronium ion after the second ionization is therefore

$$0.099 + 6.1 \times 10^{-5} \approx 0.099\ M$$

and

$$\mathbf{pH} = -\log(0.099) = \mathbf{1.00}$$

Note that the second ionization did not contribute enough to the hydronium concentration to change the value of pH.

16.87 a. The two steps in the ionization of a weak diprotic acid are:

$$H_2A(aq) + H_2O(l) \rightleftarrows H_3O^+(aq) + HA^-(aq)$$

$$HA^-(aq) + H_2O(l) \rightleftarrows H_3O^+(aq) + A^{2-}(aq)$$

The diagram that represents a weak diprotic acid is **(c)**. In this diagram, we only see the first step of the ionization, because $HA^-$ is a much weaker acid than $H_2A$.

b. Both **(b)** and **(d)** are chemically implausible situations. Because $HA^-$ is a much weaker acid than $H_2A$, you would not see a higher concentration of $A^{2-}$ compared to $HA^-$.

16.91 All the listed pairs are oxoacids that contain different central atoms whose elements are in the same group of the periodic table and have the same oxidation number. In this situation the acid with the most electronegative central atom will be the strongest.

a. $\mathbf{H_2SO_4 > H_2SeO_4}$ b. $\mathbf{H_3PO_4 > H_3AsO_4}$

16.93 **The conjugate bases are $C_6H_5O^-$ from phenol and $CH_3O^-$ from methanol. The $C_6H_5O^-$ is stabilized by resonance**:

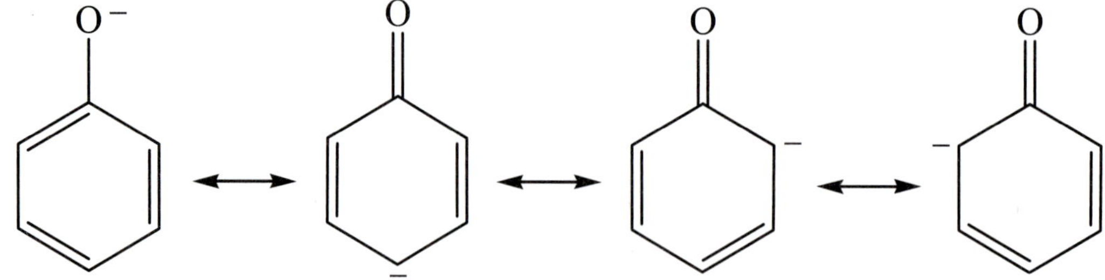

**The $CH_3O^-$ ion has no such resonance stabilization. A more stable conjugate base means an increase in the strength of the acid.**

16.99 The salt ammonium chloride completely ionizes upon dissolution, producing 0.42 $M$ [$NH_4^+$] and 0.42 $M$ [$Cl^-$] ions. $NH_4^+$ will undergo hydrolysis because it is a weak acid ($NH_4^+$ is the conjugate acid of the weak base, $NH_3$).

*Step 1:* Express the equilibrium concentrations of all species in terms of initial concentrations and a single unknown $x$, that represents the change in concentration. Let $(-x)$ be the depletion in concentration (mol/L) of $NH_4^+$. From the stoichiometry of the reaction, it follows that the increase in concentration for both $H_3O^+$ and $NH_3$ must be $x$. Complete a table that lists the initial concentrations, the change in concentrations, and the equilibrium concentrations.

| | $NH_4^+(aq)$ + | $H_2O(l)$ $\rightleftarrows$ | $NH_3(aq)$ | + $H_3O^+(aq)$ |
|---|---|---|---|---|
| Initial ($M$): | 0.42 | | 0.00 | 0.00 |
| Change ($M$): | $-x$ | | $+x$ | $+x$ |
| Equilibrium ($M$): | $(0.42 - x)$ | | $+x$ | $+x$ |

*Step 2:* You can calculate the $K_a$ value for $NH_4^+$ from the $K_b$ value of $NH_3$. The relationship is

$$K_a \times K_b = K_w$$

or

$$K_a = \frac{K_w}{K_b} = \frac{1.0 \times 10^{-14}}{1.8 \times 10^{-5}} = 5.6 \times 10^{-10}$$

*Step 3:* Write the ionization constant expression in terms of the equilibrium concentrations. Knowing the value of the equilibrium constant ($K_a$), solve for $x$.

$$K_a = \frac{[NH_3][H_3O^+]}{[NH_4^+]}$$

$$5.6 \times 10^{-10} = \frac{x^2}{0.42 - x} \approx \frac{x^2}{0.42}$$

$$x = [H^+] = 1.5 \times 10^{-5}\ M$$

$$\mathbf{pH} = -\log(1.5 \times 10^{-5}) = \mathbf{4.82}$$

Since $NH_4Cl$ is the salt of a weak base (aqueous ammonia) and a strong acid (HCl), we expect the solution to be slightly acidic, which is confirmed by the calculation.

16.101 **Strategy:** We are asked to determine the pH of a salt solution. We must first determine the identities of the ions in solution and decide which, if any, of the ions will hydrolyze. In this case, the ions in solution are $C_2H_5NH_3^+$ and $I^-$. The iodide ion, $I^-$, is the anion of the strong acid, HI. Therefore it will not hydrolyze. The $C_2H_5NH_3^+$ ion, however, is the conjugate acid of the weak base $C_2H_5NH_2$ (ethylamine). It hydrolyzes to produce $H^+$ and the weak base $C_2H_5NH_2$:

$$C_2H_5NH_3^+(aq) + H_2O(l) \rightleftarrows C_2H_5NH_2(aq) + H_3O^+(aq)$$

In order to determine the pH, we must first determine the concentration of hydronium ion produced by the hydrolysis. For this, we need the ionization constant, $K_a$, of the $C_2H_5NH_3^+$ ion.

**Solution:** Using the $K_b$ value for ethylamine and Equation 16.7, we calculate $K_a$ for $C_2H_5NH_3^+$:

$$K_a = \frac{1.0 \times 10^{-14}}{5.6 \times 10^{-4}} = 1.8 \times 10^{-11}$$

We then construct an equilibrium table and fill in what we know to find $[H_3O^+]$:

| | $C_2H_5NH_3^+(aq)$ | + | $H_2O(l)$ | $\rightleftarrows$ | $C_2H_5NH_2(aq)$ | + | $H_3O^+(aq)$ |
|---|---|---|---|---|---|---|---|
| Initial (*M*): | 0.91 | | | | 0.00 | | 0.00 |
| Change (*M*): | $-x$ | | | | $+x$ | | $+x$ |
| Equilibrium (*M*): | $(0.91 - x)$ | | | | $x$ | | $x$ |

$$1.8 \times 10^{-11} = \frac{[C_2H_5NH_2][H_3O^+]}{[C_2H_5NH_3^+]} = \frac{x^2}{(0.91 - x)}$$

Assuming that $x$ is very small compared to 0.91, we can write

$$1.8 \times 10^{-11} = \frac{x^2}{0.91}$$

and

$$x = [H_3O^+] = 4.05 \times 10^{-6}\ M$$

$$\textbf{pH} = -\log(4.05 \times 10^{-6}) = \mathbf{5.39}$$

16.103 **Strategy:** In deciding whether a salt will undergo hydrolysis, ask yourself the following questions: Is the cation a highly charged metal ion or an ammonium ion? Is the anion the conjugate base of a weak acid? If yes to either question, then hydrolysis will occur. In cases where both the cation and the anion react with water, the pH of the solution will depend on the relative magnitudes of $K_a$ for the cation and $K_b$ for the anion.

**Solution:** We first break up the salt into its cation and anion components and then examine the possible reaction of each ion with water.

a. The $Na^+$ cation does not hydrolyze. The $Br^-$ anion is the conjugate base of the strong acid HBr. Therefore, $Br^-$ will not hydrolyze either, and the solution is **neutral**.

b. The $K^+$ cation does not hydrolyze. The $SO_3^{2-}$ anion is the conjugate base of the weak acid $HSO_3^-$ and will hydrolyze to give $HSO_3^-$ and $OH^-$. The solution will be **basic**.

c. Both the $NH_4^+$ and $NO_2^-$ ions will hydrolyze. $NH_4^+$ is the conjugate acid of the weak base $NH_3$, and $NO_2^-$ is the conjugate base of the weak acid $HNO_2$. Using ionization constants from Tables 16.6 and 16.7, and Equation 16.7, we find that the $K_a$ of $NH_4^+$ ($1.0 \times 10^{-14}/1.8 \times 10^{-5} = 5.6 \times 10^{-10}$) is greater than the $K_b$ of $NO_2^-$ ($1.0 \times 10^{-14}/4.5 \times 10^{-4} = 2.2 \times 10^{-11}$). Therefore, the solution will be **acidic**.

d. $Cr^{3+}$ is a small metal cation with a high charge, which hydrolyzes to produce $H^+$ ions. The $NO_3^-$ anion does not hydrolyze. It is the conjugate base of the strong acid, $HNO_3$. The solution will be **acidic**.

16.105 There is an inverse relationship between acid strength and conjugate base strength. As acid strength decreases, the proton accepting power of the conjugate base increases. In general the weaker the acid, the stronger the conjugate base. All three of the potassium salts ionize completely to form the conjugate base of the respective acid. The greater the pH, the stronger the conjugate base, and therefore, the weaker the acid.

The order of increasing acid strength is **HZ < HY < HX**.

16.107

$$HCO_3^- \rightleftarrows H^+ + CO_3^{2-} \qquad K_a = 4.8 \times 10^{-11}$$

$$HCO_3^- + H_2O \rightleftarrows H_2CO_3 + OH^- \qquad K_b = \frac{K_w}{K_a} = \frac{1.0 \times 10^{-14}}{4.2 \times 10^{-7}} = 2.4 \times 10^{-8}$$

$HCO_3^-$ has a greater tendency to hydrolyze than to ionize ($K_b > K_a$). The solution will be basic (**pH > 7**).

16.111 The most basic oxides occur with metal ions having the lowest positive charges (or lowest oxidation numbers).

a. $\mathbf{Al_2O_3 < BaO < K_2O}$

b. $\mathbf{CrO_3 < Cr_2O_3 < CrO}$

16.113 $Al(OH)_3$ is an amphoteric hydroxide. The reaction is:

$$\mathbf{Al(OH)_3(s) + OH^-(aq) \rightarrow Al(OH)_4^-(aq)}$$

**This is a Lewis acid-base reaction.** Can you identify the acid and base?

16.117 **$AlCl_3$ is a Lewis acid with an incomplete octet of electrons and $Cl^-$ is the Lewis base donating a pair of electrons.**

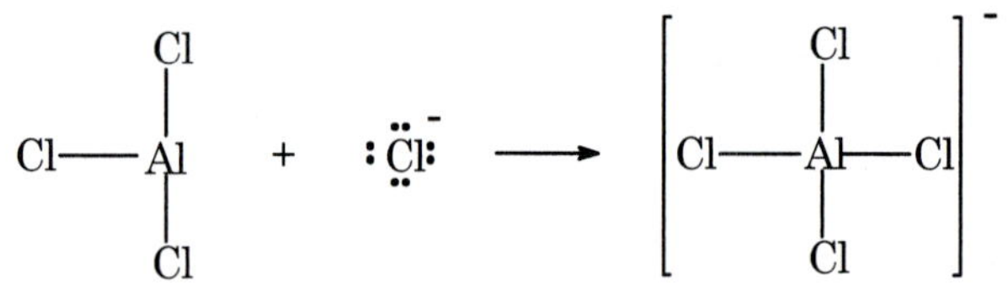

$$Cl{-}\underset{\displaystyle Cl}{\overset{\displaystyle Cl}{Al}} \;+\; :\ddot{\underset{..}{Cl}}:^- \;\longrightarrow\; \left[Cl{-}\underset{\displaystyle Cl}{\overset{\displaystyle Cl}{Al}}{-}Cl\right]^-$$

16.119 By definition Brønsted acids are proton donors, therefore such compounds must contain at least one hydrogen atom. In Problem 16.116, Lewis acids that do not contain hydrogen, and therefore are not Brønsted acids, are **$CO_2$, $SO_2$, and $BCl_3$**. Can you name others?

16.121 a. **$AlBr_3$ is the Lewis acid; $Br^-$ is the Lewis base.**

b. **Cr is the Lewis acid; CO is the Lewis base.**

c. **$Cu^{2+}$ is the Lewis acid; $CN^-$ is the Lewis base.**

16.123 In theory, the products will be $CH_3COO^-(aq)$ and $HCl(aq)$ but **this reaction will *not* occur to any measurable extent** because $Cl^-$ is the conjugate base of the strong acid, HCl. It is a negligibly weak base and has no affinity for protons.

16.125 $HNO_2$ is a weak acid with $K_a = 4.5 \times 10^{-4}$.

| | $HNO_2(aq)$ $\rightleftarrows$ | $H^+(aq)$ + | $NO_2^-(aq)$ |
|---|---|---|---|
| Initial ($M$): | 0.88 | 0.00 | 0.00 |
| Change ($M$): | $-x$ | $+x$ | $+x$ |
| Equilibrium ($M$): | $(0.88 - x)$ | $x$ | $x$ |

$$4.5 \times 10^{-4} = \frac{x^2}{0.88 - x}$$

We neglect $x$ in the denominator and find

$$x = \sqrt{(4.5 \times 10^{-4})(0.88)} = 0.0199$$

$$\textbf{pH} = -\log(0.020) = \mathbf{1.70}$$

$$\textbf{\% ionization} = \frac{0.020}{0.88} \times 100\% = \mathbf{2.3\%}$$

16.127 Choice **(c)** because 0.70 $M$ KOH has a higher pH than 0.60 $M$ NaOH. Adding an equal volume of 0.60 $M$ NaOH lowers the $[OH^-]$ to 0.65 $M$, hence lowering the pH.

16.129 a. **For the forward reaction $NH_4^+$ and $NH_3$ are the conjugate acid and base pair, respectively. For the reverse reaction $NH_3$ and $NH_2^-$ are the conjugate acid and base pair, respectively.**

b. **$H^+$ corresponds to $NH_4^+$; $OH^-$ corresponds to $NH_2^-$. For the neutral solution, $[NH_4^+] = [NH_2^-]$.**

16.131

$$K_a = \frac{[H^+][A^-]}{[HA]}$$

$$[HA] \approx 0.1\ M$$

$$[A^-] \approx 0.1\ M$$

**Therefore,**

$$K_a = [H^+] = \frac{K_w}{[OH^-]}$$

$$[OH^-] = \frac{K_w}{K_a}$$

16.133

$$HCOOH \rightleftarrows HCOO^- + H^+ \qquad K_a = 1.7 \times 10^{-4}$$

$$H^+ + OH^- \rightleftarrows H_2O \qquad K_w' = \frac{1}{K_w} = \frac{1}{1.0 \times 10^{-14}} = 1.0 \times 10^{14}$$

$$HCOOH + OH^- \rightleftarrows HCOO^- + H_2O \qquad \boldsymbol{K} = K_aK_w' = (1.7 \times 10^{-4})(1.0 \times 10^{14}) = \mathbf{1.7 \times 10^{10}}$$

16.135 a.

| $\mathbf{H^-}$ | + | $\mathbf{H_2O}$ | $\rightarrow$ | $\mathbf{OH^-}$ | + | $\mathbf{H_2}$ |
|---|---|---|---|---|---|---|
| **base$_1$** | | **acid$_2$** | | **base$_2$** | | **acid$_1$** |

b. **$H^-$ is the reducing agent and $H_2O$ is the oxidizing agent.**

16.137 The van't Hoff equation allows the calculation of an equilibrium constant at a different temperature if the value of the equilibrium constant at another temperature and $\Delta H°$ for the reaction are known.

$$\ln\frac{K_1}{K_2} = \frac{\Delta H°}{R}\left(\frac{1}{T_2} - \frac{1}{T_1}\right)$$

First, we calculate $\Delta H°$ for the ionization of water using data in Appendix 2 of the text.

$$H_2O(l) \rightleftarrows H^+(aq) + OH^-(aq)$$

$$\Delta H° = [\Delta H_f^\circ(H^+) + \Delta H_f^\circ(OH^-)] - \Delta H_f^\circ(H_2O)$$

$$\Delta H° = (0 - 229.94 \text{ kJ/mol}) - (-285.8 \text{ kJ/mol})$$

$$\Delta H° = 55.9 \text{ kJ/mol}$$

We substitute $\Delta H°$ and the equilibrium constant at 25°C (298 K) into the van't Hoff equation to solve for the equilibrium constant at 100°C (373 K).

$$\ln\frac{1.0 \times 10^{-14}}{K_2} = \frac{55.9 \times 10^3 \text{ J/mol}}{8.314 \text{ J/mol}\cdot\text{K}}\left(\frac{1}{373 \text{ K}} - \frac{1}{298 \text{ K}}\right)$$

$$\frac{1.0 \times 10^{-14}}{K_2} = e^{-4.537}$$

$$K_2 = 9.3 \times 10^{-13}$$

We substitute into the equilibrium constant expression for the ionization of water to solve for $[H^+]$ and then pH.

$$K_2 = [H^+][OH^-]$$

$$9.3 \times 10^{-13} = x^2$$

$$x = [H^+] = 9.6 \times 10^{-7}\ M$$

$$\mathbf{pH} = -\log(9.6 \times 10^{-7}) = \mathbf{6.02}$$

Note that the water is **not** acidic at 100°C because $[H^+] = [OH^-]$.

16.139 Because the P–H bond is weaker, there is a greater tendency for $PH_4^+$ to ionize. Therefore, **$PH_3$ is a weaker base than $NH_3$.**

16.141 a. **$HNO_2$**

b. **HF**

c. **$BF_3$**

d. **$NH_3$**

e. **$H_2SO_3$**

f. **$HCO_3^-$ and $CO_3^{2-}$**

The reactions for (f) are:

$$HCO_3^-(aq) + H^+(aq) \rightarrow CO_2(g) + H_2O(l)$$

$$CO_3^{2-}(aq) + 2H^+(aq) \rightarrow CO_2(g) + H_2O(l)$$

16.143 A strong acid, such as HCl, will be completely ionized, choice **(b)**.

A weak acid will only ionize to a lesser extent compared to a strong acid, choice **(c)**.

A very weak acid will remain almost exclusively as the acid molecule in solution. Choice **(d)** is the best choice.

16.145 The equations are:

$$\mathbf{Cl_2(g) + H_2O(l) \rightleftarrows HCl(aq) + HClO(aq)}$$

$$\mathbf{HCl(aq) + AgNO_3(aq) \rightleftarrows AgCl(s) + HNO_3(aq)}$$

**In the presence of $OH^-$ ions, the first equation is shifted to the right:**

$$\mathbf{H^+ \text{(from HCl)} + OH^- \rightarrow H_2O}$$

**Therefore, the concentration of HClO increases. (The 'bleaching action' is due to $ClO^-$ ions.)**

16.147 We examine the hydrolysis of the cation and anion separately.

$$NH_4CN(aq) \rightarrow NH_4^+(aq) + CN^-(aq)$$

*Cation:*

| | $NH_4^+(aq)$ | $H_2O(l)$ | $\rightleftarrows$ | $NH_3(aq)$ + | + $H_3O^+(aq)$ |
|---|---|---|---|---|---|
| Initial ($M$): | 2.00 | | | 0.00 | 0.00 |
| Change ($M$): | $-x$ | | | $+x$ | $+x$ |
| Equilibrium ($M$): | $2.00 - x$ | | | $x$ | $x$ |

$$K_a = \frac{[NH_3][H_3O^+]}{[NH_4^+]}$$

$$5.6 \times 10^{-10} = \frac{x^2}{2.00 - x} \approx \frac{x^2}{2.00}$$

$$x = 3.35 \times 10^{-5}\ M = [H_3O^+]$$

$$K_b = \frac{[HCN][OH^-]}{[CN^-]}$$

$$2.0 \times 10^{-5} = \frac{y^2}{2.00 - y} \approx \frac{y^2}{2.00}$$

$$y = 6.32 \times 10^{-3}\ M = [OH^-]$$

$CN^-$ is stronger as a base than $NH_4^+$ is as an acid. Some $OH^-$ produced from the hydrolysis of $CN^-$ will be neutralized by $H_3O^+$ produced from the hydrolysis of $NH_4^+$.

| | $H_3O^+(aq)$ + | $OH^-(aq)$ | $\rightarrow$ | $2H_2O(l)$ |
|---|---|---|---|---|
| Initial ($M$): | $3.35 \times 10^{-5}$ | $6.32 \times 10^{-3}$ | | |
| Change ($M$): | $-3.35 \times 10^{-5}$ | $-3.35 \times 10^{-5}$ | | |
| Equilibrium ($M$): | 0 | $6.29 \times 10^{-3}$ | | |

$$[OH^-] = 6.29 \times 10^{-3}\ M$$

$$pOH = 2.20$$

$$\mathbf{pH = 11.80}$$

16.149 **Loss of the first proton from a polyprotic acid is always easier than the subsequent removal of additional protons. The ease with which a proton is lost (i.e., the strength of the acid) depends on the stability of the anion that remains. An anion with a single negative charge is more easily stabilized by resonance than one with two negative charges.**

16.151 The balanced equations for the two reactions are:

$$MCO_3(s) + 2HCl(aq) \rightarrow MCl_2(aq) + CO_2(g) + H_2O(l)$$

$$HCl(aq) + NaOH(aq) \rightarrow NaCl(aq) + H_2O(l)$$

First, let's find the number of moles of excess acid from the reaction with NaOH.

$$0.03280 \text{ L} \times \frac{0.588 \text{ mol NaOH}}{1 \text{ L soln}} \times \frac{1 \text{ mol HCl}}{1 \text{ mol NaOH}} = 0.0193 \text{ mol HCl}$$

The original number of moles of acid was:

$$0.500 \text{ L} \times \frac{1.00 \text{ mol HCl}}{1 \text{ L soln}} = 0.500 \text{ mol HCl}$$

The amount of hydrochloric acid that reacted with the metal carbonate is:

$$(0.500 \text{ mol HCl}) - (0.0193 \text{ mol HCl}) = 0.4807 \text{ mol HCl}$$

The mole ratio from the balanced equation is 1 mole $MCO_3$ : 2 mole HCl. The moles of $MCO_3$ that reacted are:

$$0.4807 \text{ mol HCl} \times \frac{1 \text{ mol } MCO_3}{2 \text{ mol HCl}} = 0.2404 \text{ mol } MCO_3$$

We can now determine the molar mass of $MCO_3$, which will allow us to identify the metal.

$$\text{molar mass } MCO_3 \times \frac{20.27 \text{ g } MCO_3}{0.2404 \text{ mol } MCO_3} = 84.3 \text{ g/mol}$$

We subtract the mass of $CO_3^{2-}$ to identify the metal.

$$\text{molar mass M} = 84.3 \text{ g/mol} - 60.01 \text{ g/mol} = 24.3 \text{ g/mol}$$

The metal is **magnesium**.

16.153 When the pH is 10.00, the pOH is 4.00 and the concentration of hydroxide ion is $1.0 \times 10^{-4}$ *M*. The concentration of HCN must be the same. (Why?) If the concentration of NaCN is *x*, the table looks like:

$$CN^-(aq) + H_2O(l) \rightleftarrows HCN(aq) + OH^-(aq)$$

| | $CN^-(aq)$ | $H_2O(l)$ | $HCN(aq)$ | $OH^-(aq)$ |
|---|---|---|---|---|
| Initial (*M*): | $x$ | | 0.00 | 0.00 |
| Change (*M*): | $-1.0 \times 10^{-4}$ | | $+1.0 \times 10^{-4}$ | $+1.0 \times 10^{-4}$ |
| Equilibrium (*M*): | $(x - 1.0 \times 10^{-4})$ | | $(1.0 \times 10^{-4})$ | $(1.0 \times 10^{-4})$ |

$$K_b = \frac{[HCN][OH^-]}{[CN^-]}$$

$$2.04 \times 10^{-5} = \frac{\left(1.0 \times 10^{-4}\right)^2}{\left(x - 1.0 \times 10^{-4}\right)}$$

$$x = 5.90 \times 10^{-4}\ M = [CN^-]_0$$

$$\textbf{Amount of NaCN} = 250\text{ mL} \times \frac{5.90 \times 10^{-4}\text{ mol NaCN}}{1000\text{ mL}} \times \frac{49.01\text{ g NaCN}}{1\text{ mol NaCN}} = \mathbf{7.2 \times 10^{-3}\ g\ NaCN}$$

16.155 The equilibrium is established:

| | $CH_3COOH(aq)$ | ⇄ | $CH_3COO^-(aq)$ | $+\ H^+(aq)$ |
|---|---|---|---|---|
| Initial ($M$): | 0.150 | | 0.00 | 0.100 |
| Change ($M$): | $-x$ | | $+x$ | $+x$ |
| Equilibrium ($M$): | $(0.150 - x)$ | | $x$ | $(0.100 + x)$ |

$$K_a = \frac{[CH_3COO^-][H^+]}{[CH_3COOH]}$$

$$1.8 \times 10^{-5} = \frac{x(0.100 + x)}{0.150 - x} \approx \frac{0.100x}{0.150}$$

$$x = 2.7 \times 10^{-5}\ M$$

$2.7 \times 10^{-5}\ M$ is the $[H^+]$ contributed by $CH_3COOH$. HCl is a strong acid that completely ionizes. It contributes a $[H^+]$ of 0.100 $M$ to the solution.

$$[H^+]_{total} = [0.100 + (2.7 \times 10^{-5})]\ M \approx 0.100\ M$$

$$\mathbf{pH = 1.000}$$

To the correct number of significant digits, the contribution to the pH from $CH_3COOH$ is negligible.

16.157 a. **The pH of the solution of HA would be lower**. (Why?)

b. **The electrical conductance of the HA solution would be greater**. (Why?)

c. **The rate of hydrogen evolution from the HA solution would be greater**. Presumably, the rate of the reaction between the metal and hydrogen ion would depend on the hydrogen ion concentration (i.e., this would be part of the rate law). The hydrogen ion concentration will be greater in the HA solution.

16.159

$$HCOOH\ (aq) \rightleftarrows H^+(aq) + HCOO^-(aq)$$

| | | | |
|---|---|---|---|
| Initial ($M$): | 0.400 | 0.00 | 0.00 |
| Change ($M$): | $-x$ | $+x$ | $+x$ |
| Equilibrium ($M$): | $0.400 - x$ | $x$ | $x$ |

Total concentration of particles in solution: $(0.400 - x) + x + x = 0.400 + x$

Assuming the molarity of the solution is equal to the molality, we can write:

$$\Delta T_f = K_f m$$

$$0.758 = (1.86)(0.400 + x)$$

$$x = 0.00753 = [H^+] = [HCOO^-]$$

$$\boldsymbol{K_a} = \frac{[H^+][HCOO^-]}{[HCOOH]} = \frac{(0.00753)(0.00753)}{0.400 - 0.00753} = \mathbf{1.4 \times 10^{-4}}$$

16.161

$$pH = 10.64$$

$$pOH = 3.36$$

$$[OH^-] = 4.4 \times 10^{-4}\ M$$

$$\begin{array}{ccccccc} CH_3NH_2(aq) & + & H_2O(l) & \rightleftarrows & CH_3NH_3^+(aq) & + & OH^-(aq) \\ (x - 4.4 \times 10^{-4})\ M & & & & 4.4 \times 10^{-4}\ M & & 4.4 \times 10^{-4}\ M \end{array}$$

$$K_b = \frac{[CH_3NH_3^+][OH^-]}{[CH_3NH_2]}$$

$$4.4 \times 10^{-4} = \frac{(4.4 \times 10^{-4})(4.4 \times 10^{-4})}{x - (4.4 \times 10^{-4})}$$

$$4.4 \times 10^{-4}x - 1.9 \times 10^{-7} = 1.9 \times 10^{-7}$$

$$x = 8.6 \times 10^{-4}\ M$$

The molar mass of $CH_3NH_2$ is 31.06 g/mol.

The mass of $CH_3NH_2$ in 100.0 mL is:

$$100.0\text{ mL} \times \frac{8.6 \times 10^{-4}\text{ mol } CH_3NH_2}{1000\text{ mL}} \times \frac{31.06\text{ g } CH_3NH_2}{1\text{ mol } CH_3NH_2} = \mathbf{2.7 \times 10^{-3}\ g\ CH_3NH_2}$$

16.163 a.

$$\mathbf{NH_2^-\ (base) + H_2O\ (acid) \rightarrow NH_3 + OH^-}$$

$$\mathbf{N^{3-}\ (base) + 3H_2O\ (acid) \rightarrow NH_3 + 3OH^-}$$

b. $\mathbf{N^{3-}}$ is the stronger base since each ion produces 3 $OH^-$ ions.

16.165 From the given pH's, we can calculate the $[H^+]$ in each solution.

$$\text{Solution (1): } [H^+] = 10^{-pH} = 10^{-4.12} = 7.6 \times 10^{-5}\,M$$
$$\text{Solution (2): } [H^+] = 10^{-5.76} = 1.7 \times 10^{-6}\,M$$
$$\text{Solution (3): } [H^+] = 10^{-5.34} = 4.6 \times 10^{-6}\,M$$

We are adding solutions (1) and (2) to make solution (3). The volume of solution (2) is 0.528 L. We are going to add a given volume of solution (1) to solution (2). Let's call this volume $x$. The moles of $H^+$ in solutions (1) and (2) will equal the moles of $H^+$ in solution (3).

$$\text{mol } H^+ \text{ soln (1)} + \text{mol } H^+ \text{ soln (2)} = \text{mol } H^+ \text{ soln (3)}$$

Recall that mol = $M \times$ L. We have:

$$(7.6 \times 10^{-5}\text{ mol/L})(x\text{ L}) + (1.7 \times 10^{-6}\text{ mol/L})(0.528\text{ L}) = (4.6 \times 10^{-6}\text{ mol/L})(0.528 + x)\text{L}$$

$$(7.6 \times 10^{-5})x + (9.0 \times 10^{-7}) = (2.4 \times 10^{-6}) + (4.6 \times 10^{-6})x$$

$$(7.1 \times 10^{-5})x = 1.5 \times 10^{-6}$$

$$x = 0.021\text{ L} = \mathbf{21\ mL}$$

16.167 **When smelling salt is inhaled, some of the powder dissolves in the basic solution. The ammonium ions react with the base as follows:**

$$\mathbf{NH_4^+(aq) + OH^-(aq) \rightarrow NH_3(aq) + H_2O}$$

**It is the pungent odor of ammonia that prevents a person from fainting.**

16.169

$$K_b = 8.91 \times 10^{-6}$$

$$K_a = \frac{K_w}{K_b} = 1.1 \times 10^{-9}$$

$$\text{pH} = 7.40$$

$$[H^+] = 10^{-7.40} = 3.98 \times 10^{-8}$$

$$K_a = \frac{[H^+][\text{conjugate base}]}{[\text{acid}]}$$

Therefore,

$$\frac{[\text{conjugate base}]}{[\text{acid}]} = \frac{K_a}{[H^+]} = \frac{1.1 \times 10^{-9}}{3.98 \times 10^{-8}} = \mathbf{0.028} \text{ or } \mathbf{2.8 \times 10^{-2}}$$

16.171 **Both NaF and $SnF_2$ provide $F^-$ ions in solution.**

$$NaF \rightarrow Na^+ + F^-$$

$$SnF_2 \rightarrow Sn^{2+} + 2F^-$$

Because HF is a much stronger acid than $H_2O$, it follows that $F^-$ is a much weaker base than $OH^-$. **The $F^-$ ions replace $OH^-$ ions during the remineralization process**

$$\mathbf{5Ca^{2+} + 3PO_4^{3-} + F^- \rightarrow Ca_5(PO_4)_3F \text{ (fluorapatite)}}$$

because $OH^-$ has a much greater tendency to combine with $H^+$

$$OH^- + H^+ \rightarrow H_2O$$

than $F^-$ does.

$$F^- + H^+ \rightleftarrows HF$$

**Because $F^-$ is a weaker base than $OH^-$, fluorapatite is more resistant to attacks by acids compared to hydroxyapatite.**

16.173

$$SO_2(g) + H_2O(l) \rightleftarrows H^+(aq) + HSO_3^-(aq)$$

Recall that 0.12 ppm $SO_2$ would mean 0.12 parts $SO_2$ per 1 million ($10^6$) parts of air by volume. The number of particles of $SO_2$ per volume will be directly related to the pressure.

$$P_{SO_2} = \frac{0.12 \text{ parts } SO_2}{10^6 \text{ parts air}} \text{ atm} = 1.2 \times 10^{-7} \text{ atm}$$

We can now calculate the $[H^+]$ from the equilibrium constant expression.

$$K = \frac{[H^+][HSO_3^-]}{P_{SO_2}}$$

$$1.3 \times 10^{-2} = \frac{x^2}{1.2 \times 10^{-7}}$$

$$x^2 = (1.3 \times 10^{-2})(1.2 \times 10^{-7})$$

$$x = 3.9 \times 10^{-5}\,M = [H^+]$$

$$\mathbf{pH} = -\log(3.9 \times 10^{-5}) = \mathbf{4.41}$$

16.175 Because HA is a gas, we use the ideal gas law and the density of the compound to find the molar mass of HA.

$$PV = nRT$$

$$\frac{n}{V} = \frac{P}{RT}$$

$$\frac{n}{V} = \frac{0.982 \text{ atm}}{(0.0821 \text{ L}\cdot\text{atm/mol}\cdot\text{K})(301 \text{ K})}$$

$$\frac{n}{V} = 0.0397 \text{ mol/L}$$

Now, use the density to find molar mass.

$$\left(\frac{1.16 \text{ g}}{\text{L}}\right)\left(\frac{\text{L}}{0.0397 \text{ mol}}\right) = 29.22 \text{ g/mol}$$

Find the molarity of the solution of HA.

$$\left(\frac{2.03 \text{ g HA}}{\text{L}}\right)\left(\frac{\text{mol HA}}{29.22 \text{ g HA}}\right) = 6.95 \times 10^{-2}\ M \text{ HA}$$

The pH can be used to determine the equilibrium concentration of $H^+$ $(aq)$.

$$\text{pH} = -\log[H^+] = 5.22$$

$$[H^+]_{\text{equilibrium}} = 6.03 \times 10^{-6}$$

Now, solve the equilibrium problem.

| | HA $(aq)$ | $\rightleftarrows$ | $H^+$ $(aq)$ | + | $A^-$ $(aq)$ |
|---|---|---|---|---|---|
| Initial ($M$) | 0.0695 | | 0 | | 0 |
| Change ($M$) | $-x$ | | $+x$ | | $+x$ |
| Equilibrium ($M$) | $0.0695 - x$ | | $x$ | | $x$ |

From the pH, we know:

$$x = [H^+]_{\text{equilibrium}} = 6.03 \times 10^{-6}$$

Now, we can calculate the $K_a$.

$$\boldsymbol{K_a} = \frac{[H^+][A^-]}{[HA]} = \frac{x^2}{0.0695 - x} = \frac{(6.03 \times 10^{-6})^2}{(0.0695 - 6.03 \times 10^{-6})} = \mathbf{5.23 \times 10^{-10}}$$

16.177

$$[CO_2] = kP = (2.28 \times 10^{-3} \text{ mol/L}\cdot\text{atm})(3.20 \text{ atm}) = 7.30 \times 10^{-3}\ M$$

$$CO_2(aq) + H_2O(l) \rightleftarrows H^+(aq) + HCO_3^-(aq)$$

$(7.30 \times 10^{-3} - x)\ M$ $\qquad\qquad\qquad$ $x\ M$ $\qquad$ $x\ M$

$$K_a = \frac{[H^+][HCO_3^-]}{[CO_2]}$$

$$4.2 \times 10^{-7} = \frac{x^2}{(7.30 \times 10^{-3}) - x} \approx \frac{x^2}{7.30 \times 10^{-3}}$$

$$x = 5.5 \times 10^{-5}\ M = [H^+]$$

**pH = 4.26**

# Chapter 17

# Acid-Base Equilibria and Solubility Equilibria

17.5 a. This is a weak acid problem. Setting up the standard equilibrium table:

| | $CH_3COOH(aq)$ | $\rightleftarrows$ | $H^+(aq)$ | + | $CH_3COO^-(aq)$ |
|---|---|---|---|---|---|
| Initial ($M$): | 0.40 | | 0.00 | | 0.00 |
| Change ($M$): | $-x$ | | $+x$ | | $+x$ |
| Equilibrium ($M$): | $(0.40 - x)$ | | $x$ | | $x$ |

$$K_a = \frac{[H^+][CH_3COO^-]}{[CH_3COOH]}$$

$$1.8 \times 10^{-5} = \frac{x^2}{(0.40 - x)} \approx \frac{x^2}{0.40}$$

$$x = [H^+] = 2.7 \times 10^{-3}\ M$$

$$\mathbf{pH = 2.57}$$

b. In addition to the acetate ion formed from the ionization of acetic acid, we also have acetate ion formed from the sodium acetate dissolving.

$$CH_3COONa(aq) \rightarrow CH_3COO^-(aq) + Na^+(aq)$$

Dissolving 0.20 $M$ sodium acetate initially produces 0.20 $M$ $CH_3COO^-$ and 0.20 $M$ $Na^+$. The sodium ions are not involved in any further equilibrium (why?), but the acetate ions must be added to the equilibrium in part (a).

| | $CH_3COOH(aq)$ | $\rightleftarrows$ | $H^+(aq)$ | + | $CH_3COO^-(aq)$ |
|---|---|---|---|---|---|
| Initial ($M$): | 0.40 | | 0.00 | | 0.20 |
| Change ($M$): | $-x$ | | $+x$ | | $+x$ |
| Equilibrium ($M$): | $(0.40 - x)$ | | $x$ | | $(0.20 + x)$ |

$$K_a = \frac{[H^+][CH_3COO^-]}{[CH_3COOH]}$$

$$1.8 \times 10^{-5} = \frac{(x)(0.20 + x)}{(0.40 - x)} \approx \frac{x(0.20)}{0.40}$$

$$x = [\text{H}^+] = 3.6 \times 10^{-5}\ M$$

$$\textbf{pH} = \textbf{4.44}$$

**Think About It:** Could you have predicted whether the pH should have increased or decreased after the addition of the sodium acetate to the pure 0.40 *M* acetic acid in part (a)?

An alternate way to work part (b) of this problem is to use the Henderson-Hasselbalch equation.

$$\text{pH} = \text{p}K_\text{a} + \log\frac{[\text{conjugate base}]}{[\text{acid}]}$$

$$\textbf{pH} = -\log(1.8 \times 10^{-5}) + \log\frac{0.20\ M}{0.40\ M} = 4.74 - 0.30 = \textbf{4.44}$$

17.9 **Strategy:** The pH of a buffer system can be calculated using the Henderson-Hasselbalch equation (Equation 17.3). The $K_a$ of a conjugate acid such as $NH_4^+$ is calculated using the $K_b$ of its weak base (in this case, $NH_3$) and Equation 16.8.

**Solution:**

$$\text{NH}_4^+(aq) \rightleftarrows \text{H}_3(aq) + \text{H}^+(aq)$$

$$K_\text{a} = K_\text{w}/1.8 \times 10^{-5} = 5.56 \times 10^{-10}$$

$$\text{p}K_\text{a} = -\log K_\text{a} = 9.26$$

$$\textbf{pH} = \text{p}K_\text{a} + \log\frac{[\text{NH}_3]}{[\text{NH}_4^+]} = 9.26 + \log\frac{0.15\ M}{0.35\ M} = \textbf{8.89}$$

17.11

$$\text{H}_2\text{CO}_3(aq) \rightleftarrows \text{HCO}_3^-(aq) + \text{H}^+(aq)$$

$$K_{\text{a}_1} = 4.2 \times 10^{-7}$$

$$\text{p}K_{\text{a}_1} = 6.38$$

$$\text{pH} = \text{p}K_\text{a} + \log\frac{[\text{HCO}_3^-]}{[\text{H}_2\text{CO}_3]}$$

$$8.00 = 6.38 + \log\frac{[HCO_3^-]}{[H_2CO_3]}$$

$$\log\frac{[HCO_3^-]}{[H_2CO_3]} = 1.62$$

$$\frac{[HCO_3^-]}{[H_2CO_3]} = 41.7$$

$$\frac{[H_2CO_3]}{[HCO_3^-]} = \mathbf{0.024}$$

17.13 Using the Henderson–Hasselbalch equation:

$$pH = pK_a + \log\frac{[CH_3COO^-]}{[CH_3COOH]}$$

$$4.50 = 4.74 + \log\frac{[CH_3COO^-]}{[CH_3COOH]}$$

Thus,

$$\frac{[CH_3COO^-]}{[CH_3COOH]} = \mathbf{0.58}$$

17.15 For the first part we use $K_a$ for ammonium ion. (Why?) The Henderson–Hasselbalch equation is

$$\mathbf{pH} = -\log(5.6 \times 10^{-10}) + \log\frac{(0.20\ M)}{(0.20\ M)} = \mathbf{9.25}$$

For the second part, the acid–base reaction is

$$NH_3(g) + H^+(aq) \rightarrow NH_4^+(aq)$$

We find the number of moles of HCl added

$$10.0\text{ mL} \times \frac{0.10\text{ mol HCl}}{1000\text{ mL soln}} = 0.0010\text{ mol HCl}$$

The number of moles of $NH_3$ and $NH_4^+$ originally present are

$$65.0\text{ mL} \times \frac{0.20\text{ mol}}{1000\text{ mL soln}} = 0.013\text{ mol}$$

Using the acid-base reaction, we find the number of moles of $NH_3$ and $NH_4^+$ after addition of the HCl.

| | $NH_3(aq)$ + | $H^+(aq)$ | $\rightleftarrows$ | $NH_4^+(aq)$ |
|---|---|---|---|---|
| Initial (mol): | 0.013 | 0.0010 | | 0.013 |
| Change (mol): | −0.0010 | −0.0010 | | +0.0010 |
| Final (mol): | 0.012 | 0 | | 0.014 |

We find the new pH:

$$\mathbf{pH} = 9.25 + \log\frac{(0.012)}{(0.014)} = \mathbf{9.18}$$

17.17 **Strategy:** What constitutes a buffer system? Which of the solutions described in the problem contains a weak acid and its salt (containing the weak conjugate base)? Which contains a weak base and its salt (containing the weak conjugate acid)? Why is the conjugate base of a strong acid not able to neutralize an added acid?

**Solution:** The criteria for a buffer system are that we must have a weak acid and its salt (containing the weak conjugate base) or a weak base and its salt (containing the weak conjugate acid).

a. HCl (hydrochloric acid) is a strong acid. A buffer is a solution containing both a weak acid and a weak base. Therefore, this is *not* a buffer system.

b. $H_2SO_4$ (sulfuric acid) is a strong acid. A buffer is a solution containing both a weak acid and a weak base. Therefore, this is *not* a buffer system.

c. This solution contains both a weak acid, $H_2PO_4^-$ and its conjugate base, $HPO_4^{2-}$. Therefore, this is a buffer system.

d. $HNO_2$ (nitrous acid) is a weak acid, and its conjugate base, $NO_2^-$ (nitrite ion, the anion of the salt $KNO_2$), is a weak base. Therefore, this is a buffer system.

Only **(c)** and **(d)** are buffer systems.

17.19 In order for the buffer solution to function effectively, the $pK_a$ of the acid component must be close to the desired pH. We write

$K_{a_1} = 1.1 \times 10^{-3}$ $\quad\quad$ $pK_{a_1} = 2.96$

$K_{a_2} = 2.5 \times 10^{-6}$ $\quad\quad$ $pK_{a_2} = 5.60$

Therefore, the proper buffer system is **$Na_2A$/NaHA**.

17.21 a. In order to function as a buffer, a solution must contain species that will consume both acid and base. Solution (a) contains $HA^-$, which can consume either acid or base:

$$HA^- + H^+ \rightarrow H_2A$$

$$HA^- + OH^- \rightarrow A^{2-} + H_2O$$

and $A^{2-}$, which can consume acid:

$$A^{2-} + H+ \rightarrow HA-$$

Solution (b) contains $HA^-$, which can consume either acid or base, and $H_2A$, which can consume base:

$$H_2A + OH- \rightarrow HA^- + H_2O$$

Solution (c) contains only $H_2A$ and consequently can consume base but not acid. Solution (c) cannot function as a buffer.

Solution (d) contains $HA^-$ and $A^{2-}$.

Solutions **(a)**, **(b)**, and **(c)** can function as buffers.

b. **Solution (a)** should be the most effective buffer because it has the highest concentrations of acid- and base-consuming species.

17.27 Since the acid is monoprotic, the number of moles of KOH is equal to the number of moles of acid.

$$\text{Moles acid} = 16.4\text{ mL} \times \frac{0.08133\text{ mol}}{1000\text{ mL}} = 0.00133\text{ mol}$$

$$\textbf{Molar mass} = \frac{0.2688\text{ g}}{0.00133\text{ mol}} = \mathbf{202\ g/mol}$$

17.29 The neutralization reaction is:

$$H_2SO_4(aq) + 2NaOH(aq) \rightarrow Na_2SO_4(aq) + 2H_2O(l)$$

Since one mole of sulfuric acid combines with two moles of sodium hydroxide, we write:

$$\text{mol NaOH} = 12.5\text{ mL H}_2\text{SO}_4 \times \frac{0.500\text{ mol H}_2\text{SO}_4}{1000\text{ mL soln}} \times \frac{2\text{ mol NaOH}}{1\text{ mol H}_2\text{SO}_4} = 0.0125\text{ mol NaOH}$$

$$\textbf{concentration of NaOH} = \frac{0.0125\text{ mol NaOH}}{50.0 \times 10^{-3}\text{ L soln}} = \mathbf{0.25}\ \boldsymbol{M}$$

17.31 a. Since the acid is monoprotic, the moles of acid equals the moles of base added.

$$\text{HA}(aq) + \text{NaOH}(aq) \rightarrow \text{NaA}(aq) + \text{H}_2\text{O}(l)$$

$$\text{Moles acid} = 18.4\text{ mL} \times \frac{0.0633\text{ mol}}{1000\text{ mL soln}} = 0.00116\text{ mol}$$

We know the mass of the unknown acid in grams and the number of moles of the unknown acid.

$$\textbf{Molar mass} = \frac{0.1276\text{ g}}{0.00116\text{ mol}} = \mathbf{1.10 \times 10^2\ g/mol}$$

b. The number of moles of NaOH in 10.0 mL of solution is

$$10.0\text{ mL} \times \frac{0.0633\text{ mol}}{1000\text{ mL soln}} = 6.33 \times 10^{-4}\text{ mol}$$

The neutralization reaction is:

| | HA($aq$) | + NaOH($aq$) | $\rightarrow$ | NaA($aq$) | + | $H_2O(l)$ |
|---|---|---|---|---|---|---|
| Initial (mol): | 0.00116 | $6.33 \times 10^{-4}$ | | 0 | | |
| Change (mol): | $-6.33 \times 10^{-4}$ | $-6.33 \times 10^{-4}$ | | $+6.33 \times 10^{-4}$ | | |
| Final (mol): | $5.3 \times 10^{-4}$ | 0 | | $6.33 \times 10^{-4}$ | | |

Now, the weak acid equilibrium will be reestablished. The total volume of solution is 35.0 mL.

$$[\text{HA}] = \frac{5.3 \times 10^{-4}\text{ mol}}{0.035\text{ L}} = 0.015\ M$$

$$[\text{A}^-] = \frac{6.33 \times 10^{-4}\text{ mol}}{0.035\text{ L}} = 0.0181\ M$$

We can calculate the $[H^+]$ from the pH.

$$[H^+] = 10^{-pH} = 10^{-5.87} = 1.35 \times 10^{-6}\ M$$

| | $HA(aq)$ | $\rightleftarrows$ | $H^+(aq)$ + | $A^-(aq)$ |
|---|---|---|---|---|
| Initial ($M$): | 0.015 | | 0 | 0.0181 |
| Change ($M$): | $-1.35 \times 10^{-6}$ | | $+1.35 \times 10^{-6}$ | $+1.35 \times 10^{-6}$ |
| Equilibrium ($M$): | 0.015 | | $1.35 \times 10^{-6}$ | 0.0181 |

Substitute the equilibrium concentrations into the equilibrium constant expression to solve for $K_a$.

$$\boldsymbol{K_a} = \frac{[H^+][A^-]}{[HA]} = \frac{(1.35 \times 10^{-6})(0.0181)}{0.015} = \mathbf{1.6 \times 10^{-6}}$$

17.33 $$HCl(aq) + CH_3NH_2(aq) \rightleftarrows CH_3NH_3^+(aq) + Cl^-(aq)$$

Since the concentrations of acid and base are equal, equal volumes of each solution will need to be added to reach the equivalence point. Therefore, the solution volume is doubled at the equivalence point, and the concentration of the conjugate acid from the salt, $CH_3NH_3^+$, is:

$$\frac{0.20\ M}{2} = 0.10\ M$$

The conjugate acid undergoes hydrolysis.

| | $CH_3NH_3^+(aq)$ | $+ H_2O(l)$ | $\rightleftarrows$ | $H_3O^+(aq)$ | $+ CH_3NH_2(aq)$ |
|---|---|---|---|---|---|
| Initial ($M$): | 0.10 | | | 0 | 0 |
| Change ($M$): | $-x$ | | | $+x$ | $+x$ |
| Equilibrium ($M$): | $0.10 - x$ | | | $x$ | $x$ |

$$K_a = \frac{[H_3O^+][CH_3NH_2]}{[CH_3NH_3^+]}$$

$$2.3 \times 10^{-11} = \frac{x^2}{0.10 - x}$$

Assuming that, $0.10 - x \approx 0.10$

$$x = [H_3O^+] = 1.5 \times 10^{-6}\ M$$

$$\mathbf{pH = 5.82}$$

17.35 The reaction between $CH_3COOH$ and KOH is:

$$CH_3COOH(aq) + KOH(aq) \rightarrow CH_3COOK(aq) + H_2O(l)$$

We see that 1 mole $CH_3COOH$ is stoichiometrically equivalent to 1 mol KOH. Therefore, at every stage of titration, we can calculate the number of moles of acid reacting with base, and the pH of the solution is determined by the excess acid or base left over. At the equivalence point, however, the neutralization is complete, and the pH of the solution will depend on the extent of the hydrolysis of the salt formed, which is $CH_3COOK$.

a. No KOH has been added. This is a weak acid calculation.

| | $CH_3COOH(aq)$ | + $H_2O(l)$ ⇄ | $H_3O^+(aq)$ | + | $CH_3COO^-(aq)$ |
|---|---|---|---|---|---|
| Initial ($M$): | 0.100 | | 0 | | 0 |
| Change ($M$): | $-x$ | | $+x$ | | $+x$ |
| Equilibrium ($M$): | $0.100 - x$ | | $x$ | | $x$ |

$$K_a = \frac{[H_3O^+][CH_3COO^-]}{[CH_3COOH]}$$

$$1.8 \times 10^{-5} = \frac{(x)(x)}{0.100 - x} \approx \frac{x^2}{0.100}$$

$$x = 1.34 \times 10^{-3}\ M = [H_3O^+]$$

**pH = 2.87**

b. The number of moles of $CH_3COOH$ originally present in 25.0 mL of solution is:

$$25.0\text{ mL} \times \frac{0.100\text{ mol }CH_3COOH}{1000\text{ mL }CH_3COOH\text{ soln}} = 2.50 \times 10^{-3}\text{ mol}$$

The number of moles of KOH in 5.0 mL is:

$$5.0\text{ mL} \times \frac{0.200\text{ mol KOH}}{1000\text{ mL KOH soln}} = 1.00 \times 10^{-3}\text{ mol}$$

We work with moles at this point because when two solutions are mixed, the solution volume increases. As the solution volume increases, molarity will change, but the number of moles will remain the same. The changes in number of moles are summarized.

| | $CH_3COOH(aq)$ | $+KOH(aq)$ | $\rightarrow$ $CH_3COOK(aq)$ | $+H_2O(l)$ |
|---|---|---|---|---|
| Initial (mol): | $2.50 \times 10^{-3}$ | $1.00 \times 10^{-3}$ | 0 | |
| Change (mol): | $-1.00 \times 10^{-3}$ | $-1.00 \times 10^{-3}$ | $+1.00 \times 10^{-3}$ | |
| Final (mol): | $1.50 \times 10^{-3}$ | 0 | $1.00 \times 10^{-3}$ | |

At this stage, we have a buffer system made up of $CH_3COOH$ and $CH_3COO^-$ (from the salt, $CH_3COOK$). We use the Henderson-Hasselbalch equation to calculate the pH.

$$\text{pH} = \text{p}K_a + \log\frac{[\text{conjugate base}]}{[\text{acid}]}$$

$$\text{pH} = -\log(1.8 \times 10^{-5}) + \log\left(\frac{1.00 \times 10^{-3}}{1.50 \times 10^{-3}}\right)$$

$$\textbf{pH} = \textbf{4.56}$$

c. This part is solved similarly to part (b).

The number of moles of KOH in 10.0 mL is:

$$10.0 \text{ mL} \times \frac{0.200 \text{ mol KOH}}{1000 \text{ mL KOH soln}} = 2.00 \times 10^{-3} \text{ mol}$$

The changes in number of moles are summarized.

| | $CH_3COOH(aq)$ | $+ KOH(aq)$ | $\rightarrow$ $CH_3COOK(aq)$ | $+ H_2O(l)$ |
|---|---|---|---|---|
| Initial (mol): | $2.50 \times 10^{-3}$ | $2.00 \times 10^{-3}$ | 0 | |
| Change (mol): | $-2.00 \times 10^{-3}$ | $-2.00 \times 10^{-3}$ | $+2.00 \times 10^{-3}$ | |
| Final (mol): | $0.50 \times 10^{-3}$ | 0 | $2.00 \times 10^{-3}$ | |

At this stage, we have a buffer system made up of $CH_3COOH$ and $CH_3COO^-$ (from the salt, $CH_3COOK$). We use the Henderson-Hasselbalch equation to calculate the pH.

$$\text{pH} = \text{p}K_a + \log\frac{[\text{conjugate base}]}{[\text{acid}]}$$

$$\text{pH} = -\log(1.8 \times 10^{-5}) + \log\left(\frac{2.00 \times 10^{-3}}{0.50 \times 10^{-3}}\right)$$

**pH = 5.34**

d. We have reached the equivalence point of the titration. $2.50 \times 10^{-3}$ mole of $CH_3COOH$ reacts with $2.50 \times 10^{-3}$ mole KOH to produce $2.50 \times 10^{-3}$ mole of $CH_3COOK$. The only major species present in solution at the equivalence point is the salt, $CH_3COOK$, which contains the conjugate base, $CH_3COO^-$. Let's calculate the molarity of $CH_3COO^-$. The volume of the solution is: (25.0 mL + 12.5 mL = 37.5 mL = 0.0375 L).

$$M\ (CH_3COO^-) = \frac{2.50 \times 10^{-3}\ \text{mol}}{0.0375\ \text{L}} = 0.0667\ M$$

We set up the hydrolysis of $CH_3COO^-$, which is a weak base.

| | $CH_3COO^-(aq)$ | $+\ H_2O(l)$ | $\rightleftarrows$ | $CH_3COOH(aq)$ | $+$ | $OH^-(aq)$ |
|---|---|---|---|---|---|---|
| Initial ($M$): | 0.0667 | | | 0 | | 0 |
| Change ($M$): | $-x$ | | | $+x$ | | $+x$ |
| Equilibrium ($M$): | $0.0667 - x$ | | | $x$ | | $x$ |

$$K_b = \frac{[CH_3COOH][OH^-]}{[CH_3COO^-]}$$

$$5.56 \times 10^{-10} = \frac{(x)(x)}{0.0667 - x} \approx \frac{x^2}{0.0667}$$

$$x = 6.09 \times 10^{-6}\ M = [OH^-]$$

$$\text{pOH} = 5.22$$

**pH = 8.78**

e. We have passed the equivalence point of the titration. The excess strong base, KOH, will determine the pH at this point. The moles of KOH in 15.0 mL are:

$$15.0\ \text{mL} \times \frac{0.200\ \text{mol KOH}}{1000\ \text{mL KOH soln}} = 3.00 \times 10^{-3}\ \text{mol}$$

The changes in number of moles are summarized.

| | $CH_3COOH(aq)$ | + $KOH(aq)$ | → $CH_3COOK(aq)$ | + $H_2O(l)$ |
|---|---|---|---|---|
| Initial (mol): | $2.50 \times 10^{-3}$ | $3.00 \times 10^{-3}$ | 0 | |
| Change (mol): | $-2.50 \times 10^{-3}$ | $-2.50 \times 10^{-3}$ | $+2.50 \times 10^{-3}$ | |
| Final (mol): | 0 | $0.50 \times 10^{-3}$ | $2.50 \times 10^{-3}$ | |

Let's calculate the molarity of the KOH in solution. The volume of the solution is now 40.0 mL = 0.0400 L.

$$M\text{ (KOH)} = \frac{0.50 \times 10^{-3}\text{ mol}}{0.0400\text{ L}} = 0.0125\ M$$

KOH is a strong base. The pOH is:

$$\text{pOH} = -\log(0.0125) = 1.90$$

$$\mathbf{pH = 12.10}$$

17.37 a. HCOOH is a weak acid and NaOH is a strong base. Suitable indicators are **cresol red and phenolphthalein.**

b. HCl is a strong acid and KOH is a strong base. **Most of the indicators in Table 17.3 are suitable for a strong acid-strong base titration. Exceptions are thymol blue and, to a lesser extent, bromophenol blue and methyl orange.**

c. $HNO_3$ is a strong acid and $CH_3NH_2$ is a weak base. Suitable indicators are **bromophenol blue, methyl orange, methyl red, and chlorophenol blue.**

17.39 The weak acid equilibrium is

$$\text{HIn}(aq) \rightleftarrows \text{H}^+(aq) + \text{In}^-(aq)$$

We can write a $K_a$ expression for this equilibrium.

$$K_a = \frac{[\text{H}^+][\text{In}^-]}{[\text{HIn}]}$$

Rearranging,

$$\frac{[\text{HIn}]}{[\text{In}^-]} = \frac{[\text{H}^+]}{K_a}$$

From the pH, we can calculate the $H^+$ concentration.

$$[\text{H}^+] = 10^{-\text{pH}} = 10^{-4} = 1.0 \times 10^{-4}\ M$$

$$\frac{\mathbf{[HIn]}}{\mathbf{[In^-]}} = \frac{[\text{H}^+]}{K_a} = \frac{1.0 \times 10^{-4}}{1.0 \times 10^{-6}} = \mathbf{100}$$

Since the concentration of HIn is 100 times greater than the concentration of $In^-$, the color of the solution will be that of HIn, the nonionized formed. The color of the solution will be **red**.

17.41 a. **Diagram (c)**

b. **Diagram (b)**

c. **Diagram (d)**

d. **Diagram (a). The pH at the equivalence point is below 7 (acidic)** because the species in solution at the equivalence point is the conjugate acid of a weak base.

17.49 a. The solubility equilibrium is given by the equation

$$\text{AgI}(s) \rightleftarrows \text{Ag}^+(aq) + \text{I}^-(aq)$$

The expression for $K_{sp}$ is given by

$$K_{sp} = [\text{Ag}^+][\text{I}^-]$$

The value of $K_{sp}$ can be found in Table 17.4 of the text. If the equilibrium concentration of silver ion is the value given, the concentration of iodide ion must be

$$\mathbf{[I^-]} = \frac{K_{sp}}{[\text{Ag}^+]} = \frac{8.3 \times 10^{-17}}{9.1 \times 10^{-9}} = \mathbf{9.1 \times 10^{-9}}\ \boldsymbol{M}$$

b. The value of $K_{sp}$ for aluminum hydroxide can be found in Table 17.4 of the text. The equilibrium expressions are:

$$\text{Al(OH)}_3(s) \rightleftarrows \text{Al}^{3+}(aq) + 3\text{OH}^-(aq)$$

$$K_{sp} = [Al^{3+}][OH^-]^3$$

Using the given value of the hydroxide ion concentration, the equilibrium concentration of aluminum ion is:

$$[Al^{3+}] = \frac{K_{sp}}{[OH^-]^3} = \frac{1.8 \times 10^{-33}}{(2.9 \times 10^{-9})^3} = \mathbf{7.4 \times 10^{-8}}\ \boldsymbol{M}$$

**Think About It:** What is the pH of this solution? Will the aluminum concentration change if the pH is altered?

17.51 For $MnCO_3$ dissolving, we write

$$MnCO_3(s) \rightleftarrows Mn^{2+}(aq) + CO_3^{2-}(aq)$$

For every mole of $MnCO_3$ that dissolves, one mole of $Mn^{2+}$ will be produced and one mole of $CO_3^{2-}$ will be produced. If the molar solubility of $MnCO_3$ is $s$ mol/L, then the concentrations of $Mn^{2+}$ and $CO_3^{2-}$ are:

$$[Mn^{2+}] = [CO_3^{2-}] = s = 4.2 \times 10^{-6}\ M$$

$$\boldsymbol{K_{sp}} = [Mn^{2+}][CO_3^{2-}] = s^2 = (4.2 \times 10^{-6})^2 = \mathbf{1.8 \times 10^{-11}}$$

17.53 The charges of the M and X ions are +3 and −2, respectively (are other values possible?). We first calculate the number of moles of $M_2X_3$ that dissolve in 1.0 L of water. We carry an additional significant figure throughout this calculation to minimize rounding errors.

$$\text{Moles } M_2X_3 = (3.6 \times 10^{-17}\text{ g}) \times \frac{1\text{ mol}}{288\text{ g}} = 1.25 \times 10^{-19}\text{ mol}$$

The molar solubility, $s$, of the compound is therefore $1.3 \times 10^{-19}$ $M$. At equilibrium the concentration of $M^{3+}$ must be $2s$ and that of $X^{2-}$ must be $3s$.

$$K_{sp} = [M^{3+}]^2[X^{2-}]^3 = [2s]^2[3s]^3 = 108s^5$$

Since these are equilibrium concentrations, the value of $K_{sp}$ can be found by simple substitution

$$\boldsymbol{K_{sp}} = 108s^5 = 108(1.25 \times 10^{-19})^5 = \mathbf{3.3 \times 10^{-93}}$$

17.55 Let $s$ be the molar solubility of $Zn(OH)_2$. The equilibrium concentrations of the ions are then

$$[Zn^{2+}] = s \text{ and } [OH^-] = 2s$$

$$K_{sp} = [Zn^{2+}][OH^-]^2 = (s)(2s)^2 = 4s^3 = 1.8 \times 10^{-14}$$

$$s = \sqrt[3]{\left(\frac{1.8 \times 10^{-14}}{4}\right)} = 1.65 \times 10^{-5}$$

$$[OH^-] = 2s = 3.30 \times 10^{-5}\ M \text{ and pOH} = 4.48$$

$$\textbf{pH} = 14.00 - 4.48 = \textbf{9.52}$$

**Think About It:** If the $K_{sp}$ of $Zn(OH)_2$ were smaller by many more powers of ten, would $2s$ still be the hydroxide ion concentration in the solution?

17.57 According to the solubility rules, the only precipitate that might form is $BaCO_3$.

$$Ba^{2+}(aq) + CO_3^{2-}(aq) \rightarrow BaCO_3(s)$$

The number of moles of $Ba^{2+}$ present in the original 20.0 mL of $Ba(NO_3)_2$ solution is

$$20.0 \text{ mL} \times \frac{0.10 \text{ mol } Ba^{2+}}{1000 \text{ mL soln}} = 2.0 \times 10^{-3} \text{ mol } Ba^{2+}$$

The total volume after combining the two solutions is 70.0 mL. The concentration of $Ba^{2+}$ in 70 mL is

$$[Ba^{2+}] = \frac{2.0 \times 10^{-3} \text{ mol } Ba^{2+}}{70.0 \times 10^{-3} \text{ L}} = 2.9 \times 10^{-2}\ M$$

The number of moles of $CO_3^{2-}$ present in the original 50.0 mL $Na_2CO_3$ solution is

$$50.0 \text{ mL} \times \frac{0.10 \text{ mol } CO_3^{2-}}{1000 \text{ mL soln}} = 5.0 \times 10^{-3} \text{ mol } CO_3^{2-}$$

The concentration of $CO_3^{2-}$ in the 70.0 mL of combined solution is

$$[CO_3^{2-}] = \frac{5.0 \times 10^{-3} \text{ mol } CO_3^{2-}}{70.0 \times 10^{-3} \text{ L}} = 7.1 \times 10^{-2}\ M$$

Now we must compare $Q$ and $K_{sp}$. From Table 17.4 of the text, the $K_{sp}$ for $BaCO_3$ is $8.1 \times 10^{-9}$. As for $Q$,

$$Q = [Ba^{2+}]_0[CO_3^{2-}]_0 = (2.9 \times 10^{-2})(7.1 \times 10^{-2}) = 2.1 \times 10^{-3}$$

Since $(2.1 \times 10^{-3}) > (8.1 \times 10^{-9})$, then $Q > K_{sp}$. Therefore, **yes**, $BaCO_3$ will precipitate.

17.63 **Strategy:** In parts (b) and (c), this is a common-ion problem. In part (b), the common ion is $Br^-$, which is supplied by both $PbBr_2$ and KBr. Remember that the presence of a common ion will affect only the solubility of $PbBr_2$, but not the $K_{sp}$ value because it is an equilibrium constant. In part (c), the common ion is $Pb^{2+}$, which is supplied by both $PbBr_2$ and $Pb(NO_3)_2$.

**Solution:** a. Set up a table to find the equilibrium concentrations in pure water.

| | $PbBr_2(s)$ | $\rightleftarrows$ | $Pb^{2+}(aq)$ | + | $2Br^-(aq)$ |
|---|---|---|---|---|---|
| Initial ($M$): | | | 0 | | 0 |
| Change ($M$): | $-s$ | | $+s$ | | $+2s$ |
| Equilibrium ($M$): | | | $s$ | | $2s$ |

$$K_{sp} = [Pb^{2+}][Br^-]^2$$

$$8.9 \times 10^{-6} = (s)(2s)^2$$

$$s = \text{molar solubility} = \mathbf{0.013\ M} \text{ or } \mathbf{1.3 \times 10^{-2}\ M}$$

b. Set up a table to find the equilibrium concentrations in 0.20 $M$ KBr. KBr is a soluble salt that ionizes completely giving an initial concentration of $Br^- = 0.20\ M$.

| | $PbBr_2(s)$ | $\rightleftarrows$ | $Pb^{2+}(aq)$ | + | $2Br^-(aq)$ |
|---|---|---|---|---|---|
| Initial ($M$): | | | 0 | | 0.20 |
| Change ($M$): | $-s$ | | $+s$ | | $+2s$ |
| Equilibrium ($M$): | | | $s$ | | $0.20 + 2s$ |

$$K_{sp} = [Pb^{2+}][Br^-]^2$$

$$8.9 \times 10^{-6} = (s)(0.20 + 2s)^2$$

$$8.9 \times 10^{-6} \approx (s)(0.20)^2$$

$$s = \text{molar solubility} = \mathbf{2.2 \times 10^{-4}\ M}$$

Thus, the molar solubility of $PbBr_2$ is reduced from 0.013 $M$ to $2.2 \times 10^{-4}\ M$ as a result of the common ion ($Br^-$) effect.

c. Set up a table to find the equilibrium concentrations in 0.20 $M$ $Pb(NO_3)_2$. $Pb(NO_3)_2$ is a soluble salt that dissociates completely giving an initial concentration of $[Pb^{2+}] = 0.20\ M$.

| | $PbBr_2(s)$ | $\rightleftarrows$ | $Pb^{2+}(aq)$ | $+\ 2Br^-(aq)$ |
|---|---|---|---|---|
| Initial (*M*): | 0.20 | | 0 | |
| Change (*M*): | $-s$ | | $+s$ | $+2s$ |
| Equilibrium (*M*): | | | $0.20 + s$ | $2s$ |

$$K_{sp} = [Pb^{2+}][Br^-]^2$$

$$8.9 \times 10^{-6} = (0.20 + s)(2s)^2$$

$$8.9 \times 10^{-6} \approx (0.20)(2s)^2$$

$$s = \text{molar solubility} = \mathbf{3.3 \times 10^{-3}\ \mathit{M}}$$

Thus, the molar solubility of $PbBr_2$ is reduced from 0.013 *M* to $3.3 \times 10^{-3}$ *M* as a result of the common ion ($Pb^{2+}$) effect.

**Think About It:** You should also be able to predict the decrease in solubility due to a common-ion using Le Châtelier's principle. Adding $Br^-$ or $Pb^{2+}$ ions shifts the system to the left, thus decreasing the solubility of $PbBr_2$.

17.65 a. The equilibrium equation is:

| | $BaSO_4(s)$ | $\rightleftarrows$ | $Ba^{2+}(aq)$ | $+\ SO_4^{2-}(aq)$ |
|---|---|---|---|---|
| Initial (*M*): | | | 0 | 0 |
| Change (*M*): | $-s$ | | $+s$ | $+s$ |
| Equilibrium (*M*): | | | $s$ | $s$ |

$$K_{sp} = [Ba^{2+}][SO_4^{2-}]$$

$$1.1 \times 10^{-10} = s^2$$

$$s = \mathbf{1.0 \times 10^{-5}\ \mathit{M}}$$

The molar solubility of $BaSO_4$ in pure water is $1.0 \times 10^{-5}$ mol/L.

b. The initial concentration of $SO_4^{2-}$ is 1.0 *M*.

$$BaSO_4(s) \rightleftarrows Ba^{2+}(aq) + SO_4^{2-}(aq)$$

| | | | |
|---|---|---|---|
| Initial ($M$): | | 0 | 1.0 |
| Change ($M$): | $-s$ | $+s$ | $+s$ |
| Equilibrium ($M$): | | $s$ | $1.0 + s$ |

$$K_{sp} = [Ba^{2+}][SO_4^{2-}]$$

$$1.1 \times 10^{-10} = (s)(1.0 + s) \approx (s)(1.0)$$

$$\boldsymbol{s = 1.1 \times 10^{-10}\ M}$$

Due to the common ion effect, the molar solubility of $BaSO_4$ decreases to $1.1 \times 10^{-10}$ mol/L in 1.0 $M$ $SO_4^{2-}(aq)$ compared to $1.0 \times 10^{-5}$ mol/L in pure water.

17.67 a. $I^-$ is the conjugate base of the strong acid HI

b. $SO_4^{2-}(aq)$ is a weak base

c. $OH^-(aq)$ is a strong base

d. $C_2O_4^{2-}(aq)$ is a weak base

e. $PO_4^{3-}(aq)$ is a weak base

The solubilities of the above (**b – e**) will increase in acidic solution. Only (a), which contains an extremely weak base ($I^-$ is the conjugate base of the strong acid HI) is unaffected by the acid solution.

17.69 From Table 17.4, the value of $K_{sp}$ for iron(II) hydroxide is $1.6 \times 10^{-14}$.

a. At pH = 8.00, pOH = 14.00 – 8.00 = 6.00, and $[OH^-] = 1.0 \times 10^{-6}\ M$

$$[Fe^{2+}] = \frac{K_{sp}}{[OH^-]^2} = \frac{1.6 \times 10^{-14}}{(1.0 \times 10^{-6})^2} = 0.016\ M$$

The *molar solubility* of iron(II) hydroxide at pH = 8.00 is **0.016 *M*** or **$1.6 \times 10^{-2}$ *M***

b. At pH = 10.00, pOH = 14.00 – 10.00 = 4.00, and $[OH^-] = 1.0 \times 10^{-4}\ M$

$$[Fe^{2+}] = \frac{K_{sp}}{[OH^-]^2} = \frac{1.6 \times 10^{-14}}{(1.0 \times 10^{-4})^2} = 1.6 \times 10^{-6}\ M$$

The *molar solubility* of iron(II) hydroxide at pH = 10.00 is **$1.6 \times 10^{-6}$ *M*.**

17.71 We first determine the effect of the added ammonia. Let's calculate the concentration of $NH_3$. This is a dilution problem.

$$M_iV_i = M_fV_f$$

$$(0.60\ M)(2.00\ \text{mL}) = M_f(1002\ \text{mL})$$

$$M_f = 0.0012\ M\ NH_3$$

Ammonia is a weak base ($K_b = 1.8 \times 10^{-5}$).

$$NH_3(aq) + H_2O(l) \rightleftarrows NH_4^+(aq) + OH^-(aq)$$

| | $NH_3(aq)$ | $NH_4^+(aq)$ | $OH^-(aq)$ |
|---|---|---|---|
| Initial (*M*): | 0.0012 | 0 | 0 |
| Change (*M*): | $-x$ | $+x$ | $+x$ |
| Equilibrium (*M*): | $0.0012 - x$ | $x$ | $x$ |

$$K_b = \frac{[NH_4^+][OH^-]}{[NH_3]}$$

$$1.8 \times 10^{-5} = \frac{x^2}{(0.0012 - x)}$$

Solving the resulting quadratic equation gives $x = 0.00014$, or $[OH^-] = 0.00014\ M$

This is a solution of iron(II) sulfate, which contains $Fe^{2+}$ ions. These $Fe^{2+}$ ions could combine with $OH^-$ to precipitate $Fe(OH)_2$. Therefore, we must use $K_{sp}$ for iron(II) hydroxide. We compute the value of $Q_c$ for this solution.

$$Fe(OH)_2(s) \rightleftarrows Fe^{2+}(aq) + 2OH^-(aq)$$

$$Q = [Fe^{2+}]_0[OH^-]_0^2 = (1.0 \times 10^{-3})(0.00014)^2 = 2.0 \times 10^{-11}$$

Note that when adding 2.00 mL of $NH_3$ to 1.0 L of $FeSO_4$, the concentration of $FeSO_4$ will decrease slightly. However, rounding off to 2 significant figures, the concentration of $1.0 \times 10^{-3}$ *M* does not change. *Q* is larger than $K_{sp}$ $[Fe(OH)_2] = 1.6 \times 10^{-14}$. The concentrations of the ions in solution are greater than the equilibrium concentrations; the solution is saturated. The system will shift left to reestablish equilibrium; therefore, **a precipitate of $Fe(OH)_2$ will form.**

17.73 **Strategy:** The addition of $Cd(NO_3)_2$ to the NaCN solution results in complex ion formation. In solution, $Cd^{2+}$ ions will complex with $CN^-$ ions. The concentration of $Cd^{2+}$ will be determined by the following equilibrium

$$Cd^{2+}(aq) + 4CN^-(aq) \rightleftarrows Cd(CN)_4^{2-}$$

From Table 17.5 of the text, we see that the formation constant ($K_f$) for this reaction is very large ($K_f = 7.1 \times 10^{16}$). Because $K_f$ is so large, the reaction lies mostly to the right. At equilibrium, the concentration of $Cd^{2+}$ will be very small. As a good approximation, we can assume that essentially all the dissolved $Cd^{2+}$ ions end up as $Cd(CN)_4^{2-}$ ions. What is the initial concentration of $Cd^{2+}$ ions? A very small amount of $Cd^{2+}$ will be present at equilibrium. Set up the $K_f$ expression for the above equilibrium to solve for $[Cd^{2+}]$.

**Solution:** Calculate the initial concentration of $Cd^{2+}$ ions.

$$[Cd^{2+}]_0 = \frac{0.50 \text{ g} \times \frac{1 \text{ mol } Cd(NO_3)_2}{236.42 \text{ g } Cd(NO_3)_2} \times \frac{1 \text{ mol } Cd^{2+}}{1 \text{ mol } Cd(NO_3)_2}}{0.50 \text{ L}} = 4.2 \times 10^{-3}\ M$$

If we assume that the above equilibrium goes to completion, we can write

| | $Cd^{2+}(aq)$ + | $4CN^-(aq)$ | $\rightarrow$ $Cd(CN)_4^{2-}(aq)$ |
|---|---|---|---|
| Initial (*M*): | $4.2 \times 10^{-3}$ | 0.50 | 0 |
| Change (*M*): | $-4.2 \times 10^{-3}$ | $-4(4.2 \times 10^{-3})$ | $+4.2 \times 10^{-3}$ |
| Equilibrium (*M*): | 0 | 0.48 | $4.2 \times 10^{-3}$ |

To find the concentration of free $Cd^{2+}$ at equilibrium, use the formation constant expression.

$$K_f = \frac{[Cd(CN)_4^{2-}]}{[Cd^{2+}][CN^-]^4}$$

Rearranging,

$$[Cd^{2+}] = \frac{[Cd(CN)_4^{2-}]}{K_f[CN^-]^4}$$

Substitute the equilibrium concentrations calculated above into the formation constant expression to calculate the equilibrium concentration of $Cd^{2+}$.

$$\mathbf{[Cd^{2+}]} = \frac{[Cd(CN)_4^{2-}]}{K_f[CN^-]^4} = \frac{4.2 \times 10^{-3}}{(7.1 \times 10^{16})(0.48)^4} = \mathbf{1.1 \times 10^{-18}}\ \boldsymbol{M}$$

$$\mathbf{[Cd(CN)_4^{2-}]} = (4.2 \times 10^{-3}\ M) - (1.1 \times 10^{-18}) = \mathbf{4.2 \times 10^{-3}}\ \boldsymbol{M}$$

$$\mathbf{[CN^-]} = 0.48\ M + 4(1.1 \times 10^{-18}\ M) = \mathbf{0.48}\ \boldsymbol{M}$$

**Think About It:** Substitute the equilibrium concentrations calculated into the formation constant expression to calculate $K_f$. Also, the small value of $[Cd^{2+}]$ at equilibrium, compared to its initial concentration of $4.2 \times 10^{-3}\ M$, certainly justifies our approximation that almost all the $Cd^{2+}$ ions react.

17.75 Silver iodide is only slightly soluble. It dissociates to form a small amount of $Ag^+$ and $I^-$ ions. The $Ag^+$ ions then complex with $NH_3$ in solution to form the complex ion $Ag(NH_3)_2^+$. The balanced equations are:

$$AgI(s) \rightleftarrows Ag^+(aq) + I^-(aq) \qquad K_{sp} = [Ag^+][I^-] = 8.3 \times 10^{-17}$$

$$Ag^+(aq) + 2NH_3(aq) \rightleftarrows Ag(NH_3)_2^+(aq) \qquad K_f = \frac{[Ag(NH_3)_2^+]}{[Ag^+][NH_3]^2} = 1.5 \times 10^7$$

---

$$\text{Overall: } AgI(s) + 2NH_3(aq) \rightleftarrows Ag(NH_3)_2^+(aq) + I^-(aq) \qquad K = K_{sp} \times K_f = 1.2 \times 10^{-9}$$

If $s$ is the molar solubility of AgI then,

| | $AgI(s)$ + | $2NH_3(aq)$ ⇄ | $Ag(NH_3)_2^+(aq)$ + | $I^-(aq)$ |
|---|---|---|---|---|
| Initial ($M$): | | 1.0 | 0.0 | 0.0 |
| Change ($M$): | $-s$ | $-2s$ | $+s$ | $+s$ |
| Equilibrium ($M$): | | $(1.0 - 2s)$ | $s$ | $s$ |

Because $K_f$ is large, we can assume all of the silver ions exist as $Ag(NH_3)_2^+$. Thus,

$$[Ag(NH_3)_2^+] = [I^-] = s$$

We can write the equilibrium constant expression for the above reaction, then solve for $s$.

$$K = 1.2 \times 10^{-9} = \frac{(s)(s)}{(1.0 - 2s)^2} \approx \frac{(s)(s)}{(1.0)^2}$$

$$\mathbf{s = 3.5 \times 10^{-5}\ M}$$

At equilibrium, $3.5 \times 10^{-5}$ moles of AgI dissolves in 1 L of 1.0 $M$ $NH_3$ solution.

17.77 a. The equations are as follows:

$$CuI_2(s) \rightleftarrows Cu^{2+}(aq) + 2I^-(aq)$$

$$\mathbf{Cu^{2+}(aq) + 4NH_3(aq) \rightleftarrows [Cu(NH_3)_4]^{2+}(aq)}$$

The ammonia combines with the $Cu^{2+}$ ions formed in the first step to form the complex ion $[Cu(NH_3)_4]^{2+}$, effectively removing the $Cu^{2+}$ ions, causing the first equilibrium to shift to the right (resulting in more $CuI_2$ dissolving).

b. Similar to part (a):

$$AgBr(s) \rightleftarrows Ag^+(aq) + Br^-(aq)$$

$$\mathbf{Ag^+(aq) + 2CN^-(aq) \rightleftarrows [Ag(CN)_2]^-(aq)}$$

c. Similar to parts (a) and (b).

$$HgCl_2(s) \rightleftarrows Hg^{2+}(aq) + 2Cl^-(aq)$$

$$\mathbf{Hg^{2+}(aq) + 4Cl^-(aq) \rightleftarrows [HgCl_4]^{2-}(aq)}$$

17.81 For $Fe(OH)_3$, $K_{sp} = 1.1 \times 10^{-36}$. When $[Fe^{3+}] = 0.010\ M$, the $[OH^-]$ value is:

$$K_{sp} = [Fe^{3+}][OH^-]^3$$

or

$$[OH^-] = \left(\frac{K_{sp}}{[Fe^{3+}]}\right)^{\frac{1}{3}}$$

$$[OH^-] = \left(\frac{1.1 \times 10^{-36}}{0.010}\right)^{\frac{1}{3}} = 4.8 \times 10^{-12}\ M$$

This $[OH^-]$ corresponds to a pH of 2.68. In other words, $Fe(OH)_3$ will begin to precipitate from this solution at pH of 2.68.

For $Zn(OH)_2$, $K_{sp} = 1.8 \times 10^{-14}$. When $[Zn^{2+}] = 0.010\ M$, the $[OH^-]$ value is:

$$[OH^-] = \left(\frac{K_{sp}}{[Zn^{2+}]}\right)^{\frac{1}{2}}$$

$$[OH^-] = \left(\frac{1.8 \times 10^{-14}}{0.010}\right)^{\frac{1}{2}} = 1.3 \times 10^{-6}\ M$$

This corresponds to a pH of 8.11. In other words $Zn(OH)_2$ will begin to precipitate from the solution at pH = 8.11. These results show that $Fe(OH)_3$ will precipitate when the pH just exceeds 2.68 and that $Zn(OH)_2$ will precipitate when the pH just exceeds 8.11. Therefore, to selectively remove iron as $Fe(OH)_3$, the pH must be *greater than* **2.68** but *less than* **8.11**.

17.83 Since some $PbCl_2$ precipitates, the solution is saturated. From Table 17.4, the value of $K_{sp}$ for lead(II) chloride is $2.4 \times 10^{-4}$. The equilibrium is:

$$PbCl_2(aq) \rightleftarrows Pb^{2+}(aq) + 2Cl^-(aq)$$

We can write the solubility product expression for the equilibrium.

$$K_{sp} = [Pb^{2+}][Cl^-]^2$$

$K_{sp}$ and $[Cl^-]$ are known. Solving for the $Pb^{2+}$ concentration,

$$\mathbf{[Pb^{2+}]} = \frac{K_{sp}}{[Cl^-]^2} = \frac{2.4 \times 10^{-4}}{(0.15)^2} = \mathbf{0.011\ M}$$

17.85 **Chloride ion will precipitate $Ag^+$ but not $Cu^{2+}$. So, dissolve some solid in $H_2O$ and add HCl. If a precipitate forms, the salt was $AgNO_3$.** A flame test will also work. $Cu^{2+}$ gives a green flame test.

17.87 We can use the Henderson-Hasselbalch equation to solve for the pH when the indicator is 90% acid / 10% conjugate base and when the indicator is 10% acid / 90% conjugate base.

$$pH = pK_a + \log\frac{[\text{conjugate base}]}{[\text{acid}]}$$

Solving for the pH with 90% of the indicator in the HIn form:

$$\text{pH} = 3.46 + \log\frac{[10]}{[90]} = 3.46 - 0.95 = 2.51$$

Next, solving for the pH with 90% of the indicator in the $In^-$ form:

$$\text{pH} = 3.46 + \log\frac{[90]}{[10]} = 3.46 + 0.95 = 4.41$$

Thus the pH range varies from **2.51 to 4.41** as the [HIn] varies from 90% to 10%.

17.89 First, calculate the pH of the 2.00 *M* weak acid ($HNO_2$) solution before any NaOH is added.

$$HNO_2(aq) \rightleftarrows H^+(aq) + NO_2^-(aq)$$

| | $HNO_2(aq)$ | $H^+(aq)$ | $NO_2^-(aq)$ |
|---|---|---|---|
| Initial (*M*): | 2.00 | 0 | 0 |
| Change (*M*): | $-x$ | $+x$ | $+x$ |
| Equilibrium (*M*): | $2.00 - x$ | $x$ | $x$ |

$$K_a = \frac{[H^+][NO_2^-]}{[HNO_2]}$$

$$4.5 \times 10^{-4} = \frac{x^2}{2.00 - x} \approx \frac{x^2}{2.00}$$

$$x = [H^+] = 0.030\ M$$

$$\text{pH} = -\log(0.030) = 1.52$$

Since the pH after the addition is 1.5 pH units greater, the new pH = 1.52 + 1.50 = 3.02.

From this new pH, we can calculate the $[H^+]$ in solution.

$$[H^+] = 10^{-\text{pH}} = 10^{-3.02} = 9.55 \times 10^{-4}\ M$$

When the NaOH is added, we dilute our original 2.00 *M* $HNO_2$ solution to:

$$M_iV_i = M_fV_f$$

$$(2.00\ M)(400\text{ mL}) = M_f(600\text{ mL})$$

$$M_f = 1.33\ M$$

Since we have not reached the equivalence point, we have a buffer solution. The reaction between $HNO_2$ and NaOH is:

$$HNO_2(aq) + NaOH(aq) \rightarrow NaNO_2(aq) + H_2O(l)$$

or

$$HNO_2(aq) + OH^-(aq) \rightarrow NO_2^-(aq) + H_2O(l)$$

Since the mole ratio between $HNO_2$ and NaOH is 1:1, the decrease in $[HNO_2]$ is the same as the decrease in [NaOH].

We can calculate the decrease in $[HNO_2]$ by setting up the weak acid equilibrium. From the pH of the solution, we know that the $[H^+]$ at equilibrium is $9.55 \times 10^{-4}$ *M.*

| | $HNO_2(aq)$ | $\rightleftarrows$ | $H^+(aq)$ | + | $NO_2^-(aq)$ |
|---|---|---|---|---|---|
| Initial (*M*): | 1.33 | | 0 | | 0 |
| Change (*M*): | $-x$ | | | | $+x$ |
| Equilibrium (*M*): | $1.33 - x$ | | $9.55 \times 10^{-4}$ | | $x$ |

We can calculate $x$ from the equilibrium constant expression.

$$K_a = \frac{[H^+][NO_2^-]}{[HNO_2]}$$

$$4.5 \times 10^{-4} = \frac{(9.55 \times 10^{-4})(x)}{1.33 - x}$$

$$x = 0.426\ M$$

Thus, $x$ is the decrease in $[HNO_2]$ which equals the concentration of added $OH^-$. However, this is the concentration of NaOH after it has been diluted to 600 mL. We need to correct for the dilution from 200 mL to 600 mL to calculate the concentration of the original NaOH solution.

$$M_iV_i = M_fV_f$$

$$M_i(200\text{ mL}) = (0.426\ M)(600\text{ mL})$$

$$\mathbf{[NaOH]} = M_i = \mathbf{1.3}\ \boldsymbol{M}$$

17.91 The resulting solution is not a buffer system. There is excess NaOH and the neutralization is well past the equivalence point.

$$\text{Moles NaOH} = 0.500\ \text{L} \times \frac{0.167\ \text{mol}}{1\ \text{L}} = 0.0835\ \text{mol}$$

$$\text{Moles HCOOH} = 0.500\ \text{L} \times \frac{0.100\ \text{mol}}{1\ \text{L}} = 0.0500\ \text{mol}$$

$$\text{HCOOH}(aq) + \text{NaOH}(aq) \rightarrow \text{HCOONa}(aq)$$

| | HCOOH(*aq*) | NaOH(*aq*) | HCOONa(*aq*) |
|---|---|---|---|
| Initial (mol): | 0.0500 | 0.0835 | 0 |
| Change (mol): | −0.0500 | −0.0500 | +0.0500 |
| Final (mol): | 0 | 0.0335 | 0.0500 |

The volume of the resulting solution is 1.00 L (500 mL + 500 mL = 1000 mL).

$$[\text{Na}^+] = \frac{(0.0335 + 0.0500)\ \text{mol}}{1.00\ \text{L}} = \mathbf{0.0835}\ \boldsymbol{M}$$

$$[\text{HCOO}^-] = \frac{0.0500\ \text{mol}}{1.00\ \text{L}} = \mathbf{0.0500}\ \boldsymbol{M}$$

$$[\text{OH}^-] = \frac{0.0335\ \text{mol}}{1.00\ \text{L}} = \mathbf{0.0335}\ \boldsymbol{M}$$

$$[\text{H}^+] = \frac{K_\text{w}}{[\text{OH}^-]} = \frac{1.0 \times 10^{-14}}{0.0335} = \mathbf{3.0 \times 10^{-13}}\ \boldsymbol{M}$$

$$\text{HCOO}^-(aq) + \text{H}_2\text{O}(l) \rightleftarrows \text{HCOOH}(aq) + \text{OH}^-(aq)$$

| | HCOO⁻(*aq*) | H₂O(*l*) | HCOOH(*aq*) | OH⁻(*aq*) |
|---|---|---|---|---|
| Initial (*M*): | 0.0500 | | 0 | 0.0335 |
| Change (*M*): | −*x* | | +*x* | +*x* |
| Equilibrium (*M*): | 0.0500 − *x* | | *x* | 0.0335 + *x* |

$$K_\text{b} = \frac{[\text{HCOOH}][\text{OH}^-]}{[\text{HCOO}^-]}$$

$$5.9 \times 10^{-11} = \frac{(x)(0.0335 + x)}{(0.0500 - x)} \approx \frac{(x)(0.0335)}{(0.0500)}$$

$$x = \textbf{[HCOOH]} = \mathbf{8.8 \times 10^{-11}}\ \boldsymbol{M}$$

17.93 Most likely the increase in solubility is due to complex ion formation:

$$\mathbf{Cd(OH)_2(\mathit{s}) + 2OH^- \rightleftarrows Cd(OH)_4^{2-}(\mathit{aq})}$$

**This is a Lewis acid-base reaction.**

17.95 A solubility equilibrium is an equilibrium between a solid (reactant) and its components (products: ions, neutral molecules, etc.) in solution. Only **(d)** represents a solubility equilibrium.

**Think About It:** Consider part (b). Can you write the equilibrium constant for this reaction in terms of $K_{sp}$ for calcium phosphate?

17.97 Since equal volumes of the two solutions were used, the initial molar concentrations will be halved.

$$[Ag^+] = \frac{0.12\ M}{2} = 0.060\ M$$

$$[Cl^-] = \frac{2(0.14\ M)}{2} = 0.14\ M$$

Let's assume that the $Ag^+$ ions and $Cl^-$ ions react completely to form AgCl($s$). Then, we will reestablish the equilibrium between AgCl, $Ag^+$, and $Cl^-$.

| | $Ag^+(aq)$ | + $Cl^-(aq)$ | → AgCl($s$) |
|---|---|---|---|
| Initial ($M$): | 0.060 | 0.14 | 0 |
| Change ($M$): | −0.060 | −0.060 | +0.060 |
| Final ($M$): | 0 | 0.080 | 0.060 |

Now, setting up the equilibrium,

| | AgCl($s$) | ⇄ $Ag^+(aq)$ | + $Cl^-(aq)$ |
|---|---|---|---|
| Initial ($M$): | 0.060 | 0 | 0.080 |
| Change ($M$): | $-s$ | $+s$ | $+s$ |
| Equilibrium ($M$): | $0.060 - s$ | $s$ | $0.080 + s$ |

Set up the $K_{sp}$ expression to solve for $s$.

$$K_{sp} = [Ag^+][Cl^-]$$

$$1.6 \times 10^{-10} = (s)(0.080 + s)$$

$$s = 2.0 \times 10^{-9}\ M$$

$$[\mathbf{Ag^+}] = s = \mathbf{2.0 \times 10^{-9}}\ \boldsymbol{M}$$

$$[\mathbf{Cl^-}] = 0.080\ M + s = \mathbf{0.080}\ \boldsymbol{M}$$

$$[\mathbf{Zn^{2+}}] = \frac{0.14\ M}{2} = \mathbf{0.070}\ \boldsymbol{M}$$

$$[\mathbf{NO_3^-}] = \frac{0.12\ M}{2} = \mathbf{0.060}\ \boldsymbol{M}$$

17.99 First we find the molar solubility and then convert moles to grams. The solubility equilibrium for silver carbonate is:

$$Ag_2CO_3(s) \rightleftarrows 2Ag^+(aq) + CO_3^{2-}(aq)$$

| | | | |
|---|---|---|---|
| Initial ($M$): | | 0 | 0 |
| Change ($M$): | $-s$ | $+2s$ | $+s$ |
| Equilibrium ($M$): | | $2s$ | $s$ |

$$K_{sp} = [Ag^+]^2[CO_3^{2-}] = (2s)^2(s) = 4s^3 = 8.1 \times 10^{-12}$$

$$s = \sqrt[3]{\left(\frac{8.1 \times 10^{-12}}{4}\right)} = 1.27 \times 10^{-4}\ M$$

Converting from mol/L to g/L:

$$\textbf{concentration in g/L} = \frac{1.27 \times 10^{-4}\ \text{mol}}{1\ \text{L soln}} \times \frac{275.8\ \text{g}}{1\ \text{mol}} = \mathbf{0.035\ g/L}$$

17.101 The equilibrium reaction is:

| | $Pb(IO_3)_2(aq)$ | $\rightleftarrows$ | $Pb^{2+}(aq)$ | + | $2IO_3^-(aq)$ |
|---|---|---|---|---|---|
| Initial ($M$): | | | 0 | | 0.10 |
| Change ($M$): | $-2.4 \times 10^{-11}$ | | $+2.4 \times 10^{-11}$ | | $+2(2.4 \times 10^{-11})$ |
| Equilibrium ($M$): | | | $2.4 \times 10^{-11}$ | | $\approx 0.10$ |

Substitute the equilibrium concentrations into the solubility product expression to calculate $K_{sp}$.

$$K_{sp} = [Pb^{2+}][IO_3^-]^2$$

$$\mathbf{K_{sp}} = (2.4 \times 10^{-11})(0.10)^2 = \mathbf{2.4 \times 10^{-13}}$$

17.103 According to the solubility guidelines (Tables 4.2 and 4.3), the cation $Cd^{2+}(aq)$ will not precipitate with $NO_3^-(aq)$, $SO_4^{2-}(aq)$, or $ClO_3^-(aq)$. Likewise, the anion $S^{2-}(aq)$ will not precipitate with $Li^+(aq)$, $Na^+(aq)$, or $K^+(aq)$. However, $Cd^{2+}(aq)$ forms a complex with $CN^-$(aq) to produce $Cd(CN)_4^{2-}(aq)$. This reduces $[Cd^{2+}]$ in the solution and, by Le Chatelier's principle, promotes further dissolving of the CdS(*s*). So, **(c)** will increase the solubility.

17.105 a. The solubility product expressions for both substances have exactly the same mathematical form and are therefore directly comparable. The substance having the smaller $K_{sp}$ (**AgBr**) will precipitate first. (Why?)

b. When CuBr just begins to precipitate the solubility product expression will just equal $K_{sp}$ (saturated solution). The concentration of $Cu^+$ at this point is 0.010 $M$ (given in the problem), so the concentration of bromide ion must be:

$$K_{sp} = [Cu^+][Br^-] = (0.010)[Br^-] = 4.2 \times 10^{-8}$$

$$[Br^-] = \frac{4.2 \times 10^{-8}}{0.010} = 4.2 \times 10^{-6}\ M$$

Using this value of $[Br^-]$, we find the silver ion concentration

$$\mathbf{[Ag^+]} = \frac{K_{sp}}{[Br^-]} = \frac{7.7 \times 10^{-13}}{4.2 \times 10^{-6}} = \mathbf{1.8 \times 10^{-7}\ \mathit{M}}$$

c. The percent of silver ion remaining in solution is:

$$\% \ \mathbf{Ag^+}(\boldsymbol{aq}) = \frac{1.8 \times 10^{-7}\ M}{0.010\ M} \times 100\% = \mathbf{0.0018\%}\ \textit{or}\ \mathbf{1.8 \times 10^{-3}\%}$$

**Think About It:** Is this an effective way to separate silver from copper?

17.107 a. **Add sulfate.** $Na_2SO_4$ is soluble, $BaSO_4$ is not.

b. **Add sulfide.** $K_2S$ is soluble, PbS is not.

c. **Add iodide.** $ZnI_2$ is soluble, $HgI_2$ is not.

17.109 The amphoteric oxides cannot be used to prepare buffer solutions because **they are insoluble in water**.

17.111 a. **Mix 500 mL of 0.40 *M* $CH_3COOH$ with 500 mL of 0.40 *M* $CH_3COONa$.** Since the final volume is 1.00 L, then the concentrations of the two solutions that were mixed must be one-half of their initial concentrations.

b. **Mix 500 mL of 0.80 *M* $CH_3COOH$ with 500 mL of 0.40 *M* NaOH.** (Note: half of the acid reacts with all of the base to make a solution identical to that in part (a) above.)

$$CH_3COOH + NaOH \rightarrow CH_3COONa + H_2O$$

c. **Mix 500 mL of 0.80 *M* $CH_3COONa$ with 500 mL of 0.40 *M* HCl.** (Note: half of the salt reacts with all of the acid to make a solution identical to that in part (a) above.)

$$CH_3COO^- + H^+ \rightarrow CH_3COOH$$

17.113 a. When a strong acid $H_3O^+$ is added to a buffer solution containing the weak acid HA and its conjugate base $A^-$, the strong acid $H_3O^+$ protonates the conjugate base $A^-$. This corresponds to **figure (b)**.

b. When a strong base $OH^-$ is added to a buffer solution containing the weak acid HA and its conjugate base $A^-$, the strong base $OH^-$ de-protonates the conjugate acid HA. This corresponds to **figure (a)**.

17.115

$$pH = pK_a + \log\frac{[In^-]}{[HIn]}$$

For acid color:

$$pH = pK_a + \log\frac{1}{10}$$

$$pH = pK_a - \log 10$$

$$pH = pK_a - 1$$

For base color:

$$pH = pK_a + \log\frac{10}{1}$$

$$pH = pK_a + 1$$

Combining these two equations:

$$\mathbf{pH = pK_a \pm 1}$$

17.117 a. **The $pK_b$ value can be determined at the half-equivalence point of the titration (half the volume of added acid needed to reach the equivalence point). At this point in the titration $pH = pK_a$, where $K_a$ refers to the acid ionization constant of the conjugate acid of the weak base. The Henderson-Hasselbalch equation reduces to $pH = pK_a$ when [acid] = [conjugate base]. Once the $pK_a$ value is determined, the $pK_b$ value can be calculated as follows:**

$$\mathbf{pK_a + pK_b = 14.00}$$

b. Let B represent the base, and $BH^+$ represents its conjugate acid.

$$B(aq) + H_2O(l) \rightleftarrows BH^+(aq) + OH^-(aq)$$

$$K_b = \frac{[BH^+][OH^-]}{[B]}$$

$$[OH^-] = \frac{K_b[B]}{[BH^+]}$$

Taking the negative logarithm of both sides of the equation gives:

$$-\log[OH^-] = -\log K_b - \log\frac{[B]}{[BH^+]}$$

$$\mathbf{pOH = p}K_{\mathbf{b}}\mathbf{ + \log\frac{[BH^+]}{[B]}}$$

**The titration curve would look very much like Figure 17.4 of the text, except the *y*-axis would be pOH and the *x*-axis would be volume of strong acid added. The p$K_b$ value can be determined at the half-equivalence point of the titration (half the volume of added acid needed to reach the equivalence point). At this point in the titration, the concentrations of the buffer components, [B] and [$BH^+$], are equal, and hence pOH = p$K_b$.**

17.119 **Strategy:** Based on the fact that the containers in each figure are the same size, we assume that the volumes of all four solutions are equal. Further, define the concentration of each species as the number of corresponding circles per box. In this case, we can define the equilibrium quotient $Q$ as

$$Q = [\text{number of } M^+ \text{ circles per box}][\text{number of } X^- \text{ circles per box}].$$

According to the information given in part **(a)** of the problem, the $K_{sp}$ is:

$$K_{sp} = (5\ M^+ \text{ circles/box})(5\ X^- \text{ circles/box}) = 25 \text{ (omit the units).}$$

Calculate $Q$ for each of **(b)-(d)** and compare it to $K_{sp}$.

**Solution:** a. **saturated**

b. $Q = (8\ M^+ \text{ circles/box})(3\ X^- \text{ circles/box}) = 24 < K_{sp}$, **unsaturated**.

c. $Q = (6\ M^+ \text{ circles/box})(5\ X^- \text{ circles/box}) = 30 > K_{sp}$, **supersaturated**.

d. $Q = (5\ M^+ \text{ circles/box})(3\ X^- \text{ circles/box}) = 15 < K_{sp}$, **unsaturated**.

17.121 The initial number of moles of $Ag^+$ is

$$\text{mol } Ag^+ = 50.0 \text{ mL} \times \frac{0.010 \text{ mol } Ag^+}{1000 \text{ mL soln}} = 5.0 \times 10^{-4} \text{ mol } Ag^+$$

We can use the counts of radioactivity as being proportional to concentration. Thus, we can use the ratio to determine the quantity of $Ag^+$ still in solution. However, since our original 50 mL of solution has been diluted to 500 mL, the counts per mL will be reduced by ten. Our diluted solution would then produce 7402.5 counts per minute if no removal of $Ag^+$ had occurred.

The number of moles of $Ag^+$ that correspond to 44.4 counts are:

$$44.4 \text{ counts} \times \frac{5.0 \times 10^{-4} \text{ mol Ag}^+}{7402.5 \text{ counts}} = 3.0 \times 10^{-6} \text{ mol Ag}^+$$

$$\text{Original mol of IO}_3^- = 100 \text{ mL} \times \frac{0.030 \text{ mol IO}_3^-}{1000 \text{ mL soln}} = 3.0 \times 10^{-3} \text{ mol}$$

The quantity of $IO_3^-$ remaining after reaction with $Ag^+$:

$$(\text{original moles} - \text{moles reacted with Ag}^+) = (3.0 \times 10^{-3} \text{ mol}) - [(5.0 \times 10^{-4} \text{ mol}) - (3.0 \times 10^{-6} \text{ mol})]$$

$$= 2.5 \times 10^{-3} \text{ mol IO}_3^-$$

The total final volume is 500 mL or 0.50 L.

$$[\text{Ag}^+] = \frac{3.0 \times 10^{-6} \text{ mol Ag}^+}{0.50 \text{ L}} = 6.0 \times 10^{-6}\,M$$

$$[\text{IO}_3^-] = \frac{2.5 \times 10^{-3} \text{ mol IO}_3^-}{0.50 \text{ L}} = 5.0 \times 10^{-3}\,M$$

$$\text{AgIO}_3(s) \rightleftarrows \text{Ag}^+(aq) + \text{IO}_3^-(aq)$$

$$K_{sp} = [\text{Ag}^+][\text{IO}_3^-] = (6.0 \times 10^{-6})(5.0 \times 10^{-3}) = \mathbf{3.0 \times 10^{-8}}$$

17.123

$$\text{BaSO}_4(s) \rightleftarrows \text{Ba}^{2+}(aq) + \text{SO}_4^{2-}(aq)$$

$$K_{sp} = [\text{Ba}^{2+}][\text{SO}_4^{2-}] = 1.1 \times 10^{-10}$$

$$\mathbf{[Ba^{2+}] = 1.0 \times 10^{-5}\,M}$$

In 5.0 L, the number of moles of $Ba^{2+}$ is

$$(5.0 \text{ L})\left(\frac{1.0 \times 10^{-5} \text{ mol Ba}^{2+}}{1 \text{ L}}\right) = 5 \times 10^{-5} \text{ mol Ba}^{2+}.$$

In practice, even less $BaSO_4$ will dissolve because the $BaSO_4$ is not in contact with the entire volume of blood. **$Ba(NO_3)_2$ is too soluble to be used for this purpose.**

17.125 Because oxalate is the anion of a weak acid, increasing the hydrogen ion concentration (decreasing the pH) would consume oxalate ion to produce hydrogen oxalate and oxalic acid:

$$\text{C}_2\text{O}_4^{2-}(aq) + \text{H}^+(aq) \rightarrow \text{HC}_2\text{O}_4^-(aq)$$

$$HC_2O_4^-(aq) + H^+(aq) \rightarrow H_2C_2O_4(aq)$$

Decreasing the concentration of oxalate ion would, via Le Châtelier's principle, increase the solubility of calcium oxalate. **Decreasing the pH would increase the solubility of calcium oxalate and should help minimize the formation of calcium oxalate kidney stones.**

Note: Although this makes sense from the standpoint of principles presented in Chapter 17, the actual mechanism of kidney-stone formation is more complex than can be described using only these equilibria.

17.127

$$\text{pH} = \text{p}K_\text{a} + \log\frac{[\text{conjugate base}]}{[\text{acid}]}$$

**At pH = 1.0,**

–COOH

$$1.0 = 2.3 + \log\frac{[\text{–COO}^-]}{[\text{–COOH}]}$$

$$\frac{[\text{–COOH}]}{[\text{–COO}^-]} = 20$$

$-NH_3^+$

$$1.0 = 9.6 + \log\frac{[\text{–NH}_2]}{[\text{–NH}_3^+]}$$

$$\frac{[\text{–NH}_3^+]}{[\text{–NH}_2]} = 4 \times 10^8$$

Therefore the **predominant species** is: $^+$**NH$_3$ – CH$_2$ – COOH**

**At pH = 7.0,**

–COOH

$$7.0 = 2.3 + \log\frac{[\text{–COO}^-]}{[\text{–COOH}]}$$

$$\frac{[\text{–COOH}]}{[\text{–COO}^-]} = 5 \times 10^4$$

$-NH_3^+$

$$7.0 = 9.6 + \log\frac{[\text{–NH}_2]}{[\text{–NH}_3^+]}$$

$$\frac{[\text{–NH}_3^+]}{[\text{–NH}_2]} = 4 \times 10^2$$

**Predominant species:** $^+$**NH$_3$ – CH$_2$ – COO$^-$**

**At pH = 12.0,**

–COOH

$$12.0 = 2.3 + \log\frac{[\text{–COO}^-]}{[\text{–COOH}]}$$

$$\frac{[\text{–COO}^-]}{[\text{–COOH}]} = 5 \times 10^9$$

$\text{–NH}_3^+$

$$12.0 = 9.6 + \log\frac{[\text{–NH}_2]}{[\text{–NH}_3^+]}$$

$$\frac{[\text{–NH}_2]}{[\text{–NH}_3^+]} = 2.5 \times 10^2$$

**Predominant species: $NH_2 - CH_2 - COO^-$**

17.129 Assuming the density of water to be 1.00 g/mL, 0.05 g $Pb^{2+}$ per $10^6$ g water is equivalent to $5 \times 10^{-5}$ g $Pb^{2+}$/L

$$\frac{0.05 \text{ g Pb}^{2+}}{1 \times 10^6 \text{ g H}_2\text{O}} \times \frac{1 \text{ g H}_2\text{O}}{1 \text{ mL H}_2\text{O}} \times \frac{1000 \text{ mL H}_2\text{O}}{1 \text{ L H}_2\text{O}} = 5 \times 10^{-5} \text{ g Pb}^{2+}\text{/L}$$

$$\text{PbSO}_4(s) \rightleftarrows \text{Pb}^{2+}(aq) + \text{SO}_4^{2-}(aq)$$

| | $PbSO_4(s)$ | $Pb^{2+}(aq)$ | $SO_4^{2-}(aq)$ |
|---|---|---|---|
| Initial ($M$): | | 0 | 0.10 |
| Change ($M$): | $-s$ | $+s$ | $+s$ |
| Equilibrium ($M$): | | $s$ | $s$ |

$$K_{sp} = [\text{Pb}^{2+}][\text{SO}_4^{2-}]$$

$$1.6 \times 10^{-8} = s^2$$

$$s = 1.3 \times 10^{-4}\ M$$

The solubility of $PbSO_4$ in g/L is:

$$\frac{1.3 \times 10^{-4} \text{ mol}}{1 \text{ L}} \times \frac{303.3 \text{ g}}{1 \text{ mol}} = 4.0 \times 10^{-2} \text{ g/L}$$

**Yes**. The $[Pb^{2+}]$ exceeds the safety limit of $5 \times 10^{-5}$ g $Pb^{2+}$/L.

17.131 **The ionized polyphenols have a dark color. In the presence of citric acid from lemon juice, the anions**

**are converted to the lighter-colored acids.**

17.133 **(c)** has the highest $[H^+]$

$$F^- + SbF_5 \rightarrow SbF_6^-$$

Removal of $F^-$ promotes further ionization of HF.

17.135 **Strategy:** Let $s$ be the molar solubility of $Ag_2CO_3$. We wish to calculate

$$K_{sp} = [Ag^+]^2[CO_3^{2-}] = (2s)^2(s) = 4s^3 .$$

Each mole of $CO_3^{2-}(aq)$ produces one mole of $CO_2(g)$:

$$CO_3^{2-}(aq) + 2HCl(aq) \rightarrow H_2O(l) + 2Cl^-(aq) + CO_2(g),$$

so the number of moles of $CO_3^{2-}(aq)$ in the saturated solution is equal to the number of moles of $CO_2\ (g)$ collected. Use the ideal gas law and the given information to calculate $n(CO_2) = n(CO_3^{2-})$, then use this result to calculate $s$ and $K_{sp}$.

**Solution:** The amount of $CO_2$ is:

$$n(CO_2) = \frac{PV}{RT} = \frac{(114\ \text{mmHg})\left(\dfrac{1\ \text{atm}}{760\ \text{mmHg}}\right)(0.019\ \text{L})}{(0.0821\ \text{L}\cdot\text{atm/mol}\cdot\text{K})(273+5)(\text{K})} = 1.25 \times 10^{-4}\ \text{mol}.$$

Since $n(CO_3^{2-}) = n(CO_2) = 1.25 \times 10^{-4}$ mol and the volume of the saturated solution is 1.0 L, then

$$s = \frac{1.25\times10^{-4}\ \text{mol}\ CO_3^{2-}}{1.0\ \text{L}} = 1.3\times10^{-4}\ M.$$

Thus, $\boldsymbol{K_{sp}} = 4\left(1.3 \times 10^{-4}\right)^3 = \mathbf{8.8 \times 10^{-12}}$

17.137 a. $H^+ + OH^- \rightarrow H_2O \qquad K = \mathbf{1.0 \times 10^{14}}$

b. $H^+ + NH_3 \rightarrow NH_4^+ \qquad K = \dfrac{1}{K_a} = \dfrac{1}{5.6 \times 10^{-10}} = \mathbf{1.8 \times 10^{9}}$

c. $$CH_3COOH + OH^- \rightarrow CH_3COO^- + H_2O$$

Broken into 2 equations:

$$CH_3COOH \rightarrow CH_3COO^- + H^+ \qquad K_a$$

$$H^+ + OH^- \rightarrow H_2O \qquad 1/K_w$$

$$K = \frac{K_a}{K_w} = \frac{1.8 \times 10^{-5}}{1.0 \times 10^{-14}} = \mathbf{1.8 \times 10^9}$$

d. $CH_3COOH + NH_3 \rightarrow CH_3COONH_4$

Broken into 2 equations:

$$CH_3COOH \rightarrow CH_3COO^- + H^+ \qquad K_a$$

$$NH_3 + H^+ \rightarrow NH_4^+ \qquad \frac{1}{K_a'}$$

$$K = \frac{K_a}{K_a'} = \frac{1.8 \times 10^{-5}}{5.6 \times 10^{-10}} = \mathbf{3.2 \times 10^4}$$

# Chapter 18

# Entropy, Free Energy, and Equilibrium

18.7 **Strategy:** According to equation 18.2, the number of ways of arranging $N$ particles in $X$ cells is $W = X^N$. Use this equation to calculate the number of arrangements and equation 18.1 to calculate the entropy.

**Setup:** For the setup in the figure, $X = 4$ when the barrier is in place and $X = 8$ when it is absent.

**Solution:** a. **With barrier:** $N = 2$; $\boldsymbol{W} = 4^2 = \mathbf{16}$; **Without barrier**: $N = 2$; $\boldsymbol{W} = 8^2 = \mathbf{64}$

b. From part (a), we know that **16** of the 64 arrangements have both particles in the left side of the container. Similarly, there are **16** ways for the particles to be found on the right-hand side. The number of arrangements with one particle per side is 64 – 16 – 16 = **32**.
**Both particles on one side:** $\boldsymbol{S} = k \ln W^N = (1.38 \times 10^{-23}$ J/K)ln(32) = **3.83 × 10$^{-23}$ J/K; particles on opposite sides:** $\boldsymbol{S} = (1.38 \times 10^{-23}$ J/K)ln(32) = **4.78 × 10$^{-23}$. The most probable state is the one with the larger entropy; that is, the state in which the particles are on opposite sides.**

18.13 **Strategy:** Equation 18.4 gives the entropy change for the isothermal expansion of an ideal gas. Substitute the given values into the equation and compute $\Delta S$.

**Solution:** a.

$$\boldsymbol{\Delta S_{sys}} = (0.0050 \text{ mol})(8.314 \text{ J/mol}\cdot\text{K})\ln\left(\frac{52.5 \text{ mL}}{112 \text{ mL}}\right) = \mathbf{-0.031\ J/K}$$

b.

$$\boldsymbol{\Delta S_{sys}} = (0.015 \text{ mol})(8.314 \text{ J/mol}\cdot\text{K})\ln\left(\frac{22.5 \text{ mL}}{225 \text{ mL}}\right) = \mathbf{-0.29\ J/K}$$

c.

$$\boldsymbol{\Delta S_{sys}} = (22.1 \text{ mol})(8.314 \text{ J/mol}\cdot\text{K})\ln\left(\frac{275 \text{ L}}{122 \text{ L}}\right) = \mathbf{1.5\times10^2\ J/K}$$

18.15 **Strategy:** To calculate the standard entropy change of a reaction, we look up the standard entropies of reactants and products in Appendix 2 of the text and apply Equation 18.7. As in the calculation of enthalpy of reaction, the stoichiometric coefficients have no units, so $\Delta S^\circ_{rxn}$ is expressed in units of J/K·mol.

**Solution:** The standard entropy change for a reaction can be calculated using the following equation.

$$\Delta S^\circ_{rxn} = \Sigma nS^\circ(\text{products}) - \Sigma mS^\circ(\text{reactants})$$

a.

$$\mathbf{\Delta S^\circ_{rxn}} = S^\circ(\text{Cu}) + S^\circ(\text{H}_2\text{O}(g)) - [S^\circ(\text{H}_2) + S^\circ(\text{CuO})]$$

$$= (1)(33.3\text{ J/K·mol}) + (1)(188.7\text{ J/K·mol}) - [(1)(131.0\text{ J/K·mol}) + (1)(43.5\text{ J/K·mol})]$$

$$= \mathbf{47.5\ J/K{\cdot}mol}$$

b.

$$\mathbf{\Delta S^\circ_{rxn}} = S^\circ(\text{Al}_2\text{O}_3) + 3S^\circ(\text{Zn}) - [2S^\circ(\text{Al}) + 3S^\circ(\text{ZnO})]$$

$$= (1)(50.99\text{ J/K·mol}) + (3)(41.6\text{ J/K·mol}) - [(2)(28.3\text{ J/K·mol}) + (3)(43.9\text{ J/K·mol})]$$

$$= \mathbf{-12.5\ J/K{\cdot}mol}$$

c.

$$\mathbf{\Delta S^\circ_{rxn}} = S^\circ(\text{CO}_2) + 2S^\circ(\text{H}_2\text{O}(l)) - [S^\circ(\text{CH}_4) + 2S^\circ(\text{O}_2)]$$

$$= (1)(213.6\text{ J/K·mol}) + (2)(69.9\text{ J/K·mol}) - [(1)(186.2\text{ J/K·mol}) + (2)(205.0\text{ J/K·mol})]$$

$$= \mathbf{-242.8\ J/K{\cdot}mol}$$

Why was the entropy value for water different in parts (a) and (c)?

18.17 In order of increasing entropy per mole at 25°C:

**(c) < (d) < (e) < (a) < (b)**

a. **Ne(*g*): a monatomic gas of higher molar mass than $H_2$. (For gas phase molecules, increasing molar mass tends to increase the molar entropy. While Ne(*g*) is less complex structurally than $H_2(g)$, its much larger molar mass more than offsets the complexity difference.)**

b. **$SO_2(g)$: a polyatomic gas of higher complexity and higher molar mass than Ne(*g*)** (see the explanation for (a) above).

c. **Na(*s*): highly ordered, crystalline material.**

d. **NaCl(*s*): highly ordered crystalline material, but with more particles per mole than Na(s).**

e. **$H_2$: a diatomic gas, hence of higher entropy than a solid.**

18.21 **Strategy:** Assume all reactants and products are in their standard states. The entropy change in the surroundings is related to the enthalpy change of the system by Equation 18.7. For each reaction in Exercise 18.17, use Appendix 2 to calculate $\Delta H_{sys}$ ( = $\Delta H_{rxn}$) (Section 5.3) and then use

Equation 18.7 to calculate $\Delta S_{surr}$. Finally, use Equation 18.8 to compute $\Delta S_{univ}$. If $\Delta S_{univ}$ is positive, then according to the second law of thermodynamics, the reaction is spontaneous. The values for $\Delta S_{sys}$ are found in the solution for problem 18.16.

**Solution:** a.

$$\Delta H_{sys} = \Delta H^\circ_f\ (Cu) + \Delta H^\circ_f\ (H_2O(g)) - [\Delta H^\circ_f\ (H_2) + \Delta H^\circ_f\ (CuO)]$$

$$\Delta H_{sys} = 0 + (1)(-241.8\text{ kJ/mol}) - [0 + (1)(-155.2\text{ kJ/mol})]$$

$$\Delta H_{sys} = -86.6\text{ kJ/mol}$$

$$\Delta S_{surr} = -\frac{\Delta H_{sys}}{T} = -\frac{(-86.6\text{ kJ/mol})}{298\text{ K}} = 0.291\text{ kJ/mol}\cdot\text{K}$$

$$\mathbf{\Delta S_{surr} = 291\ J/mol\cdot K}$$

$$\Delta S_{univ} = \Delta S_{sys} + \Delta S_{surr} = 47.5\text{ J/mol}\cdot\text{K} + 291\text{ J/mol}\cdot\text{K}$$

$$\Delta S_{univ} = 339\text{ J/mol}\cdot\text{K}$$

$$\Delta S_{univ} > 0$$

**spontaneous**

b.

$$\Delta H_{sys} = \Delta H^\circ_f\ (Al_2O_3) + 3\Delta H^\circ_f\ (Zn) - [2\Delta H^\circ_f\ (Al) + 3\Delta H^\circ_f\ (ZnO)]$$

$$\Delta H_{sys} = (1)(-1669.8\text{ kJ/mol}) + 0 - [0 + (3)(-348.0\text{ kJ/mol})]$$

$$\Delta H_{sys} = -626\text{ kJ/mol}$$

$$\Delta S_{surr} = -\frac{\Delta H_{sys}}{T} = -\frac{(-626\text{ kJ/mol})}{298\text{ K}} = 2.10\text{ kJ/mol}\cdot\text{K}$$

$$\mathbf{\Delta S_{surr} = 2.10\times 10^3\ J/mol\cdot K}$$

$$\Delta S_{univ} = \Delta S_{sys} + \Delta S_{surr} = -12.5\text{ J/mol}\cdot\text{K} + 2.10\times 10^3\text{ J/mol}\cdot\text{K}$$

$$\Delta S_{univ} = 2.09\times 10^3\text{ J/mol}\cdot\text{K}$$

$$\Delta S_{univ} > 0$$

**spontaneous**

c. $$\Delta H_{sys} = \Delta H^\circ_f(CO_2) + 2\Delta H^\circ_f(H_2O(l)) - [\Delta H^\circ_f(CH_4) + 2\Delta H^\circ_f(O_2)]$$

$$\Delta H_{sys} = (1)(-393.5\text{ kJ/mol}) + (2)(-285.8\text{ kJ/mol}) - [(1)(-74.85\text{ kJ/mol}) + 0]$$

$$\Delta H_{sys} = -890.3\text{ kJ/mol}$$

$$\Delta S_{surr} = -\frac{\Delta H_{sys}}{T} = -\frac{(-890.3\text{ kJ/mol})}{298\text{ K}} = 2.99\text{ kJ/mol}\cdot\text{K}$$

$$\mathbf{\Delta S_{surr} = 2.99\times 10^3\text{ J/mol}\cdot\text{K}}$$

$$\Delta S_{univ} = \Delta S_{sys} + \Delta S_{surr} = -242.8\text{ J/mol}\cdot\text{K} + 2.99\times 10^3\text{ J/mol}\cdot\text{K}$$

$$\Delta S_{univ} = 2.75\times 10^3\text{ J/mol}\cdot\text{K}$$

$$\Delta S_{univ} > 0$$

**spontaneous**

18.23 **Strategy:** Assume all reactants and products are in their standard states. According to Equations 18.7, the entropy change in the surroundings for an isothermal process is :

$$\Delta S_{surr} = -\frac{\Delta H_{sys}}{T}$$

Also, Equation 18.8 states that the entropy change of the universe is:

$$\Delta S_{univ} = \Delta S_{sys} + \Delta S_{surr}$$

Use Appendix 2 to calculate $\Delta S_{sys}$ ( = $\Delta S_{rxn}$) and $\Delta H_{sys}$ ( = $\Delta H_{rxn}$). Substitute these results into the above equations and determine the sign of $\Delta S_{univ}$. If $\Delta S_{univ}$ is positive, then according to the second law of thermodynamics, the reaction is spontaneous.

**Solution:** a. $$\Delta S_{sys} = S^\circ(PCl_5) - [S^\circ(PCl_3(l)) + S^\circ(Cl_2)]$$

$$\Delta S_{sys} = 364.5\text{ J/mol}\cdot\text{K} - [217.1\text{ J/mol}\cdot\text{K} + 223.0\text{ J/mol}\cdot\text{K}]$$

$$\mathbf{\Delta S_{sys} = -75.6\text{ J / mol}\cdot\text{K}}$$

$$\Delta H_{sys} = \Delta H_f^\circ\ (PCl_5) - [\Delta H_f^\circ\ (PCl_3(l)) + \Delta H_f^\circ\ (Cl_2)]$$

$$\Delta H_{sys} = -374.9\ \text{kJ/mol} - [-319.7\ \text{kJ/mol} + 0]$$

$$\Delta H_{sys} = -55.2\ \text{kJ/mol}$$

$$\Delta S_{surr} = -\frac{\Delta H_{sys}}{T} = -\frac{(-55.2\ \text{kJ/mol})}{298\ \text{K}} = 0.185\ \text{kJ/mol}\cdot\text{K}$$

$$\mathbf{\Delta S_{surr} = 185\ J/mol \cdot K}$$

$$\Delta S_{univ} = \Delta S_{sys} + \Delta S_{surr} = -75.6\ \text{J/mol}\cdot\text{K} + 185\ \text{J/mol}\cdot\text{K}$$

$$\Delta S_{univ} = 109\ \text{J/mol}\cdot\text{K}$$

$$\Delta S_{univ} > 0$$

**spontaneous**

b.

$$\Delta S_{sys} = 2S^\circ(Hg) + S^\circ(O_2) - 2S^\circ(HgO)$$

$$\Delta S_{sys} = (2)(77.4\ \text{J/mol}\cdot\text{K}) + 205.0\ \text{J/mol}\cdot\text{K} - (2)(72.0\ \text{J/mol}\cdot\text{K})$$

$$\mathbf{\Delta S_{sys} = 215.8\ J/mol \cdot K}$$

$$\Delta H_{sys} = 2\Delta H_f^\circ\ (Hg) + \Delta H_f^\circ\ (O_2) - 2\Delta H_f^\circ\ (HgO)$$

$$\Delta H_{sys} = 0 + 0 - (-90.7\ \text{kJ/mol})$$

$$\Delta H_{sys} = 90.7\ \text{kJ/mol}$$

$$\Delta S_{surr} = -\frac{\Delta H_{sys}}{T} = -\frac{(90.7\ \text{kJ/mol})}{298\ \text{K}} = -0.304\ \text{kJ/mol}\cdot\text{K}$$

$$\mathbf{\Delta S_{surr} = -304\ J/mol \cdot K}$$

$$\Delta S_{univ} = \Delta S_{sys} + \Delta S_{surr} = 215.8\ \text{J/mol}\cdot\text{K} + (-304\ \text{J/mol}\cdot\text{K})$$

$$\Delta S_{univ} = -88\ \text{J/mol}\cdot\text{K}$$

$$\Delta S_{univ} < 0$$

**not spontaneous**

c. $$\Delta S_{sys} = 2S°(\text{H}(g)) - S°(\text{H}_2(g)) = (2)(114.6\ \text{J/mol}\cdot\text{K}) - (1)(131.0\ \text{J/mol}\cdot\text{K})$$

$$\mathbf{\Delta S_{sys} = 98.2\ J / mol \cdot K}$$

To calculate $\Delta H_{rxn}$ for $H_2 \rightarrow 2H$, use the bond enthalpy from Table 8.6.

$$\mathbf{\Delta H_{sys} = 436.4\ kJ / mol}$$

Calculate $\Delta S_{surr}$ and $\Delta S_{univ}$:

$$\Delta S_{surr} = -\frac{\Delta H_{sys}}{T} = -\frac{(436.4\ \text{kJ/mol})}{298\ \text{K}} = -1.46\ \text{kJ/mol}\cdot\text{K}$$

$$\mathbf{\Delta S_{surr} = -1.46 \times 10^3\ J / mol \cdot K}$$

$$\Delta S_{univ} = \Delta S_{sys} + \Delta S_{surr} = 98.2\ \text{J/mol}\cdot\text{K} + \left(-1.46 \times 10^3\ \text{J/mol}\cdot\text{K}\right)$$

$$\Delta S_{univ} = -1.36 \times 10^3\ \text{J/mol}\cdot\text{K}$$

$$\Delta S_{univ} < 0$$

**not spontaneous**

d. $$\Delta S_{sys} = S°(\text{UF}_6) - [S°(\text{U}) + 3S°(\text{F}_2)]$$

$$\Delta S_{sys} = 378\ \text{J/mol}\cdot\text{K} - [50.21\ \text{J/mol}\cdot\text{K} + (3)(203.34\ \text{J/mol}\cdot\text{K})]$$

$$\mathbf{\Delta S_{sys} = -282\ J / mol \cdot K}$$

$$\Delta H_{sys} = \Delta H^\circ_f\ (\text{UF}_6) - [\Delta H^\circ_f\ (\text{U}) + 3\Delta H^\circ_f\ (\text{F}_2)]$$

$$\Delta H_{sys} = -2147\ \text{kJ/mol} - [0 + 0]$$

$$\Delta H_{sys} = -2147 \text{ kJ/mol}$$

$$\Delta S_{surr} = -\frac{\Delta H_{sys}}{T} = -\frac{(-2147 \text{ kJ/mol})}{298 \text{ K}} = 7.20 \text{ kJ/mol}\cdot\text{K}$$

$$\mathbf{\Delta S_{surr} = 7.20 \times 10^3 \ J / mol \cdot K}$$

$$\Delta S_{univ} = \Delta S_{sys} + \Delta S_{surr} = -282 \text{ J/mol}\cdot\text{K} + 7.20 \times 10^3 \text{ J/mol}\cdot\text{K}$$

$$\Delta S_{univ} = 6.92 \times 10^3 \text{ J/mol}\cdot\text{K}$$

$$\Delta S_{univ} > 0$$

**spontaneous**

**Think About It:** We could have assumed that the reaction in (c) occurred at constant volume instead of constant pressure. Would this have changed the conclusion about the spontaneity of the reaction?

18.29 **Strategy:** To calculate the standard free-energy change of a reaction, we look up the standard free energies of formation of reactants and products in Appendix 2 of the text and apply Equation 18.13. Note that all the stoichiometric coefficients have no units so $\Delta G^\circ_{rxn}$ is expressed in units of kJ/mol. The standard free energy of formation of any element in its stable allotropic form at 1 atm and 25°C is zero.

**Solution:** The standard free energy change for a reaction can be calculated using the following equation.

$$\Delta G^\circ_{rxn} = \Sigma n\Delta G^\circ_f(\text{products}) - \Sigma m\Delta G^\circ_f(\text{reactants})$$

a.

$$\Delta G^\circ_{rxn} = 2\Delta G^\circ_f(\text{MgO}) - [2\Delta G^\circ_f(\text{Mg}) + \Delta G^\circ_f(\text{O}_2)]$$

$$\mathbf{\Delta G^\circ_{rxn}} = (2)(-569.6 \text{ kJ/mol}) - [(2)(0) + (1)(0)] = \mathbf{-1139 \ kJ / mol}$$

b.

$$\Delta G^\circ_{rxn} = 2\Delta G^\circ_f(\text{SO}_3) - [2\Delta G^\circ_f(\text{SO}_2) + \Delta G^\circ_f(\text{O}_2)]$$

$$\mathbf{\Delta G^\circ_{rxn}} = (2)(-370.4 \text{ kJ/mol}) - [(2)(-300.4 \text{ kJ/mol}) + (1)(0)] = \mathbf{-140.0 \ kJ / mol}$$

c.

$$\mathbf{\Delta G^\circ_{rxn}} = (4)(-394.4 \text{ kJ/mol}) + (6)(-237.2 \text{ kJ/mol})$$

$$\Delta G^\circ_{rxn} = 4\Delta G^\circ_f(CO_2(g)) + 6\Delta G^\circ_f(H_2O(l)) - \left[2\Delta G^\circ_f(C_2H_6(g)) + 7\Delta G^\circ_f(O_2(g))\right]$$

$$- [(2)(-32.89\text{ kJ/mol}) + (7)(0)] = \mathbf{-2935\text{ kJ / mol}}$$

18.31 a. Calculate $\Delta G$ from $\Delta H$ and $\Delta S$.

$$\Delta G = \Delta H - T\Delta S = -126{,}000\text{ J/mol} - (298\text{ K})(84\text{ J/K·mol}) = -151{,}000\text{ J/mol}$$

The free energy change is negative so the reaction is spontaneous at 298 K. Since $\Delta H$ is negative and $\Delta S$ is positive, **the reaction is spontaneous at all temperatures**.

b. Calculate $\Delta G$.

$$\Delta G = \Delta H - T\Delta S = -11{,}700\text{ J/mol} - (298\text{ K})(-105\text{ J/K·mol}) = +19{,}600\text{ J}$$

The free energy change is positive at 298 K which means the reaction is not spontaneous at that temperature. The positive sign of $\Delta G$ results from the large negative value of $\Delta S$. At lower temperatures, the $-T\Delta S$ term will be smaller thus allowing the free energy change to be negative.

$\Delta G$ will equal zero when $\Delta H = T\Delta S$.

Rearranging,

$$\boldsymbol{T} = \frac{\Delta H}{\Delta S} = \frac{-11700\text{ J/mol}}{-105\text{ J/K}\cdot\text{mol}} = \mathbf{111\text{ K}}$$

At temperatures **below 111 K**, $\Delta G$ will be negative and the reaction will be spontaneous.

18.33 **Strategy:** Equation 18.7 from the text relates the entropy change of the surroundings to the enthalpy change of the system and the temperature at which a phase change occurs.

$$\Delta S_{surr} = \frac{-\Delta H_{sys}}{T}$$

We know that $\Delta S_{sys} = -\Delta S_{surr}$, so we can rewrite Equation 18.7 as

$$\Delta S_{sys} = \frac{\Delta H_{sys}}{T}$$

**Solution:**

$$\boldsymbol{\Delta S_{fus}} = \frac{\Delta H_{fus}}{T} = \frac{23.4\text{ kJ/mol}}{(234.3\text{ K})} \times \frac{1000\text{ J}}{1\text{ kJ}} = \mathbf{99.9\text{ J/K·mol}}$$

$$\Delta S_{vap} = \frac{\Delta H_{vap}}{T} = \frac{59.0 \text{ kJ/mol}}{(630 \text{ K})} \times \frac{1000 \text{ J}}{1 \text{ kJ}} = \mathbf{93.6 \text{ J/K·mol}}$$

**Think About It:** Remember that Celsius temperatures must be converted to Kelvin.

18.35 Using Equation 18.12 from the text,

$$\Delta G^\circ_{rxn} = 2\Delta G^\circ_f(C_2H_5OH) + 2\Delta G^\circ_f(CO_2) - \Delta G^\circ_f(C_6H_{12}O_6)$$

$$\Delta G^\circ_{rxn} = (2)(-174.18 \text{ kJ/mol}) + (2)(-394.4 \text{ kJ/mol}) - (-910.56 \text{ kJ/mol}) = \mathbf{-226.6 \text{ kJ/mol}}$$

18.37 Using Equation 18.2 from the text,

$$\Delta G^\circ_{rxn} = \Delta G^\circ_f(NO_3^-) - [\frac{1}{2}\Delta G^\circ_f(O_2) + \Delta G^\circ_f(NO_2^-)]$$

$$= (-110.5 \text{ kJ/mol}) - [(0 \text{ kJ/mol}) + (-34.6 \text{ kJ/mol}) = -75.9 \text{ kJ/mol}$$

**75.9 kJ** of Gibbs free energy are released.

18.41 **Strategy:** According to Equation 18.14 of the text, the equilibrium constant for the reaction is related to the standard free energy change; that is, $\Delta G^\circ = -RT \ln K$. Since we are given $\Delta G^\circ$ in the problem, we can solve for the equilibrium constant. What temperature unit should be used?

**Solution:** Solving Equation 18.14 for $K$ gives

$$K_p = e^{\frac{-\Delta G^\circ}{RT}} = e^{\frac{-2.60 \times 10^3 \text{ J/mol}}{(8.314 \text{ J/K·mol})(298 \text{ K})}} = e^{-1.05} = \mathbf{0.35}$$

18.43
$$K_{sp} = [Fe^{2+}][OH^-]^2 = 1.6 \times 10^{-14}$$

$$\Delta G^\circ = -RT\ln K_{sp} = -(8.314 \text{ J/K·mol})(298 \text{ K})\ln(1.6 \times 10^{-14}) = \mathbf{7.9 \times 10^4 \text{ J/mol}} = \mathbf{79 \text{ kJ/mol}}$$

18.45 a. We first find the standard free energy change of the reaction.

$$\Delta G^\circ_{rxn} = \Delta G^\circ_f(PCl_3(g)) + \Delta G^\circ_f(Cl_2(g)) - \Delta G^\circ_f(PCl_5(g))$$
$$= (1)(-269.6 \text{ kJ/mol}) + (1)(0) - (1)(-305.0 \text{ kJ/mol}) = \mathbf{35.4 \text{ kJ/mol}}$$

We can calculate $K_P$ by rearranging Equation 18.14 of the text.

$$K_P = e^{\frac{-\Delta G^\circ}{RT}} = e^{\frac{-35.4 \times 10^3\ \text{J/mol}}{(8.314\ \text{J/K·mol})(298\ \text{K})}} = e^{-14.3} = \mathbf{6.2 \times 10^{-7}}$$

b. We are finding the free energy difference between the reactants and the products at their nonequilibrium values. The result tells us the direction of and the potential for further chemical change. We use the given nonequilibrium pressures to compute $Q_P$.

$$Q_P = \frac{P_{PCl_3} P_{Cl_2}}{P_{PCl_5}} = \frac{(0.27)(0.40)}{0.0029} = 37$$

The value of $\Delta G$ (notice that this is not the standard free energy difference) can be found using Equation 18.13 of the text and the result from part (a).

$$\mathbf{\Delta G} = \Delta G^\circ + RT\ln Q = (35.4 \times 10^3\ \text{J/mol}) + (8.314\ \text{J/K·mol})(298\ \text{K})\ln(37) = \mathbf{44.6\ kJ/mol}$$

**Think About It:** Which way is the direction of spontaneous change for this system? What would be the value of $\Delta G$ if the given data were equilibrium pressures? What would be the value of $Q_P$ in that case?

18.47 The expression of $K_P$ is: $K_P = P_{CO_2}$

Thus you can predict the equilibrium pressure directly from the value of the equilibrium constant. The only task at hand is computing the values of $K_P$ using Equations 18.10 and 18.14 of the text.

a. At 25°C, $\Delta G^\circ = \Delta H^\circ - T\Delta S^\circ = (177.8 \times 10^3\ \text{J/mol}) - (298\ \text{K})(160.5\ \text{J/K·mol})$

$$= 130.0 \times 10^3\ \text{J/mol}$$

$$P_{CO_2} = K_P = e^{\frac{-\Delta G^\circ}{RT}} = e^{\frac{-130.0 \times 10^3\ \text{J/mol}}{(8.314\ \text{J/K·mol})(298\ \text{K})}} = e^{-52.47} = \mathbf{1.6 \times 10^{-23}\ atm}$$

b. At 800°C, $\Delta G^\circ = \Delta H^\circ - T\Delta S^\circ = (177.8 \times 10^3\ \text{J/mol}) - (1073\ \text{K})(160.5\ \text{J/K·mol})$

$$= 5.58 \times 10^3\ \text{J/mol}$$

$$P_{CO_2} = K_P = e^{\frac{-\Delta G^\circ}{RT}} = e^{\frac{-5.58 \times 10^3\ \text{J/mol}}{(8.314\ \text{J/K·mol})(1073\ \text{K})}} = e^{-0.625} = \mathbf{0.535\ atm}$$

**Think About It:** What assumptions are made in the second calculation?

18.49 The equilibrium constant expression is: $K_P = P_{H_2O}$

We are actually finding the equilibrium vapor pressure of water (compare to Problem 18.47). We use Equation 18.14 of the text.

$$P_{H_2O} = K_P = e^{\frac{-\Delta G^\circ}{RT}} = e^{\frac{-8.6 \times 10^3 \text{ J/mol}}{(8.314 \text{ J/K·mol})(298 \text{ K})}} = e^{-3.47} = \mathbf{3.1 \times 10^{-2}\ atm} \text{ or } \mathbf{23.6\ mmHg}$$

**Think About It:** The positive value of $\Delta G^\circ$ indicates that reactants are favored at equilibrium at 25°C. Is that what you would expect?

18.53

$$C_6H_{12}O_6 + 6O_2 \rightarrow 6CO_2 + 6H_2O \qquad \Delta G^\circ = -2880 \text{ kJ/mol}$$

$$ADP + H_3PO_4 \rightarrow ATP + H_2O \qquad \Delta G^\circ = +31 \text{ kJ/mol}$$

Maximum number of ATP molecules synthesized:

$$2880 \text{ kJ/mol} \times \frac{1 \text{ ATP molecule}}{31 \text{ kJ/mol}} = \mathbf{93\ ATP\ molecules}$$

18.55 **Strategy:** Melting is an endothermic process, $\mathbf{\Delta H_{fus} > 0}$. Also, a liquid generally has a higher entropy than the solid at the same temperature, so $\mathbf{\Delta S_{fus} > 0}$. To determine the sign of $\Delta G_{fus}$, use the fact that melting is spontaneous ($\Delta G_{fus} < 0$) for temperatures above the freezing point and is not spontaneous ($\Delta G_{fus} > 0$) for temperatures below the freezing point.

**Solution:**

a. The temperature is above the freezing point, so melting is spontaneous and $\mathbf{\Delta G_{fus} < 0}$.

b. The temperature is at the freezing point, so the solid and the liquid are in equilibrium and $\mathbf{\Delta G_{fus} = 0}$.

c. The temperature is below the freezing point, so melting is not spontaneous and $\mathbf{\Delta G_{fus} > 0}$.

18.57 **Only *U* and *H* are associated with the first law alone.**

18.59 We can calculate $\Delta S_{sys}$ from standard entropy values in Appendix 2 of the text. We can calculate $\Delta S_{surr}$ from the $\Delta H_{sys}$ value given in the problem. Finally, we can calculate $\Delta S_{univ}$ from the $\Delta S_{sys}$ and $\Delta S_{surr}$ values.

$$\mathbf{\Delta S_{sys}} = (2)(69.9 \text{ J/K·mol}) - [(2)(131.0 \text{ J/K·mol}) + (1)(205.0 \text{ J/K·mol})] = \mathbf{-327\ J/K·mol}$$

$$\mathbf{\Delta S_{surr}} = \frac{-\Delta H_{sys}}{T} = \frac{-(-571.6 \times 10^3 \text{ J/mol})}{298 \text{ K}} = \mathbf{1918\ J/K \cdot mol}$$

$$\mathbf{\Delta S_{univ}} = \Delta S_{sys} + \Delta S_{surr} = (-327 + 1918) \text{ J/K·mol} = \mathbf{1591\ J/K·mol}$$

18.61 If the process is *spontaneous* as well as *endothermic*, the signs of $\Delta G$ and $\Delta H$ must be negative and positive, respectively. Since $\Delta G = \Delta H - T\Delta S$, the sign of **$\Delta S$ must be positive ($\Delta S > 0$)** for $\Delta G$ to be negative.

18.63 a. Using the relationship:

$$\frac{\Delta H_{vap}}{T_{bp}} = \Delta S_{vap} \approx 90\ \text{J/K} \cdot \text{mol}$$

| | |
|---|---|
| **benzene** | $\boldsymbol{\Delta S_{vap} = 87.8\ \text{J/K·mol}}$ |
| **hexane** | $\boldsymbol{\Delta S_{vap} = 90.1\ \text{J/K·mol}}$ |
| **mercury** | $\boldsymbol{\Delta S_{vap} = 93.7\ \text{J/K·mol}}$ |
| **toluene** | $\boldsymbol{\Delta S_{vap} = 91.8\ \text{J/K·mol}}$ |

**Trouton's rule is a statement about $\Delta S^{o}_{vap}$. In most substances, the molecules are in constant and random motion in both the liquid and gas phases, so $\Delta S^{\circ}_{vap}$ = 90 J/K·mol.**

b. Using the data in Table 11.6 of the text, we find:

| | |
|---|---|
| **ethanol** | $\boldsymbol{\Delta S_{vap} = 111.9\ \text{J/K·mol}}$ |
| **water** | $\boldsymbol{\Delta S_{vap} = 109.4\ \text{J/K·mol}}$ |

**In ethanol and water, there are fewer possible arrangements of the molecules due to the network of H-bonds, so $\Delta S^{o}_{vap}$ is greater.**

18.65 ***q*, and *w* are *not* state functions**. Recall that state functions represent properties that are determined by the state of the system, regardless of how that condition is achieved. Heat and work are not state functions because they are not properties of the system. They manifest themselves only during a process (during a change). Thus their values depend on the path of the process and vary accordingly.

18.67 For a phase transition, $\Delta G = 0$. We write:

$$\Delta G = \Delta H - T\Delta S$$

$$0 = \Delta H - T\Delta S$$

$$\Delta S_{sub} = \frac{\Delta H_{sub}}{T}$$

Substituting $\Delta H$ and the temperature, $(-78° + 273°)\text{K} = 195\ \text{K}$, gives

$$\Delta S_{sub} = \frac{\Delta H_{sub}}{T} = \frac{25.2 \times 10^3 \text{ J}}{195 \text{ K}} = 129 \text{ J/K} \cdot \text{mol}$$

This value of $\Delta S_{sub}$ is for the sublimation of 1 mole of $CO_2$. We convert to the $\Delta S$ value for the sublimation of 84.8 g of $CO_2$.

$$84.8 \text{ g } CO_2 \times \frac{1 \text{ mol } CO_2}{44.01 \text{ g } CO_2} \times \frac{129 \text{ J}}{\text{K} \cdot \text{mol}} = \mathbf{249 \text{ J / K}}$$

18.69 **Equation 18.10 represents the standard free-energy change for a reaction, and not for a particular compound like $CO_2$. The correct form is:**

$$\Delta G° = \Delta H° - T\Delta S°$$

**For a given reaction, $\Delta G°$ and $\Delta H°$ would need to be calculated from standard formation values (graphite, oxygen, and carbon dioxide) first, before plugging into the equation. Also, $\Delta S°$ would need to be calculated from standard entropy values.**

$$\mathbf{C(graphite) + O_2(g) \rightarrow CO_2(g)}$$

18.71 a. Each CO molecule has two possible orientations in the crystal,

CO or OC

If there is no preferred orientation, then for one molecule there are two, or $2^1$, choices of orientation. Two molecules have four or $2^2$ choices, and for 1 mole of CO there are $2^{N_A}$ choices. From Equation 18.1 of the text:

$$S = k \ln W$$

$$S = (1.38 \times 10^{-23} \text{ J/K}) \ln 2^{6.022 \times 10^{23}}$$

$$S = (1.38 \times 10^{-23} \text{ J/K})(6.022 \times 10^{23} \text{ /mol}) \ln 2$$

$$\boldsymbol{S = 5.76 \text{ J/K} \cdot \text{mol}}$$

b. **The fact that the actual residual entropy is 4.2 J/K·mol means that the orientation is not totally random.**

18.73 For the reaction: $CaCO_3(s) \rightleftarrows CaO(s) + CO_2(g)$ $K_P = P_{CO_2}$

Using the equation from Problem 18.66:

$$\ln\frac{K_2}{K_1} = \frac{\Delta H^\circ}{R}\left(\frac{1}{T_1} - \frac{1}{T_2}\right) = \frac{\Delta H^\circ}{R}\left(\frac{T_2 - T_1}{T_1 T_2}\right)$$

Substituting,

$$\ln\frac{1829}{22.6} = \frac{\Delta H^\circ}{8.314\text{ J/K}\cdot\text{mol}}\left(\frac{1223\text{ K} - 973\text{ K}}{(973\text{ K})(1223\text{ K})}\right)$$

Solving,

$$\mathbf{\Delta H^\circ = 1.74 \times 10^5\ J/mol = 174\ kJ/mol}$$

18.75 a. **$\Delta S$ is positive** b. **$\Delta S$ is negative** c. **$\Delta S$ is positive** d. **$\Delta S$ is positive**

18.77 At the temperature of the normal boiling point the free energy difference between the liquid and gaseous forms of mercury (or any other substances) is zero, i.e. the two phases are in equilibrium. We can therefore use Equation 18.10 of the text to find this temperature. For the equilibrium,

$$\text{Hg}(l) \rightleftarrows \text{Hg}(g)$$

$$\Delta G = \Delta H - T\Delta S = 0$$

$$\Delta H = \Delta H_f^\circ[\text{Hg}(g)] - \Delta H_f^\circ[\text{Hg}(l)] = 60{,}780\text{ J/mol} - 0 = 60{,}780\text{ J/mol}$$

$$\Delta S = S^\circ[\text{Hg}(g)] - S^\circ[\text{Hg}(l)] = 174.7\text{ J/K·mol} - 77.4\text{ J/K·mol} = 97.3\text{ J/K·mol}$$

$$T_{bp} = \frac{\Delta H}{\Delta S} = \frac{60780\text{ J/mol}}{97.3\text{ J/K}\cdot\text{mol}} = \mathbf{625\ K}$$

**Think About It:** What assumptions are made? Notice that the given enthalpies and entropies are at standard conditions, namely 25°C and 1.00 atm pressure. In performing this calculation **we assume that $\Delta H^\circ$ and $\Delta S^\circ$ do not depend on temperature**. The actual normal boiling point of mercury is 356.58°C. Is the assumption of the temperature independence of these quantities reasonable?

18.79 **No. A negative $\Delta G^\circ$ tells us that a reaction has the potential to happen, but gives no indication of the rate.**

18.81 a.

$$\Delta G^\circ = 2\Delta G_f^\circ(\text{HBr}) - \Delta G_f^\circ(\text{H}_2) - \Delta G_f^\circ(\text{Br}_2) = (2)(-53.2\text{ kJ/mol}) - (1)(0) - (1)(0)$$

$$\mathbf{\Delta G^\circ = -106.4\ kJ/mol}$$

$$\ln K_P = \frac{-\Delta G^\circ}{RT} = \frac{106.4 \times 10^3\text{ J/mol}}{(8.314\text{ J/K}\cdot\text{mol})(298\text{ K})} = 42.9$$

$$\mathbf{K_P = 4 \times 10^{18}}$$

b.
$$\Delta G° = \Delta G_f^\circ(\text{HBr}) - \tfrac{1}{2}\Delta G_f^\circ(\text{H}_2) - \tfrac{1}{2}\Delta G_f^\circ(\text{Br}_2) = (1)(-53.2\text{ kJ/mol}) - (\tfrac{1}{2})(0) - (\tfrac{1}{2})(0)$$

$$\mathbf{\Delta G° = -53.2\ kJ/mol}$$

$$\ln K_P = \frac{-\Delta G°}{RT} = \frac{53.2 \times 10^3\text{ J/mol}}{(8.314\text{ J/K}\cdot\text{mol})(298\text{ K})} = 21.5$$

$$\mathbf{K_P = 2 \times 10^9}$$

**The $K_P$ in (a) is the square of the $K_P$ in (b). Both $\Delta G°$ and $K_P$ depend on the number of moles of reactants and products specified in the balanced equation.**

18.83 **Talking involves various biological processes (to provide the necessary energy) that lead to a increase in the entropy of the universe. Since the overall process (talking) is spontaneous, the entropy of the universe must increase.**

18.85 a. If $\Delta G°$ for the reaction is 173.4 kJ/mol,

then,

$$\Delta G_f^\circ = \frac{173.4\text{ kJ/mol}}{2} = \mathbf{86.7\ kJ/mol}$$

b.
$$\Delta G° = -RT\ln K_P$$

$$173.4 \times 10^3\text{ J/mol} = -(8.314\text{ J/K}\cdot\text{mol})(298\text{ K})\ln K_P$$

$$\mathbf{K_P = 4 \times 10^{-31}}$$

c. $\Delta H°$ for the reaction is $2 \times \Delta H_f^\circ(\text{NO}) = (2)(90.4\text{ kJ/mol}) = 180.8\text{ kJ/mol}$

Using the equation in Problem 18.66:

$$\ln\frac{K_2}{4 \times 10^{-31}} = \frac{180.8 \times 10^3\text{ J/mol}}{8.314\text{ J/mol}\cdot\text{K}}\left(\frac{1373\text{ K} - 298\text{ K}}{(1373\text{ K})(298\text{ K})}\right)$$

$$\mathbf{K_2 = 3 \times 10^{-6}}$$

d. **Lightning supplies the energy necessary to drive this reaction, converting the two most abundant gases in the atmosphere into NO(*g*). The NO gas dissolves in the rain, which carries it into the soil where it is converted into nitrate and nitrite by bacterial action. This "fixed" nitrogen is a necessary nutrient for plants.**

18.87 As discussed in Chapter 18 of the text for the decomposition of calcium carbonate, a reaction favors the formation of products at equilibrium when

$$\Delta G° = \Delta H° - T\Delta S° < 0$$

If we can calculate $\Delta H°$ and $\Delta S°$, we can solve for the temperature at which decomposition begins to favor products. We use data in Appendix 2 of the text to solve for $\Delta H°$ and $\Delta S°$.

$$\Delta H° = \Delta H_f^\circ[\text{MgO}(s)] + \Delta H_f^\circ[\text{CO}_2(g)] - \Delta H_f^\circ[\text{MgCO}_3(s)]$$

$$\Delta H° = -601.8 \text{ kJ/mol} + (-393.5 \text{ kJ/mol}) - (-1112.9 \text{ kJ/mol}) = 117.6 \text{ kJ/mol}$$

$$\Delta S° = S°[\text{MgO}(s)] + S°[\text{CO}_2(g)] - S°[\text{MgCO}_3(s)]$$

$$\Delta S° = 26.78 \text{ J/K·mol} + 213.6 \text{ J/K·mol} - 65.69 \text{ J/K·mol} = 174.7 \text{ J/K·mol}$$

For the reaction to begin to favor products,

$$\Delta H° - T\Delta S° < 0$$

or

$$T > \frac{\Delta H°}{\Delta S°}$$

$$T > \frac{117.6 \times 10^3 \text{ J/mol}}{174.7 \text{ J/K·mol}}$$

$$\mathbf{T > 673.2\ K}$$

18.89 a.

$$\Delta G° = \Delta G_f^\circ(\text{H}_2) + \Delta G_f^\circ(\text{Fe}^{2+}) - \Delta G_f^\circ(\text{Fe}) - 2\Delta G_f^\circ(\text{H}^+)]$$

$$\Delta G° = (1)(0) + (1)(-84.9 \text{ kJ/mol}) - (1)(0) - (2)(0)$$

$$\Delta G° = -84.9 \text{ kJ/mol}$$

$$\Delta G° = -RT\ln K$$

$$-84.9 \times 10^3 \text{ J/mol} = -(8.314 \text{ J/mol·K})(298 \text{ K})\ln K$$

$$\mathbf{K = 7.6 \times 10^{14}}$$

b.
$$\Delta G° = \Delta G_f°(H_2) + \Delta G_f°(Cu^{2+}) - \Delta G_f°(Cu) - 2\Delta G_f°(H^+)]$$

$$\Delta G° = 64.98 \text{ kJ/mol}$$

$$\Delta G° = -RT\ln K$$

$$64.98 \times 10^3 \text{ J/mol} = -(8.314 \text{ J/mol·K})(298 \text{ K})\ln K$$

$$\mathbf{K = 4.1 \times 10^{-12}}$$

**The activity series is correct. The very large value of *K* for reaction (a) indicates that *products* are highly favored; whereas, the very small value of *K* for reaction (b) indicates that *reactants* are highly favored.**

18.91 First convert to moles of ice.

$$74.6 \text{ g } H_2O(s) \times \frac{1 \text{ mol } H_2O(s)}{18.02 \text{ g } H_2O(s)} = 4.14 \text{ mol } H_2O(s)$$

For a phase transition:

$$\Delta S_{sys} = \frac{\Delta H_{sys}}{T}$$

$$\Delta S_{sys} = \frac{(4.14 \text{ mol})(6010 \text{ J/mol})}{273 \text{ K}} = \mathbf{91.1 \text{ J/K}}$$

$$\Delta S_{surr} = \frac{-\Delta H_{sys}}{T}$$

$$\Delta S_{surr} = \frac{-(4.14 \text{ mol})(6010 \text{ J/mol})}{273 \text{ K}} = \mathbf{-91.1 \text{ J/K}}$$

$$\Delta S_{univ} = \Delta S_{sys} + \Delta S_{surr} = \mathbf{0}$$

**The system is at equilibrium.**

**Think About It:** We ignored the freezing point depression of the salt water. How would inclusion of this effect change our conclusion?

18.93 Since the adsorption is spontaneous, **$\Delta G$ must be negative**. When hydrogen bonds to the surface of the catalyst, there is a decrease in the number of possible arrangements of atoms in the system, **$\Delta S$ must be negative**. Since there is a decrease in entropy, the adsorption must be exothermic for the process to be spontaneous, **$\Delta H$ must be negative**.

18.95 We can calculate $\Delta G°$ at 872 K from the equilibrium constant, $K_1$.

$$\Delta G° = -RT\ln K$$

$$\Delta G° = -(8.314\text{ J/mol}\cdot\text{K})(872\text{ K})\ln(1.80\times10^{-4})$$

$$\boldsymbol{\Delta G°} = 6.25\times10^{4}\text{ J/mol} = \mathbf{62.5\ kJ/mol}$$

We use the equation derived in Problem 18.66 to calculate $\Delta H°$.

$$\ln\frac{K_2}{K_1} = \frac{\Delta H°}{R}\left(\frac{1}{T_1}-\frac{1}{T_2}\right)$$

$$\ln\frac{0.0480}{1.80\times10^{-4}} = \frac{\Delta H°}{8.314\text{ J/mol}\cdot\text{K}}\left(\frac{1}{872\text{ K}}-\frac{1}{1173\text{ K}}\right)$$

$$\boldsymbol{\Delta H°} = \mathbf{157.8\ kJ/mol}$$

Now that both $\Delta G°$ and $\Delta H°$ are known, we can calculate $\Delta S°$ at 872 K.

$$\Delta G° = \Delta H° - T\Delta S°$$

$$62.5\times10^{3}\text{ J/mol} = (157.8\times10^{3}\text{ J/mol}) - (872\text{ K})\Delta S°$$

$$\boldsymbol{\Delta S°} = \mathbf{109\ J/K{\cdot}mol}$$

18.97 The reaction mixtures are at equilibrium if $K = Q$. Use Equation 18.14 to calculate $K$.

$$\Delta G° = -RT\ln K$$

$$-3400\text{ J/mol} = -(8.314\text{ J/mol}\cdot\text{K})(298\text{ K})\ln K$$

$$1.372 = \ln K$$

Taking the anti-ln of both sides,

$$e^{-1.372} = K$$

$$\boldsymbol{K} = \mathbf{3.9 \approx 4}$$

To find $Q$ of each reaction mixture, take the concentration of products over concentration of reactants raised to the appropriate power.

For the reaction, $A_2(g) + B_2(g) \rightleftarrows 2AB(g)$

$$Q = \frac{[AB]^2}{[A_2][B_2]}$$

The partial pressure is equal to the number of molecules times 0.10 atm. Therefore, the partial pressure for 3 molecules of AB is 3 × 0.10 atm = 0.30 atm.

$$Q_i = \frac{[0.30]^2}{[0.30][0.20]} = 1.5$$

$$Q_{ii} = \frac{[0.60]^2}{[0.20][0.30]} = 6.0$$

$$Q_{iii} = \frac{[0.40]^2}{[0.20][0.20]} = 4.0$$

a. Since $K = Q$ for reaction (iii), then **reaction (iii) is at equilibrium**.

b. Use Equation 18.13 to calculate $\Delta G$.

$$\Delta G = \Delta G° + RT \ln Q$$

$$\Delta G_i = -3400 \text{ J/mol} + (8.314 \text{ J/K} \cdot \text{mol})(298 \text{ K})\ln(1.5) = -2.4\times 10^3 \text{ J/mol}$$

$$\Delta G_{ii} = -3400 \text{ J/mol} + (8.314 \text{ J/K} \cdot \text{mol})(298 \text{ K})\ln(6) = 1.0 \times 10^3 \text{ J/mol}$$

$$\Delta G_{iii} = -3400 \text{ J/mol} + (8.314 \text{ J/K} \cdot \text{mol})(298 \text{ K})\ln(4) = 35 \text{ J/mol}$$

**Reaction *i*** has a negative $\Delta G$ value.

c. **Reactions *ii* and *iii*** have positive $\Delta G$ values.

18.99 a. It is the reverse of a **disproportionation redox reaction**.

b. $$\Delta G° = (2)(-228.6 \text{ kJ/mol}) - (2)(-33.0 \text{ kJ/mol}) - (1)(-300.4 \text{ kJ/mol})$$

$$\Delta G° = -90.8 \text{ kJ/mol}$$

$$-90.8 \times 10^3 \text{ J/mol} = -(8.314 \text{ J/mol·K})(298 \text{ K})\ln K$$

$$\mathbf{K = 8.2 \times 10^{15}}$$

Because of the large value of $K$, **this method is feasible for removing $SO_2$.**

c.
$$\Delta H^\circ = (2)(-241.8 \text{ kJ/mol}) + (3)(0) - (2)(-20.15 \text{ kJ/mol}) - (1)(-296.4 \text{ kJ/mol})$$

$$\Delta H^\circ = -146.9 \text{ kJ/mol}$$

$$\Delta S^\circ = (2)(188.7 \text{ J/K·mol}) + (3)(31.88 \text{ J/K·mol}) - (2)(205.64 \text{ J/K·mol}) - (1)(248.5 \text{ J/K·mol})$$

$$\Delta S^\circ = -186.7 \text{ J/K·mol}$$

$$\Delta G^\circ = \Delta H^\circ - T\Delta S^\circ$$

Due to the negative entropy change, $\Delta S^\circ$, the free energy change, $\Delta G^\circ$, will become positive at higher temperatures. Therefore, the reaction will be **less effective** at high temperatures.

18.101 We can calculate $K_P$ from $\Delta G^\circ$.

$$\Delta G^\circ = (1)(-394.4 \text{ kJ/mol}) + (0) - (1)(-137.3 \text{ kJ/mol}) - (1)(-255.2 \text{ kJ/mol})$$

$$\Delta G^\circ = -1.9 \text{ kJ/mol}$$

$$-1.9 \times 10^3 \text{ J/mol} = -(8.314 \text{ J/mol·K})(1173 \text{ K})\ln K_P$$

$$K_P = 1.2$$

Now, from $K_P$, we can calculate the mole fractions of CO and $CO_2$.

$$K_P = \frac{P_{CO_2}}{P_{CO}} = 1.2 \qquad P_{CO_2} = 1.2P_{CO}$$

$$\boldsymbol{X_{CO}} = \frac{P_{CO}}{P_{CO} + P_{CO_2}} = \frac{P_{CO}}{P_{CO} + 1.2P_{CO}} = \frac{1}{2.2} = \mathbf{0.45}$$

$$\boldsymbol{X_{CO_2}} = 1 - 0.45 = \mathbf{0.55}$$

**We assumed that $\Delta G^\circ$ calculated from $\Delta G^\circ_f$ values was temperature independent**. The $\Delta G^\circ_f$ values in Appendix 2 of the text are measured at 25°C, but the temperature of the reaction is 900°C.

18.103 Assuming that both $\Delta H^\circ$ and $\Delta S^\circ$ are temperature independent, we can calculate both $\Delta H^\circ$ and $\Delta S^\circ$.

$$\Delta H^\circ = \Delta H^\circ_f(CO) + \Delta H^\circ_f(H_2) - [\Delta H^\circ_f(H_2O(g)) + \Delta H^\circ_f(C)]$$

$$\Delta H^\circ = (1)(-110.5 \text{ kJ/mol}) + (1)(0)] - [(1)(-241.8 \text{ kJ/mol}) + (1)(0)]$$

$$\Delta H° = 131.3 \text{ kJ/mol}$$

$$\Delta S° = S°(CO) + S°(H_2) - [S°(H_2O) + S°(C)]$$

$$\Delta S° = [(1)(197.9 \text{ J/K·mol}) + (1)(131.0 \text{ J/K·mol})] - [(1)(188.7 \text{ J/K·mol}) + (1)(5.69 \text{ J/K·mol})]$$

$$\Delta S° = 134.5 \text{ J/K·mol}$$

It is obvious from the given conditions that the reaction must take place at a fairly high temperature (in order to have red–hot coke). Setting $\Delta G° = 0$

$$0 = \Delta H° - T\Delta S°$$

$$\boldsymbol{T} = \frac{\Delta H°}{\Delta S°} = \frac{131.3 \text{ kJ/mol} \times \dfrac{1000 \text{ J}}{1 \text{ kJ}}}{134.5 \text{ J/K} \cdot \text{mol}} = \mathbf{976\ K} = \mathbf{703°C}$$

The temperature must be greater than 703°C for the reaction to be spontaneous.

18.105 Setting $\Delta G$ equal to zero and solving Equation 18.10for $T$ gives

$$\text{T} = \frac{\Delta H}{\Delta S} = \frac{1.25 \times 10^5 \text{ J/mol}}{397 \text{ J/K} \cdot \text{mol}} = 314.9 \text{ K} = \mathbf{42°C}$$

18.107 **Since we are dealing with the same ion ($K^+$)**, Equation 18.13 of the text can be written as:

$$\Delta G = \Delta G° + RT\ln Q$$

$$\Delta G = 0 + (8.314 \text{ J/mol} \cdot \text{K})(310 \text{ K})\ln\left(\frac{400 \text{ m}M}{15 \text{ m}M}\right)$$

$$\mathbf{\Delta G = 8.5 \times 10^3\ J/mol = 8.5\ kJ/mol}$$

18.109 Begin by writing the equation for the process described in the problem. Hydrogen ions must be transported from a region where their concentration is $10^{-7.4}$ ($4 \times 10^{-8}$ $M$) to a region where their concentration is $10^{-1}$ (0.1 $M$).

$$H^+(aq, 4 \times 10^{-8}\ M) \rightleftarrows H^+(aq, 0.1\ M)$$

$\Delta G°$ for this process is 0. We use Equation 18.13 to calculate $\Delta G$. ($T = 37°C + 273 = 310$ K.)

$$\Delta G = \Delta G° + RT\ln Q = (8.314 \times 10^{-3} \text{ kJ/mol})(310 \text{ K})\left(\ln\frac{0.1\ M}{4 \times 10^{-8}\ M}\right) = 38 \text{ kJ/mol}$$

Therefore, the Gibbs free energy required for the secretion of 1 mole of $H^+$ ions from the blood plasma to the stomach is **38 kJ**.

18.111 a.

$$\mathbf{2CO + 2NO \rightarrow 2CO_2 + N_2}$$

b. **The oxidizing agent is NO; the reducing agent is CO.**

c.

$$\Delta G^\circ = 2\Delta G_f^\circ(CO_2) + \Delta G_f^\circ(N_2) - 2\Delta G_f^\circ(CO) - 2\Delta G_f^\circ(NO)$$

$$\Delta G^\circ = (2)(-394.4 \text{ kJ/mol}) + (0) - (2)(-137.3 \text{ kJ/mol}) - (2)(86.7 \text{ kJ/mol}) = -687.6 \text{ kJ/mol}$$

$$\Delta G^\circ = -RT\ln K_P$$

$$\ln K_P = \frac{6.876 \times 10^5 \text{ J/mol}}{(8.314 \text{ J/K}\cdot\text{mol})(298 \text{ K})} = 277.5$$

$$\boldsymbol{K_P = 3 \times 10^{120}}$$

d.

$$\boldsymbol{Q_P} = \frac{(P_{N_2})(P_{CO_2})^2}{(P_{CO})^2(P_{NO})^2} = \frac{(0.80)(3.0 \times 10^{-4})^2}{(5.0 \times 10^{-5})^2(5.0 \times 10^{-7})^2} = \mathbf{1.2 \times 10^{14}}$$

Since $Q_P << K_P$, the reaction will proceed **to the right.**

e.

$$\Delta H^\circ = 2\Delta H_f^\circ(CO_2) + \Delta H_f^\circ(N_2) - 2\Delta H_f^\circ(CO) - 2\Delta H_f^\circ(NO)$$

$$\Delta H^\circ = (2)(-393.5 \text{ kJ/mol}) + (0) - (2)(-110.5 \text{ kJ/mol}) - (2)(90.4 \text{ kJ/mol}) = -746.8 \text{ kJ/mol}$$

Since $\Delta H^\circ$ is negative, raising the temperature will decrease $K_P$ (Le Chatelier's principle), thereby increasing the amount of reactants and decreasing the amount of products. **No**, the formation of $N_2$ and $CO_2$ is not favored by raising the temperature.

18.113 a. $\Delta G^\circ$ for $CH_3COOH$:

$$\Delta G^\circ = -(8.314 \text{ J/mol}\cdot\text{K})(298 \text{ K})\ln(1.8 \times 10^{-5})$$

$$\boldsymbol{\Delta G^\circ} = 2.7 \times 10^4 \text{ J/mol} = \mathbf{27\ kJ/mol}$$

$\Delta G^\circ$ for $CH_2ClCOOH$:

$$\Delta G^\circ = -(8.314 \text{ J/mol}\cdot\text{K})(298 \text{ K})\ln(1.4 \times 10^{-3})$$

$$\boldsymbol{\Delta G^\circ} = 1.6 \times 10^4 \text{ J/mol} = \mathbf{16\ kJ/mol}$$

b. The $T\Delta S°$ term determines the value of $\Delta G°$. **The *system's* entropy change dominates.**

c. **The breaking and making of specific O–H bonds. Other contributions include solvent separation and ion solvation.**

d. **The $CH_3COO^-$ ion, which is smaller than $CH_2ClCOO^-$, can participate in hydration to a greater extent, leading to solutions with fewer possible arrangements.**

# Chapter 19

# Electrochemistry

19.1 **Strategy:** We follow the stepwise procedure for balancing redox reactions presented in Section 19.1 of the text.

**Solution:** a. In Step 1, we separate the half-reactions.

| | |
|---|---|
| oxidation: | $Fe^{2+} \rightarrow Fe^{3+}$ |
| reduction: | $H_2O_2 \rightarrow H_2O$ |

Step 2 is unnecessary because the half-reactions are already balanced with respect to iron.

Step 3: We balance each half-reaction for O by adding $H_2O$. The oxidation half-reaction is already balanced in this regard (it contains no O atoms). The reduction half-reaction requires the addition of one $H_2O$ on the product side.

$$H_2O_2 \rightarrow H_2O + H_2O \quad \text{or simply} \quad H_2O_2 \rightarrow 2H_2O$$

Step 4: We balance each half reaction for H by adding $H^+$. Again, the oxidation half-reaction is already balanced in this regard. The reduction half-reaction requires the addition of two $H^+$ ions on the reactant side.

$$H_2O_2 + 2H^+ \rightarrow 2H_2O$$

Step 5: We balance both half-reactions for charge by adding electrons. In the oxidation half-reaction, there is a total charge of +2 on the left and a total charge of +3 on the right. Adding one electron to the product side makes the total charge on each side +2.

$$Fe^{2+} \rightarrow Fe^{3+} + e^-$$

In the reduction half-reaction, there is a total charge of +2 on the left and a total charge of 0 on the right. Adding two electrons to the reactant side makes the total charge on each side 0.

$$H_2O_2 + 2H^+ + 2e^- \rightarrow 2H_2O$$

Step 6: Because the number of electrons is not the same in both half-reactions, we multiply the oxidation half-reaction by 2.

$$2\times(Fe^{2+} \rightarrow Fe^{3+} + e^-) \quad = \quad 2Fe^{2+} \rightarrow 2Fe^{3+} + 2e^-$$

Step 7: Finally, we add the resulting half-reactions, cancelling electrons and any other identical species to get the overall balanced equation.

$$2Fe^{2+} \rightarrow 2Fe^{3+} + 2e^-$$

$$H_2O_2 + 2H^+ + 2e^- \rightarrow 2H_2O$$

$$\mathbf{2H^+ + H_2O_2 + 2Fe^{2+} \rightarrow 2Fe^{3+} + 2H_2O}$$

b. In Step 1, we separate the half-reactions.

oxidation: $Cu \rightarrow Cu^{2+}$

reduction: $HNO_3 \rightarrow NO$

Step 2 is unnecessary because the half-reactions are already balanced with respect to copper and nitrogen.

Step 3: We balance each half-reaction for O by adding $H_2O$. The oxidation half-reaction is already balanced in this regard (it contains no O atoms). The reduction half-reaction requires the addition of two $H_2O$ molecules on the product side.

$$HNO_3 \rightarrow NO + 2H_2O$$

Step 4: We balance each half reaction for H by adding $H^+$. Again, the oxidation half-reaction is already balanced in this regard. The reduction half-reaction requires the addition of three $H^+$ ions on the reactant side.

$$3H^+ + HNO_3 \rightarrow NO + 2H_2O$$

Step 5: We balance both half-reactions for charge by adding electrons. In the oxidation half-reaction, there is a total charge of 0 on the left and a total charge of +2 on the right. Adding two electrons to the product side makes the total charge on each side 0.

$$Cu \rightarrow Cu^{2+} + 2e^-$$

In the reduction half-reaction, there is a total charge of +3 on the left and a total charge of 0 on the right. Adding three electrons to the reactant side makes the total charge on each side 0.

$$3H^+ + HNO_3 + 3e^- \rightarrow NO + 2H_2O$$

Step 6: Because the number of electrons is not the same in both half-reactions, we multiply the oxidation half-reaction by 3, and the reduction half-reaction by 2.

$$3\times(Cu \rightarrow Cu^{2+} + 2e^-) = 3Cu \rightarrow 3Cu^{2+} + 6e^-$$

$$2\times(3H^+ + HNO_3 + 3e^- \rightarrow NO + 2H_2O) = 6H^+ + 2HNO_3 + 6e^- \rightarrow 2NO + 4H_2O$$

Step 7: Finally, we add the resulting half-reactions, cancelling electrons and any other identical species to get the overall balanced equation.

$$3Cu \rightarrow 3Cu^{2+} + 6e^-$$

$$6H^+ + 2HNO_3 + 6e^- \rightarrow 2NO + 4H_2O$$

$$\mathbf{6H^+ + 2HNO_3 + 3Cu \rightarrow 3Cu^{2+} + 2NO + 4H_2O}$$

c. In Step 1, we separate the half-reactions.

oxidation: $CN^- \rightarrow CNO^-$

reduction: $MnO_4^- \rightarrow MnO_2$

Step 2 is unnecessary because the half-reactions are already balanced with respect to carbon, nitrogen, and manganese.

Step 3: We balance each half-reaction for O by adding $H_2O$. The oxidation half-reaction requires the addition of one $H_2O$ molecule to the reactant side. The reduction half-reaction requires the addition of two $H_2O$ molecules on the product side.

$$H_2O + CN^- \rightarrow CNO^-$$

$$MnO_4^- \rightarrow MnO_2 + 2H_2O$$

Step 4: We balance each half reaction for H by adding $H^+$. The oxidation half-reaction requires the addition of 2 $H^+$ ions to the product side. The reduction half-reaction requires the addition of four $H^+$ ions on the reactant side.

$$H_2O + CN^- \rightarrow CNO^- + 2H^+$$

$$4H^+ + MnO_4^- \rightarrow MnO_2 + 2H_2O$$

Step 5: We balance both half-reactions for charge by adding electrons. In the oxidation half-reaction, there is a total charge of −1 on the left and a total charge of +1 on the right. Adding two electrons to the product side makes the total charge on each side −1.

$$H_2O + CN^- \rightarrow CNO^- + 2H^+ + 2e^-$$

In the reduction half-reaction, there is a total charge of +3 on the left and a total charge of 0 on the right. Adding three electrons to the reactant side makes the total charge on each side 0.

$$3e^- + 4H^+ + MnO_4^- \rightarrow MnO_2 + 2H_2O$$

Step 6: Because the number of electrons is not the same in both half-reactions, we multiply the oxidation half-reaction by 3, and the reduction half-reaction by 2.

$$3\times(H_2O + CN^- \rightarrow CNO^- + 2H^+ + 2e^-) = 3H_2O + 3CN^- \rightarrow 3CNO^- + 6H^+ + 6e^-$$

$$2\times(3e^- + 4H^+ + MnO_4^- \rightarrow MnO_2 + 2H_2O) = 6e^- + 8H^+ + 2MnO_4^- \rightarrow 2MnO_2 + 4H_2O$$

Step 7: Finally, we add the resulting half-reactions, cancelling electrons and any other identical species to get the overall balanced equation.

$$3H_2O + 3CN^- \rightarrow 3CNO^- + 6H^+ + 6e^-$$
$$6e^- + 8H^+ + 2MnO_4^- \rightarrow 2MnO_2 + 4H_2O$$
$$3CN^- + 2MnO_4^- + 2H^+ \rightarrow 3CNO^- + 2MnO_2 + H_2O$$

Balancing a redox reaction in basic solution requires two additional steps.

Step 8: For each $H^+$ ion in the final equation, we add one $OH^-$ ion to each side of the equation, combining the $H^+$ and $OH^-$ ions to produce $H_2O$.

$$3CN^- + 2MnO_4^- + 2H^+ \rightarrow 3CNO^- + 2MnO_2 + H_2O$$
$$+2OH^- \qquad\qquad +2OH^-$$

$$2H_2O + 2MnO_4^- + 3CN^- \rightarrow 2MnO_2 + 3CNO^- + H_2O + 2OH^-$$

Step 9: Lastly, we cancel $H_2O$ molecules that result from Step 8.

$$\mathbf{3CN^- + 2MnO_4^- + H_2O \rightarrow 3CNO^- + 2MnO_2 + 2OH^-}$$

Parts (d) and (e) are solved using the methods outlined in (a) through (c).

d. $\mathbf{6OH^- + 3Br_2 \rightarrow BrO_3^- + 3H_2O + 5Br^-}$

e. $\mathbf{2S_2O_3^{2-} + I_2 \rightarrow S_4O_6^{2-} + 2I^-}$

19.11 **Strategy:** At first, it may not be clear how to assign the electrodes in the galvanic cell. From Table 19.1 of the text, we write the standard reduction potentials of Al and Ag and compare their values to determine which is the anode half-reaction and which is the cathode half-reaction.

**Solution:** The standard reduction potentials are:

$$Ag^{+}(1.0\ M) + e^{-} \rightarrow Ag(s) \qquad E° = 0.80\ V$$
$$Al^{3+}(1.0\ M) + 3e^{-} \rightarrow Al(s) \qquad E° = -1.66\ V$$

The silver half-reaction, with the more positive $E°$ value, will occur at the cathode (as a reduction). The aluminum half-reaction will occur at the anode (as an oxidation). In order to balance the numbers of electrons in the two half-reactions, we multiply the reduction by 3 before summing the two half-reactions. Note that multiplying a half-reaction by 3 does not change its $E°$ value because reduction potential is an intensive property.

$$Al(s) \rightarrow Al^{3+}(1.0\ M) + 3e^{-}$$
$$3Ag^{+}(1.0\ M) + 3e^{-} \rightarrow 3Ag(s)$$

Overall: $\mathbf{Al(s) + 3Ag^{+}(1.0\ M) \rightarrow Al^{3+}(1.0\ M) + 3Ag(s)}$

We find the emf of the cell using Equation 19.1.

$$E^{\circ}_{cell} = E^{\circ}_{cathode} - E^{\circ}_{anode} = E^{\circ}_{Ag^{+}/Ag} - E^{\circ}_{Al^{3+}/Al}$$

$$E^{\circ}_{cell} = 0.80\ V - (-1.66\ V) = \mathbf{2.46\ V}$$

**Check:** The positive value of $E°$ shows that the forward reaction is favored.

19.13 The half–reaction for oxidation is:

$$2H_2O(l) \xrightarrow{\text{oxidation (anode)}} O_2(g) + 4H^{+}(aq) + 4e^{-} \qquad E^{\circ}_{anode} = +1.23\ V$$

The species that can oxidize water to molecular oxygen must have an $E^{\circ}_{red}$ more positive than +1.23 V. From Table 19.1 of the text we see that only $\mathbf{Cl_2(g)}$ and $\mathbf{MnO_4^{-}(aq)}$ in acid solution can oxidize water to oxygen.

19.15 **Strategy:** In each case, we can calculate the standard cell emf from the potentials for the two half-reactions.

$$E^{\circ}_{cell} = E^{\circ}_{cathode} - E^{\circ}_{anode}$$

**Solution:** a. $E° = -0.40\ V - (-2.87\ V) = \mathbf{2.47\ V}$. The reaction is **spontaneous.**

b. $E° = -0.14\ V - 1.07\ V = \mathbf{-1.21\ V}$. The reaction is **not spontaneous.**

c. $E° = -0.25\ V - 0.80\ V = \mathbf{-1.05\ V}$. The reaction is **not spontaneous.**

d. $E° = 0.77\ V - 0.15\ V = \mathbf{0.62\ V}$. The reaction is **spontaneous.**

19.17 **Strategy:** The greater the tendency for the substance to be oxidized, the stronger its tendency to act as a reducing agent. The species that has a stronger tendency to be oxidized will have a smaller reduction potential.

**Solution:** In each pair, look for the one with the smaller reduction potential. This indicates a greater tendency for the substance to be oxidized.

a. From Table 19.1, we have the following reduction potentials:

Na: −2.71 V
Li: −3.05 V

Li has the smaller (more negative) reduction potential and is therefore more easily oxidized. The more easily oxidized substance is the better reducing agent. **Li** is the better reducing agent.

Following the same logic for parts (b) through (d) gives

b. $H_2$

c. $Fe^{2+}$

d. $Br^-$

19.21 **Strategy:** The relationship between the equilibrium constant, *K*, and the standard emf is given by Equation 19.5 of the text: $E^\circ_{cell} = \frac{0.0592\text{ V}}{n}\log K$. Thus, knowing $E^\circ_{cell}$ and *n* (the moles of electrons transferred) we can the determine equilibrium constant. We find the standard reduction potentials in Table 19.1 of the text.

$$E^\circ_{cell} = E^\circ_{cathode} - E^\circ_{anode} = -0.76\text{ V} - (-2.37\text{ V}) = 1.61\text{ V}$$

**Solution:** We must rearrange Equation 19.5 to solve for *K*:

$$E^\circ_{cell} = \frac{0.0592\text{ V}}{n}\log K$$

$$\log K = \frac{nE^\circ_{cell}}{0.0592\text{ V}}$$

$$K = 10^{nE^\circ_{cell}/0.0592\text{ V}}$$

We see in the reaction that Mg goes to $Mg^{2+}$ and $Zn^{2+}$ goes to Zn. Therefore, two moles of electrons are transferred during the redox reaction; i.e., $n = 2$.

$$K = 10^{(2)(1.61\text{ V})/0.0592\text{ V}}$$

$$\boldsymbol{K} = 10^{54.4} = \mathbf{3 \times 10^{54}}$$

19.23 In each case we use standard reduction potentials from Table 19.1 together with Equation 19.5 of the text.

a. We break the equation into two half–reactions:

$$Br_2(l) + 2e^- \xrightarrow{\text{reduction (cathode)}} 2Br^-(aq) \qquad E^\circ_{\text{cathode}} = 1.07\text{ V}$$

$$2I^-(aq) \xrightarrow{\text{oxidation (anode)}} I_2(s) + 2e^- \qquad E^\circ_{\text{anode}} = 0.53\text{ V}$$

The standard emf is

$$E^\circ_{\text{cell}} = E^\circ_{\text{cathode}} - E^\circ_{\text{anode}} = 1.07\text{ V} - 0.53\text{ V} = 0.54\text{ V}$$

Next, we can calculate $K$ using Equation 19.5 of the text.

$$K = 10^{nE^\circ_{\text{cell}}/0.0592\text{ V}}$$

$$K = 10^{(2)(0.54\text{ V})/0.0592\text{ V}}$$

$$\boldsymbol{K} = 10^{18.2} = \mathbf{2 \times 10^{18}}$$

b. We break the equation into two half–reactions:

$$2Ce^{4+}(aq) + 2e^- \xrightarrow{\text{reduction (cathode)}} 2Ce^{3+}(aq) \qquad E^\circ_{\text{cathode}} = 1.61\text{ V}$$

$$2Cl^-(aq) \xrightarrow{\text{oxidation (anode)}} Cl_2(g) + 2e^- \qquad E^\circ_{\text{anode}} = 1.36\text{ V}$$

The standard emf is

$$E^\circ_{\text{cell}} = E^\circ_{\text{cathode}} - E^\circ_{\text{anode}} = 1.61\text{ V} - 1.36\text{ V} = 0.25\text{ V}$$

$$K = 10^{(2)(0.25\text{ V})/0.0592\text{ V}}$$

$$\boldsymbol{K} = 10^{8.44} = \mathbf{3 \times 10^{8}}$$

c. We break the equation into two half–reactions:

$$MnO_4^-(aq) + 8H^+(aq) + 5e^- \xrightarrow{\text{reduction (cathode)}} Mn^{2+}(aq) + 4H_2O(l) \qquad E^\circ_{\text{cathode}} = 1.51\text{ V}$$

$$5Fe^{2+}(aq) \xrightarrow{\text{oxidation (anode)}} 5Fe^{3+}(aq) + 5e^- \qquad E^\circ_{\text{anode}} = 0.77\text{ V}$$

The standard emf is

$$E^\circ_{\text{cell}} = E^\circ_{\text{cathode}} - E^\circ_{\text{anode}} = 1.51\text{ V} - 0.77\text{ V} = 0.74\text{ V}$$

$$K = 10^{(5)(0.74\text{ V})/0.0592\text{ V}}$$

$$\boldsymbol{K} = 10^{62.5} = \mathbf{3 \times 10^{62}}$$

19.25 **Strategy:** The spontaneous reaction that occurs must include one reduction and one oxidation. We examine the reduction potentials of the species present to determine which half reaction will occur as the reduction and which will occur as the oxidation. The relationship between the standard free energy change and the standard emf of the cell is given by Equation 19.3 of the text: $\Delta G^\circ = -nFE^\circ_{\text{cell}}$. The relationship between the equilibrium constant, *K*, and the standard emf is given by Equation 19.5 of the text: $E^\circ_{\text{cell}} = \frac{0.0592\text{ V}}{n}\log K$. Thus, once we determine $E^\circ_{\text{cell}}$, we can calculate $\Delta G^\circ$ and *K*.

**Solution:** The half-reactions (both written as reductions) are:

$$Fe^{3+}(aq) + e^- \rightarrow Fe^{2+}(aq) \qquad E^\circ_{\text{anode}} = 0.77\text{ V}$$

$$Ce^{4+}(aq) + e^- \rightarrow Ce^{3+}(aq) \qquad E^\circ_{\text{cathode}} = 1.61\text{ V}$$

Because it has the larger reduction potential, $Ce^{4+}$ is the more easily reduced and will oxidize $Fe^{2+}$ to $Fe^{3+}$. This makes the $Fe^{2+}/Fe^{3+}$ half-reaction the anode. The spontaneous reaction is

$$\mathbf{Ce^{4+}(aq) + Fe^{2+}(aq) \rightarrow Ce^{3+}(aq) + Fe^{3+}(aq)}$$

The standard cell emf is found using Equation 19.1 of the text.

$$E^\circ_{\text{cell}} = E^\circ_{\text{cathode}} - E^\circ_{\text{anode}} = 1.61\text{ V} - 0.77\text{ V} = 0.84\text{ V}$$

The values of $\Delta G^\circ$ and $K_c$ are found using Equations 19.3 and 19.5 of the text.

$$\boldsymbol{\Delta G^\circ} = -nFE^\circ_{\text{cell}} = -(1)(96500\text{ J/V}\cdot\text{mol})(0.84\text{ V}) = \mathbf{-81\text{ kJ/mol}}$$

$$\log K = \frac{nE^\circ_{\text{cell}}}{0.0592\text{ V}}$$

$$K = 10^{nE^\circ_{\text{cell}}/0.0592\text{ V}}$$

$$\boldsymbol{K} = 10^{(1)(0.84\text{ V})/0.0592\text{ V}} = \mathbf{2 \times 10^{14}}$$

**Check:** The negative value of $\Delta G^\circ$ and the large positive value of $K$, both indicate that the reaction favors products at equilibrium. The result is consistent with the fact that $E^\circ$ for the galvanic cell is positive.

19.29 **Strategy:** The standard emf ($E^\circ$) can be calculated using the standard reduction potentials in Table 19.1 of the text. Because the reactions are not run under standard-state conditions (concentrations are not 1 *M*), we need the Nernst equation (Equation 19.7) of the text to calculate the emf ($E$) of a hypothetical galvanic cell. Remember that solids do not appear in the reaction quotient ($Q$) term in the Nernst equation. We can calculate $\Delta G$ from $E$ using Equation 19.2 of the text: $\Delta G = -nFE_{\text{cell}}$.

**Solution:** The half-cell reactions are:

$$Cu^{2+}(aq) + 2e^- \rightarrow Cu(s)$$
$$Zn^{2+}(aq) + 2e^- \rightarrow Zn(s)$$

$$E = E^\circ - \frac{0.0592\text{ V}}{n}\log Q$$

$$E = 1.10\text{ V} - \frac{0.0592\text{ V}}{2}\log\frac{[Zn^{2+}]}{[Cu^{2+}]}$$

$$\boldsymbol{E} = 1.10\text{ V} - \frac{0.0592\text{ V}}{2}\log\frac{0.25}{0.15} = \mathbf{1.09\text{ V}}$$

19.31 The overall reaction is: $Zn(s) + 2H^+(aq) \rightarrow Zn^{2+}(aq) + H_2(g)$

$$\boldsymbol{E^\circ_{\text{cell}}} = E^\circ_{\text{cathode}} - E^\circ_{\text{anode}} = 0.00\text{ V} - (-0.76\text{ V}) = \mathbf{0.76\text{ V}}$$

$$E = E^\circ - \frac{0.0592\text{ V}}{n}\log\frac{[Zn^{2+}]P_{H_2}}{[H^+]^2}$$

$$\boldsymbol{E} = 0.76\text{ V} - \frac{0.0592\text{ V}}{2}\log\frac{(0.45)(2.0)}{(1.8)^2} = \mathbf{0.78\text{ V}}$$

19.33 As written, the reaction is not spontaneous under standard state conditions; the cell emf is negative.

$$E^\circ_{\text{cell}} = E^\circ_{\text{cathode}} - E^\circ_{\text{anode}} = -0.76\text{ V} - 0.34\text{ V} = -1.10\text{ V}$$

The reaction will become spontaneous when the concentrations of zinc(II) and copper(II) ions are such as to make the emf positive. The turning point is when the emf is zero. We solve the Nernst equation (Equation 19.6) for the $[Cu^{2+}]/[Zn^{2+}]$ ratio at this point.

$$E_{cell} = E^\circ - \frac{RT}{nF}\ln Q$$

At 25°C:

$$0 = -1.10\ \text{V} - \frac{(8.314\ \text{J/mol}\cdot\text{K})(298\ \text{K})}{(2\ \text{mol}\ e^-)(96{,}500\ \text{J/V}\cdot\text{mol}\ e^-)}\ln\frac{[Cu^{2+}]}{[Zn^{2+}]}$$

$$\ln\frac{[Cu^{2+}]}{[Zn^{2+}]} = -85.7$$

$$\frac{[Cu^{2+}]}{[Zn^{2+}]} = e^{-85.7} = \mathbf{6.0 \times 10^{-38}}$$

In other words for the reaction to be spontaneous, the $[Cu^{2+}]/[Zn^{2+}]$ ratio must be less than $6.0 \times 10^{-38}$.

19.39 We can calculate the standard free energy change, $\Delta G^\circ$, from the standard free energies of formation, $\Delta G_f^\circ$ using Equation 18.12 of the text. Then, we can calculate the standard cell emf, $E_{cell}^\circ$, from $\Delta G^\circ$.

The overall reaction is:

$$C_3H_8(g) + 5O_2(g) \longrightarrow 3CO_2(g) + 4H_2O(l)$$

$$\Delta G_{rxn}^\circ = 3\Delta G_f^\circ(CO_2(g)) + 4\Delta G_f^\circ(H_2O(l)) - \left[\Delta G_f^\circ(C_3H_8(g)) + 5\Delta G_f^\circ(O_2(g))\right]$$

$$\Delta G_{rxn}^\circ = (3)(-394.4\ \text{kJ/mol}) + (4)(-237.2\ \text{kJ/mol}) - [(1)(-23.5\ \text{kJ/mol}) + (5)(0)] = -2108.5\ \text{kJ/mol}$$

We can now calculate the standard emf using the following equation:

$$\Delta G^\circ = -nFE_{cell}^\circ$$

or

$$E_{cell}^\circ = \frac{-\Delta G^\circ}{nF}$$

Check the half-reactions of the text to determine that 20 moles of electrons are transferred during this redox reaction.

$$E^\circ_{\text{cell}} = \frac{-(-2108.5 \times 10^3 \text{ J/mol})}{(20)(96500 \text{ J/V} \cdot \text{mol})} = \mathbf{1.09\ V}$$

19.43 **Strategy:** A faraday is a mole of electrons. Knowing how many moles of electrons are needed to reduce a mole of magnesium ions, we can determine how many moles of magnesium will be produced. The half-reaction shows that two moles of $e^-$ are required per mole of Mg.

$$Mg^{2+}(aq) + 2e^- \longrightarrow Mg(s)$$

**Solution:**

$$\textbf{Mass Mg} = 1.00\ F \times \frac{1 \text{ mol Mg}}{2 \text{ mol } e^-} \times \frac{24.31 \text{ g Mg}}{1 \text{ mol Mg}} = \mathbf{12.2\ g\ Mg}$$

19.45 The half-reactions are:

$$Na^+ + e^- \rightarrow Na$$
$$Al^{3+} + 3e^- \rightarrow Al$$

As long as we are comparing equal masses, we can use any mass that is convenient. In this case, we will use 1 g.

$$1 \text{ g Na} \times \frac{1 \text{ mol}}{22.99 \text{ g Na}} \times 1\ e^- = 0.043 \text{ mol } e^-$$

$$1 \text{ g Al} \times \frac{1 \text{ mol}}{26.98 \text{ g Al}} \times 3\ e^- = 0.11 \text{ mol } e^-$$

It is cheaper to prepare 1 ton of sodium by electrolysis.

19.47 Find the amount of oxygen using the ideal gas equation

$$n = \frac{PV}{RT} = \frac{\left(755 \text{ mmHg} \times \frac{1 \text{ atm}}{760 \text{ mmHg}}\right)(0.076 \text{ L})}{(0.0821 \text{ L} \cdot \text{atm/K} \cdot \text{mol})(298 \text{ K})} = 3.1 \times 10^{-3} \text{ mol } O_2$$

Since the half-reaction shows that one mole of oxygen requires four faradays of electric charge, we write

$$(3.1 \times 10^{-3} \text{ mol } O_2) \times \frac{4\ F}{1 \text{ mol } O_2} = \mathbf{0.012}\ \boldsymbol{F}$$

19.49 The half-reactions are:

$$Cu^{2+}(aq) + 2e^- \rightarrow Cu(s)$$

$$2Br^-(aq) \rightarrow Br_2(l) + 2e^-$$

The mass of copper produced is:

$$4.50\text{ A} \times 1\text{ h} \times \frac{3600\text{ s}}{1\text{ h}} \times \frac{1\text{ C}}{1\text{ A}\cdot\text{s}} \times \frac{1\text{ mol } e^-}{96500\text{ C}} \times \frac{1\text{ mol Cu}}{2\text{ mol } e^-} \times \frac{63.55\text{ g Cu}}{1\text{ mol Cu}} = \mathbf{5.33\text{ g Cu}}$$

The mass of bromine produced is:

$$4.50\text{ A} \times 1\text{ h} \times \frac{3600\text{ s}}{1\text{ h}} \times \frac{1\text{ C}}{1\text{ A}\cdot\text{s}} \times \frac{1\text{ mol } e^-}{96500\text{ C}} \times \frac{1\text{ mol Br}_2}{2\text{ mol } e^-} \times \frac{159.8\text{ g Br}_2}{1\text{ mol Br}_2} = \mathbf{13.4\text{ g Br}_2}$$

19.51 The half-reaction is: $Co^{2+} + 2e^- \rightarrow Co$

The half-reaction tells us that 2 moles of electrons are needed to reduce 1 mol of $Co^{2+}$ to Co metal. We can set up the following strategy to calculate the quantity of electricity (in C) needed to deposit 2.35 g of Co.

$$2.35\text{ g Co} \times \frac{1\text{ mol Co}}{58.93\text{ g Co}} \times \frac{2\text{ mol } e^-}{1\text{ mol Co}} \times \frac{96500\text{ C}}{1\text{ mol } e^-} = \mathbf{7.70 \times 10^3\text{ C}}$$

19.53 The half-reaction for the oxidation of chloride ion is:

$$2Cl^-(aq) \rightarrow Cl_2(g) + 2e^-$$

First, let's calculate the moles of $e^-$ flowing through the cell in one hour.

$$1500\text{ A} \times \frac{1\text{ C}}{1\text{ A}\cdot\text{s}} \times \frac{3600\text{ s}}{1\text{ h}} \times \frac{1\text{ mol } e^-}{96500\text{ C}} = 55.96\text{ mol e}^-$$

Next, let's calculate the hourly production rate of chlorine gas (in kg). Note that the anode efficiency is 93.0%.

$$55.96\text{ mol } e^- \times \frac{1\text{ mol Cl}_2}{2\text{ mol } e^-} \times \frac{0.07090\text{ kg Cl}_2}{1\text{ mol Cl}_2} \times \frac{93.0\%}{100\%} = \mathbf{1.84\text{ kg Cl}_2\text{/h}}$$

19.55 The quantity of charge passing through the solution is:

$$0.750\text{ A} \times \frac{1\text{ C}}{1\text{ A}\cdot\text{s}} \times \frac{60\text{ s}}{1\text{ min}} \times \frac{1\text{ mol } e^-}{96500\text{ C}} \times 25.0\text{ min} = 1.166 \times 10^{-2}\text{ mol } e^-$$

Since the charge of the copper ion is +2, the number of moles of copper formed must be:

$$(1.166 \times 10^{-2}\text{ mol } e^-) \times \frac{1\text{ mol Cu}}{2\text{ mol } e^-} = 5.83 \times 10^{-3}\text{ mol Cu}$$

The units of molar mass are grams per mole. The molar mass of copper is:

$$\frac{0.369\text{ g}}{5.83 \times 10^{-3}\text{ mol}} = \mathbf{63.3\text{ g/mol}}$$

19.57 The number of faradays supplied is:

$$1.44\text{ g Ag}\times\frac{1\text{ mol Ag}}{107.9\text{ g Ag}}\times\frac{1\text{ mol }e^-}{1\text{ mol Ag}}=0.01335\text{ mol }e^-$$

Since we need three faradays to reduce one mole of $X^{3+}$, the molar mass of X must be:

$$\frac{0.120\text{ g X}}{0.01335\text{ mol }e^-}\times\frac{3\text{ mol }e^-}{1\text{ mol X}}=\mathbf{27.0\ g/mol}$$

19.63 a. The half-reactions are:

$$\mathbf{H_2(\mathit{g})\ \rightarrow\ 2H^+(\mathit{aq})+2\mathit{e}^-}$$
$$\mathbf{Ni^{2+}(\mathit{aq})+2\mathit{e}^-\ \rightarrow\ Ni(\mathit{s})}$$

The complete balanced equation is:

$$\mathbf{H_2(\mathit{g})+Ni^{2+}(\mathit{aq})\rightarrow\ 2H^+(\mathit{aq})+Ni(\mathit{s})}$$

Ni(*s*) is below and to the right of $H^+$(*aq*) in Table 19.1 of the text (see the half-reactions at –0.25 and 0.00 V). Therefore, the spontaneous reaction is the reverse of the above reaction; therefore, **the reaction will proceed to the left.**

b. The half-reactions are:

$$\mathbf{5\mathit{e}^-+8H^+(\mathit{aq})+MnO_4^-(\mathit{aq})\rightarrow\ Mn^{2+}(\mathit{aq})+4H_2O}$$
$$\mathbf{2Cl^-(\mathit{aq})\ \rightarrow\ Cl_2(\mathit{g})+2\mathit{e}^-}$$

The complete balanced equation is:

$$\mathbf{16H^+(\mathit{aq})+2MnO_4^-(\mathit{aq})+10Cl^-(\mathit{aq})\ \rightarrow\ 2Mn^{2+}(\mathit{aq})+8H_2O+5Cl_2(\mathit{g})}$$

In Table 19.1 of the text, $Cl^-$(*aq*) is below and to the right of $MnO_4^-$(*aq*); therefore the spontaneous reaction is as written. **The reaction will proceed to the right**.

c. The half-reactions are:

$$\mathbf{Cr(\mathit{s})\ \rightarrow\ Cr^{3+}(\mathit{aq})+3\mathit{e}^-}$$
$$\mathbf{Zn^{2+}(\mathit{aq})+2\mathit{e}^-\ \rightarrow\ Zn(\mathit{s})}$$

The complete balanced equation is:

$$\mathbf{2Cr(\mathit{s})+3Zn^{2+}(\mathit{aq})\ \rightarrow\ 2Cr^{3+}(\mathit{aq})+3Zn(\mathit{s})}$$

In Table 19.1 of the text, Zn(*s*) is below and to the right of $Cr^{3+}$(*aq*); therefore the spontaneous reaction is the reverse of the reaction as written. **The reaction will proceed to the left**.

19.65 The reduction potential of a half-cell is temperature-dependent. **A** small **non-zero emf will appear if the temperatures of the two half-cells are different.**

19.67 a. The balanced equation is:

$$\mathbf{2MnO_4^-+6H^++5H_2O_2\rightarrow\ 2Mn^{2+}+8H_2O+5O_2}$$

b. The number of moles of potassium permanganate in 36.44 mL of the solution is

$$36.44 \text{ mL} \times \frac{0.01652 \text{ mol}}{1000 \text{ mL soln}} = 6.020 \times 10^{-4} \text{ mol of } KMnO_4$$

From the balanced equation it can be seen that in this particular reaction 2 moles of permanganate is stoichiometrically equivalent to 5 moles of hydrogen peroxide. The number of moles of $H_2O_2$ oxidized is therefore

$$(6.020 \times 10^{-4} \text{ mol } MnO_4^-) \times \frac{5 \text{ mol } H_2O_2}{2 \text{ mol } MnO_4^-} = 1.505 \times 10^{-3} \text{ mol } H_2O_2$$

The molar concentration of $H_2O_2$ is:

$$\mathbf{[H_2O_2]} = \frac{1.505 \times 10^{-3} \text{ mol}}{25.0 \times 10^{-3} \text{ L}} = 0.0602 \text{ mol/L} = \mathbf{0.0602}\ \boldsymbol{M}$$

19.69 It might appear that because the sum of the first two half-reactions gives Equation (3), $E_3^\circ$ is given by $E_1^\circ + E_2^\circ = 0.33$ V. This is not the case, however, because emf is not an extensive property. We cannot set $E_3^\circ = E_1^\circ + E_2^\circ$. On the other hand, the Gibbs energy is an extensive property, so we can add the separate Gibbs energy changes to obtain the overall Gibbs energy change.

$$\Delta G_3^\circ = \Delta G_1^\circ + \Delta G_2^\circ$$

Substituting the relationship $\Delta G^\circ = -nFE^\circ$, we obtain

$$n_3FE_3^\circ = n_1FE_1^\circ + n_2FE_2^\circ$$

$$E_3^\circ = \frac{n_1E_1^\circ + n_2E_2^\circ}{n_3}$$

$n_1 = 2$, $n_2 = 1$, and $n_3 = 3$.

$$\boldsymbol{E_3^\circ} = \frac{(2)(-0.44 \text{ V}) + (1)(0.77 \text{ V})}{3} = \mathbf{-0.037\ V}$$

19.71 The solubility equilibrium of AgBr is: $AgBr(s) \rightleftarrows Ag^+(aq) + Br^-(aq)$

By reversing the first given half-reaction and adding it to the first, we obtain:

$Ag(s) \rightarrow Ag^+(aq) + e^-$ $\quad E^\circ_{anode} = 0.80$ V

$AgBr(s) + e^- \rightarrow Ag(s) + Br^-(aq)$ $\quad E^\circ_{cathode} = 0.07$ V

---

$AgBr(s) \rightleftarrows Ag^+(aq) + Br^-(aq)$ $\quad E^\circ_{cell} = E^\circ_{cathode} - E^\circ_{anode} = 0.07\text{ V} - 0.80\text{ V} = -0.73\text{ V}$

At equilibrium, we have:

$$E = E^\circ - \frac{0.0592\text{ V}}{n}\log[\text{Ag}^+][\text{Br}^-]$$

$$0 = -0.73\text{ V} - \frac{0.0592\text{ V}}{1}\log K_{sp}$$

$$\log K_{sp} = -12.33$$

$$\boldsymbol{K_{sp}} = 10^{-12.33} = \mathbf{5 \times 10^{-13}}$$

(Note that this value differs from that given in Table 17.4 of the text, since the data quoted here were obtained from a student's lab report.)

19.73 a. If this were a standard cell, the concentrations would all be 1.00 *M*, and the voltage would just be the standard emf calculated from Table 19.1 of the text. Since cell emf's depend on the concentrations of the reactants and products, we must use the Nernst equation (Equation 19.7 of the text) to find the emf of a nonstandard cell.

$$E = E^\circ - \frac{0.0592\text{ V}}{n}\log Q$$

$$E = 3.17\text{ V} - \frac{0.0592\text{ V}}{2}\log\frac{[\text{Mg}^{2+}]}{[\text{Ag}^+]^2}$$

$$\boldsymbol{E} = 3.17\text{ V} - \frac{0.0592\text{ V}}{2}\log\frac{0.10}{(0.10)^2} = \mathbf{3.14\text{ V}}$$

b. First we calculate the concentration of silver ion remaining in solution after the deposition of 1.20 g of silver metal

Ag originally in solution: $\frac{0.100\text{ mol Ag}^+}{1\text{ L}} \times 0.346\text{ L} = 3.46 \times 10^{-2}\text{ mol Ag}^+$

Ag deposited: $1.20\text{ g Ag} \times \frac{1\text{ mol}}{107.9\text{ g}} = 1.11 \times 10^{-2}\text{ mol Ag}$

Ag remaining in solution: $(3.46 \times 10^{-2}\text{ mol Ag}) - (1.11 \times 10^{-2}\text{ mol Ag}) = 2.35 \times 10^{-2}\text{ mol Ag}$

$$[\text{Ag}^+] = \frac{2.35 \times 10^{-2}\text{ mol}}{0.346\text{ L}} = 6.79 \times 10^{-2}\ M$$

The overall reaction is: $\text{Mg}(s) + 2\text{Ag}^+(aq) \rightarrow \text{Mg}^{2+}(aq) + 2\text{Ag}(s)$

We use the balanced equation to find the amount of magnesium metal suffering oxidation and dissolving.

$$(1.11 \times 10^{-2} \text{ mol Ag}) \times \frac{1 \text{ mol Mg}}{2 \text{ mol Ag}} = 5.55 \times 10^{-3} \text{ mol Mg}$$

The amount of magnesium originally in solution was

$$0.288 \text{ L} \times \frac{0.100 \text{ mol}}{1 \text{ L}} = 2.88 \times 10^{-2} \text{ mol}$$

The new magnesium ion concentration is:

$$\frac{[(5.55 \times 10^{-3}) + (2.88 \times 10^{-2})] \text{mol}}{0.288 \text{ L}} = 0.119\ M$$

The new cell emf is:

$$E = E^{\circ} - \frac{0.0592 \text{ V}}{n} \log Q$$

$$\boldsymbol{E} = 3.17 \text{ V} - \frac{0.0592 \text{ V}}{2} \log \frac{0.119}{(6.79 \times 10^{-2})^2} = \mathbf{3.13\ V}$$

19.75 The cell voltage is given by:

$$E_{\text{cell}} = E^{\circ} - \frac{0.0592 \text{ V}}{2} \log \frac{[Cu^{2+}]_{\text{dilute}}}{[Cu^{2+}]_{\text{concentrated}}}$$

$$\boldsymbol{E_{\text{cell}}} = 0 \text{ V} - \frac{0.0592 \text{ V}}{2} \log \frac{0.080}{1.2} = \mathbf{0.035\ V}$$

19.77 Since this is a concentration cell, the standard emf is zero. (Why?) Using Equation 19.6, we can write equations to calculate the cell voltage for the two cells.

$$(1) \quad E_{\text{cell}} = -\frac{RT}{nF} \ln Q = -\frac{RT}{2F} \ln \frac{[Hg_2^{2+}]\text{soln A}}{[Hg_2^{2+}]\text{soln B}}$$

$$(2) \quad E_{\text{cell}} = -\frac{RT}{nF} \ln Q = -\frac{RT}{1F} \ln \frac{[Hg^{+}]\text{soln A}}{[Hg^{+}]\text{soln B}}$$

In the first case, two electrons are transferred per mercury ion ($n = 2$), while in the second only one is transferred ($n = 1$). Note that the concentration ratio will be 1:10 in both cases. The voltages calculated at 18°C are:

$$(1)\qquad E_{\text{cell}} = \frac{-(8.314\ \text{J/K}\cdot\text{mol})(291\ \text{K})}{2(96500\ \text{J}\cdot\text{V}^{-1}\text{mol}^{-1})}\ln 10^{-1} = 0.0289\ \text{V}$$

$$(2)\qquad E_{\text{cell}} = \frac{-(8.314\ \text{J/K}\cdot\text{mol})(291\ \text{K})}{1(96500\ \text{J}\cdot\text{V}^{-1}\text{mol}^{-1})}\ln 10^{-1} = 0.0577\ \text{V}$$

Since the calculated cell potential for cell (1) agrees with the measured cell emf, we conclude that the **mercury(I) is $Hg_2^{2+}$**.

19.79 We begin by treating this like an ordinary stoichiometry problem (see Chapter 3).

*Step 1:* Calculate the number of moles of Mg and $Ag^+$.

The number of moles of magnesium is:

$$1.56\ \text{g Mg} \times \frac{1\ \text{mol Mg}}{24.31\ \text{g Mg}} = 0.0642\ \text{mol Mg}$$

The number of moles of silver ion in the solution is:

$$\frac{0.100\ \text{mol Ag}^+}{1\ \text{L}} \times 0.1000\ \text{L} = 0.0100\ \text{mol Ag}^+$$

*Step 2:* Calculate the mass of Mg remaining by determining how much Mg reacts with $Ag^+$.

The balanced equation for the reaction is:

$$2\text{Ag}^+(aq) + \text{Mg}(s) \longrightarrow 2\text{Ag}(s) + \text{Mg}^{2+}(aq)$$

Since you need twice as much $Ag^+$ compared to Mg for complete reaction, $Ag^+$ is the limiting reagent. The amount of Mg consumed is:

$$0.0100\ \text{mol Ag}^+ \times \frac{1\ \text{mol Mg}}{2\ \text{mol Ag}^+} = 0.00500\ \text{mol Mg}$$

The amount of magnesium remaining is:

$$(0.0642 - 0.00500)\ \text{mol Mg} \times \frac{24.31\ \text{g Mg}}{1\ \text{mol Mg}} = \mathbf{1.44\ g\ Mg}$$

*Step 3:* Assuming complete reaction, calculate the concentration of $Mg^{2+}$ ions produced.

Since the mole ratio between Mg and $Mg^{2+}$ is 1:1, the mol of $Mg^{2+}$ formed will equal the mol of Mg reacted. The concentration of $Mg^{2+}$ is:

$$[Mg^{2+}]_0 = \frac{0.00500 \text{ mol}}{0.100 \text{ L}} = 0.0500\ M$$

*Step 4:* We can calculate the equilibrium constant for the reaction from the standard cell emf.

$$E^\circ_{\text{cell}} = E^\circ_{\text{cathode}} - E^\circ_{\text{anode}} = 0.80 \text{ V} - (-2.37 \text{ V}) = 3.17 \text{ V}$$

We can then compute the equilibrium constant.

$$K = e^{\frac{nE^\circ_{\text{cell}}}{0.0257}}$$

$$K = e^{\frac{(2)(3.17)}{0.0257}} = 1\times 10^{107}$$

*Step 5:* To find equilibrium concentrations of $Mg^{2+}$ and $Ag^+$, we have to solve an equilibrium problem.

Let $x$ be the small amount of $Mg^{2+}$ that reacts to achieve equilibrium. The concentration of $Ag^+$ will be $2x$ at equilibrium. Assume that essentially all $Ag^+$ has been reduced so that the initial concentration of $Ag^+$ is zero.

| | $2Ag^+(aq)$ | + | $Mg(s)$ | ⇄ | $2Ag(s)$ | + | $Mg^{2+}(aq)$ |
|---|---|---|---|---|---|---|---|
| Initial ($M$): | 0.0000 | | | | | | 0.0500 |
| Change ($M$): | $+2x$ | | | | | | $-x$ |
| Equilibrium ($M$): | $2x$ | | | | | | $(0.0500 - x)$ |

$$K = \frac{[Mg^{2+}]}{[Ag^+]^2}$$

$$1\times 10^{107} = \frac{(0.0500 - x)}{(2x)^2}$$

We can assume $0.0500 - x \approx 0.0500$.

$$1\times 10^{107} \approx \frac{0.0500}{(2x)^2}$$

$$(2x)^2 = \frac{0.0500}{1\times 10^{107}} = 0.0500\times 10^{-107}$$

$$(2x)^2 = 5.00\times 10^{-109} = 50.0\times 10^{-110}$$

$$2x = 7\times 10^{-55}\ M$$

$$[\mathbf{Ag^+}] = \mathbf{2}x = \mathbf{7 \times 10^{-55}}\ M$$

$$[\mathbf{Mg^{2+}}] = 0.0500 - x = \mathbf{0.0500}\ M$$

19.81 a. Since this is an acidic solution, the gas must be hydrogen gas from the reduction of hydrogen ion. The two electrode reactions and the overall cell reaction are:

$$\begin{array}{ll} \text{anode:} & Cu(s) \longrightarrow Cu^{2+}(aq) + 2e^- \\ \text{cathode:} & 2H^+(aq) + 2e^- \longrightarrow H_2(g) \\ \hline & Cu(s) + 2H^+(aq) \longrightarrow Cu^{2+}(aq) + H_2(g) \end{array}$$

Since 0.584 g of copper was consumed, the amount of hydrogen gas produced is:

$$0.584\text{ g Cu} \times \frac{1\text{ mol Cu}}{63.55\text{ g Cu}} \times \frac{1\text{ mol }H_2}{1\text{ mol Cu}} = 9.20 \times 10^{-3}\text{ mol }H_2$$

At STP, 1 mole of an ideal gas occupies a volume of 22.41 L. Thus, the volume of hydrogen gas at STP is:

$$V_{H_2} = (9.20 \times 10^{-3}\text{ mol }H_2) \times \frac{22.41\text{ L}}{1\text{ mol}} = \mathbf{0.206\ L}$$

b. From the current and the time, we can calculate the amount of charge:

$$1.18\text{ A} \times \frac{1\text{ C}}{1\text{ A}\cdot\text{s}} \times (1.52 \times 10^3\text{ s}) = 1.79 \times 10^3\text{ C}$$

Since we know the charge of an electron, we can compute the number of electrons.

$$(1.79 \times 10^3\text{ C}) \times \frac{1\ e^-}{1.6022 \times 10^{-19}\text{ C}} = 1.12 \times 10^{22}\ e^-$$

Using the amount of copper consumed in the reaction and the fact that 2 mol of $e^-$ are produced for every 1 mole of copper consumed, we can calculate Avogadro's number.

$$\frac{1.12 \times 10^{22}\ e^-}{9.20 \times 10^{-3}\text{ mol Cu}} \times \frac{1\text{ mol Cu}}{2\text{ mol }e^-} = \mathbf{6.09 \times 10^{23}\ e^-/mol\ e^-}$$

In practice, Avogadro's number can be determined by electrochemical experiments like this. The charge of the electron can be found independently by Millikan's experiment.

19.83 a. We can calculate $\Delta G°$ from standard free energies of formation.

$$\Delta G° = 2\Delta G_f^\circ(N_2) + 6\Delta G_f^\circ(H_2O) - [4\Delta G_f^\circ(NH_3) + 3\Delta G_f^\circ(O_2)]$$

$$\Delta G = 0 + (6)(-237.2\ \text{kJ/mol}) - [(4)(-16.6\ \text{kJ/mol}) + 0]$$

$$\mathbf{\Delta G = -1356.8\ kJ/mol}$$

b. The half-reactions are:

$$4NH_3(g) \longrightarrow 2N_2(g) + 12H^+(aq) + 12e^-$$

$$3O_2(g) + 12H^+(aq) + 12e^- \longrightarrow 6H_2O(l)$$

The overall reaction is a 12-electron process. We can calculate the standard cell emf from the standard free energy change, $\Delta G°$.

$$\Delta G° = -nFE_{\text{cell}}^\circ$$

$$E_{\text{cell}}^\circ = \frac{-\Delta G°}{nF} = \frac{-\left(\dfrac{-1356.8\ \text{kJ}}{1\ \text{mol}} \times \dfrac{1000\ \text{J}}{1\ \text{kJ}}\right)}{(12)(96500\ \text{J/V}\cdot\text{mol})} = \mathbf{1.17\ V}$$

19.85 The reduction of $Ag^+$ to Ag metal is:

$$Ag^+(aq) + e^- \longrightarrow Ag$$

We can calculate both the moles of Ag deposited and the moles of Au deposited.

$$?\ \text{mol Ag} = 2.64\ \text{g Ag} \times \frac{1\ \text{mol Ag}}{107.9\ \text{g Ag}} = 2.45 \times 10^{-2}\ \text{mol Ag}$$

$$?\ \text{mol Au} = 1.61\ \text{g Au} \times \frac{1\ \text{mol Au}}{197.0\ \text{g Au}} = 8.17 \times 10^{-3}\ \text{mol Au}$$

We do not know the oxidation state of Au ions, so we will represent the ions as $Au^{n+}$. If we divide the mol of Ag by the mol of Au, we can determine the ratio of $Ag^+$ reduced compared to $Au^{n+}$ reduced.

$$\frac{2.45 \times 10^{-2}\ \text{mol Ag}}{8.17 \times 10^{-3}\ \text{mol Au}} = 3$$

That is, the same number of electrons that reduced the $Ag^+$ ions to Ag reduced only one-third the number of moles of the $Au^{n+}$ ions to Au. Thus, each $Au^{n+}$ required three electrons per ion for every one electron for

$Ag^+$. The oxidation state for the gold ion is **+3**; the ion is $Au^{3+}$.

$$Au^{3+}(aq) + 3e^- \longrightarrow Au$$

19.87 We reverse the first half–reaction and add it to the second to come up with the overall balanced equation

$$Hg_2^{2+} \longrightarrow 2Hg^{2+} + 2e^- \qquad E^\circ_{anode} = +0.92\ V$$

$$Hg_2^{2+} + 2e^- \longrightarrow 2Hg \qquad E^\circ_{cathode} = +0.85\ V$$

$$2Hg_2^{2+} \longrightarrow 2Hg^{2+} + 2Hg \qquad E^\circ_{cell} = 0.85\ V - 0.92\ V = -0.07\ V$$

Since the standard cell potential is an intensive property,

$$Hg_2^{2+}(aq) \longrightarrow Hg^{2+}(aq) + Hg(l) \qquad E^\circ_{cell} = -0.07\ V$$

We calculate $\Delta G°$ from $E°$.

$$\boldsymbol{\Delta G°} = -nFE° = -(1)(96500\ J/V{\cdot}mol)(-0.07\ V) = \mathbf{6.8\ kJ/mol}$$

The corresponding equilibrium constant is:

$$K = \frac{[Hg^{2+}]}{[Hg_2^{2+}]}$$

We calculate $K$ from $\Delta G°$.

$$\Delta G° = -RT\ln K$$

$$\ln K = \frac{-6.8 \times 10^3\ J/mol}{(8.314\ J/K \cdot mol)(298\ K)}$$

$$\boldsymbol{K} = \mathbf{0.064}$$

19.89 The reactions for the electrolysis of NaCl($aq$) are:

Anode: $2Cl^-(aq) \longrightarrow Cl_2(g) + 2e^-$

Cathode: $2H_2O(l) + 2e^- \longrightarrow H_2(g) + 2OH^-(aq)$

Overall: $2H_2O(l) + 2Cl^-(aq) \longrightarrow H_2(g) + Cl_2(g) + 2OH^-(aq)$

From the pH of the solution, we can calculate the $OH^-$ concentration. From the $[OH^-]$, we can calculate the moles of $OH^-$ produced. Then, from the moles of $OH^-$ we can calculate the average current used.

$$pH = 12.24$$

$$\text{pOH} = 14.00 - 12.24 = 1.76$$

$$[\text{OH}^-] = 1.74 \times 10^{-2}\ M$$

The moles of $OH^-$ produced are:

$$\frac{1.74 \times 10^{-2}\ \text{mol}}{1\ \text{L}} \times 0.300\ \text{L} = 5.22 \times 10^{-3}\ \text{mol OH}^-$$

From the balanced equation, it takes 1 mole of $e^-$ to produce 1 mole of $OH^-$ ions.

$$(5.22 \times 10^{-3}\ \text{mol OH}^-) \times \frac{1\ \text{mol}\ e^-}{1\ \text{mol OH}^-} \times \frac{96500\ \text{C}}{1\ \text{mol}\ e^-} = 504\ \text{C}$$

Recall that 1 C = 1 A·s

$$504\ \text{C} \times \frac{1\ \text{A} \cdot \text{s}}{1\ \text{C}} \times \frac{1\ \text{min}}{60\ \text{s}} \times \frac{1}{6.00\ \text{min}} = \mathbf{1.4\ A}$$

19.91 The reaction is:

$$\text{Pt}^{n+} + ne^- \longrightarrow \text{Pt}$$

Thus, we can calculate the charge of the platinum ions by realizing that $n$ mol of $e^-$ are required per mol of Pt formed.

The moles of Pt formed are:

$$9.09\ \text{g Pt} \times \frac{1\ \text{mol Pt}}{195.1\ \text{g Pt}} = 0.0466\ \text{mol Pt}$$

Next, calculate the charge passed in C.

$$\text{C} = 2.00\ \text{h} \times \frac{3600\ \text{s}}{1\ \text{h}} \times \frac{2.50\ \text{C}}{1\ \text{s}} = 1.80 \times 10^4\ \text{C}$$

Convert to moles of electrons.

$$?\ \text{mol e}^- = (1.80 \times 10^4\ \text{C}) \times \frac{1\ \text{mol}\ e^-}{96500\ \text{C}} = 0.187\ \text{mol}\ e^-$$

We now know the number of moles of electrons (0.187 mol $e^-$) needed to produce 0.0466 mol of Pt metal. We can calculate the number of moles of electrons needed to produce 1 mole of Pt metal.

$$\frac{0.187 \text{ mol } e^-}{0.0466 \text{ mol Pt}} = 4.01 \text{ mol } e^-/\text{mol Pt}$$

Since we need 4 moles of electrons to reduce 1 mole of Pt ions, the charge on the Pt ions must be **+4**.

19.93 From Table 19.1 of the text.

$$H_2O_2(aq) + 2H^+(aq) + 2e^- \rightarrow 2H_2O(l) \qquad E^\circ_{\text{cathode}} = 1.77 \text{ V}$$
$$H_2O_2(aq) \rightarrow O_2(g) + 2H^+(aq) + 2e^- \qquad E^\circ_{\text{anode}} = 0.68 \text{ V}$$

$$2H_2O_2(aq) \rightarrow 2H_2O(l) + O_2(g) \qquad E^\circ_{\text{cell}} = E^\circ_{\text{cathode}} - E^\circ_{\text{anode}} = 1.77 \text{ V} - (0.68 \text{ V}) = 1.09 \text{ V}$$

Because $E^\circ$ is positive, the decomposition is **spontaneous**.

19.95 Cells of higher voltage require very reactive oxidizing and reducing agents, which are difficult to handle. (From Table 19.1 of the text, we see that 5.92 V is the theoretical limit of a cell made up of $Li^+/Li$ and $F_2/F^-$ electrodes under standard-state conditions.) Batteries made up of several cells in series are easier to use.

19.97 The half-reactions are:

$$Zn(s) + 4OH^-(aq) \rightarrow Zn(OH)_4^{2-}(aq) + 2e^- \qquad E^\circ_{\text{anode}} = -1.36 \text{ V}$$
$$Zn^{2+}(aq) + 2e^- \rightarrow Zn(s) \qquad E^\circ_{\text{cathode}} = -0.76 \text{ V}$$

$$Zn^{2+}(aq) + 4OH^-(aq) \rightarrow Zn(OH)_4^{2-}(aq) \qquad E^\circ_{\text{cell}} = -0.76 \text{ V} - (-1.36 \text{ V}) = 0.60 \text{ V}$$

$$E^\circ_{\text{cell}} = -\frac{0.0592 \text{ V}}{n} \log K_f$$

$$\boldsymbol{K_f} = 10^{nE^\circ/0.0592 \text{ V}} = 10^{(2)(0.60 \text{ V})/0.0592 \text{ V}} = \mathbf{2 \times 10^{20}}$$

19.99 a. Since electrons flow from X to SHE, **$E^\circ_{\text{red}}$ for X is negative**. Thus **$E^\circ_{\text{red}}$ for Y is positive**.

b.

$$Y^{2+} + 2e^- \rightarrow Y \qquad E^\circ_{\text{cathode}} = 0.34 \text{ V}$$
$$X \rightarrow X^{2+} + 2e^- \qquad E^\circ_{\text{anode}} = -0.25 \text{ V}$$

$$X + Y^{2+} \rightarrow X^{2+} + Y \qquad \boldsymbol{E^\circ_{\text{cell}}} = 0.34 \text{ V} - (-0.25 \text{ V}) = \mathbf{0.59 \text{ V}}$$

19.101 a. Gold does not tarnish in air because the reduction potential for oxygen is not sufficiently positive to result in the oxidation of gold.

$$O_2 + 4H^+ + 4e^- \rightarrow 2H_2O \qquad E^\circ_{\text{cathode}} = 1.23 \text{ V}$$

That is, $E^\circ_{\text{cell}} = E^\circ_{\text{cathode}} - E^\circ_{\text{anode}} < 0$, for either oxidation by $O_2$ to $Au^+$ or $Au^{3+}$.

$$E^\circ_{\text{cell}} = 1.23\text{ V} - 1.50\text{ V} < 0$$

or

$$E^\circ_{\text{cell}} = 1.23\text{ V} - 1.69\text{ V} < 0$$

b.

| | |
|---|---|
| $3(Au^+ + e^- \rightarrow Au)$ | $E^\circ_{\text{cathode}} = 1.69\text{ V}$ |
| $Au \rightarrow Au^{3+} + 3e^-$ | $E^\circ_{\text{anode}} = 1.50\text{ V}$ |
| $3Au^+ \rightarrow 2Au + Au^{3+}$ | $E^\circ_{\text{cell}} = 1.69\text{ V} - 1.50\text{ V} = 0.19\text{ V}$ |

Calculating $\Delta G$,

$$\Delta G^\circ = -nFE^\circ = -(3)(96{,}500\text{ J/V·mol})(0.19\text{ V}) = -55.0\text{ kJ/mol}$$

For spontaneous electrochemical equations, $\Delta G^\circ$ must be negative. **Yes**, the disproportionation occurs spontaneously.

c. Since the most stable oxidation state for gold is $Au^{3+}$, the predicted reaction is:

$$\mathbf{2Au + 3F_2 \rightarrow 2AuF_3}$$

19.103 We can calculate $\Delta G^\circ_{\text{rxn}}$ using the following equation (Equation 18.12).

$$\Delta G^\circ_{\text{rxn}} = \Sigma n\Delta G^\circ_f(\text{products}) - \Sigma m\Delta G^\circ_f(\text{reactants})$$

$$\Delta G^\circ_{\text{rxn}} = 0 + 0 - [(1)(-293.8\text{ kJ/mol}) + 0] = 293.8\text{ kJ/mol}$$

Next, we can calculate $E^\circ$ using the equation

$$\Delta G^\circ = -nFE^\circ$$

We use a more accurate value for Faraday's constant.

$$293.8 \times 10^3\text{ J/mol} = -(1)(96485.338\text{ J/V·mol})E^\circ$$

$$\mathbf{E^\circ = -3.05\text{ V}}$$

19.105 First, we need to calculate $E^\circ_{\text{cell}}$, then we can calculate $K$ from the cell potential.

| | |
|---|---|
| $H_2(g) \rightarrow 2H^+(aq) + 2e^-$ | $E^\circ_{\text{anode}} = 0.00\text{ V}$ |
| $2H_2O(l) + 2e^- \rightarrow H_2(g) + 2OH^-$ | $E^\circ_{\text{cathode}} = -0.83\text{ V}$ |
| $2H_2O(l) \rightarrow 2H^+(aq) + 2OH^-(aq)$ | $E^\circ_{\text{cell}} = -0.83\text{ V} - 0.00\text{ V} = -0.83\text{ V}$ |

We want to calculate $K$ for the reaction: $H_2O(l) \rightarrow H^+(aq) + OH^-(aq)$. The cell potential for this reaction will be the same as the above reaction, but the moles of electrons transferred, $n$, will equal one.

$$E^\circ_{\text{cell}} = \frac{0.0592\text{ V}}{n}\log K_w$$

$$\log K_w = \frac{nE^\circ_{\text{cell}}}{0.0592\text{ V}}$$

$$K_w = 10^{nE^\circ_{\text{cell}}/0.0592\text{ V}}$$

$$\boldsymbol{K_w = 10^{(1)(-0.83\text{ V})/0.0592\text{ V}} = 1 \times 10^{-14}}$$

19.107 a. unchanged b. unchanged c. squared d. doubled e. doubled

19.109

$$F_2\,(g) + 2H^+\,(aq) + 2e^- \rightarrow 2HF\,(g)$$

$$E = E^\circ - \frac{RT}{2F}\ln\frac{P_{HF}^2}{P_{F_2}[H^+]^2}$$

With increasing $[H^+]$, $E$ will be larger. As $[H^+]$ increases, $F_2$ does become a **stronger oxidizing agent**.

19.111

| | |
|---|---|
| $Pb \rightarrow Pb^{2+} + 2e^-$ | $E^\circ_{\text{anode}} = -0.13\text{ V}$ |
| $2H^+ + 2e^- \rightarrow H_2$ | $E^\circ_{\text{cathode}} = 0.00\text{ V}$ |
| $Pb + 2H^+ \rightarrow Pb^{2+} + H_2$ | $E^\circ_{\text{cell}} = 0.00\text{ V} - (-0.13\text{ V}) = 0.13\text{ V}$ |

$$\text{pH} = 1.60$$

$$[H^+] = 10^{-1.60} = 0.025\,M$$

$$E = E^\circ - \frac{RT}{nF}\ln\frac{[Pb^{2+}]P_{H_2}}{[H^+]^2}$$

$$0 = 0.13 - \frac{0.0592\text{ V}}{2}\log\frac{(0.035)P_{H_2}}{(0.025)^2}$$

$$4.39 = \log \frac{(0.035)P_{H_2}}{(0.025)^2}$$

$$P_{H_2} = \mathbf{4.4 \times 10^2\ atm}$$

19.113 a. $Au(s) + 3HNO_3(aq) + 4HCl(aq) \rightarrow HAuCl_4(aq) + 3H_2O(l) + 3NO_2(g)$

b. The function of HCl is to increase the acidity and to form the stable complex ion, $AuCl_4^-$.

19.115 The surface area of an open cylinder is $2\pi rh$. The surface area of the culvert is

$$2\pi(0.900\text{ m})(40.0\text{ m}) \times 2 \text{ (for both sides of the iron sheet)} = 452\text{ m}^2$$

Converting to units of cm$^2$,

$$452\text{ m}^2 \times \left(\frac{100\text{ cm}}{1\text{ m}}\right)^2 = 4.52 \times 10^6\text{ cm}^2$$

The volume of the Zn layer is

$$0.200\text{ mm} \times \frac{1\text{ cm}}{10\text{ mm}} \times (4.52 \times 10^6\text{ cm}^2) = 9.04 \times 10^4\text{ cm}^3$$

The mass of Zn needed is

$$(9.04 \times 10^4\text{ cm}^3) \times \frac{7.14\text{ g}}{1\text{ cm}^3} = 6.45 \times 10^5\text{ g Zn}$$

$$Zn^{2+} + 2e^- \rightarrow Zn$$

$$Q = (6.45 \times 10^5\text{ g Zn}) \times \frac{1\text{ mol Zn}}{65.41\text{ g Zn}} \times \frac{2\text{ mol } e^-}{1\text{ mol Zn}} \times \frac{96500\text{ C}}{1\text{ mol } e^-} = 1.90 \times 10^9\text{ C}$$

$$1\text{ J} = 1\text{ C} \times 1\text{ V}$$

$$\text{Total energy} = \frac{(1.90 \times 10^9\text{ C})(3.26\text{ V})}{0.95 \leftarrow \text{(efficiency)}} = 6.52 \times 10^9\text{ J}$$

$$\text{Cost} = (6.52 \times 10^9\text{ J}) \times \frac{1\text{ kW}}{1000\ \frac{\text{J}}{\text{s}}} \times \frac{1\text{ h}}{3600\text{ s}} \times \frac{\$0.12}{1\text{ kWh}} = \mathbf{\$217}$$

19.117 a. $$1\text{A}\cdot\text{h} = 1\text{A} \times 3600\text{s} = \mathbf{3600\ C}$$

b. Anode: $Pb + SO_4^{2-} \rightarrow PbSO_4 + 2e^-$

Two moles of electrons are produced by 1 mole of Pb. Recall that the charge of 1 mol $e^-$ is 96,500 C. We can set up the following conversions to calculate the capacity of the battery.

$$\text{mol Pb} \rightarrow \text{mol } e^- \rightarrow \text{coulombs} \rightarrow \text{ampere hour}$$

$$406\text{ g} \times \frac{1\text{ mol Pb}}{207.2\text{ g Pb}} \times \frac{2\text{ mol } e^-}{1\text{ mol Pb}} \times \frac{96500\text{ C}}{1\text{ mol } e^-} = (3.78 \times 10^5\text{ C}) \times \frac{1\text{ h}}{3600\text{ s}} = \mathbf{105\ A\cdot h}$$

This ampere·hour cannot be fully realized because the concentration of $H_2SO_4$ keeps decreasing.

c. $$\boldsymbol{E^\circ_{\textbf{cell}}} = 1.70\text{ V} - (-0.31\text{ V}) = \mathbf{2.01\ V} \qquad \text{(From Table 19.1 of the text)}$$

$$\Delta G^\circ = -nFE^\circ$$

$$\boldsymbol{\Delta G^\circ} = -(2)(96500\text{ J/V}\cdot\text{mol})(2.01\text{ V}) = \mathbf{-3.88 \times 10^5\ J/mol = -388\ kJ/mol}$$

19.119 a. The half–reactions are:

(i) $MnO_4^-(aq) + 8H^+(aq) + 5e^- \longrightarrow Mn^{2+}(aq) + 4H_2O(l)$

(ii) $C_2O_4^{2-}(aq) \longrightarrow 2CO_2(g) + 2e^-$

We combine the half-reactions to cancel electrons, that is, [2 × equation (i)] + [5 × equation (ii)]

$$\mathbf{2MnO_4^-(\mathit{aq}) + 16H^+(\mathit{aq}) + 5C_2O_4^{2-}(\mathit{aq}) \longrightarrow 2Mn^{2+}(\mathit{aq}) + 10CO_2(\mathit{g}) + 8H_2O(\mathit{l})}$$

b. We can calculate the moles of $KMnO_4$ from the molarity and volume of solution.

$$24.0\text{ mL KMnO}_4 \times \frac{0.0100\text{ mol KMnO}_4}{1000\text{ mL soln}} = 2.40 \times 10^{-4}\text{ mol KMnO}_4$$

We can calculate the mass of oxalic acid from the stoichiometry of the balanced equation. The mole ratio between oxalate ion and permanganate ion is 5:2.

$$(2.40 \times 10^{-4}\text{ mol KMnO}_4) \times \frac{5\text{ mol H}_2\text{C}_2\text{O}_4}{2\text{ mol KMnO}_4} \times \frac{90.04\text{ g H}_2\text{C}_2\text{O}_4}{1\text{ mol H}_2\text{C}_2\text{O}_4} = 0.0540\text{ g H}_2\text{C}_2\text{O}_4$$

Finally, the percent by mass of oxalic acid in the sample is:

$$\textbf{\% oxalic acid} = \frac{0.0540\text{ g}}{1.00\text{ g}} \times 100\% = \textbf{5.40\%}$$

19.121 The balanced equation is:

$$2MnO_4^- + 5C_2O_4^{2-} + 16H^+ \longrightarrow 2Mn^{2+} + 10CO_2 + 8H_2O$$

Therefore, 2 mol $MnO_4^-$ reacts with 5 mol $C_2O_4^{2-}$

$$\text{Moles of MnO reacted} = 24.2\text{ mL} \times \frac{9.56\times10^{-4}\text{ mol MnO}_4^-}{1000\text{ mL soln}} = 2.314\times10^{-5}\text{ mol MnO}_4^-$$

Recognize that the mole ratio of $Ca^{2+}$ to $C_2O_4^{2-}$ is 1:1 in $CaC_2O_4$. The mass of $Ca^{2+}$ in 10.0 mL is:

$$(2.314\times10^{-5}\text{ mol MnO}_4^-) \times \frac{5\text{ mol Ca}^{2+}}{2\text{ mol MnO}_4^-} \times \frac{40.08\text{ g Ca}^{2+}}{1\text{ mol Ca}^{2+}} = 2.32\times10^{-3}\text{ g Ca}^{2+}$$

Finally, converting to mg/mL, we have:

$$\frac{2.32\times10^{-3}\text{ g Ca}^{2+}}{10.0\text{ mL}} \times \frac{1000\text{ mg}}{1\text{ g}} = \textbf{0.232 mg Ca}^{2+}\textbf{/mL blood}$$

19.123 **Strategy:** According to Section 18.7 of the text, 31 kJ of free energy are required to convert 1 mole ADP to ATP:

$$\text{ADP} + H_3PO_4 \rightarrow \text{ATP} + H_2O \quad \Delta G° = 31\text{ kJ/mol}$$

To determine the number of moles of ATP that can be produced, calculate the amount of free energy from the oxidation of 1 mole of nitrite and divide this energy by 31 kJ/mol.

**Solution:** Assume that the nitrite, nitrate, ADP, and ATP are present in their standard states. The overall reaction for converting the nitrite to nitrate is

$$2NO_2^-(aq) + O_2(g) \rightarrow 2NO_3^-(aq).$$

Use Equation 19.3 to relate $\Delta E°$ to $\Delta G°$:

$$\Delta G° = -nF\Delta E° = -(4\text{ mol e}^-)(96500\text{ J}\cdot\text{V/mol e}^-)(1.23\text{ V} - 0.42\text{ V})$$

$$\Delta G° = -310\text{ kJ}$$

Since the oxidation of 2 moles of nitrite provide 310 kJ of free energy, then

$$\text{yield} = \left(\frac{310\text{ kJ}}{2\text{ mol NO}_2^-(aq)}\right)\left(\frac{1\text{ mol ATP}}{31\text{ kJ}}\right) = \mathbf{5\ mol\ ATP / mol\ NO_2^-}.$$

19.125 The balanced equation is:

$$5SO_2(g) + 2MnO_4^-(aq) + 2H_2O(l) \rightarrow 5SO_4^{2-}(aq) + 2Mn^{2+}(aq) + 4H^+(aq)$$

The mass of $SO_2$ in the water sample is given by

$$7.37\text{ mL} \times \frac{0.00800\text{ mol}}{1000\text{ mL soln}} \times \frac{5\text{ mol SO}_2}{2\text{ mol KMnO}_4} \times \frac{64.07\text{ g SO}_2}{1\text{ mol SO}_2} = \mathbf{0.00944\ g\ SO_2}$$

19.127 The half–reaction for the oxidation of water to oxygen is:

$$2H_2O(l) \xrightarrow{\text{oxidation (anode)}} O_2(g) + 4H^+(aq) + 4e^-$$

Knowing that one mole of any gas at STP occupies a volume of 22.41 L, we find the number of moles of oxygen.

$$4.26\text{ L O}_2 \times \frac{1\text{ mol}}{22.41\text{ L}} = 0.190\text{ mol O}_2$$

Since four electrons are required to form one oxygen molecule, the number of electrons must be:

$$0.190\text{ mol O}_2 \times \frac{4\text{ mol e}^-}{1\text{ mol O}_2} \times \frac{6.022\times10^{23}\ e^-}{1\text{ mol}} = 4.58\times10^{23}\ e^-$$

The amount of charge passing through the solution is:

$$6.00\text{ A} \times \frac{1\text{ C}}{1\text{ A}\cdot\text{s}} \times \frac{3600\text{ s}}{1\text{ h}} \times 3.40\text{ h} = 7.34\times10^4\text{ C}$$

We find the electron charge by dividing the amount of charge by the number of electrons.

$$\frac{7.34\times10^4\text{ C}}{4.58\times10^{23}\ e^-} = \mathbf{1.60\times10^{-19}\ C/e^-}$$

In actual fact, this sort of calculation can be used to find Avogadro's number, not the electron charge. The latter can be measured independently, and one can use this charge together with electrolytic data like the above to calculate the number of objects in one mole. See also Problem 19.81.

19.129 The balanced equation is: $5Fe^{2+} + MnO_4^- + 8H^+ \longrightarrow Mn^{2+} + 5Fe^{3+} + 4\ H_2O$

Calculate the amount of iron(II) in the original solution using the mole ratio from the balanced equation.

$$23.0 \text{ mL} \times \frac{0.0200 \text{ mol KMnO}_4}{1000 \text{ mL soln}} \times \frac{5 \text{ mol Fe}^{2+}}{1 \text{ mol KMnO}_4} = 0.00230 \text{ mol Fe}^{2+}$$

The concentration of iron(II) must be:

$$\mathbf{[Fe^{2+}]} = \frac{0.00230 \text{ mol}}{0.0250 \text{ L}} = \mathbf{0.0920}\ \boldsymbol{M}$$

The total iron concentration can be found by simple proportion because the same sample volume (25.0 mL) and the same $KMnO_4$ solution were used.

$$[\text{Fe}]_{\text{total}} = \frac{40.0 \text{ mL KMnO}_4}{23.0 \text{ mL KMnO}_4} \times 0.0920\ M = 0.160\ M$$

$$\mathbf{[Fe^{3+}]} = [\text{Fe}]_{\text{total}} - [\text{Fe}^{2+}] = \mathbf{0.0680}\ \boldsymbol{M}$$

# Chapter 20

# Nuclear Chemistry

20.5 **Strategy:** In balancing nuclear equations, note that the sum of atomic numbers and that of mass numbers must match on both sides of the equation.

**Solution:**

a. On the left side of this equation the atomic number sum is 13 (12 + 1) and the mass number sum is 27 (26 + 1). These sums must be the same on the right side. Remember that the atomic and mass numbers of an alpha particle are 2 and 4, respectively. The atomic number of X is therefore 11 (13 – 2) and the mass number is 23 (27 – 4). X is sodium–23 ($^{23}_{11}\textbf{Na}$).

b. On the left side of this equation the atomic number sum is 28 (27 + 1) and the mass number sum is 61 (59 + 2). These sums must be the same on the right side. The atomic number of X is therefore 1 (28 – 27) and the mass number is also 1 (61 – 60). X is a proton ($^{1}_{1}\textbf{p}$ or $^{1}_{1}\textbf{H}$).

c. On the left side of this equation the atomic number sum is 92 (92 + 0) and the mass number sum is 236 (235 + 1). These sums must be the same on the right side. The atomic number of X is therefore 0 $\left[\dfrac{92-(36+56)}{3}\right]$ and the mass number is 1 $\left[\dfrac{236-(94+139)}{3}\right]$. X is a neutron ($^{1}_{0}\textbf{n}$).

d. On the left side of this equation the atomic number sum is 26 (24 + 2) and the mass number sum is 57 (53 + 4). These sums must be the same on the right side. The atomic number of X is therefore 26 (26 – 0) and the mass number is 56 (57 – 1). X is iron–56 ($^{56}_{26}\textbf{Fe}$).

e. On the left side of this equation the atomic number sum is 8 and the mass number sum is 20. The sums must be the same on the right side. The atomic number of X is therefore –1 (8 – 9) and the mass number is 0 (20 – 20). X is a β particle ($^{0}_{-1}\boldsymbol{\beta}$).

20.13 **Strategy:** We first convert the mass in amu to grams. Then, assuming the nucleus to be spherical, we calculate its volume. Dividing mass by volume gives density.

**Solution:** The mass is:

$$235\ \text{amu} \times \frac{1\ \text{g}}{6.022 \times 10^{23}\ \text{amu}} = 3.902 \times 10^{-22}\ \text{g}$$

The volume is, $V = 4/3\pi r^3$.

$$V = \frac{4}{3}\pi\left(\left(7.0 \times 10^{-3}\ \text{pm}\right) \times \frac{1\ \text{cm}}{1 \times 10^{10}\ \text{pm}}\right)^3 = 1.437 \times 10^{-36}\ \text{cm}^3$$

The density is:

$$\frac{3.902 \times 10^{-22}\ \text{g}}{1.437 \times 10^{-36}\ \text{cm}^3} = \mathbf{2.72 \times 10^{14}\ g/cm^3}$$

20.15 Nickel (**Ni**), selenium (**Se**), and cadmium (**Cd**) have more stable isotopes. All three have even atomic numbers (see Table 20.2 of the text).

20.17 **Strategy:** We can use Equation 20.1, $\Delta E = (\Delta m)c^2$, to solve the problem. Recall the following conversion factor:

$$1\ \text{J} = \frac{1\ \text{kg}\cdot\text{m}^2}{\text{s}^2}$$

**Solution:** The mass change is:

$$\Delta m = \frac{\Delta E}{c^2} = \frac{-436400\ \text{J/mol}}{(3.00 \times 10^8\ \text{m/s})^2} = \mathbf{-4.85 \times 10^{-12}\ kg/mol\ H_2}$$

**Think About It:** Is this mass measurable with ordinary laboratory analytical balances?

20.19 **Strategy:** To calculate the nuclear binding energy, we first determine the difference between the mass of the nucleus and the mass of all the protons and neutrons, which gives us the mass defect. Next, we apply Einstein's mass-energy relationship [$\Delta E = (\Delta m)c^2$], (Equation 20.1); and use the procedure shown in Sample Problem 20.2.

**Solution:** a. There are 4 neutrons and 3 protons in a Li–7 nucleus. The predicted mass is:

$$(3)(\text{mass of proton}) + (4)(\text{mass of neutron}) = (3)(1.007825\ \text{amu}) + (4)(1.008665\ \text{amu})$$

$$\text{predicted mass} = 7.058135\ \text{amu}$$

The mass defect, that is the difference between the predicted mass and the measured mass is:

$$\Delta m = 7.01600\ \text{amu} - 7.058135\ \text{amu} = -0.042135\ \text{amu}$$

The mass that is converted in energy, that is the energy released is:

$$\Delta E = (\Delta m)c^2 = \left(-0.042135 \text{ amu} \times \frac{1 \text{ kg}}{6.022 \times 10^{26} \text{ amu}}\right)(3.00 \times 10^8 \text{ m/s})^2$$

$$= -6.30 \times 10^{-12} \text{ kg·m}^2\text{/s}^2 = -6.30 \times 10^{-12} \text{ J}$$

Note that we report the binding energy without referring to its sign. Therefore, the nuclear binding energy is $\mathbf{6.30 \times 10^{-12}}$ **J**. When comparing the stability of any two nuclei we must account for the fact that they have different numbers of nucleons. For this reason, it is more meaningful to use the *nuclear binding energy per nucleon*, defined as

$$\text{nuclear binding energy per nucleon} = \frac{\text{nuclear binding energy}}{\text{number of nucleons}}$$

The binding energy per nucleon for $^7$Li is:

$$\frac{6.30 \times 10^{-12} \text{ J}}{7 \text{ nucleons}} = \mathbf{9.00 \times 10^{-13} \text{ J / nucleon}}$$

b. Using the same procedure as in (a), using 1.007825 amu for $^1_1\text{H}$ and 1.008665 amu for $^1_0\text{n}$, we can show that for chlorine–35:

Nuclear binding energy = $\mathbf{4.78 \times 10^{-11}}$ **J**

Nuclear binding energy per nucleon = $\mathbf{1.37 \times 10^{-12}}$ **J/nucleon**

20.21 **Strategy:** The mass of one $^{48}$Cr atom is the sum of the masses of its constituents (24 protons, 24 neutrons, and 24 electrons) plus the mass defect. We can use the given nuclear binding energy and Equation 20.1 to compute the mass defect. Remember that 1 J = 1 kg·m/s$^2$.

**Solution:** Calculate the mass defect:

$$\Delta E = \Delta mc^2$$

$$(-1.37340 \times 10^{-12} \text{ (kg} \cdot \text{m/s}^2\text{)/nucleon})(48 \text{ nucleons}) = (3.00 \times 10^8 \text{ m/s})^2 \Delta m$$

$$\Delta m = -7.32 \times 10^{-28} \text{ kg}$$

Calculate the mass of the $^{48}$Cr atom (ignore the mass of the electrons):

**mass $^{48}$Cr atom** = (24 protons)( 1.007825 amu/proton) + (24 neutrons)(1.008665 amu) $- 7.32 \times 10^{-28}$ kg

$$= 48.39576 \text{ amu} - 7.32 \times 10^{-28} \text{ kg}$$

$$= (48.39576 \text{ amu})(1 \text{ kg}/6.022 \times 10^{26} \text{ amu}) - 7.32 \times 10^{-28} \text{ kg}$$

$$= 8.036 \times 10^{-26} \text{ kg} - 7.32 \times 10^{-28} \text{ kg}$$

$$= \mathbf{7.963 \times 10^{-26}\ kg}$$

20.25 **Strategy:** Alpha emission decreases the atomic number by two and the mass number by four. Beta emission increases the atomic number by one and has no effect on the mass number.

**Solution:** a.

$$^{232}_{90}\text{Th} \xrightarrow{\alpha} {}^{228}_{88}\text{Ra} \xrightarrow{\beta} {}^{228}_{89}\text{Ac} \xrightarrow{\beta} {}^{228}_{90}\text{Th}$$

b.

$$^{235}_{92}\text{U} \xrightarrow{\alpha} {}^{231}_{90}\text{Th} \xrightarrow{\beta} {}^{231}_{91}\text{Pa} \xrightarrow{\alpha} {}^{227}_{89}\text{Ac}$$

c.

$$^{237}_{93}\text{Np} \xrightarrow{\alpha} {}^{233}_{91}\text{Pa} \xrightarrow{\beta} {}^{233}_{92}\text{U} \xrightarrow{\alpha} {}^{229}_{90}\text{Th}$$

20.27 The number of atoms decreases by half for each half-life. For ten half-lives we have:

$$(5.00 \times 10^{22} \text{ atoms}) \times \left(\frac{1}{2}\right)^{10} = \mathbf{4.88 \times 10^{19}\ atoms}$$

20.29 **Strategy:** We use the equation from Section 20.3 (a variation of Equation 14.3) to solve for $t$, the age of the artifact. Recall that the ratio in this equation can be expressed as the number of radioactive nuclei or as the activity (disintegrations per unit time) of the sample.

$$\ln\frac{N_t}{N_0} = -kt$$

The problem gives the half-life. We must first solve for $k$ using the equation

$$t_{1/2} = \frac{0.693}{k}$$

$$k = \frac{0.693}{t_{1/2}} = \frac{0.693}{5715 \text{ yr}} = 1.213 \times 10^{-4} \text{ yr}^{-1}$$

**Solution:** Rearranging the equation to solve for $t$ gives

$$t = -\frac{\left(\ln\frac{N_t}{N_0}\right)}{k} = -\frac{\left(\ln\frac{18.9\text{ dis/min}}{27.5\text{ dis/min}}\right)}{1.213\times10^{-4}\text{ yr}^{-1}} = \mathbf{3.09\times10^3\text{ yr}}$$

**Think About It:** Note that because the activity is *more* than half the original activity, we expect the age of the artifact to be *less* than a half-life.

20.31 Let's consider the decay of A first.

$$k = \frac{0.693}{t_{1/2}} = \frac{0.693}{4.50\text{ s}} = 0.154\text{ s}^{-1}$$

Let's convert $k$ to units of day$^{-1}$.

$$0.154\frac{1}{\text{s}}\times\frac{3600\text{ s}}{1\text{ h}}\times\frac{24\text{ h}}{1\text{ d}} = 1.33\times10^4\text{ d}^{-1}$$

Next, use the first-order rate equation to calculate the amount of A left after 30 days.

$$\ln\frac{N_t}{N_0} = -kt$$

Let $x$ be the amount of A left after 30 days.

$$\ln\frac{x}{1.00} = -(1.33\times10^4\text{ d}^{-1})(30\text{ d}) = -3.99\times10^5$$

$$\frac{x}{1.00} = e^{(-3.99\times10^5)}$$

$$x \approx 0$$

Thus, **no A remains.**

For B: As calculated above, all of A is converted to B in less than 30 days. In fact, essentially all of A is gone in less than 1 day! This means that at the beginning of the 30 day period, there is 1.00 mol of B present. The half life of B is 15 days, so that after two half-lives (30 days), there should be **0.25 mole of B** left.

For C: As in the case of A, the half-life of C is also very short. Therefore, at the end of the 30–day period, **no C is left**.

For D: D is not radioactive. 0.75 mol of B reacted in 30 days; therefore, due to a 1:1 mole ratio between B and D, there should be **0.75 mole of D** present after 30 days.

20.33 **Strategy:** We use the integrated rate law to solve for the activity at time $t$, given that time $t = 8.4 \times 10^3$ yr.

$$\ln\frac{N_t}{N_0} = -kt$$

**Solution:** Taking the inverse ln of both sides of the equation, we get

$$\frac{N_t}{N_0} = e^{-kt} = e^{-\left(1.213 \times 10^{-4}\ \text{yr}^{-1}\right)\left(8.4 \times 10^3\ \text{yr}\right)} = e^{-1.019} = 0.361$$

$$N_t = N_0(0.361) = (15.3\ \text{dpm})(0.361) = \mathbf{5.5\ dpm}$$

20.35 **Strategy:** We begin by using the integrated rate law and the rate constant, $k$ (determined in Problem 20.34), to determine the ratio of $^{238}$U at $t = 0$ to $^{238}$U at $t = 1.7 \times 10^8$ yr.

**Solution:**

$$\ln\frac{^{238}\text{U}_t}{^{238}\text{U}_0} = -\left(1.54 \times 10^{-10}\ \text{yr}^{-1}\right)\left(1.7 \times 10^8\ \text{yr}\right) = -0.02618$$

It simplifies the problem to put the larger mass in the numerator. Doing this simply changes the sign of the result:

$$\ln\frac{^{238}\text{U}_0}{^{238}\text{U}_t} = 0.02618$$

Taking the inverse natural log of both sides gives

$$\frac{^{238}\text{U}_0}{^{238}\text{U}_t} = e^{0.02618} = 1.027$$

The mass ratio of $^{238}$U at $t = 0$ to $^{238}$U now (at $t = 1.7 \times 10^8$ yr) is 1.027. Since an infinite number of mass combinations that would give this ratio, we must assume a value for one of the masses. It makes the mass easier if we assume that the mass of $^{238}$U now is 1.000 g. This gives a mass of 1.027 g $^{238}$U at $t = 0$. The difference between these masses is the mass of $^{238}$U that has decayed to $^{206}$Pb. However, because a nucleus loses mass as it decays, the mass of $^{206}$Pb will be less than the mass of $^{238}$U that decayed. We determine the mass of $^{206}$Pb as follows:

$$0.027\ \text{g}\ ^{238}\text{U} \times \frac{206\ \text{g}\ ^{206}\text{Pb}}{238\ \text{g}\ ^{238}\text{U}} = 0.023\ \text{g}\ ^{206}\text{Pb}$$

Finally, if there currently a mass of 1.000 g $^{238}$U and a mass of 0.023 g $^{206}$Pb, the ratio of $^{238}$U to $^{206}$Pb is

$$\frac{1.000 \text{ g } ^{238}\text{U}}{0.023 \text{ g } ^{206}\text{Pb}} = \mathbf{43:1}$$

20.39 a. **$^{14}N(\alpha,p)^{17}O$** b. **$^{9}Be(\alpha,n)^{12}C$** c. **$^{238}U(d,2n)^{238}Np$**

20.41 a. **$^{40}Ca(d,p)^{41}Ca$** b. **$^{32}S(n,p)^{32}P$** c. **$^{239}Pu(\alpha,n)^{242}Cm$**

Remember that it is unnecessary to include the subscripted atomic number. The elemental symbol and atomic mass is sufficient to identify an isotope unambiguously.

20.43 Upon bombardment with neutrons, mercury–198 is first converted to mercury–199, which then emits a proton. The reaction is:

$$^{198}_{80}\text{Hg} + ^{1}_{0}\text{n} \longrightarrow ^{199}_{80}\text{Hg} \longrightarrow ^{198}_{79}\text{Au} + ^{1}_{1}\text{p}$$

20.55 **The fact that the radioisotope appears only in the $I_2$ shows that the $IO_3^-$ is formed only from the $IO_4^-$.** Does this result rule out the possibility that $I_2$ could be formed from $IO_4^-$ as well? Can you suggest an experiment to answer the question?

20.57 **Add iron-59 to the person's diet, and allow a few days for the iron–59 isotope to be incorporated into the person's body. Isolate red blood cells from a blood sample and monitor radioactivity from the hemoglobin molecules present in the red blood cells.**

20.63 We start with the integrated first-order rate law (Equation 14.3):

$$\ln\frac{N_t}{N_0} = -kt$$

We can calculate the rate constant, *k*, from the half-life using Equation 14.5 of the text, and then substitute into Equation 14.3 to solve for the time.

$$t_{1/2} = \frac{0.693}{k}$$

$$k = \frac{0.693}{t_{1/2}} = \frac{0.693}{28.1 \text{ yr}} = 0.02466 \text{ yr}^{-1}$$

Substituting:

$$\ln\left(\frac{0.200}{1.00}\right) = -\left(0.02466 \text{ yr}^{-1}\right)t$$

$$t = \mathbf{65.3\ yr}$$

20.65 a. The balanced equation is:

$$\mathbf{{}^{3}_{1}H \rightarrow {}^{3}_{2}He + {}^{0}_{-1}\beta}$$

b. The number of tritium (T) atoms in 1.00 kg of water is:

$$(1.00\times10^{3}\text{ g }H_2O)\times\frac{1\text{ mol }H_2O}{18.02\text{ g }H_2O}\times\frac{6.022\times10^{23}\text{ molecules }H_2O}{1\text{ mol }H_2O}\times\frac{2\text{ H atoms}}{1\text{ }H_2O}\times\frac{1\text{ T atom}}{1.0\times10^{17}\text{ H atoms}}$$

$$= 6.68\times10^{8}\text{ T atoms}$$

The number of disintegrations per minute (dpm) will be:

$$\text{rate} = k\text{ (number of T atoms)} = kN = \frac{0.693}{t_{1/2}}N$$

$$\text{rate} = \left(\frac{0.693}{12.5\text{ yr}}\times\frac{1\text{ yr}}{365\text{ day}}\times\frac{1\text{ day}}{24\text{ h}}\times\frac{1\text{ h}}{60\text{ min}}\right)(6.68\times10^{8}\text{ T atoms})$$

$$\mathbf{rate} = 70.5\text{ T atoms/min} = \mathbf{70.5\ dpm}$$

20.67 a. ${}^{235}_{92}U + {}^{1}_{0}n \rightarrow {}^{140}_{56}Ba + 3{}^{1}_{0}n + \mathbf{{}^{93}_{36}Kr}$

b. ${}^{235}_{92}U + {}^{1}_{0}n \rightarrow {}^{144}_{55}Cs + {}^{90}_{37}Rb + \mathbf{2{}^{1}_{0}n}$

c. ${}^{235}_{92}U + {}^{1}_{0}n \rightarrow {}^{87}_{35}Br + \mathbf{{}^{146}_{57}La} + 3{}^{1}_{0}n$

d. ${}^{235}_{92}U + {}^{1}_{0}n \rightarrow {}^{160}_{62}Sm + {}^{72}_{30}Zn + \mathbf{4{}^{1}_{0}n}$

20.69 The balanced nuclear equations are:

a. $\mathbf{{}^{3}_{1}H \rightarrow {}^{3}_{2}He + {}^{0}_{-1}\beta}$

b. $\mathbf{{}^{242}_{94}Pu \rightarrow {}^{4}_{2}\alpha + {}^{238}_{92}U}$

c. $\mathbf{{}^{131}_{53}I \rightarrow {}^{131}_{54}Xe + {}^{0}_{-1}\beta}$

d. $\mathbf{{}^{251}_{98}Cf \rightarrow {}^{247}_{96}Cm + {}^{4}_{2}\alpha}$

20.71 a. $\mathbf{{}^{209}_{83}Bi + {}^{4}_{2}\alpha \longrightarrow {}^{211}_{85}At + 2\,{}^{1}_{0}n}$

b. $\mathbf{{}^{209}_{83}Bi(\alpha, 2n){}^{211}_{85}At}$

20.73 a. The volume of a sphere is

$$V = \frac{4}{3}\pi r^3$$

Volume is proportional to the number of nucleons. Therefore,

$$V \propto \text{A (mass number)}$$

$$r^3 \propto \text{A}$$

$$\mathbf{r = r_0A^{1/3}, \text{ where } r_0 \text{ is a proportionality constant.}}$$

b. We can calculate the volume of the $^{238}U$ nucleus by substituting the equation derived in part (a) into the equation for the volume of a sphere.

$$V = \frac{4}{3}\pi r^3 = \frac{4}{3}\pi r_0^{\,3}\text{A}$$

$$V = \frac{4}{3}\pi(1.2 \times 10^{-15}\text{ m})^3(238) = \mathbf{1.7 \times 10^{-42}\ m^3}$$

20.75 a.

$$^{238}_{92}\text{U} \rightarrow {}^{234}_{90}\text{Th} + {}^{4}_{2}\alpha$$

$$\Delta m = 234.03596 + 4.002603 - 238.05078 = -0.01222\text{ amu} \times \frac{1\text{ kg}}{6.022\times10^{26}\text{ amu}} = 2.029\times10^{-29}\text{ kg}$$

$$\Delta E = 2.029 \times 10^{-29}\text{ kg }(3.00 \times 10^8\text{ m/s})^2 = \mathbf{1.83 \times 10^{-12}\ J}$$

b. **The α particle will move away faster because it is smaller.**

20.77 One curie represents $3.70 \times 10^{10}$ disintegrations/s. The rate of decay of the isotope is given by the rate law: rate $= kN$, where $N$ is the number of atoms in the sample and $k$ is the first-order rate constant. We find the value of $k$ in units of $s^{-1}$:

$$k = \frac{0.693}{t_{1/2}} = \frac{0.693}{1.6 \times 10^3\text{ yr}} = 4.33 \times 10^{-4}\text{ yr}^{-1}$$

$$\frac{4.33 \times 10^{-4}\text{ yr}^{-1}}{1\text{ yr}} \times \frac{1\text{ yr}}{365\text{ d}} \times \frac{1\text{ d}}{24\text{ h}} \times \frac{1\text{ h}}{3600\text{ s}} = 1.37 \times 10^{-11}\text{ s}^{-1}$$

Now, we can calculate $N$, the number of Ra atoms in the sample.

$$\text{rate} = kN$$

$$3.7 \times 10^{10} \text{ disintegrations/s} = (1.37 \times 10^{-11} \text{ s}^{-1})N$$

$$N = 2.70 \times 10^{21} \text{ Ra atoms}$$

By definition, 1 curie corresponds to exactly $3.7 \times 10^{10}$ nuclear disintegrations per second which is the decay rate equivalent to that of *1 g of radium*. Thus, the mass of $2.70 \times 10^{21}$ Ra atoms is 1 g.

$$\frac{2.70 \times 10^{21} \text{ Ra atoms}}{1.0 \text{ g Ra}} \times \frac{226.03 \text{ g Ra}}{1 \text{ mol Ra}} = \mathbf{6.1 \times 10^{23} \text{ atoms/mol} = N_A}$$

20.79

$$\ln\frac{N_t}{N_0} = -kt$$

$$\ln\frac{\text{mass of old sample}}{\text{mass of fresh sample}} = -\left(1.21 \times 10^{-4} \text{ yr}^{-1}\right)\left(60000 \text{ yr}\right)$$

$$\ln\frac{1.0 \text{ g}}{x \text{ g}} = -7.26$$

$$\frac{1.0}{x} = e^{-7.26}$$

$$x = 1422$$

$$\textbf{Percent of C-14 left} = \frac{1.0}{1422} \times 100\% = \mathbf{0.070\%}$$

20.81 **The nuclear submarine can be submerged for a long period without refueling; Conventional diesel engines receive an input of oxygen. A nuclear reactor does not.**

20.83 First, let's calculate the number of disintegrations/s to which 7.4 mCi corresponds.

$$7.4 \text{ mCi} \times \frac{1 \text{ Ci}}{1000 \text{ mCi}} \times \frac{3.7 \times 10^{10} \text{ disintegrations/s}}{1 \text{ Ci}} = 2.74 \times 10^{8} \text{ disintegrations/s}$$

This is the rate of decay. We can now calculate the number of iodine-131 atoms to which this radioactivity corresponds. First, we calculate the half-life in seconds:

$$t_{\frac{1}{2}} = 8.1 \text{ d} \times \frac{24 \text{ h}}{1 \text{ d}} \times \frac{3600 \text{ h}}{1 \text{ h}} = 7.00 \times 10^{5} \text{ s}$$

$$k = \frac{0.693}{t_{1/2}} = \frac{0.693}{7.00 \times 10^{5}\ \text{s}} = 9.90 \times 10^{-7}\ \text{s}^{-1}$$

$$\text{rate} = kN$$

Therefore,

$$2.74 \times 10^{8}\ \text{disintegrations/s} = (9.90 \times 10^{-7}\ \text{s}^{-1})N$$

$$N = \mathbf{2.8 \times 10^{14}\ iodine\text{-}131\ atoms}$$

20.85 **A small scale chain reaction (fission of $^{235}U$) took place. Copper played the crucial role of reflecting neutrons from the splitting uranium-235 atoms back into the uranium sphere to trigger the chain reaction. Note that a sphere has the most appropriate geometry for such a chain reaction. In fact, during the implosion process prior to an atomic explosion, fragments of uranium-235 are pressed roughly into a sphere for the chain reaction to occur (see Section 20.5 of the text).**

20.87 In this problem, we are asked to calculate the molar mass of a radioactive isotope. Grams of sample are given in the problem, so if we can find moles of sample we can calculate the molar mass. The rate constant can be calculated from the half-life. Then, from the rate of decay and the rate constant, the number of radioactive nuclei can be calculated. The number of radioactive nuclei can be converted to moles.

First, we convert the half-life to units of minutes because the rate is given in dpm (disintegrations per minute). Then, we calculate the rate constant from the half-life.

$$(1.3 \times 10^{9}\ \text{yr}) \times \frac{365\ \text{days}}{1\ \text{yr}} \times \frac{24\ \text{h}}{1\ \text{day}} \times \frac{60\ \text{min}}{1\ \text{h}} = 6.8 \times 10^{14}\ \text{min}$$

$$k = \frac{0.693}{t_{1/2}} = \frac{0.693}{6.8 \times 10^{14}\ \text{min}} = 1.0 \times 10^{-15}\ \text{min}^{-1}$$

Next, we calculate the number of radioactive nuclei from the rate and the rate constant.

$$\text{rate} = kN$$

$$2.9 \times 10^{4}\ \text{dpm} = (1.0 \times 10^{-15}\ \text{min}^{-1})N$$

$$N = 2.9 \times 10^{19}\ \text{nuclei}$$

Convert to moles of nuclei, and then determine the molar mass.

$$(2.9 \times 10^{19}\ \text{nuclei}) \times \frac{1\ \text{mol}}{6.022 \times 10^{23}\ \text{nuclei}} = 4.8 \times 10^{-5}\ \text{mol}$$

$$\textbf{molar mass} = \frac{\text{g of substance}}{\text{mol of substance}} = \frac{0.0100\text{ g}}{4.8 \times 10^{-5}\text{ mol}} = \mathbf{2.1 \times 10^2\ g/mol}$$

20.89 **Using A for element 110, D for element 111, E for element 112, G for element 114, J for element 115, L for element 116, M for element 117, and Q for element 118:**

$$^{208}_{82}\text{Pb} + {}^{62}_{28}\text{Ni} \rightarrow {}^{270}_{110}\text{A}$$

$$^{209}_{83}\text{Bi} + {}^{64}_{28}\text{Ni} \rightarrow {}^{273}_{111}\text{D}$$

$$^{208}_{82}\text{Pb} + {}^{66}_{30}\text{Zn} \rightarrow {}^{274}_{112}\text{E}$$

$$^{244}_{94}\text{Pu} + {}^{48}_{20}\text{Ca} \rightarrow {}^{289}_{114}\text{G} + 3\,{}^{1}_{0}\text{n}$$

$$^{243}_{95}\text{Am} + {}^{48}_{20}\text{Ca} \rightarrow {}^{291}_{115}\text{J}$$

$$^{248}_{96}\text{Cm} + {}^{48}_{20}\text{Ca} \rightarrow {}^{296}_{116}\text{L}$$

$$^{249}_{97}\text{Bk} + {}^{48}_{20}\text{Ca} \rightarrow {}^{297}_{117}\text{M}$$

$$^{249}_{98}\text{Cf} + {}^{48}_{20}\text{Ca} \rightarrow {}^{297}_{118}\text{Q}$$

**A and D are transition metals. E resembles Zn, Cd, and Hg. G is in the carbon family, J is in the nitrogen family and L is in the oxygen family. M is a halide and Q is a noble gas and likely a metalloid.**

20.91 **Since the new particle's mass exceeds the sum of the masses of the electron and positron, the process violates the law of conservation of mass. But, it does not violate Einstein's more general law of mass-energy conservation, $\Delta E = \Delta mc^2$. The large mass of the new particle reflects that fact that the process is extremely endothermic.**

20.93 **Only $^3$H has a suitable half-life. The other half-lives are either too long or too short to determine the time span of 6 years accurately.**

20.95 *Step 1:* The half-life of carbon-14 is 5715 years. From the half-life, we can calculate the rate constant, $k$.

$$k = \frac{0.693}{t_{1/2}} = \frac{0.693}{5715\text{ yr}} = 1.21 \times 10^{-4}\text{ yr}^{-1}$$

*Step 2:* The age of the object can now be calculated using the following equation.

$$\ln\frac{N_t}{N_0} = -kt$$

$N$ = the number of radioactive nuclei. In the problem, we are given disintegrations per second per gram. The number of disintegrations is directly proportional to the number of radioactive nuclei. We can write,

$$\ln\frac{\text{decay rate of old sample}}{\text{decay rate of fresh sample}} = -kt$$

$$\ln\frac{0.186\ \text{dps/g C}}{0.260\ \text{dps/g C}} = -(1.21\times10^{-4}\ \text{yr}^{-1})t$$

$$\boldsymbol{t = 2.77\times10^{3}\ \text{yr}}$$

20.97 **Normally the human body concentrates iodine in the thyroid gland. The purpose of the large doses of KI is to displace radioactive iodine from the thyroid and allow its excretion from the body.**

20.99

$$\textbf{U–238, } t_{1/2} = \mathbf{4.5\times10^{9}\ yr}\ \textbf{ and Th–232, } t_{1/2} = \mathbf{1.4\times10^{10}\ yr}$$

**They are still present because of their long half-lives.**

20.101 1 Ci = $3.7\times10^{10}$ decays/s

Let $R_0$ be the activity of the injected 20.0 mCi $^{99m}$Tc.

$$R_0 = (20.0\times10^{-3}\ \text{Ci})\times\frac{3.70\times10^{10}\ \text{decays/s}}{1\ \text{Ci}} = 7.4\times10^{8}\ \text{decays/s}$$

$R_0 = kN_0$, where $N_0$ =number of $^{99m}$Tc nuclei present.

$$k = \frac{0.693}{t_{1/2}} = \frac{0.693}{6.0\ \text{h}} = 0.1155\ \text{h}^{-1}\times\frac{1\ \text{h}}{3600\ \text{s}} = 3.208\times10^{-5}\ \text{s}^{-1}$$

$$N_0 = \frac{R_0}{k} = \frac{7.4\times10^{8}\ \text{decays/s}}{3.208\times10^{-5}\ /\text{s}} = 2.307\times10^{13}\ \text{decays} = 2.307\times10^{13}\ \text{nuclei}$$

Each of the nuclei emits a photon of energy $2.29\times10^{-14}$ J. The total energy absorbed by the patient is

$$\boldsymbol{E} = \frac{2}{3}(2.307\times10^{13}\ \text{nuclei})\times\left(\frac{2.29\times10^{-14}\ \text{J}}{1\ \text{nuclei}}\right) = \mathbf{0.352\ J}$$

The rad is:

$$\frac{0.352\ \text{J}/10^{-2}\ \text{J}}{70} = \mathbf{0.503\ rad}$$

Given that RBE = 0.98, the rem is:

$$(0.503)(0.98) = \mathbf{0.49\ rem}$$

20.103 **Strategy:** We are asked to find the volume of helium formed at STP from the radioactive decay of radium-226. The volume is easily calculated using the ideal gas law:

$$V = n_{\text{He}}RT/P$$

The overall decay process produces 5 mol He for each mole of $^{226}$Ra that decays:

$$\text{mol He} = 5(\text{mol }^{226}\text{Ra decayed}) = 5(\text{initial mol }^{226}\text{Ra} - \text{final mol }^{226}\text{Ra})$$

The initial amount of $^{226}$Ra can be found easily from the initial mass (1.00 g) and the molar mass (226 g/mol) of radium-226. But, calculating the final amount of $^{226}$Ra does not appear to be straightforward since the decay sequence from radium-226 to lead-206 involves several intermediate radioisotopes, each with its own half-life. Fortunately, though, the half-life of the first step in the decay sequence (radium-226 → radon-222, half-life 1600 years) is significantly longer than any other half-life in the sequence, so we can approximate the overall decay process using first-order kinetics and the rate constant of the slow (first) step:

$$^{226}\text{Ra} \rightarrow {}^{206}\text{Pb} + 5\,{}^{4}\text{He} \qquad t_{1/2} \approx 1600\text{ yr}$$

$$\ln\frac{\text{final mol }^{226}\text{Ra}}{\text{initial mol }^{226}\text{Ra}} \approx -kt$$

Plugging the known values into the above equation, we can solve it for the final moles of $^{226}$Ra, which in turn gives us the moles of He produced and the volume of He produced.

**Solution:** Calculate the initial moles of $^{226}$Ra:

$$\text{initial mol }^{226}\text{Ra} = \left(1.00\text{ g }^{226}\text{Ra}\right)\left(\frac{1\text{ mol}}{226\text{ g }^{226}\text{Ra}}\right) = 4.42 \times 10^{-3}\text{ mol }^{226}\text{Ra}$$

Next, find the final moles of $^{226}$Ra:

$$\ln\frac{\text{final mol }^{226}\text{Ra}}{0.00442\text{ mol }^{226}\text{Ra}} \approx -\left(\frac{0.693}{1600\text{ yr}}\right)(125\text{ yr})$$

$$\text{final mol }^{226}\text{Ra} \approx \left(0.00442\text{ mol }^{226}\text{Ra}\right)e^{-(0.693)(125)/1600}$$

$$\text{final mol }^{226}\text{Ra} \approx 0.00419\text{ mol }^{226}\text{Ra}$$

The number of moles of He produced is:

$$\text{mol He} = 5(0.00422\ \text{mol} - 0.00419\ \text{mol}) = 1.5\times10^{-4}\ \text{mol He}$$

Finally, the volume of He at STP is:

$$V = \frac{(1.5 \times 10^{-4}\ \text{mol})(0.0821\ \text{L}\cdot\text{atm/mol}\cdot\text{K})(273\ \text{K})}{1\ \text{atm}}$$

$$V = 3.4 \times 10^{-3}\ \text{L} = \mathbf{3.4\ mL}$$

# Chapter 21

# Environmental Chemistry

21.5 For ideal gases, mole fraction is the same as volume fraction. From Table 21.1 of the text, $CO_2$ is 0.033% of the composition of dry air, by volume. The value 0.033% means 0.033 volumes (or moles, in this case) out of 100 or

$$X_{CO_2} = \frac{0.033}{100} = \mathbf{3.3 \times 10^{-4}}$$

To change to parts per million (ppm), we multiply the mole fraction by one million.

$$(3.3 \times 10^{-4})(1 \times 10^{6}) = \mathbf{330\ ppm}$$

21.7 **In the stratosphere, the air temperature rises with altitude. This warming effect is the result of exothermic reactions triggered by UV radiation from the sun**. For further discussion, see Sections 21.2 and 21.3 of the text.

21.11 **Strategy:** We are given the bond enthalpy of the OH radical and are asked to determine the minimum wavelength of light required to break the bond. The bond enthalpy is given in kJ/mol, so we first must determine the energy required to break a single bond. We then use Equation 6.2 to calculate the frequency, and Equation 6.1 to convert from frequency to wavelength.

**Solution:** Determine the energy required to break a single OH bond.

$$\frac{460\ \text{kJ}}{1\ \text{mol}} \times \frac{1000\ \text{J}}{1\ \text{kJ}} \times \frac{1\ \text{mol}}{6.022 \times 10^{23}\ \text{photons}} = 7.64 \times 10^{-19}\ \text{J}$$

The frequency can now be calculated using Equation 6.2.

$$E = h\nu$$

$$\nu = \frac{E}{h} = \frac{7.64 \times 10^{-19}\ \text{J}}{6.63 \times 10^{-34}\ \text{J} \cdot \text{s}} = 1.15 \times 10^{15}\ \text{s}^{-1}$$

Finally, using Equation 6.1, we convert from frequency to wavelength.

$$\lambda = \frac{3.00 \times 10^{8}\ \text{m/s}}{1.15 \times 10^{15}\ \text{s}^{-1}} = 2.6 \times 10^{-7}\ \text{m} = \mathbf{260\ nm}$$

21.21 **Strategy:** We are asked to calculate the number of $O_3$ molecules in the stratosphere and their mass in kg. We can simplify the calculation of volume by multiplying the thickness of the layer by the surface area of a sphere the radius of Earth. Then we can use the molar volume of an ideal gas to determine the amount of ozone.

**Solution:** The formula for the volume is $4\pi r^2 h$, where $r = 6.371 \times 10^6$ m and $h = 3.0 \times 10^{-3}$ m (or 3.0 mm).

$$V = 4\pi(6.371 \times 10^6\text{ m})^2(3.0 \times 10^{-3}\text{ m}) = 1.5 \times 10^{12}\text{ m}^3 \times \frac{1000\text{ L}}{1\text{ m}^3} = 1.5 \times 10^{15}\text{ L}$$

Recall that at STP, one mole of gas occupies 22.41 L.

$$\text{moles O}_3 = (1.5 \times 10^{15}\text{ L}) \times \frac{1\text{ mol}}{22.41\text{ L}} = 6.7 \times 10^{13}\text{ mol O}_3$$

$$\textbf{molecules O}_3 = (6.7 \times 10^{13}\text{ mol O}_3) \times \frac{6.022 \times 10^{23}\text{ molecules}}{1\text{ mol}} = \mathbf{4.0 \times 10^{37}\ molecules}$$

$$\textbf{mass O}_3\textbf{ (kg)} = (6.7 \times 10^{13}\text{ mol O}_3) \times \frac{48.00\text{ g O}_3}{1\text{ mol O}_3} \times \frac{1\text{ kg}}{1000\text{ g}} = \mathbf{3.2 \times 10^{12}\ kg\ O_3}$$

21.23 The formula for Freon-11 is $CFCl_3$ and for Freon-12 is $CF_2Cl_2$. The equations are:

$$\mathbf{CCl_4 + HF \rightarrow HCl + CFCl_3\ (Freon\text{-}11)}$$

$$\mathbf{CFCl_3 + HF \rightarrow HCl + CF_2Cl_2\ (Freon\text{-}12)}$$

A catalyst is necessary for both reactions.

21.25 $\lambda = 250$ nm

$$\upsilon = \frac{3.00 \times 10^8\text{ m/s}}{250 \times 10^{-9}\text{ m}} = 1.20 \times 10^{15}\text{ /s}$$

$$E = h\nu = (6.63 \times 10^{-34}\text{ J·s})(1.20 \times 10^{15}\text{ /s}) = 7.96 \times 10^{-19}\text{ J}$$

Converting to units of kJ/mol:

$$\frac{7.96 \times 10^{-19}\text{ J}}{1\text{ photon}} \times \frac{6.022 \times 10^{23}\text{ photons}}{1\text{ mol}} \times \frac{1\text{ kJ}}{1000\text{ J}} = \mathbf{479\ kJ / mol}$$

Solar radiation preferentially breaks the C–Cl bond. **There is not enough energy to break the C–F bond.**

21.27 The Lewis structures for chlorine nitrate and chlorine monoxide are:

$$:\ddot{\underset{..}{Cl}}-\ddot{\underset{..}{O}}-\overset{+}{\underset{\underset{:O:}{||}}{N}}-\ddot{\underset{..}{O}}:^{-} \qquad\qquad :\ddot{\underset{..}{Cl}}-\ddot{\underset{..}{O}}\cdot$$

21.39 The equation is: $2ZnS + 3O_2 \rightarrow 2ZnO + 2SO_2$

$$(4.0 \times 10^4 \text{ ton ZnS}) \times \frac{1 \text{ ton} \cdot \text{mol ZnS}}{97.46 \text{ ton ZnS}} \times \frac{1 \text{ ton} \cdot \text{mol } SO_2}{1 \text{ ton} \cdot \text{mol ZnS}} \times \frac{64.07 \text{ ton } SO_2}{1 \text{ ton} \cdot \text{mol } SO_2} = \mathbf{2.6 \times 10^4 \text{ tons } SO_2}$$

21.53 The volume a gas occupies is directly proportional to the number of moles of gas. Therefore, 0.42 ppm by volume can also be expressed as a mole fraction.

$$X_{O_3} = \frac{n_{O_3}}{n_{\text{total}}} = \frac{0.42}{1 \times 10^6} = 4.2 \times 10^{-7}$$

The partial pressure of ozone can be calculated from the mole fraction and the total pressure.

$$P_{O_3} = X_{O_3} P_T = (4.2 \times 10^{-7})(748 \text{ mmHg}) = (3.14 \times 10^{-4} \text{ mmHg}) \times \frac{1 \text{ atm}}{760 \text{ mmHg}} = \mathbf{4.1 \times 10^{-7} \text{ atm}}$$

Substitute into the ideal gas equation to calculate moles of ozone.

$$n_{O_3} = \frac{P_{O_3} V}{RT} = \frac{(4.1 \times 10^{-7} \text{ atm})(1 \text{ L})}{(0.0821 \text{ L} \cdot \text{atm/mol} \cdot \text{K})(293 \text{ K})} = 1.7 \times 10^{-8} \text{ mol}$$

Number of $O_3$ molecules per liter:

$$\left(1.7 \times 10^{-8} \text{ mol } O_3\right) \times \frac{6.022 \times 10^{23} \text{ molecules}}{1 \text{ mol } O_3} = \mathbf{1 \times 10^{16} \text{ molecules/L}}$$

21.59 The room volume is:

$$17.6 \text{ m} \times 8.80 \text{ m} \times 2.64 \text{ m} = 4.09 \times 10^2 \text{ m}^3$$

Since $1 \text{ m}^3 = 1 \times 10^3$ L, then the volume of the container is $4.09 \times 10^5$ L. The quantity, $8.00 \times 10^2$ ppm is:

$$\frac{8.00 \times 10^2}{1 \times 10^6} = 8.00 \times 10^{-4} = \text{mole fraction of CO}$$

The pressure of the CO(atm) is:

$$P_{CO} = X_{CO}P_T = (8.00 \times 10^{-4})(756\text{ mmHg}) \times \frac{1\text{ atm}}{760\text{ mmHg}} = 7.96 \times 10^{-4}\text{ atm}$$

The moles of CO is:

$$n = \frac{PV}{RT} = \frac{(7.96 \times 10^{-4}\text{ atm})(4.09 \times 10^5\text{ L})}{(0.0821\text{ L}\cdot\text{atm/K}\cdot\text{mol})(293\text{ K})} = 13.5\text{ mol}$$

The mass of CO in the room is:

$$\textbf{mass} = 13.5\text{ mol} \times \frac{28.01\text{ g CO}}{1\text{ mol CO}} = \mathbf{378\text{ g CO}}$$

21.61 **$O_3$: greenhouse gas, toxic to humans, attacks rubber; $SO_2$: toxic to humans, forms acid rain; $NO_2$: forms acid rain, destroys ozone; CO: toxic to humans; PAN: a powerful lachrymator, causes breathing difficulties; Rn: causes lung cancer.**

21.63 a. **Its small concentration is the result of the high reactivity of the OH radical.**

b. **OH has an unpaired electron; free radicals are always good oxidizing agents.**

c.

$$\mathbf{OH + NO_2 \rightarrow HNO_3}$$

d.

$$\mathbf{OH + SO_2 \rightarrow HSO_3}$$

$$\mathbf{HSO_3 + O_2 + H_2O \rightarrow H_2SO_4 + HO_2}$$

21.65 **Most water molecules contain oxygen-16, but a small percentage of water molecules contain oxygen-18. The ratio of the two isotopes in the ocean is essentially constant, but the ratio in the water vapor evaporated from the oceans is temperature-dependent, with the vapor becoming slightly enriched with oxygen-18 as temperature increases. The water locked up in ice cores provides a historical record of this oxygen-18 enrichment, and thus ice cores contain information about past global temperatures.**

21.67 In one second, the energy absorbed by $CO_2$ is 6.7 J. If we can calculate the energy of one photon of light with a wavelength of 14993 nm, we can then calculate the number of photons absorbed per second.

The energy of one photon with a wavelength of 14993 nm is:

$$E = \frac{hc}{\lambda} = \frac{(6.63 \times 10^{-34}\text{ J}\cdot\text{s})(3.00 \times 10^8\text{ m/s})}{14993 \times 10^{-9}\text{ m}} = 1.3266 \times 10^{-20}\text{ J}$$

The number of photons absorbed by $CO_2$ per second is:

$$6.7\text{ J} \times \frac{1\text{ photon}}{1.3266 \times 10^{-20}\text{ J}} = \mathbf{5.1 \times 10^{20}\text{ photons}}$$

21.69 From $\Delta H_f^\circ$ values given in Appendix 2 of the text, we can calculate $\Delta H^\circ$ for the reaction

$$NO_2 \rightarrow NO + O$$

Then, we can calculate $\Delta E^\circ$ from $\Delta H^\circ$. The $\Delta E^\circ$ calculated will have units of kJ/mol. If we can convert this energy to units of J/molecule, we can calculate the wavelength required to decompose $NO_2$.

We use the $\Delta H_f^\circ$ values in Appendix 2 and Equation 5.18 of the text.

$$\Delta H_{rxn}^\circ = \Sigma n\Delta H_f^\circ(\text{products}) - \Sigma m\Delta H_f^\circ(\text{reactants})$$

Consider reaction (1):

$$\Delta H^\circ = \Delta H_f^\circ(NO) + \Delta H_f^\circ(O) - \Delta H_f^\circ(NO_2)$$

$$\Delta H^\circ = (1)(90.4\text{ kJ/mol}) + (1)(249.4\text{ kJ/mol}) - (1)(33.85\text{ kJ/mol})$$

$$\Delta H^\circ = 306.0\text{ kJ/mol}$$

From Equation 5.10 of the text,

$$\Delta E^\circ = \Delta H^\circ - P\Delta V$$

Using the ideal gas equation to substitute for $P\Delta V$, we get

$$\Delta E^\circ = \Delta H^\circ - RT\Delta n$$

$$\Delta E^\circ = (306.0 \times 10^3\text{ J/mol}) - (8.314\text{ J/mol}\cdot\text{K})(298\text{ K})(1)$$

$$\Delta E^\circ = 304 \times 10^3\text{ J/mol}$$

This is the energy needed to dissociate 1 mole of $NO_2$. We need the energy required to dissociate *one molecule* of $NO_2$.

$$\frac{304 \times 10^3\text{ J}}{1\text{ mol }NO_2} \times \frac{1\text{ mol }NO_2}{6.022 \times 10^{23}\text{ molecules }NO_2} = 5.05 \times 10^{-19}\text{ J/molecule}$$

The longest wavelength that can dissociate $NO_2$ is:

$$\lambda = \frac{hc}{E} = \frac{(6.63 \times 10^{-34}\text{ J}\cdot\text{s})(3.00 \times 10^8\text{ m/s})}{5.05 \times 10^{-19}\text{ J}} = \mathbf{3.94 \times 10^{-7}\text{ m} = 394\text{ nm}}$$

21.71 A Lewis base is a substance that can donate an electron pair. A Lewis acid is a substance that can accept a lone pair. In this case, **the lone pair on the S in $SO_2$ functions as the Lewis base and the Ca in CaO functions as a Lewis acid**.

21.73 a. The reactions representing the formation of acid rain [$H_2SO_4(aq)$] and the damage that acid rain causes to marble ($CaCO_3$) statues are:

$$2SO_2(g) + O_2(g) \rightarrow 2SO_3(g)$$

$$SO_3(g) + H_2O(l) \rightarrow H_2SO_4(aq)$$

$$CaCO_3(s) + H_2SO_4(aq) \rightarrow CaSO_4(s) + H_2O(l) + CO_2(g)$$

First, we convert the mass of $SO_2$ to moles of $SO_2$. Next, we convert to moles of $H_2SO_4$ that are produced (20% of $SO_2$ is converted to $H_2SO_4$). Then, we convert to the moles of $CaCO_3$ damaged per statue (5% of 1000 lb statue is damaged). And finally, we can calculate the number of marble statues that are damaged.

$$(50 \times 10^6 \text{ tons SO}_2) \times \frac{2000 \text{ lb}}{1 \text{ ton}} \times \frac{453.6 \text{ g}}{1 \text{ lb}} \times \frac{1 \text{ mol SO}_2}{64.07 \text{ g SO}_2} = 7.1 \times 10^{11} \text{ mol SO}_2$$

$$(0.20) \times (7.1 \times 10^{11} \text{ mol SO}_2) \times \frac{1 \text{ mol H}_2\text{SO}_4}{1 \text{ mol SO}_2} = 1.4 \times 10^{11} \text{ mol H}_2\text{SO}_4$$

The moles of $CaCO_3$ damaged per stature are:

$$(0.05) \times (1000 \text{ lb CaCO}_3) \times \frac{453.6 \text{ g}}{1 \text{ lb}} \times \frac{1 \text{ mol CaCO}_3}{100.1 \text{ g CaCO}_3} = 226.6 \text{ mol CaCO}_3/\text{statue}$$

The number of statues damaged by $1.4 \times 10^{11}$ moles of $H_2SO_4$ is:

$$(1.4 \times 10^{11} \text{ mol H}_2\text{SO}_4) \times \frac{1 \text{ mol CaCO}_3}{1 \text{ mol H}_2\text{SO}_4} \times \frac{1 \text{ statue}}{226.6 \text{ mol CaCO}_3} = \mathbf{6.2 \times 10^8 \text{ statues}}$$

Of course we don't have $6.2 \times 10^8$ marble statues around. This figure just shows that any outdoor objects/statues made of marble are susceptible to attack by acid rain.

b. **The $CO_2$ liberated from limestone contributes to global warming.**

21.75 **The use of the aerosol liberates CFC's that destroy the ozone layer.**

21.77 **The size of tree rings can be related to $CO_2$ content, where the number of rings indicates the age of the tree. The amount of $CO_2$ in ice can be directly measured from portions of polar ice in different layers obtained by drilling. The "age" of $CO_2$ can be determined by radiocarbon dating and other methods.**

21.79 a.

$$\mathbf{N_2O + O \rightleftarrows 2NO}$$

$$\mathbf{2NO + 2O_3 \rightleftarrows 2NO_2 + 2O_2}$$

$$\mathbf{Overall: N_2O + O + 2O_3 \rightleftarrows 2NO_2 + 2O_2}$$

b. **$N_2O$ is a more effective greenhouse gas than $CO_2$ because it has a permanent dipole.**

c. The moles of adipic acid are:

$$(2.2 \times 10^9 \text{ kg adipic acid}) \times \frac{1000 \text{ g}}{1 \text{ kg}} \times \frac{1 \text{ mol adipic acid}}{146.1 \text{ g adipic acid}} = 1.5 \times 10^{10} \text{ mol adipic acid}$$

The number of moles of adipic acid is given as being equivalent to the moles of $N_2O$ produced, and from the overall balanced equation, one mole of $N_2O$ will react with two moles of $O_3$. Thus,

$1.5 \times 10^{10}$ mol adipic acid $\rightarrow$ $1.5 \times 10^{10}$ mol $N_2O$ which reacts with $\mathbf{3.0 \times 10^{10}}$ **mol $O_3$**.

21.81 There is one C–Br bond per $CH_3Br$ molecule. The energy needed to break one C–Br bond is:

$$E = \frac{293 \times 10^3 \text{ J}}{1 \text{ mol}} \times \frac{1 \text{ mol}}{6.022 \times 10^{23} \text{ molecules}} = 4.865 \times 10^{-19} \text{ J}$$

Using Equations 6.1 and 6.2 of the text, we can now calculate the wavelength associated with this energy.

$$E = \frac{hc}{\lambda}$$

$$\lambda = \frac{hc}{E} = \frac{(6.63 \times 10^{-34} \text{ J}\cdot\text{s})(3.00 \times 10^8 \text{ m/s})}{4.865 \times 10^{-19} \text{ J}} = 4.09 \times 10^{-7} \text{ m} = 409 \text{ nm}$$

C—Cl = 340 kJ/mol, so **the photons that photolyze C—Cl bonds could easily photolyze the C—Br bonds as well. Light of wavelength 409 nm (visible) or shorter will break the C—Br bond.**

21.83 Total amount of heat absorbed is:

$$(1.8 \times 10^{20} \text{ mol}) \times \frac{29.1 \text{ J}}{\text{K} \times \text{mol}} \times 3 \text{ K} = 1.6 \times 10^{22} \text{ J} = \mathbf{1.6 \times 10^{19} \text{ kJ}}$$

The heat of fusion of ice in units of J/kg is:

$$\frac{6.01 \times 10^3 \text{ J}}{1 \text{ mol}} \times \frac{1 \text{ mol}}{18.02 \text{ g}} \times \frac{1000 \text{ g}}{1 \text{ kg}} = 3.3 \times 10^5 \text{ J/kg}$$

The amount of ice melted by the temperature rise:

$$(1.6 \times 10^{22}\ \text{J}) \times \frac{1\ \text{kg}}{3.3 \times 10^{5}\ \text{J}} = \mathbf{4.8 \times 10^{16}\ kg\ ice}$$

21.85

$$(3.1 \times 10^{10}\ \text{g}) \times \frac{2.4}{100} \times \frac{1\ \text{mol S}}{32.07\ \text{g S}} \times \frac{1\ \text{mol SO}_2}{1\ \text{mol S}} = 2.3 \times 10^{7}\ \text{mol SO}_2$$

$$V = \frac{nRT}{P} = \frac{(2.3 \times 10^{7}\ \text{mol})(0.0821\ \text{L} \cdot \text{atm/mol} \cdot \text{K})(273\ \text{K})}{1\ \text{atm}} = \mathbf{5.2 \times 10^{8}\ L\ SO_2}$$

# Chapter 22

# Coordination Chemistry

22.11 a. The oxidation number of Cr is **+3**.

b. The coordination number of Cr is **6**.

c. **Oxalate ion** ( $C_2O_4^{2-}$ ) is a bidentate ligand.

22.13 **Strategy:** The oxidation number of the metal atom is equal to its charge. First we look for known charges in the species. Recall that alkali metals are +1 and alkaline earth metals are +2. Also determine if the ligand is a charged or neutral species. From the known charges, we can deduce the net charge of the metal and hence its oxidation number.

**Solution:** a. Since **Na** is always **+1** and the oxygens are −2, **Mo** must have an oxidation number of **+6**.

b. **Magnesium** is **+2** and oxygen −2; therefore **W** is **+6**.

c. CO ligands are neutral species, so the iron atom bears no net charge. The oxidation number of **Fe** is **0**.

22.15 **Strategy:** We follow the procedure for naming coordination compounds outlined in Section 22.1 of the text and refer to Tables 22.4 and 22.5 of the text for names of ligands and anions containing metal atoms.

**Solution:** a. Ethylenediamine is a neutral ligand, and each chloride has a −1 charge. Therefore, cobalt has a oxidation number of +3. The correct name for the ion is ***cis*–dichlorobis(ethylenediamine)cobalt(III)**. The prefix *bis* means two; we use this instead of *di* because *di* already appears in the name ethylenediamine.

b. There are four chlorides each with a −1 charge; therefore, Pt has a +4 charge. The correct name for the compound is **pentaamminechloroplatinum(IV) chloride**.

c. There are three chlorides each with a −1 charge; therefore, Co has a +3 charge. The correct name for the compound is **pentaamminechlorocobalt(III) chloride**.

22.17 **Strategy:** We follow the procedure in Section 22.1 of the text and refer to Tables 22.4 and 22.5 of the text for names of ligands and anions containing metal atoms.

**Solution:** a. There are two ethylenediamine ligands and two chloride ligands. The correct formula is $\mathbf{[Cr(en)_2Cl_2]^+}$.

b. There are five carbonyl (CO) ligands. The correct formula is $\mathbf{Fe(CO)_5}$.

c. There are four cyanide ligands each with a −1 charge. Therefore, the complex ion has a −2 charge, and two $K^+$ ions are needed to balance the −2 charge of the anion. The correct formula is $\mathbf{K_2[Cu(CN)_4]}$.

d. There are four $NH_3$ ligands, an aquo ($H_2O$) ligand, and a chloride ligand. Two chloride ions are needed to balance the +2 charge of the complex ion. The correct formula is $\mathbf{[Co(NH_3)_4(H_2O)Cl]Cl_2}$.

22.23 a. In general for any $MA_2B_4$ octahedral molecule, only **two** geometric isomers are possible. The only real distinction is whether the two A–ligands are *cis* or *trans*.

b. A model or a careful drawing is very helpful to understand the $MA_3B_3$ octahedral structure. There are only **two** possible geometric isomers. The first has all A's (and all B's) *cis*; this is called the facial isomer. The second has two A's (and two B's) at opposite ends of the molecule (*trans*). Try to make or draw other possibilities. What happens?

22.25 a. There are *cis* and *trans* geometric isomers (See Problem 22.23). No optical isomers.

Cl
$H_3N$ — Co — $NH_3$
$H_3N$ — Co — $NH_3$
Cl
*trans*

Cl
$H_3N$ — Co — Cl
$H_3N$ — Co — $NH_3$
$NH_3$
*cis*

b. There are two optical isomers. The three bidentate ethylenediamine ligands are represented by the curved lines. (See Figure 22.3 of the text.)

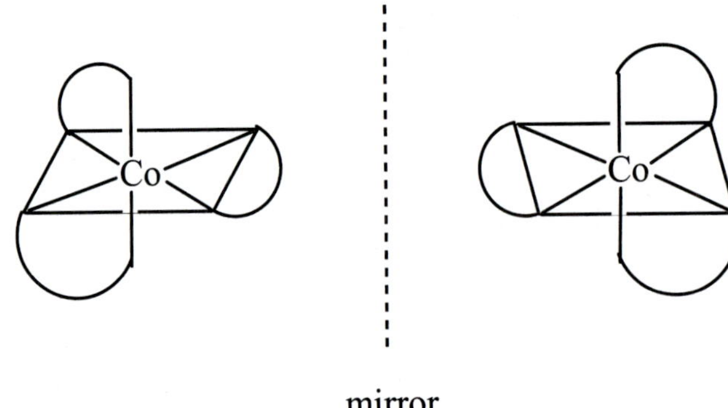


22.31 a. Wavelengths of 470 nm fall between blue and blue-green, corresponding to an observed color in the **orange** part of the spectrum.

b. We convert wavelength to photon energy using the Planck relationship.

$$\Delta E = \frac{hc}{\lambda} = \frac{(6.63 \times 10^{-34}\ \text{J}\cdot\text{s})(3.00 \times 10^{8}\ \text{m/s})}{470 \times 10^{-9}\ \text{m}} = 4.23 \times 10^{-19}\ \text{J}$$

$$\frac{4.23 \times 10^{-19}\ \text{J}}{1\ \text{photon}} \times \frac{6.022 \times 10^{23}\ \text{photons}}{1\ \text{mol}} \times \frac{1\ \text{kJ}}{1000\ \text{J}} = \mathbf{255\ kJ/mol}$$

22.33 *Step 1:* The equation for freezing-point depression is

$$\Delta T_f = K_f m$$

Solve this equation algebraically for molality ($m$), then substitute $\Delta T_f$ and $K_f$ into the equation to calculate the molality.

$$m = \frac{\Delta T_f}{K_f} = \frac{0.56°\text{C}}{1.86°\text{C}/m} = 0.30\ m$$

*Step 2:* Multiplying the molality by the mass of solvent (in kg) gives moles of unknown solute. Then, dividing the mass of solute (in g) by the moles of solute, gives the molar mass of the unknown solute.

$$?\ \text{mol of unknown solute} = \frac{0.30\ \text{mol solute}}{1\ \text{kg water}} \times 0.0250\ \text{kg water} = 0.0075\ \text{mol solute}$$

$$\text{molar mass of unknown} = \frac{0.875\ \text{g}}{0.0075\ \text{mol}} = 117\ \text{g/mol}$$

The molar mass of $Co(NH_3)_4Cl_3$ is 233.4 g/mol, which is twice the computed molar mass. This implies dissociation into two ions in solution; hence, there are **two moles** of ions produced per one mole of

$Co(NH_3)_4Cl_3$. The formula must be:

$$\mathbf{[Co(NH_3)_4Cl_2]Cl}$$

which contains the complex ion $[Co(NH_3)_4Cl_2]^+$ and a chloride ion, $Cl^-$. **Refer to Problem 22.25 (a) for a diagram of the structure** of the complex ion.

22.35 **Δ would be greater for the higher oxidation state.**

22.39 **Use a radioactive label such as $^{14}CN^-$ (in NaCN). Add NaCN to a solution of $K_3Fe(CN)_6$. Isolate some of the $K_3Fe(CN)_6$ and check its radioactivity. If the complex shows radioactivity, then it must mean that the $CN^-$ ion has participated in the exchange reaction.**

22.41 **$Cu(CN)_2$ is the white precipitate.**

$$Cu^{2+}(aq) + 2CN^-(aq) \rightarrow Cu(CN)_2(s)$$

**It is soluble in KCN(*aq*), due to formation of $[Cu(CN)_4]^{2-}$,**

$$Cu(CN)_2(s) + 2CN^-(aq) \rightarrow [Cu(CN)_4]^{2-}(aq)$$

**so the concentration of $Cu^{2+}$ is to small for $Cu^{2+}$ ions to precipitate with sulfide.**

22.43 The formation constant expression is:

$$K_f = \frac{\left[[Fe(H_2O)_5NCS]^{2+}\right]}{\left[[Fe(H_2O)_6]^{3+}\right]\left[SCN^-\right]}$$

Notice that the original volumes of the Fe(III) and $SCN^-$ solutions were both 1.0 mL and that the final volume is 10.0 mL. This represents a tenfold dilution, and the concentrations of Fe(III) and $SCN^-$ become 0.020 *M* and $1.0 \times 10^{-4}$ *M*, respectively.

We make a table.

| | $[Fe(H_2O)_6]^{3+}$ + | $SCN^-$ | ⇄ | $[Fe(H_2O)_5NCS]^{2+}$ | + $H_2O$ |
|---|---|---|---|---|---|
| Initial (*M*): | 0.020 | $1.0 \times 10^{-4}$ | | 0 | 0.00 |
| Change (*M*): | $-7.3 \times 10^{-5}$ | $-7.3 \times 10^{-5}$ | | $+7.3 \times 10^{-5}$ | $+x$ |
| Equilibrium (*M*): | 0.020 | $2.7 \times 10^{-5}$ | | $7.3 \times 10^{-5}$ | $x$ |

$$K_f = \frac{7.3 \times 10^{-5}}{(0.020)(2.7 \times 10^{-5})} = \mathbf{1.4 \times 10^2}$$

22.45 The square planar complex shown in the problem has **3** geometric isomers. They are:

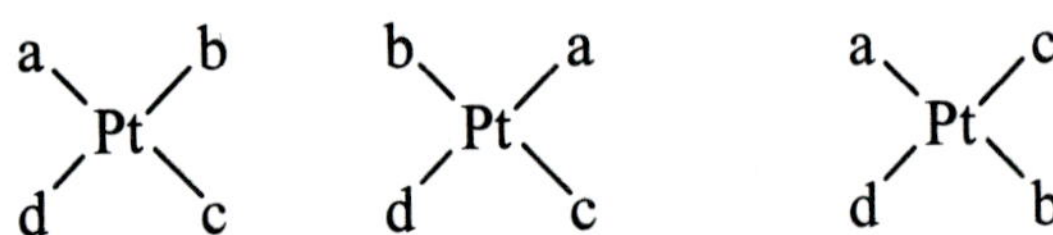


Note that in the first structure a is *trans* to c, in the second a is *trans* to d, and in the third a is *trans* to b. Make sure you realize that if we switch the positions of b and d in structure 1, we do not obtain another geometric isomer. A 180° rotation about the a–Pt–c axis gives structure 1.

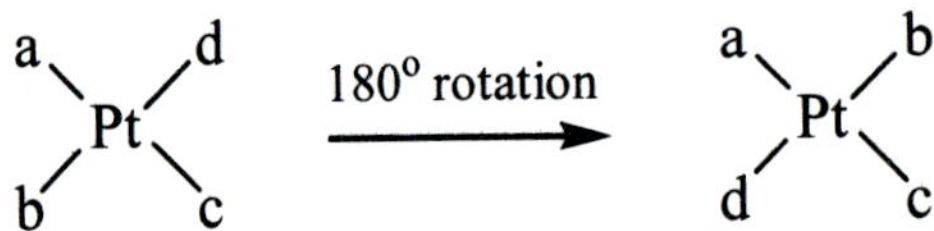


22.47 **$Ti^{3+}/Fe^{3+}$.**

22.49 **$Mn^{3+}$ is $3d^4$ and $Cr^{3+}$ is $3d^5$. Therefore, $Mn^{3+}$ has a greater tendency to accept an electron and is a stronger oxidizing agent. The $3d^5$ electron configuration of $Cr^{3+}$ is a stable configuration.**

22.51 When a substance appears blue, it is absorbing light from the orange part of the spectrum. Substances that appear yellow are absorbing light from the higher energy blue-violet part of the spectrum. Thus, **Y** causes a larger crystal field splitting than X, so it is a stronger field ligand.

22.53 To find the concentration at equilibrium, we make the following table.

| | $Pb^{2+}$ + | $EDTA^{4-}$ | $\rightleftarrows$ | $[Pb(EDTA)]^{2-}$ |
|---|---|---|---|---|
| Initial ($M$): | $1.0 \times 10^{-3}$ | $2.0 \times 10^{-3}$ | | 0 |
| Change ($M$): | $-x$ | $-x$ | | $+x$ |
| Equilibrium ($M$): | $1.0 \times 10^{-3} - x$ | $2.0 \times 10^{-3} - x$ | | $x$ |

$$K_f = \frac{\left[Pb[EDTA]^{2-}\right]}{\left[Pb^{2+}\right]\left[EDTA^{4-}\right]} = 1.0\times10^{18}$$

$$1.0 \times 10^{18} = \frac{x}{(1.0\times10^{-3} - x)(2.0 \times 10^{-3} - x)}$$

Because $K_f$ is very large, we can either assume that the reaction goes to completion or solve the equation. **If the reaction goes to completion, $Pb^{2+}$ is the limiting reagent**. The equilibrium concentration of $[Pb(EDTA)]^{2-}$ would be $1.0 \times10^{-3}$ $M$. Solving the equation would be done as follows.

$$1.0 \times 10^{18} = \frac{x}{(2.0 \times 10^{-6}) - (3.0 \times 10^{-3})x + x^2}$$

$$0 = (1.0 \times 10^{18})x^2 - (3.0 \times 10^{15})x + (2.0 \times 10^{12})$$

Use the quadratic equation to solve for $x$.

$$x = \frac{-b \pm \sqrt{b^2 - 4ac}}{2a}$$

$$= \frac{(3.0 \times 10^{15}) \pm (1.0 \times 10^{15})}{2.0 \times 10^{18}}$$

$$= 2.0 \times 10^{-3} \quad or \quad 1.0 \times 10^{-3}$$

Since the equilibrium concentration of $Pb^{2+}$ is $1.0 \times 10^{-3} - x$, the first answer is not possible. Thus, the equilibrium concentration of $[Pb(EDTA)]^{2-}$ is $1.0 \times 10^{-3}$ *M.*

22.55 a. **$[Cr(H_2O)_6]Cl_3$, number of ions: 4**

b. **$[Cr(H_2O)_5Cl]Cl_2 \cdot H_2O$, number of ions: 3**

c. **$[Cr(H_2O)_4Cl_2]Cl \cdot 2H_2O$, number of ions: 2**

**Compare the compounds with equal molar amounts of NaCl, $MgCl_2$, and $FeCl_3$ in an electrical conductance experiment. The solution that has similar conductance to the NaCl solution contains (c); the solution with the conductance similar to $MgCl_2$ contains (b); and the solution with conductance similar to $FeCl_3$ contains (a).**

22.57

$$Zn(s) \rightarrow Zn^{2+}(aq) + 2e^- \qquad E^o_{anode} = -0.76 \text{ V}$$

$$2[Cu^{2+}(aq) + e^- \rightarrow Cu^+(aq)] \qquad E^o_{cathode} = 0.15 \text{ V}$$

$$Zn(s) + 2Cu^{2+}(aq) \rightarrow Zn^{2+}(aq) + 2Cu^+(aq)$$

$$E^o_{cell} = E^o_{cathode} - E^o_{anode} = 0.15 \text{ V} - (-0.76 \text{ V}) = 0.91 \text{ V}$$

We carry additional significant figures throughout the remainder of this calculation to minimize rounding errors.

$$\Delta G° = -nFE° = -(2)(96500 \text{ J/V·mol})(0.91 \text{ V}) = -1.756 \times 10^5 \text{ J/mol} = \mathbf{-1.8 \times 10^2 \text{ kJ/mol}}$$

$$\Delta G° = -RT \ln K$$

$$\ln K = \frac{-\Delta G°}{RT} = \frac{-(-1.756 \times 10^5 \text{ J/mol})}{(8.314 \text{ J/K·mol})(298 \text{ K})}$$

$$\ln K = 70.88$$

$$K = e^{70.88} = \mathbf{6 \times 10^{30}}$$

22.59 **Iron is much more abundant than cobalt.**

22.61 The oxidation number of the metal atom is equal to its charge. Low-spin means that the crystal field splitting is large, so the lower orbitals will fill completely before the upper orbitals are populated.

**$[Mn(CN)_6]^{5-}$**

Each cyanide ligand is −1, so **Mn is +1**. Its electron configuration is $[Ar]4s^13d^5$. Thus, it has **one unpaired *d* electron.**

$$\frac{}{d_{z^2}} \quad \frac{}{d_{x^2-y^2}}$$

$$\frac{\uparrow\downarrow}{d_{xy}} \quad \frac{\uparrow\downarrow}{d_{xz}} \quad \frac{\uparrow}{d_{yz}}$$

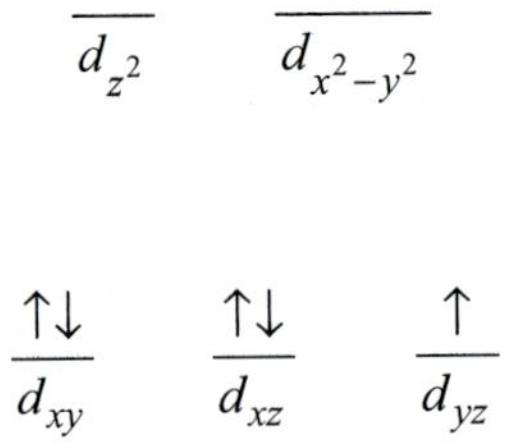

**$[Mn(CN)_6]^{4-}$**

Each cyanide ligand is −1, so **Mn is +2**. Its electron configuration is $[Ar]3d^5$. Thus, it has **one unpaired *d* electron.**

$$\frac{}{d_{z^2}} \quad \frac{}{d_{x^2-y^2}}$$

$$\frac{\uparrow\downarrow}{d_{xy}} \quad \frac{\uparrow\downarrow}{d_{xz}} \quad \frac{\uparrow}{d_{yz}}$$

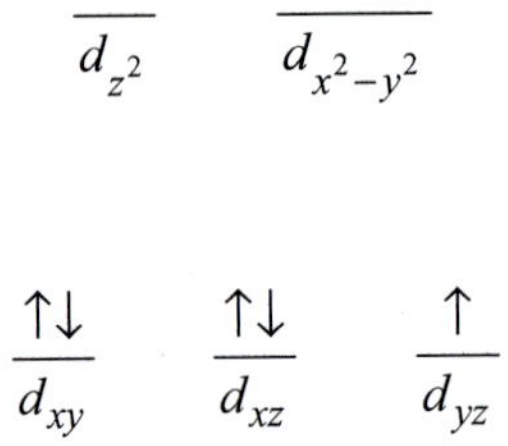

**$[Mn(CN)_6]^{3-}$**

Each cyanide ligand is −1, so **Mn is +3**. Its electron configuration is $[Ar]3d^4$. Thus, it has **two unpaired *d* electrons.**

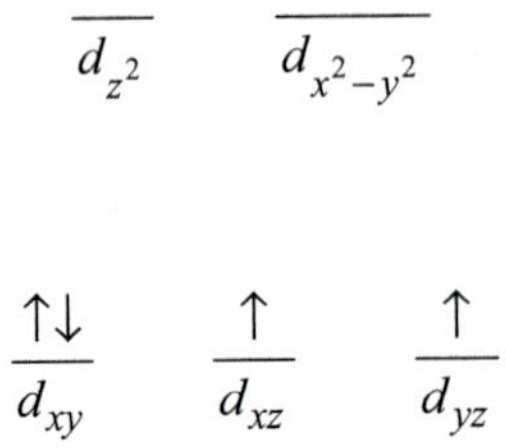

$$\frac{}{d_{z^2}} \quad \frac{}{d_{x^2-y^2}}$$

$$\frac{\uparrow\downarrow}{d_{xy}} \quad \frac{\uparrow}{d_{xz}} \quad \frac{\uparrow}{d_{yz}}$$

22.63 **Complexes are expected to be colored when the highest occupied orbitals have between one and nine *d* electrons. Such complexes can therefore have $d \rightarrow d$ transitions (that are usually in the visible part of the electromagnetic radiation spectrum). $Zn^{2+}$, $Cu^{+}$, and $Pb^{2+}$ are $d^{10}$ ions. $V^{5+}$, $Ca^{2+}$, and $Sc^{3+}$ are $d^0$ ions.**

22.65 **Dipole moment measurement. Only the *cis* isomer has a dipole moment.**

22.67 **EDTA sequesters metal ions (like $Ca^{2+}$ and $Mg^{2+}$) which are essential for the growth and function of bacteria.**

22.69 a. **Tc**

b. **W**

c. Argon has 18 $e^-$. Add 3 $e^-$ for the $3d^3$ and 4 $e^-$ to produce a neutral atom. The result is 25 $e^-$ on the neutral atom. Thus, the atom is Mn and the ion is $\mathbf{Mn^{4+}}$.

d. Xenon has 54 $e^-$. Add 22 $e^-$ for the $4f^{14}5d^8$ and 3 $e^-$ to produce a neutral atom. The result is 79 $e^-$ on the neutral atom. Thus, the atom is Au and the ion is $\mathbf{Au^{3+}}$.

22.71 **The purple color is caused by the build-up of deoxyhemoglobin. When either oxyhemoglobin or deoxyhemoglobin takes up CO, the carbonylhemoglobin takes on a red color, the same as oxyhemoglobin.**

22.73 A 100.00 g sample of hemoglobin contains 0.34 g of iron. In moles this is:

$$0.34 \text{ g Fe} \times \frac{1 \text{ mol}}{55.85 \text{ g}} = 6.1 \times 10^{-3} \text{ mol Fe}$$

The amount of hemoglobin that contains one mole of iron must be:

$$\frac{100.00 \text{ g hemoglobin}}{6.1 \times 10^{-3} \text{ mol Fe}} = \mathbf{1.6 \times 10^4 \text{ g hemoglobin / mol Fe}}$$

We compare this to the actual molar mass of hemoglobin:

$$\frac{6.5 \times 10^4 \text{ g hemoglobin}}{1 \text{ mol hemoglobin}} \times \frac{1 \text{ mol Fe}}{1.6 \times 10^4 \text{ g hemoglobin}} = 4 \text{ mol Fe/1 mol hemoglobin}$$

**The discrepancy between our minimum value and the actual value can be explained by realizing that there are four iron atoms per mole of hemoglobin.**

22.75 **Oxyhemoglobin absorbs higher energy light than deoxyhemoglobin. Oxyhemoglobin is diamagnetic (low spin), while deoxyhemoglobin is paramagnetic (high spin). These differences occur because oxygen ($O_2$) is a strong–field ligand.** The crystal field splitting diagrams are:

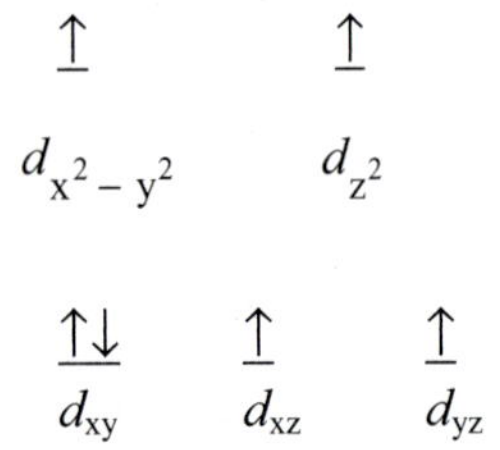

| | | |
|---|---|---|
| ↑ | ↑ | |
| $d_{x^2-y^2}$ | $d_{z^2}$ | |
| ↑↓ | ↑ | ↑ |
| $d_{xy}$ | $d_{xz}$ | $d_{yz}$ |

deoxyhemoglobin

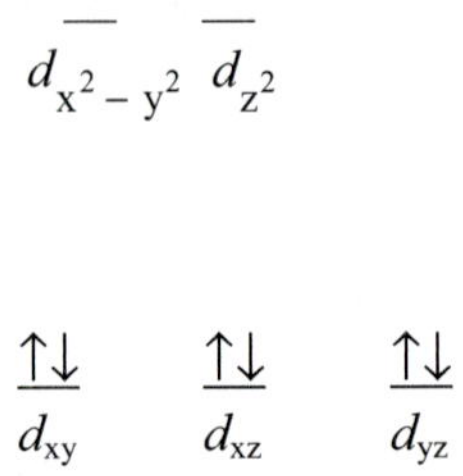

| | | |
|---|---|---|
| — | — | |
| $d_{x^2-y^2}$ | $d_{z^2}$ | |
| ↑↓ | ↑↓ | ↑↓ |
| $d_{xy}$ | $d_{xz}$ | $d_{yz}$ |

oxyhemoglobin

22.77 The reaction is: $Ag^+(aq) + 2CN^-(aq) \rightleftarrows [Ag(CN)_2]^-(aq)$

$$K_f = 1.0 \times 10^{21} = \frac{[[Ag(CN)_2]^-]}{[Ag^+][CN^-]^2}$$

First, we calculate the initial concentrations of $Ag^+$ and $CN^-$. Then, because $K_f$ is so large, we assume that the reaction goes to completion. This assumption will allow us to solve for the concentration of $Ag^+$ at equilibrium. The initial concentrations of $Ag^+$ and $CN^-$ are:

$$[CN^-] = \frac{\frac{5.0\text{ mol}}{1\text{ L}} \times 9.0\text{ L}}{99.0\text{ L}} = 0.455\ M$$

$$[Ag^+] = \frac{\frac{0.20\text{ mol}}{1\text{ L}} \times 90.0\text{ L}}{99.0\text{ L}} = 0.182\ M$$

We set up a table to determine the concentrations after complete reaction.

| | $Ag^+(aq)$ | + | $2CN^-(aq)$ | $\rightleftarrows$ | $[Ag(CN)_2]^-(aq)$ |
|---|---|---|---|---|---|
| Initial ($M$): | 0.182 | | 0.455 | | 0 |
| Change ($M$): | −0.182 | | −(2)(0.182) | | +0.182 |
| Equilibrium ($M$): | 0 | | 0.0910 | | 0.182 |

$$K_f = \frac{[[Ag(CN)_2]^-]}{[Ag^+][CN^-]^2}$$

$$1.0 \times 10^{21} = \frac{0.182\ M}{[Ag^+](0.0910\ M)^2}$$

$$\mathbf{[Ag^+] = 2.2 \times 10^{-20}\ M}$$

22.79 a. The equilibrium constant can be calculated from $\Delta G°$. We can calculate $\Delta G°$ from the cell potential.

From Table 19.1 of the text,

$$Cu^{2+} + 2e^- \rightarrow Cu \qquad E° = 0.34\text{ V and } \Delta G° = -(2)(96500\text{ J/V·mol})(0.34\text{ V}) = -6.562 \times 10^4\text{ J/mol}$$

$$Cu^{2+} + e^- \rightarrow Cu^+ \qquad E° = 0.15\text{ V and } \Delta G° = -(1)(96500\text{ J/V·mol})(0.15\text{ V}) = -1.448 \times 10^4\text{ J/mol}$$

These two equations need to be arranged to give the disproportionation reaction in the problem. We keep the first equation as written and reverse the second equation and multiply by two.

$$Cu^{2+} + 2e^- \rightarrow Cu \qquad \Delta G° = -6.562 \times 10^4\text{ J/mol}$$

$$2Cu^+ \rightarrow 2Cu^{2+} + 2e^- \qquad \Delta G° = +(2)(1.448 \times 10^4\text{ J/mol})$$

$$2Cu^+ \rightarrow Cu^{2+} + Cu \qquad \Delta G° = -6.562 \times 10^4\text{ J/mol} + 2.896 \times 10^4\text{ J/mol} = -3.666 \times 10^4\text{ J/mol}$$

We use Equation 18.15 of the text to calculate the equilibrium constant.

$$\Delta G° = -RT\ln K$$

$$K = e^{-\Delta G°/RT}$$

$$K = e^{-(-3.666 \times 10^4\text{ J/mol})/(8.314\text{ J/mol·K})(298\text{ K})}$$

$$\mathbf{K = 2.7 \times 10^6}$$

b. **Free $Cu^+$ ions are unstable in solution [as shown in part (a)]. Therefore, the only stable compounds containing $Cu^+$ ions are insoluble.**

# Chapter 23

# Metallurgy and the Chemistry of Metals

23.3 a. $\mathbf{CaCO_3}$

b. $\mathbf{CaCO_3 \cdot MgCO_3}$

c. $\mathbf{CaF_2}$

d. **NaCl**

e. $\mathbf{Al_2O_3}$

f. $\mathbf{Fe_3O_4}$

g. $\mathbf{Be_3Al_2Si_6O_{18}}$

h. **PbS**

i. $\mathbf{MgSO_4 \cdot 7H_2O}$

j. $\mathbf{CaSO_4}$

23.13 For the given reaction we can calculate the standard free energy change from the standard free energies of formation. Then, we can calculate the equilibrium constant, $K_p$, from the standard free energy change.

$$\Delta G° = \Delta G_f^o[\text{Ni(CO)}_4] - [4\Delta G_f^o(\text{CO}) + \Delta G_f^o(\text{Ni})]$$

$$\Delta G° = (1)(-587.4 \text{ kJ/mol}) - [(4)(-137.3 \text{ kJ/mol}) + (1)(0)] = -38.2 \text{ kJ/mol} = -3.82 \times 10^4 \text{ J/mol}$$

Substitute $\Delta G°$, $R$, and $T$ (in K) into the following equation to solve for $K_p$.

$$\Delta G° = -RT\ln K_p$$

$$\ln K_p = \frac{-\Delta G°}{RT} = \frac{-(-3.82 \times 10^4 \text{ J/mol})}{(8.314 \text{ J/K}\cdot\text{mol})(353 \text{ K})}$$

$$\mathbf{K_p = 4.5 \times 10^5}$$

23.15 a. We first find the mass of ore containing $2.0 \times 10^8$ kg of copper.

$$(2.0 \times 10^8 \text{ kg Cu}) \times \frac{100\% \text{ ore}}{0.80\% \text{ Cu}} = 2.5 \times 10^{10} \text{ kg ore}$$

We can then compute the volume from the density of the ore.

$$(2.5 \times 10^{10} \text{ kg}) \times \frac{1000 \text{ g}}{1 \text{ kg}} \times \frac{1 \text{ cm}^3}{2.8 \text{ g}} = \mathbf{8.9 \times 10^{12} \text{ cm}^3}$$

b. From the formula of chalcopyrite it is clear that two moles of sulfur dioxide will be formed per mole of copper. The mass of sulfur dioxide formed will be:

$$(2.0 \times 10^8 \text{ kg Cu}) \times \frac{1 \text{ mol Cu}}{0.06355 \text{ kg Cu}} \times \frac{2 \text{ mol } SO_2}{1 \text{ mol Cu}} \times \frac{0.06407 \text{ kg } SO_2}{1 \text{ mol } SO_2} = \mathbf{4.0 \times 10^8 \text{ kg } SO_2}$$

23.17 **Ag, Pt, and Au will not be oxidized but the other metals will.** (See Table 19.1 of the text.)

23.19 Very electropositive metals (i.e., very strong reducing agents) can only be isolated from their compounds by electrolysis. No chemical reducing agent is strong enough. In the given list, preparation of **$Al_2O_3$, NaCl, and $CaCl_2$** would require electrolysis.

23.33 All of these reactions are discussed in Section 23.5 of the text.

a. $$\mathbf{2K(s) + 2H_2O(l) \rightarrow 2KOH(aq) + H_2(g)}$$

b. $$\mathbf{NaH(s) + H_2O(l) \rightarrow NaOH(aq) + H_2(g)}$$

c. $$\mathbf{2Na(s) + O_2(g) \rightarrow Na_2O_2(s)}$$

d. $$\mathbf{K(s) + O_2(g) \rightarrow KO_2(s)}$$

23.35 $\mathbf{NaH + H_2O \rightarrow NaOH + H_2}$

23.39 First magnesium is treated with concentrated nitric acid (redox reaction) to obtain magnesium nitrate.

$$\mathbf{3Mg(s) + 8HNO_3(aq) \rightarrow 3Mg(NO_3)_2(aq) + 4H_2O(l) + 2NO(g)}$$

**The magnesium nitrate is recovered from solution by evaporation, dried, and heated in air to obtain magnesium oxide:**

$$\mathbf{2Mg(NO_3)_2(s) \rightarrow 2MgO(s) + 4NO_2(g) + O_2(g)}$$

23.41 **The electron configuration of magnesium is [Ne]$3s^2$. The 3*s* electrons are outside the neon core (shielded), so they have relatively low ionization energies. Removing the third electron means separating an electron from the neon (closed shell) core, which requires a great deal more energy.**

23.43 **Even though helium and the Group 2A metals have $ns^2$ outer electron configurations, helium has a**

**closed shell noble gas configuration and the Group 2A metals do not. The electrons in He are much closer to and more strongly attracted by the nucleus. Hence, the electrons in He are not easily removed. Helium is inert.**

23.45 a. quicklime: **CaO(*s*)** b. slaked lime: **$Ca(OH)_2$(*s*)**

23.49 a. The relationship between cell voltage and free energy difference is:

$$\Delta G = -nFE$$

In the given reaction $n = 6$. We write:

$$E = \frac{-\Delta G}{nF} = \frac{-594 \times 10^3 \text{ J/mol}}{(6)(96500 \text{ J/V}\times\text{mol})} = -1.03 \text{ V}$$

The balanced equation shows *two* moles of aluminum. Is this the voltage required to produce *one* mole of aluminum? If we divide everything in the equation by two, we obtain:

$$\tfrac{1}{2}Al_2O_3(s) + \tfrac{3}{2}C(s) \rightarrow Al(l) + \tfrac{3}{2}CO(g)$$

For the new equation $n = 3$ and $\Delta G$ is $\left(\frac{1}{2}\right)$(594 kJ/mol) = 297 kJ/mol. We write:

$$E = \frac{-\Delta G}{nF} = \frac{-297 \times 10^3 \text{ J/mol}}{(3)(96500 \text{ J/V}\times\text{mol})} = -1.03 \text{ V}$$

The minimum voltage that must be applied is **1.03 V** (a negative sign in the answers above means that 1.03 V is required to produce the Al). The voltage required to produce one mole or one thousand moles of aluminum is the same; the amount of *current* will be different in each case.

b. First we convert 1.00 kg (1000 g) of Al to moles.

$$(1.00 \times 10^3 \text{ g Al}) \times \frac{1 \text{ mol Al}}{26.98 \text{ g Al}} = 37.1 \text{ mol Al}$$

The reaction in part (a) shows us that three moles of electrons are required to produce one mole of aluminum. The voltage is three times the minimum calculated above (namely, –3.09 V or –3.09 J/C). We can find the electrical energy by using the same equation with the other voltage.

$$\Delta G = -nFE = -(37.1 \text{ mol Al})\left(\frac{3 \text{ mol e}^-}{1 \text{ mol Al}} \times \frac{96500 \text{ C}}{1 \text{ mol e}^-}\right)\left(\frac{-3.09 \text{ J}}{1 \text{ C}}\right)$$

$$= \mathbf{3.32 \times 10^7\ J/mol\ =\ 3.32 \times 10^4\ kJ/mol}$$

This equation can be used because electrical work can be calculated by multiplying the voltage by the amount of charge transported through the circuit (joules = volts × coulombs). The $nF$ term in Equation 19.2 of the text used above represents the amount of charge.

What is the significance of the positive sign of the free energy change? Would the manufacturing of aluminum be a different process if the free energy difference were negative?

23.51 The two complex ions can be classified as $AB_4$ and $AB_6$ structures (no unshared electron pairs on Al and 4 or 6 attached atoms, respectively). Their VSEPR geometries are **tetrahedral** and **octahedral.**

**The accepted explanation for the nonexistence of $AlCl_6^{3-}$ is that the chloride ion is too big to form an octahedral cluster around a very small $Al^{3+}$ ion.**

23.53 $\mathbf{4Al(NO_3)_3(\mathit{s}) \rightarrow 2Al_2O_3(\mathit{s}) + 12NO_2(\mathit{g}) + 3O_2(\mathit{g})}$

23.55 **The "bridge" bonds in $Al_2Cl_6$ break at high temperature: $Al_2Cl_6(g) \rightleftarrows 2AlCl_3(g)$.**

**This increases the number of molecules in the gas phase and causes the pressure to be higher than expected for pure $Al_2Cl_6$.**

If you know the equilibrium constants for the above reaction at higher temperatures, could you calculate the expected pressure of the $AlCl_3$–$Al_2Cl_6$ mixture?

23.57 **In $Al_2Cl_6$, each aluminum atom is surrounded by 4 bonding pairs of electrons ($AB_4$–type molecule), and therefore each aluminum atom is $sp^3$ hybridized. VSEPR analysis shows $AlCl_3$ to be an $AB_3$–type molecule (no lone pairs on the central atom). The geometry should be trigonal planar, and the aluminum atom should therefore be $sp^2$ hybridized.**

23.59 The formulas of the metal oxide and sulfide are MO and MS (why?). The balanced equation must therefore be:

$$2MS(s) + 3O_2(g) \rightarrow 2MO(s) + 2SO_2(g)$$

The number of moles of MO and MS are equal. We let $x$ be the molar mass of metal. The number of moles of metal oxide is:

$$0.972\ \text{g} \times \frac{1\ \text{mol}}{(x + 16.00)\,\text{g}}$$

The number of moles of metal sulfide is:

$$1.164\ \text{g} \times \frac{1\ \text{mol}}{(x + 32.07)\,\text{g}}$$

The moles of metal oxide equal the moles of metal sulfide.

$$\frac{0.972}{(x + 16.00)} = \frac{1.164}{(x + 32.07)}$$

We solve for *x*.

$$0.972(x + 32.07) = 1.164(x + 16.00)$$

$$\mathbf{x = 65.4\ g/mol}$$

23.61 **Copper(II) ion is more easily reduced than either water or hydrogen ion** (How can you tell? See Section 19.3 of the text.) **Copper metal is more easily oxidized than water. Water should not be affected by the copper purification process under standard conditions.**

23.63 Using Equation 18.12 from the text:

a.

$$\Delta G^{\circ}_{rxn} = 4\Delta G^{\circ}_{f}(Fe) + 3\Delta G^{\circ}_{f}(O_2) - 2\Delta G^{\circ}_{f}(Fe_2O_3)$$

$$\mathbf{\Delta G^{\circ}_{rxn}} = (4)(0) + (3)(0) - (2)(-741.0\ kJ/mol) = \mathbf{1482\ kJ / mol}$$

b.

$$\Delta G^{\circ}_{rxn} = 4\Delta G^{\circ}_{f}(Al) + 3\Delta G^{\circ}_{f}(O_2) - 2\Delta G^{\circ}_{f}(Al_2O_3)$$

$$\mathbf{\Delta G^{\circ}_{rxn}} = (4)(0) + (3)(0) - (2)(-1576.4\ kJ/mol) = \mathbf{3152.8\ kJ / mol}$$

23.65 **Mg(*s*) reacts with $N_2$(*g*) at high temperatures to produce $Mg_3N_2$(*s*). Ti(*s*) also reacts with $N_2$(*g*) at high temperatures to produce TiN(*s*).**

23.67 a. **In water the aluminum(III) ion causes an increase in the concentration of hydrogen ion (lower pH). This results from the effect of the small diameter and high charge (3+) of the aluminum ion on surrounding water molecules. The aluminum ion draws electrons in the O–H bonds to itself, thus allowing easy formation of $H^+$ ions.**

b. **$Al(OH)_3$ is an amphoteric hydroxide. It will dissolve in strong base with the formation of a complex ion.**

$$\mathbf{Al(OH)_3(s) + OH^-(aq) \rightarrow Al(OH)_4^-\ (aq)}$$

**The concentration of $OH^-$ in aqueous ammonia is too low for this reaction to occur.**

23.69 Calcium oxide is a base. The reaction is a neutralization.

$$\mathbf{CaO(s) + 2HCl(aq) \rightarrow CaCl_2(aq) + H_2O(l)}$$

23.71 **Metals have closely spaced energy levels and (referring to Figure 23.10 of the text) a very small energy gap between filled and empty levels. Consequently, many electronic transitions can take place with absorption and subsequent emission of light continually occurring. Some of these transitions fall in the visible region of the spectrum and give rise to the flickering appearance**.

23.73 **NaF: cavity prevention.** ($F^-$)

**$Li_2CO_3$: antidepressant.** ($Li^+$)

**$Mg(OH)_2$: laxative** (Milk of magnesia®).

**$CaCO_3$: calcium supplement; antacid**.

**$BaSO_4$: radiocontrast agent**.

23.75 **Both Li and Mg form oxides ($Li_2O$ and MgO). Other Group 1A metals (Na, K, etc.) also form peroxides and superoxides. In Group 1A, only Li forms nitride ($Li_3N$), like Mg ($Mg_3N_2$).**

**Li resembles Mg in that its carbonate, fluoride, and phosphate have low solubilities**.

23.77 We know that Ag, Cu, Au, and Pt are found as free elements in nature, which leaves **Zn** by process of elimination. Table 19.1 of the text indicates that the standard oxidation potential of Zn is +0.76 V. The positive value indicates that Zn is easily oxidized to $Zn^{2+}$ and will not exist as a free element in nature.

23.79 There are 10.00 g of Na in 13.83 g of the mixture (the mass of sodium in the reactants and products is equal). This amount of Na is equal to the mass of Na in $Na_2O$ plus the mass of Na in $Na_2O_2$.

$$10.00 \text{ g Na} = \text{mass of Na in } Na_2O + \text{mass of Na in } Na_2O_2$$

To calculate the mass of Na in each compound, the grams of compound need to be converted to grams of Na using the mass percentage of Na in the compound. If $x$ equals the mass of $Na_2O$, then the mass of $Na_2O_2$ is $13.83 - x$. We set up the following expression and solve for $x$. We carry an additional significant figure throughout the calculation to minimize rounding errors.

$$10.00 \text{ g Na} = \text{mass of Na in } Na_2O + \text{mass of Na in } Na_2O_2$$

$$10.00 \text{ g Na} = \left[x \text{ g } Na_2O \times \frac{(2)(22.99 \text{ g Na})}{61.98 \text{ g } Na_2O}\right] + \left[(13.83 - x) \text{ g } Na_2O_2 \times \frac{(2)(22.99 \text{ g Na})}{77.98 \text{ g } Na_2O_2}\right]$$

$$10.00 = 0.74185x + 8.1547 - 0.58964x$$

$$0.15221x = 1.8453$$

$x = 12.123$ g, which equals the mass of $Na_2O$.

The mass of $Na_2O_2$ is $13.83 - x$, which equals 1.707g.

The mass percent of each compound in the mixture is:

$$\textbf{\% Na}_2\textbf{O} = \frac{12.123\text{ g}}{13.83\text{ g}} \times 100 = \textbf{87.66\%}$$

$$\textbf{\% Na}_2\textbf{O}_2 = 100\% - 87.66\% = \textbf{12.34\%}$$

23.81 First, we calculate the density of $O_2$ in $KO_2$ using the mass percentage of $O_2$ in the compound.

$$\frac{32.00\text{ g O}_2}{71.10\text{ g KO}_2} \times \frac{2.15\text{ g KO}_2}{1\text{ cm}^3} = 0.968\text{ g O}_2/\text{cm}^3$$

$$\frac{0.968\text{ g O}_2}{1\text{ cm}^3} \times \frac{1000\text{ cm}^3}{1\text{ L}} = 968\text{ g O}_2/\text{L}$$

Now, we can use Equation 11.7 of the text to calculate the pressure of oxygen gas that would have the same density as that provided by $KO_2$.

$$d = \frac{PM}{RT}$$

$$P = \frac{dRT}{M} = \frac{\left(\frac{968\text{ g}}{1\text{ L}}\right)\left(0.0821\frac{\text{L}\cdot\text{atm}}{\text{mol}\cdot\text{K}}\right)(293\text{ K})}{\left(\frac{32.00\text{ g O}_2}{1\text{ mol}}\right)} = \textbf{727 atm}$$

Obviously, using $O_2$ instead of $KO_2$ is not practical.

# Chapter 24

# Nonmetallic Elements and Their Compounds

24.11 Element number 17 is the halogen, chlorine. Since it is a nonmetal, chlorine will form the molecular compound **HCl**. Element 20 is the alkaline earth metal, calcium, which will form an ionic hydride, **$CaH_2$**.

**A water solution of HCl is called hydrochloric acid. Calcium hydride will react according to the equation** (see Section 24.2 of the text):

$$CaH_2(s) + 2H_2O(l) \rightarrow Ca(OH)_2(aq) + 2H_2(g)$$

24.13 **NaH: Ionic compound, reacts with water as follows:**

$$NaH(s) + H_2O(l) \rightarrow NaOH(aq) + H_2(g)$$

**$CaH_2$: Ionic compound, reacts with water as follows:**

$$CaH_2(s) + 2H_2O(l) \rightarrow Ca(OH)_2(aq) + 2H_2(g)$$

**$CH_4$: Covalent compound, unreactive, burns in air or oxygen:**

$$CH_4(g) + 2O_2(g) \rightarrow CO_2(g) + 2H_2O(l)$$

**$NH_3$: Covalent compound, weak base in water:**

$$NH_3(aq) + H_2O(l) \rightleftarrows NH_4^+(aq) + OH^-(aq)$$

**$H_2O$: Covalent compound, forms strong intermolecular hydrogen bonds, good solvent for both ionic compounds and substances capable of forming hydrogen bonds.**

**HCl: Covalent compound (polar), acts as a strong acid in water:**

$$HCl(g) + H_2O(l) \rightarrow H_3O^+(aq) + Cl^-(aq)$$

24.15 The equation is: $$CaH_2(s) + 2H_2O(l) \rightarrow Ca(OH)_2(aq) + 2H_2(g)$$

First, let's calculate the moles of $H_2$ using the ideal gas law.

$$n = \frac{PV}{RT}$$

$$\text{mol } H_2 = \frac{\left(746 \text{ mmHg} \times \dfrac{1 \text{ atm}}{760 \text{ mmHg}}\right)(26.4 \text{ L})}{(0.0821 \text{ L}\cdot\text{atm/mol}\cdot\text{K})(293 \text{ K})} = 1.08 \text{ mol } H_2$$

Now, we can calculate the mass of $CaH_2$ using the correct mole ratio from the balanced equation.

$$\textbf{Mass CaH}_\mathbf{2} = 1.08 \text{ mol } H_2 \times \frac{1 \text{ mol } CaH_2}{2 \text{ mol } H_2} \times \frac{42.10 \text{ g}}{1 \text{ mol } CaH_2} = \mathbf{22.7 \text{ g } CaH_2}$$

24.17 According to Table 19.1 of the text, **$H_2$ can reduce $Cu^{2+}$, but not $Na^+$**. (How can you tell?) The reaction is:

$$\mathbf{CuO(\mathit{s}) + H_2(\mathit{g}) \rightarrow Cu(\mathit{s}) + H_2O(\mathit{l})}$$

24.25 The Lewis structure is:

$$\left[:C\equiv C:\right]^{2-}$$

24.27 a. The reaction is: $\mathbf{2NaHCO_3(\mathit{s}) \rightarrow Na_2CO_3(\mathit{s}) + H_2O(\mathit{g}) + CO_2(\mathit{g})}$

Is this an endo- or an exothermic process?

b. The hint provides the identities of the reactants. The reaction is:

$$\mathbf{Ca(OH)_2(\mathit{aq}) + CO_2(\mathit{g}) \rightarrow CaCO_3(\mathit{s}) + H_2O(\mathit{l})}$$

The visual proof is the formation of a white precipitate of $CaCO_3$. Why would a water solution of NaOH be unsuitable to qualitatively test for carbon dioxide?

24.29 **Heat causes bicarbonates to decompose according to the reaction:**

$$\mathbf{2HCO_3^- \rightarrow CO_3^{2-} + H_2O + CO_2}$$

**Generation of carbonate ion causes precipitation of the insoluble $MgCO_3$.**

Do you think there is much chance of finding natural mineral deposits of calcium or magnesium bicarbonates?

24.31 The wet sodium hydroxide is first converted to sodium carbonate:

$$2NaOH(\mathit{aq}) + CO_2(\mathit{g}) \rightarrow Na_2CO_3(\mathit{aq}) + H_2O(\mathit{l})$$

The sodium carbonate reacts with water to produce sodium hydrogen carbonate:

$$Na_2CO_3(aq) + H_2O(l) + CO_2(g) \rightarrow 2NaHCO_3(aq)$$

Eventually, the sodium hydrogen carbonate precipitates (the water solvent evaporates since $NaHCO_3$ is not hygroscopic). Thus, most of the white solid is **$NaHCO_3$ plus some $Na_2CO_3$.**

24.33 **Yes.** Carbon monoxide and molecular nitrogen are isoelectronic. Both have 14 electrons. What other diatomic molecules discussed in these problems are isoelectronic with CO?

24.39 The balanced equation is: $\mathbf{KNO_3(s) + C(s) \rightarrow KNO_2(s) + CO(g)}$

The maximum amount of potassium nitrite (theoretical yield) is:

$$57.0\text{ g KNO}_3 \times \frac{1\text{ mol KNO}_3}{101.1\text{ g KNO}_3} \times \frac{1\text{ mol KNO}_2}{1\text{ mol KNO}_3} \times \frac{85.11\text{ g KNO}_2}{1\text{ mol KNO}_2} = \mathbf{48.0\text{ g KNO}_2}$$

24.41 a.

$$\Delta G^{\circ}_{\text{rxn}} = 2\Delta G^{\circ}_{\text{f}}(\text{NO}) - [0 + 0]$$

$$173.4\text{ kJ/mol} = 2\Delta G^{\circ}_{\text{f}}(\text{NO})$$

$$\Delta G^{\circ}_{\text{f}}(\text{NO}) = \mathbf{86.7\text{ kJ / mol}}$$

b. From Equation 18.14 of the text:

$$\Delta G^{\circ} = -RT\ln K_{\text{p}}$$

$$173.4 \times 10^3\text{ J/mol} = -(8.314\text{ J/K·mol})(298\text{ K})\ln K_{\text{p}}$$

$$\mathbf{K_p = 4 \times 10^{-31}}$$

c. Using Equation 15.4 of the text [$K_p = K_c(0.0821\,T)^{\Delta n}$], $\Delta n = 0$, then $K_p = \mathbf{K_c = 4 \times 10^{-31}}$

24.43

$$\Delta T_{\text{b}} = K_{\text{b}}m = 0.409°\text{C}$$

$$\text{molality} = \frac{0.409°\text{C}}{2.34°\text{C}/m} = 0.175\ m$$

The number of grams of white phosphorus in 1 kg of solvent is:

$$\frac{1.645\text{ g phosphorus}}{75.5\text{ g CS}_2} \times \frac{1000\text{ g}}{1\text{ kg}} = 21.8\text{ g phosphorus/kg CS}_2$$

The molar mass of white phosphorus is:

$$\frac{21.8\text{ g phosphorus/kg CS}_2}{0.175\text{ mol phosphorus/kg CS}_2} = \mathbf{125\ g/mol}$$

Let the molecular formula of white phosphorus be $P_n$, so that:

$$n \times 30.97\text{ g/mol} = 125\text{ g/mol}$$

$$n = 4$$

The molecular formula of white phosphorus is $\mathbf{P_4}$.

24.45 a. $\mathbf{2NaNO_3(\mathit{s}) \rightarrow 2NaNO_2(\mathit{s}) + O_2(\mathit{g})}$ b. $\mathbf{NaNO_3(\mathit{s}) + C(\mathit{s}) \rightarrow NaNO_2(\mathit{s}) + CO(\mathit{g})}$

24.47 The balanced equation is: $\mathbf{2NH_3(\mathit{g}) + CO_2(\mathit{g}) \rightarrow (NH_2)_2CO(\mathit{s}) + H_2O(\mathit{l})}$

If pressure increases, the position of equilibrium will shift in the direction with the smallest number of molecules in the gas phase, that is, to the right. Therefore, **the reaction should be run at high pressure.**

Write the expression for $Q_p$ for this reaction. Does increasing pressure cause $Q_p$ to increase or decrease? Is this consistent with the above prediction?

24.49 **The oxidation state of N in nitric acid is +5, the highest oxidation state for N. N can be reduced to an oxidation state of −3.**

24.51 a. $\mathbf{NH_4NO_3(\mathit{s}) \rightarrow N_2O(\mathit{g}) + 2H_2O(\mathit{l})}$

b. $\mathbf{2KNO_3(\mathit{s}) \rightarrow 2KNO_2(\mathit{s}) + O_2(\mathit{g})}$

c. $\mathbf{Pb(NO_3)_2(\mathit{s}) \rightarrow PbO(\mathit{s}) + 2NO_2(\mathit{g}) + O_2(\mathit{g})}$

24.53 The atomic radius (see Figure 7.6 in the text) of P (110 pm) is considerably larger than that of N (75 pm); consequently, the 3*p* orbital on a P atom cannot overlap effectively with a 3*p* orbital on a neighboring P atom to form a pi bond. Simply stated, **the phosphorus is too large to allow effective overlap of the 3*p* orbitals to form π bonds.**

24.55 $PH_4^+$ is similar to $NH_4^+$. The hybridization of phosphorus in $PH_4^+$ is $\mathbf{\mathit{sp}^3}$.

24.63
$$\Delta G^\circ = \Delta G_f^\circ(NO_2) + \Delta G_f^\circ(O_2) - [\Delta G_f^\circ(NO) + \Delta G_f^\circ(O_3)]$$

$$\Delta G^\circ = (1)(51.8\text{ kJ/mol}) + (0) - [(1)(86.7\text{ kJ/mol}) + (1)(163.4\text{ kJ/mol})] = \mathbf{-198.3\ kJ/mol}$$

$$\Delta G^\circ = -RT\ln K_p$$

$$\ln K_p = \frac{-\Delta G°}{RT} = \frac{198.3 \times 10^3 \text{ J/mol}}{(8.314 \text{ J/K}\cdot\text{mol})(298 \text{ K})}$$

$$\boldsymbol{K_p = 6 \times 10^{34}}$$

Since there is no change in the number of moles of gases, $\boldsymbol{K_c}$ **is *equal* to** $\boldsymbol{K_p}$.

24.65 First we convert gallons of water to grams of water.

$$(2.0 \times 10^2 \text{ gal}) \times \frac{3.785 \text{ L}}{1 \text{ gal}} \times \frac{1000 \text{ mL}}{1 \text{ L}} \times \frac{1.00 \text{ g } H_2O}{1 \text{ mL}} = 7.6 \times 10^5 \text{ g } H_2O$$

An $H_2S$ concentration of 22 ppm indicates that in 1 million grams of water, there will be 22 g of $H_2S$. First, let's calculate the number of moles of $H_2S$ in $7.6 \times 10^5$ g of $H_2O$:

$$(7.6 \times 10^5 \text{ g } H_2O) \times \frac{22 \text{ g } H_2S}{1.0 \times 10^6 \text{ g } H_2O} \times \frac{1 \text{ mol } H_2S}{34.09 \text{ g } H_2S} = 0.49 \text{ mol } H_2S$$

The mass of chlorine required to react with 0.49 mol of $H_2S$ is:

$$0.49 \text{ mol } H_2S \times \frac{1 \text{ mol } Cl_2}{1 \text{ mol } H_2S} \times \frac{70.90 \text{ g } Cl_2}{1 \text{ mol } Cl_2} = \mathbf{35 \text{ g } Cl_2}$$

24.67 **To form $OF_6$ there would have to be six bonds (twelve electrons) around the oxygen atom. This would violate the octet rule. Since oxygen does not have *d* orbitals, it cannot have an expanded octet.**

24.69 Each reaction uses $H_2SO_4(l)$ as a reagent.

a. $\mathbf{HCOOH}(l) \rightleftarrows \mathbf{CO}(g) + \mathbf{H_2O}(l)$

b. $\mathbf{4H_3PO_4}(l) \rightleftarrows \mathbf{P_4O_{10}}(s) + \mathbf{6H_2O}(l)$

c. $\mathbf{2HNO_3}(l) \rightleftarrows \mathbf{N_2O_5}(g) + \mathbf{H_2O}(l)$

d. $\mathbf{2HClO_3}(l) \rightleftarrows \mathbf{Cl_2O_5}(l) + \mathbf{H_2O}(l)$

24.71 a. **To exclude light**. As stated in the problem, the decomposition of hydrogen peroxide is accelerated by light.

b. The STP volume of oxygen gas formed is:

$$15.0 \text{ g soln} \times \frac{7.50\% \; H_2O_2}{100\% \text{ soln}} \times \frac{1 \text{ mol } H_2O_2}{34.02 \text{ g } H_2O_2} \times \frac{1 \text{ mol } O_2}{2 \text{ mol } H_2O_2} \times \frac{22.41 \text{ L } O_2}{1 \text{ mol } O_2} = \mathbf{0.371 \text{ L } O_2}$$

24.73 Following the rules given in Section 4.4 of the text, we assign hydrogen an oxidation number of +1 and **F** an oxidation number of **−1**. Since HFO is a neutral molecule, the oxidation number of **O** is **0 (zero)**. Can you think of other compounds in which oxygen has this oxidation number?

24.75 A check of Table 19.1 of the text shows that sodium ion cannot be reduced by any of the substances mentioned in this problem; it is a "spectator ion". We focus on the substances that are actually undergoing oxidation or reduction and write half-reactions for each.

$$2I^-(aq) \rightarrow I_2(s)$$

$$H_2SO_4(aq) \rightarrow H_2S(g)$$

Balancing the oxygen, hydrogen, and charge gives:

$$2I^-(aq) \rightarrow I_2(s) + 2e^-$$

$$H_2SO_4(aq) + 8H^+(aq) + 8e^- \rightarrow H_2S(g) + 4H_2O(l)$$

Multiplying the iodine half-reaction by four and combining gives the balanced redox equation.

$$H_2SO_4(aq) + 8I^-(aq) + 8H^+(aq) \rightarrow H_2S(g) + 4I_2(s) + 4H_2O(l)$$

The hydrogen ions come from extra sulfuric acid. We add one sodium ion for each iodide ion to obtain the final equation.

$$\mathbf{9H_2SO_4(\mathit{aq}) + 8NaI(\mathit{aq}) \rightarrow 4I_2(\mathit{s}) + H_2S(\mathit{g}) + 4H_2O(\mathit{l}) + 8NaHSO_4(\mathit{aq})}$$

24.79 The reaction is: $2Br^-(aq) + Cl_2(g) \rightarrow 2Cl^-(aq) + Br_2(l)$

The number of moles of chlorine needed is:

$$167\text{ g Br}^- \times \frac{1\text{ mol Br}^-}{79.90\text{ g Br}^-} \times \frac{1\text{ mol Cl}_2}{2\text{ mol Br}^-} = 1.05\text{ mol Cl}_2(g)$$

Use the ideal gas equation to calculate the volume of $Cl_2$ needed.

$$V_{Cl_2} = \frac{nRT}{P} = \frac{(1.05\text{ mol})(0.0821\text{ L}\cdot\text{atm/K}\cdot\text{mol})(293\text{ K})}{(1\text{ atm})} = \mathbf{25.3\text{ L}}$$

24.81 a. $I_3^-$ $AB_2E_3$ **Linear**

b. $SiCl_4$ $AB_4$ **Tetrahedral**

c. $PF_5$ $AB_5$ **Trigonal bipyramidal**

d. $SF_4$ $AB_4E$ **See-saw**

24.83 The structures are:

$$H{-}\ddot{\underset{..}{F}}{:}{\cdots}H{-}\ddot{\underset{..}{F}}{:} \qquad [{:}\ddot{\underset{..}{F}}{\cdots}H{\cdots}\ddot{\underset{..}{F}}{:}]^-$$

The $HF_2^-$ ion has the strongest known hydrogen bond. More complex hydrogen bonded HF clusters are also known.

24.85 **As with iodide salts, a redox reaction occurs between sulfuric acid and sodium bromide.**

$$\mathbf{2H_2SO_4(\mathit{aq}) + 2NaBr\ (\mathit{aq}) \rightarrow SO_2(\mathit{g}) + Br_2(\mathit{l}) + 2H_2O(\mathit{l}) + Na_2SO_4(\mathit{aq})}$$

24.87 The balanced equation is:

$$\mathbf{I_2O_5(\mathit{s}) + 5CO(\mathit{g}) \rightarrow 5CO_2(\mathit{g}) + I_2(\mathit{s})}$$

The oxidation number of iodine changes from +5 to 0 and the oxidation number of carbon changes from +2 to +4. **Iodine** is **reduced**; **carbon** is **oxidized.**

24.89 The balanced equations are:

a. $$\mathbf{2H_3PO_3(\mathit{aq}) \rightarrow H_3PO_4(\mathit{aq}) + PH_3(\mathit{g}) + O_2(\mathit{g})}$$

b. $$\mathbf{Li_4C(\mathit{s}) + 4HCl(\mathit{aq}) \rightarrow 4LiCl(\mathit{aq}) + CH_4(\mathit{g})}$$

c. $$\mathbf{2HI(\mathit{g}) + 2HNO_2(\mathit{aq}) \rightarrow I_2(\mathit{s}) + 2NO(\mathit{g}) + 2H_2O(\mathit{l})}$$

d. $$\mathbf{H_2S(\mathit{g}) + 2Cl_2(\mathit{g}) \rightarrow 2HCl(\mathit{g}) + SCl_2(\mathit{l})}$$

24.91 a. $\mathbf{SiCl_4}$ b. $\mathbf{F^-}$ c. **F** d. $\mathbf{CO_2}$

24.93 **There is no change in oxidation number; it is zero for both compounds.**

24.95 The Lewis structures are shown below. In $PCl_4^+$, which is tetrahedral, the hybridization of the phosphorus atom is $\boldsymbol{sp^3}$ hybridized. In $PCl_6^-$, which is octahedral, the hybridization of the phosphorus atom is $\boldsymbol{sp^3d^2}$.

$$\left[\begin{array}{ccc} & Cl & \\ & | & \\ Cl{-} & P & {-}Cl \\ & | & \\ & Cl & \end{array}\right]^+ \qquad \left[\begin{array}{c} Cl\diagdown\ \overset{Cl}{|}\ \diagup Cl \\ P \\ Cl\diagup\ \underset{Cl}{|}\ \diagdown Cl \end{array}\right]^-$$

24.97 We know that $\Delta G° = -RT\ln K$ and $\Delta G° = \Delta H° - T\Delta S°$. We can first calculate $\Delta H°$ and $\Delta S°$ using data in Appendix 2 of the text. Then, we can calculate $\Delta G°$ and lastly $K$.

$$\Delta H^\circ = 2\Delta H_f^\circ[CO(g)] - \{\Delta H_f^\circ[C(s)] + \Delta H_f^\circ[CO_2(g)]\}$$

$$\Delta H^\circ = (2)(-110.5\ \text{kJ/mol}) - (0 + -393.5\ \text{kJ/mol}) = 172.5\ \text{kJ/mol}$$

$$\Delta S^\circ = 2S^\circ[CO(g)] - \{S^\circ[C(s)] + S^\circ[CO_2(g)]\}$$

$$\Delta S^\circ = (2)(197.9\ \text{J/K·mol}) - (5.69\ \text{J/K·mol} + 213.6\ \text{J/K·mol}) = 176.5\ \text{J/K·mol}$$

At 298 K (25°C),

$$\Delta G^\circ = \Delta H^\circ - T\Delta S^\circ = (172.5 \times 10^3\ \text{J/mol}) - (298\ \text{K})(176.5\ \text{J/K}\cdot\text{mol}) = 1.199 \times 10^5\ \text{J/mol}$$

$$\Delta G^\circ = -RT\ln K$$

$$\boldsymbol{K} = e^{\frac{-\Delta G^\circ}{RT}} = e^{\frac{-(1.199 \times 10^5\ \text{J/mol})}{(8.314\ \text{J/K·mol})(298\ \text{K})}} = \mathbf{9.61 \times 10^{-22}}$$

At 373 K (100°C),

$$\Delta G^\circ = (172.5 \times 10^3\ \text{J/mol}) - (373\ \text{K})(176.5\ \text{J/K·mol}) = 1.07 \times 10^5\ \text{J/mol}$$

$$\boldsymbol{K} = e^{\frac{-\Delta G^\circ}{RT}} = e^{\frac{-(1.07\times 10^5\ \text{J/mol})}{(8.314\ \text{J/mol})(373\ \text{K})}} = \mathbf{1.2 \times 10^{-15}}$$

24.99 **The glass is etched (dissolved) by the reaction**

$$\mathbf{6HF(\mathit{aq}) + SiO_2(\mathit{s}) \rightarrow H_2SiF_6(\mathit{aq}) + 2H_2O(\mathit{l})}$$

**This process gives the glass a frosted appearance.**

24.101 The reactions are:

$$P_4(s) + 5O_2(g) \rightarrow P_4O_{10}(s)$$
$$P_4O_{10}(s) + 6H_2O(l) \rightarrow 4H_3PO_4(aq)$$

First, we calculate the moles of $H_3PO_4$ produced. Next, we can calculate the molarity of the phosphoric acid solution. Finally, we can determine the pH of the $H_3PO_4$ solution (a weak acid).

$$10.0\ \text{g}\ P_4 \times \frac{1\ \text{mol}\ P_4}{123.9\ \text{g}\ P_4} \times \frac{1\ \text{mol}\ P_4O_{10}}{1\ \text{mol}\ P_4} \times \frac{4\ \text{mol}\ H_3PO_4}{1\ \text{mol}\ P_4O_{10}} = 0.323\ \text{mol}\ H_3PO_4$$

$$\text{Molarity} = \frac{0.323\ \text{mol}}{0.500\ \text{L}} = 0.646\ M$$

We set up the ionization of the weak acid, $H_3PO_4$. The $K_a$ value for $H_3PO_4$ can be found in Table 16.8 of the text.

| | $H_3PO_4(aq)$ + | $H_2O(aq)$ | $\rightleftarrows$ | $H_3O^+(aq)$ | + $H_2PO_4^-(aq)$ |
|---|---|---|---|---|---|
| Initial (*M*): | 0.646 | | | 0 | 0 |
| Change (*M*): | $-x$ | | | $+x$ | $+x$ |
| Equilibrium (*M*): | $0.646 - x$ | | | $x$ | $x$ |

$$K_a = \frac{[H_3O^+][H_2PO_4^-]}{[H_3PO_4]}$$

$$7.5 \times 10^{-3} = \frac{(x)(x)}{(0.646 - x)}$$

$$x^2 + 7.5 \times 10^{-3}x - 4.85 \times 10^{-3} = 0$$

Solving the quadratic equation,

$$x = 0.066\ M = [H_3O^+]$$

Following the procedure in Problem 16.140 and the discussion in Section 16.8 of the text, we can neglect the contribution to the hydronium ion concentration from the second and third ionization steps. Thus,

$$\mathbf{pH} = -\log(0.066) = \mathbf{1.18}$$

24.103 At the normal boiling point, the pressure of HF is 1 atm. We use Equation 10.10 of the text to calculate the density of HF.

$$\boldsymbol{d} = \frac{P\mathcal{M}}{RT} = \frac{(1\ \text{atm})(20.01\frac{\text{g}}{\text{mol}})}{\left(0.0821\frac{\text{L}\cdot\text{atm}}{\text{mol}\cdot\text{K}}\right)(273 + 19.5)\text{K}} = \mathbf{0.833\ g/L}$$

**The molar mass derived from the observed density is 74.41, which suggests that the molecules are associated to some extent in the gas phase. This makes sense due to strong hydrogen bonding in HF.**

# Chapter 25

# Organic Chemistry

25.7 a. **amine** b. **aldehyde** c. **ketone** d. **carboxylic acid** e. **alcohol**

25.9 a.

| | |
|---|---|
| 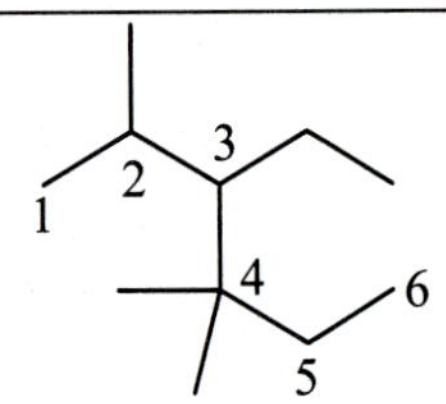 1 2 3 4 5 6 (skeletal structure) **3-ethyl-2,4,4-trimethyllhexane** | The longest continuous chain has 6 carbons, so the alkane is named as a derivative of *hexane*. Its substituents, in alphabetical order, are an ethyl group and three methyl groups. It is an ethyl trimethyl derivative of hexane. When the chain is numbered beginning at the end nearest the first branch, the substituted carbons are C-2, 3, 4, and 4. |

b.

| | |
|---|---|
| 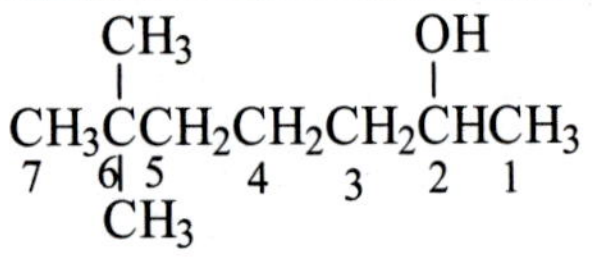 $CH_3CCH_2CH_2CH_2CHCH_3$ with $CH_3$ above and below C-6 and OH on C-2 (numbered 7 6 5 4 3 2 1) **6,6-dimethyl-2-heptanol** | This compound is named as an alcohol. The longest continuous chain that bears the –OH group has 7 carbons. Numbering starts from the end nearest the –OH group. The compound is a 6,6-dimethyl derivative of 2-heptanol. |

c.

| | |
|---|---|
| 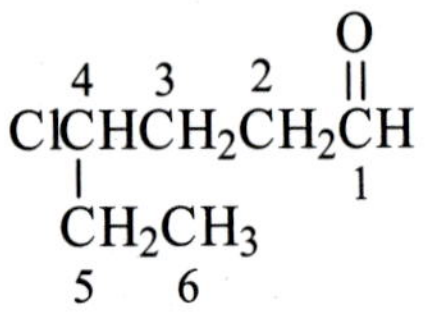 $ClCHCH_2CH_2CH$ (=O) with $CH_2CH_3$ on C-4 (numbered 4 3 2 1; 5 6) **4-chlorohexanal** | The compound is an aldehyde. When naming aldehydes, we drop the –e ending of the alkane name and replace it by –al. The chain is numbered beginning at the carbonyl group and is 6 carbons in length. This *hexanal* derivative has a chlorine substituent at C-4. |

25.11 Notice that the longest continuous chain has 8 carbons, not seven, and that the chain is numbered right-to-left so that the first-appearing substituent is at C-3, not C-4. The alkane is **3,5-dimethyloctane**.

25.13 a. 2,2,4-trimethylpentane ("isooctane") $\mathbf{(CH_3)_3CCH_2CH(CH_3)_2}$

b. 3-methyl-1-butanol ("isoamyl alcohol") $\mathbf{HO(CH_2)_2CH(CH_3)_2}$

c. hexanamide ("caproamide") **$CH_3(CH_2)_4C(O)NH_2$**

d. 2,2,2-trichloroethanal ("chloral") **$Cl_3CCHO$**

25.15

$H_3C$ $CH_3$ $CH_3$

OH **C**

**A** O $CH_3$ O OH **B**

**A** Carbonyl (Ketone)
**B** Carboxy (Carboxylic acid)
**C** Hydroxy (Alcohol)

25.17

H
Primary amino N
H

O
C Carboxy
OH

25.19 a. $CH_3CH_2CHCH_2CH_2CH_3$
$CH_3$

b. $CH_3$
$CH_3CHCHCH_2CH_3$
$CH_3$

c. Br
$CH_3CHCH_2CHCH_3$
$C_6H_5$

d. $CH_3$ $CH_3$
$CH_3CH_2CHCHCHCH_2CH_2CH_3$
$CH_3$

25.21 a.

**Kekule**

```
    H   H   H   H   H   O   H   O
    |   |   |   |   |   ||  |   ||
H—C—C—C—C—C—C—C—C—O—H
    |   |   |   |   |       |
    H   H   H   H   H       H
```

**Skeletal (line)**

O O OH

b. **Condensed** $\mathbf{(C_2H_5)_2CHCH_2CO_2C(CH_3)_3}$

**Skeletal (line)**

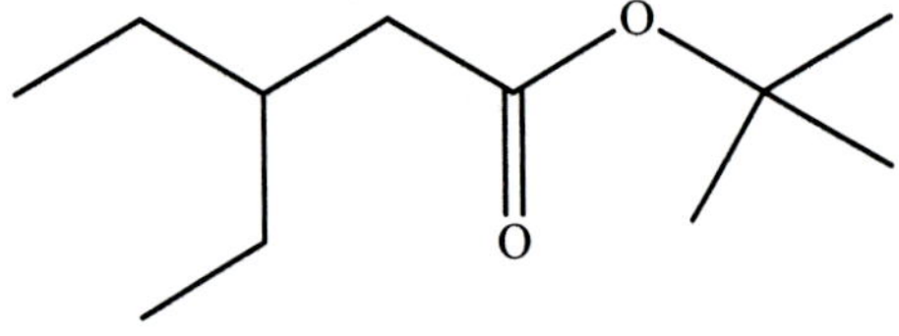


c. **Condensed** $\mathbf{(CH_3)_2CHCH_2NHCH(CH_3)_2}$

**Kekule**

```
    H   H   H   H   H   H
    |   |   |   |   |   |
H—C—C—C—N—C—C—H
    |   |   |       |   |
    H   |   H       |   H
        |           |
      H—C—H       H—C—H
        |           |
        H           H
```

25.23 a.

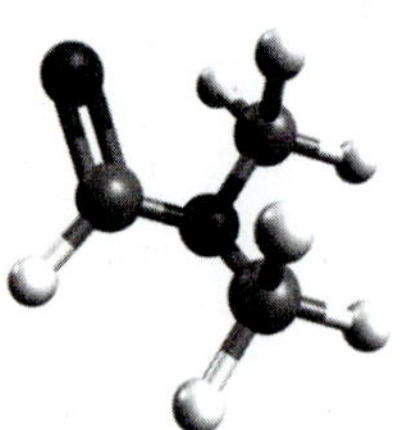

$C_3H_7NO$: DMF is

**Condensed structural:** $\mathbf{(CH_3)_2NCHO}$

**Kekule:**

**Line:**

b. $C_6H_8O_7$: citric acid is

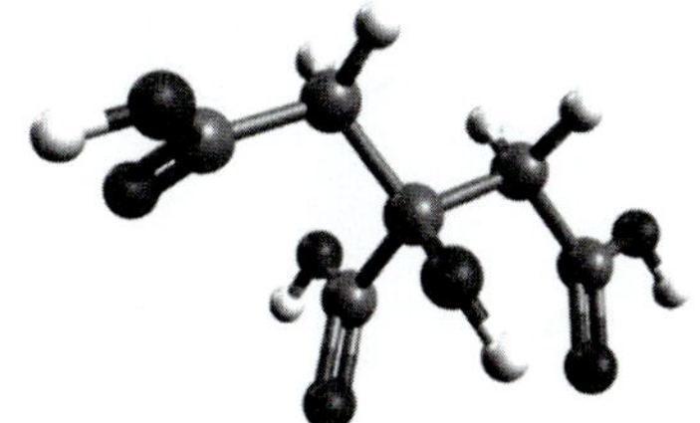

**Condensed structural:**

**$(CH_2COOH)_2C(OH)COOH$**

**Kekule:**

**Line:**

c.

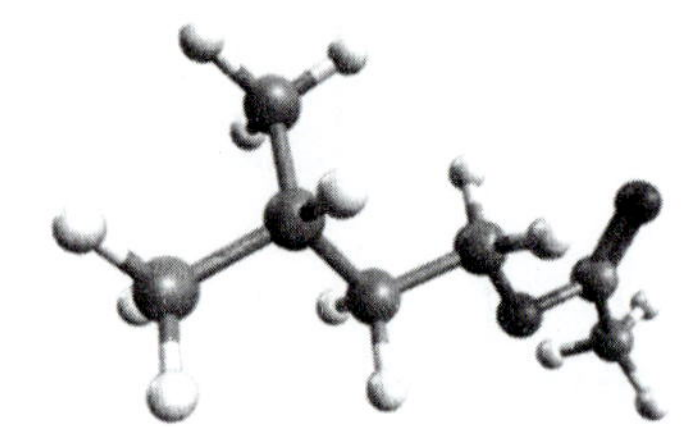

$C_6H_8O_6$: isoamyl acetate is

**Condensed structural:**

$(CH_3)_2CH(CH_2)_2OC(O)CH_3$

**Kekule:**

**Line:** O, O

25.25 a.

**line:** OH

b. **structural:** $(CH_3)_3CCH_2CHCH_2CHO$ (Br on the CH)

c.

**line:** O, O

25.27 a. $CH_3—C\equiv N\colon \longleftrightarrow CH_3—\overset{+}{C}=\ddot{N}\colon^{-}$

b. $CH_3$–C$^+$(–$CH_3$)–Ö–H ⟷ $CH_3$–C(–$CH_3$)=O$^+$–H

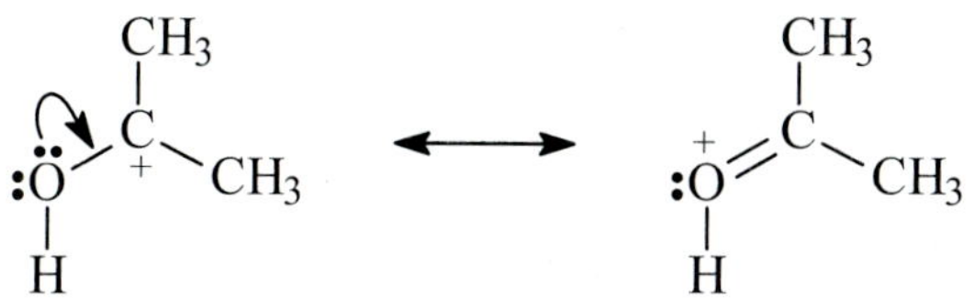

c. $CH_3$–C(–O$^-$)=$CH_2$ ⟷ $CH_3$–C(=O)–$CH_2^-$

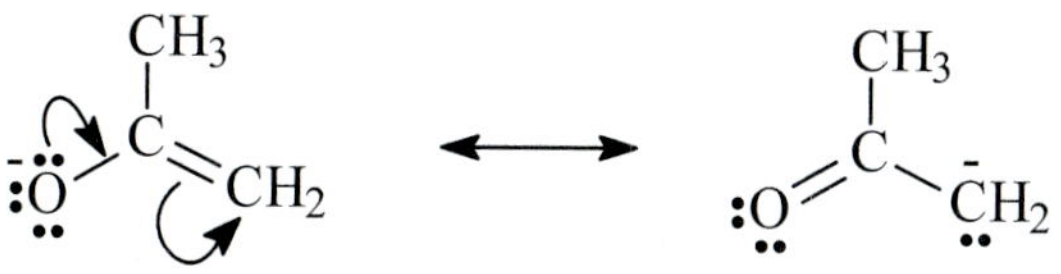

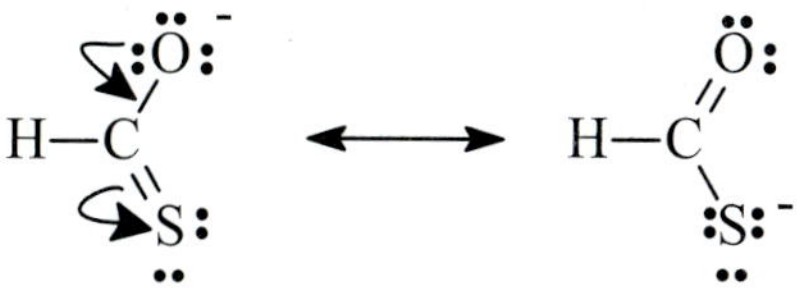

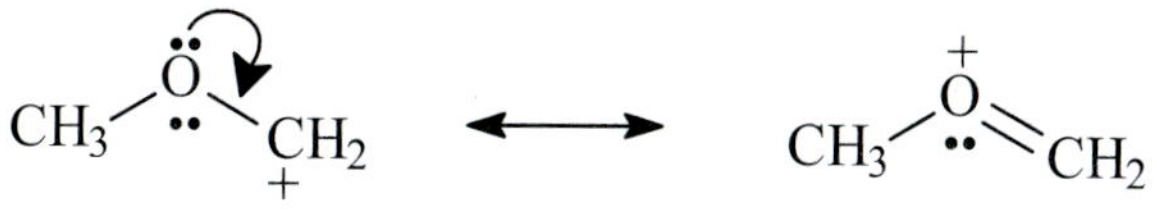

25.29 a. H–C(–O$^-$)=S ⟷ H–C(=O)–S$^-$

b. $CH_3$–O–$CH_2^+$ ⟷ $CH_3$–O$^+$=$CH_2$

c. H H H H
H– (benzene ring) –$\overset{+}{C}H_2$ ⟷ H–(+)=$CH_2$
H H H H

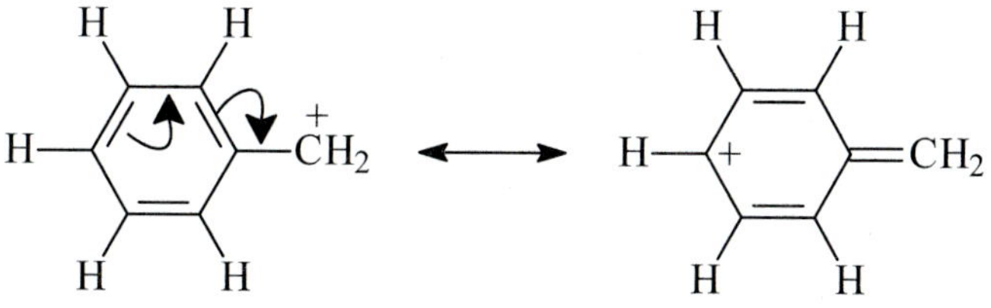

25.37 There are 9 possible isomers:

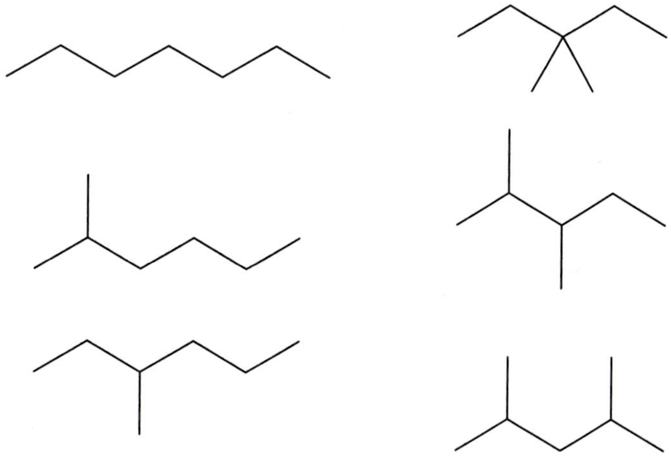

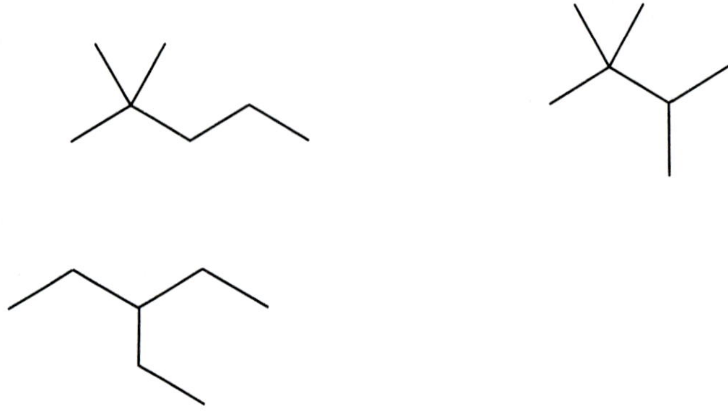

25.39 There are 5 possible isomers:

25.41 There are 4 possible isomers

25.43 a.

$$\mathrm{CH_3{-}CH_2{-}\underset{}{CH}(CH_3){-}CH(NH_2){-}C(=O){-}NH_2} \qquad \mathrm{CH_3{-}CH_2{-}\overset{*}{C}H(CH_3){-}\overset{*}{C}H(NH_2){-}C(=O){-}NH_2}$$

b.

Although the cyclopropane has two asymmetric carbon atoms, the molecule is not chiral because it has a plane of symmetry.

25.45

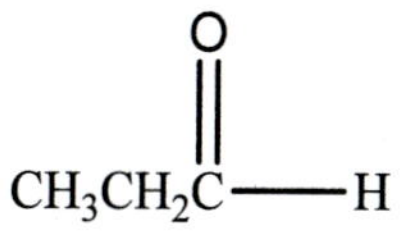

$CH_3CH_2C(=O)—H$

25.47 a. 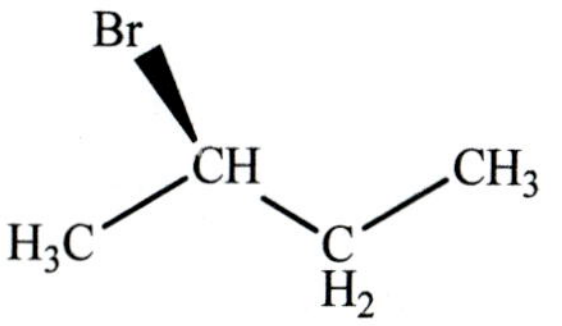 Br, CH, $CH_3$, $H_3C$, C, $H_2$

b. OH, $H_2$, C, C, HO, CH, O, $O_2N$

c. HO, $H_2$, C, CH, $H_3C$, CH, $CH_3$, $H_3C$

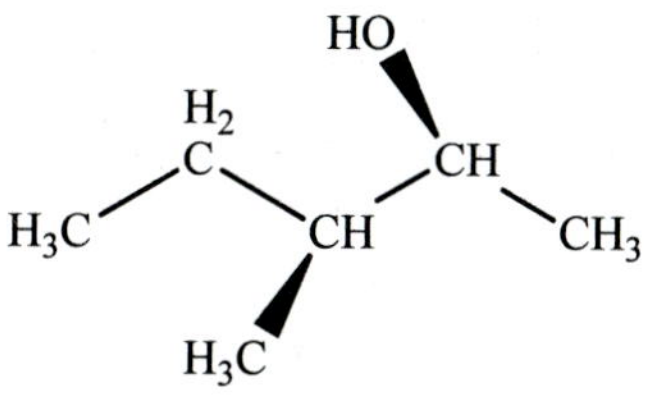

d. 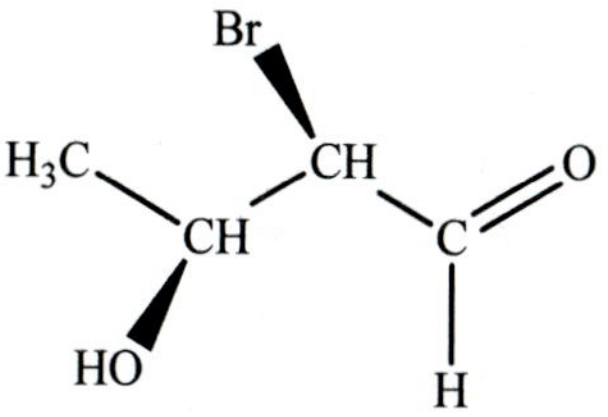 Br, $H_3C$, CH, O, CH, C, HO, H

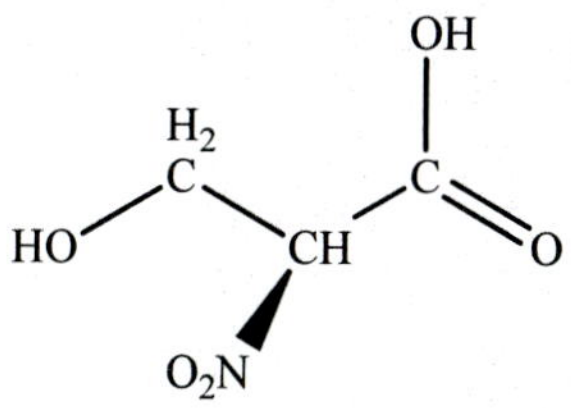

25.55

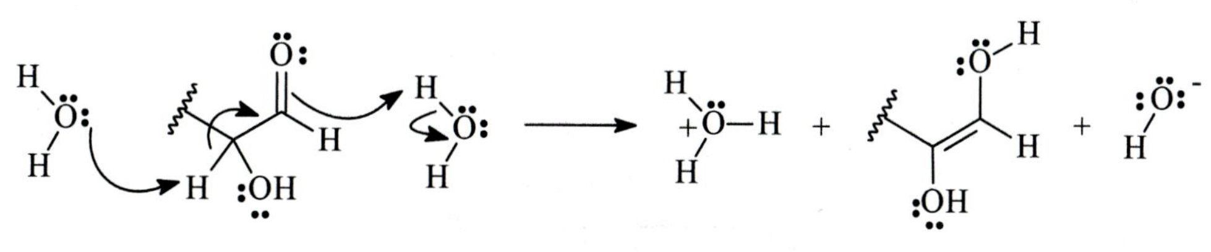

25.57 a. The aldehyde C has a partial positive charge since O is more electronegative than C; this positive charge attracts the negative end of the acetylide (nucleophilic attack), and as the C--C coordinate covalent bond forms, one of the C-O bond pairs flows onto the O atom and becomes a lone pair. The now negative O atom then attracts the positive (H) end of a water molecule, and as the O--H coordinate covalent bond forms, the H-O bond pair flows onto the O atom of the water molecule to form the hydroxyl ion leaving group.

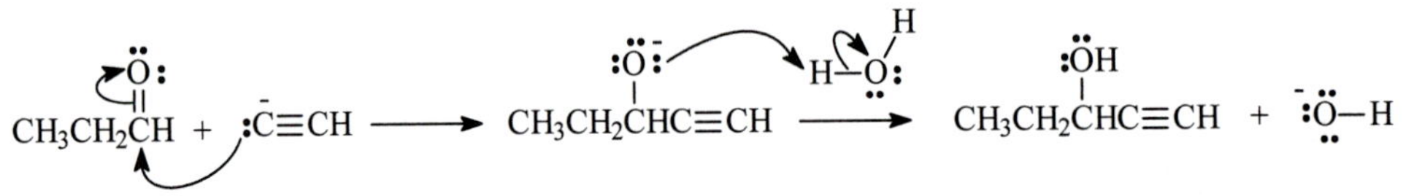

$$CH_3CH_2CH{=}O + {:}C{\equiv}CH^- \longrightarrow CH_3CH_2CH(O^-)C{\equiv}CH \xrightarrow{H_2O} CH_3CH_2CH(OH)C{\equiv}CH + {:}\ddot{O}{-}H^-$$

b. The C atom attached to Br has a partial positive charge because Br is more electronegative than C. As the flexible $C_4$ chain wiggles and bends, the negative S atom can approach and be attracted to the positive C atom (nucleophilic attack), and as the S—C coordinate covalent bond forms, the C—Br bond pair flows onto the Br atom to produce the bromide leaving group.

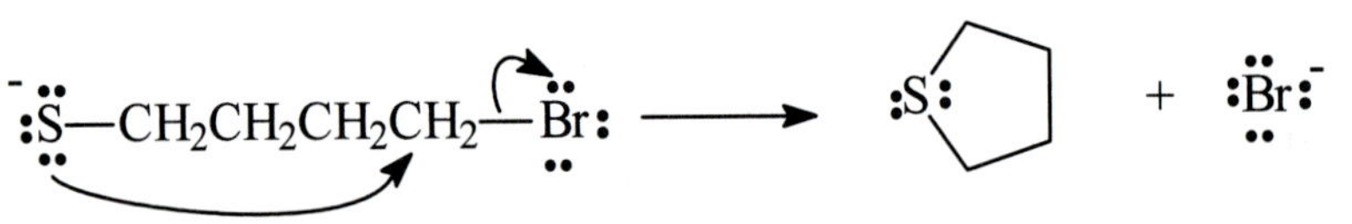

$$^-{:}\ddot{S}{-}CH_2CH_2CH_2CH_2{-}\ddot{Br}{:} \longrightarrow \text{(cyclic thioether, } {:}S{:}\text{ in five-membered ring)} + {:}\ddot{Br}{:}^-$$

25.59 Cl in an aromatic ring is an electron-withdrawing group, causing parital positive charges to appear on the H atom(s) next to it. The amide anion is attracted to this charge (nucleophilic attack), and as the N—H coordinate covalent bond forms, the C—H bond pair is forced onto the ring as a pi-bond, which in turn forces the Cl—C bond pair onto the Cl to form the chloride leaving group. This is an **elimination** reaction.

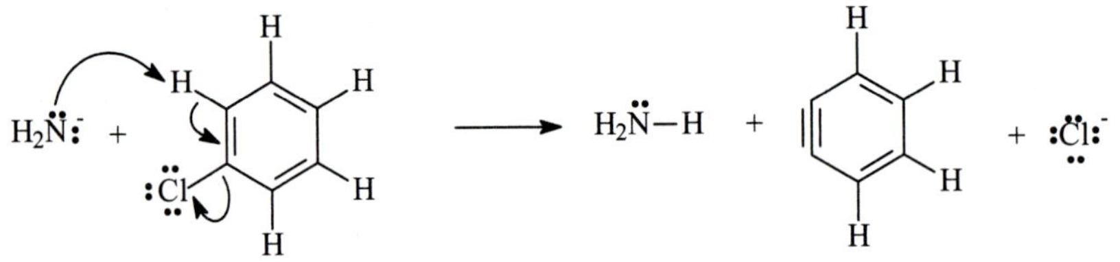

$$H_2\ddot{N}{:}^- + C_6H_5Cl \longrightarrow H_2\ddot{N}{-}H + C_6H_4 + {:}\ddot{Cl}{:}^-$$

25.61 **No. This is not an oxidation-reduction reaction. There are no changes in oxidation states for any of the atoms.**

25.63 a. **The sulfuric acid releases protons which then protonate the terminal $CH_2$ group. This creates a cation; nucleophilic attack by water produces the alcohol with release of another proton. Sulfuric acid is thus a catalyst.**

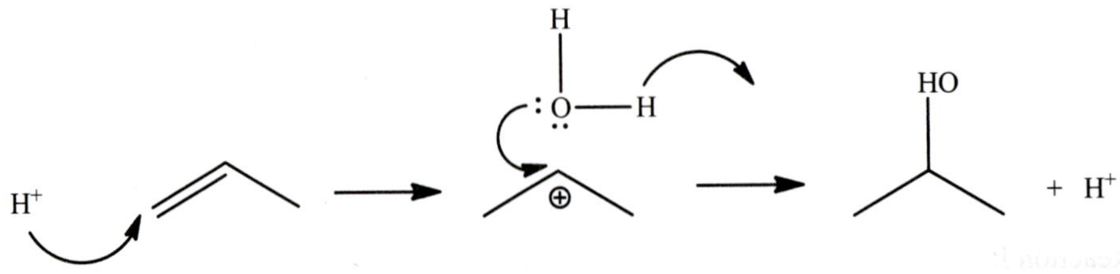

$$H^+ + CH_2{=}CHCH_3 \longrightarrow CH_3\overset{\oplus}{C}HCH_3 \xrightarrow{H_2O} (CH_3)_2CHOH + H^+$$

b.
**n-propanol**:

OH

c. **Isopropanol is not chiral since the central C atom is bonded to two identical ($CH_3$) groups.**

25.71 $\left[ -\overset{H}{\underset{H}{C}}-\overset{H}{\underset{Cl}{C}}-\overset{H}{\underset{H}{C}}-\overset{Cl}{\underset{Cl}{C}}- \right]_n$

25.73 a. $CH_2{=}CHCH{=}CH_2$

b. $HO-C(=O)-CH_2-CH_2-CH_2-CH_2-CH_2-CH_2-NH_2$

25.75 There are two, glycine-lysine and lysine-glycine.

$H_2N-CH(H)-C(=O)-N(H)-CH((CH_2)_4NH_2)-C(=O)-OH$ GlyLys

$H_2N-CH((CH_2)_4NH_2)-C(=O)-N(H)-CH(H)-C(=O)-OH$ LysGly

25.77 The structures and names of the constitutionally isomeric $C_4H_9$ alkyl groups are:

| $CH_3CH_2CH_2CH_2-$ | $CH_3\underset{\vert}{C}HCH_2CH_3$ | $CH_3\underset{CH_3}{C}HCH_2-$ | $CH_3\overset{CH_3}{\underset{CH_3}{C}}-$ |
|---|---|---|---|
| Butyl | *sec*-Butyl | Isobutyl | *tert*-Butyl |

25.79 a.

Reaction I: $H-\ddot{O}:^- + :\ddot{O}=C=\ddot{O}: \longrightarrow H-\ddot{O}-C(=O)-\ddot{O}:^-$

b.

Reaction II: $H-\ddot{O}:^-$ + $H-\ddot{O}-C(=O)-\ddot{O}:^-$ ⟶ $^-:\ddot{O}-C(=O)-\ddot{O}:^-$ + $H-\ddot{O}-H$

c. Reaction I is **nucleophilic addition** of hydroxide to C=O.

d. Reaction II is **acid-base.** Hydroxide abstracts a proton from hydrogencarbonate ion.

25.81 a. $CH_3CH_2CH(CH_3)CH=CH_2$

b. $H_3C$, H, $C_2H_5$, $C_2H_3$     H, $CH_3$, $C_2H_5$, $C_2H_3$

25.83 a. ***cis/trans* stereoisomers**

b. **constitutional isomers**

c. **resonance structures**

d. **different representations of the same structure**

25.85

25.87 **(b):** Two enantiomers of 2-chlorobutane were formed in equal amounts.

25.89 There is one non-cyclic isomer of $C_4H_6$ with two double bonds (1,3-butadiene) and two isomers with one triple bond (1-butyne and 2-butyne).

| | | |
|---|---|---|
| *sp*$^3$ *sp*$^3$ | *sp*$^3$ *sp*$^3$ | |
| $CH_3CH_2C{\equiv}CH$ | $CH_3C{\equiv}CCH_3$ | $H_2C{=}CHCH{=}CH_2$ |
| *sp* *sp* | *sp* *sp* | all *sp*$^2$ |

25.91 **Since N is less electronegative than O, electron donation in the amide would be more pronounced.**

25.93

$$-\overset{O}{\overset{\|}{C}}-C_6H_4-\overset{O}{\overset{\|}{C}}-NH-C_6H_4-NH-\overset{O}{\overset{\|}{C}}-C_6H_4-\overset{O}{\overset{\|}{C}}-NH-C_6H_4-NH-$$

$$\left[HO-\overset{O}{\overset{\|}{C}}-C_6H_4-\overset{O}{\overset{\|}{C}}-OH\right]_n + \left[NH_2-C_6H_4-NH_2\right]_n$$

$$\rightarrow \left[\overset{O}{\overset{\|}{C}}-C_6H_4-\overset{O}{\overset{\|}{C}}-NH-C_6H_4-NH\right]_n + nH_2O$$

25.95 **The N atom in the amide bond is protonated, then nucleophilic addition of water to the carbonyl cleaves the amide bond to produce the original acid and amine.**

$$-\overset{O}{\overset{\|}{C}}-\overset{H}{\overset{|}{\ddot{N}}}- \xrightarrow{H^+} -\overset{O}{\overset{\|}{C}}-\overset{H}{\overset{|}{\underset{H}{\underset{|}{N^{\oplus}}}}}- \xrightarrow{H_2O} -\overset{O}{\overset{\|}{C}}-\underset{H^{\oplus}}{\underset{|}{O}}H + H_2N-$$

25.97 The two common a-amino acids with R = $C_4H_9$ are leucine (Leu, isobutyl) and isoleucine (Ile, 2-n-butyl)

$$\underset{\text{Leucine}}{CH_3\overset{CH_3}{\overset{|}{C}}HCH_2\underset{NH_2}{\underset{|}{C}}HCO_2H} \qquad \underset{\text{Isoleucine}}{CH_3CH_2\overset{CH_3}{\overset{|}{C}}H\underset{NH_2}{\underset{|}{C}}HCO_2H}$$

25.99

$$\left[\overset{O}{\overset{\|}{C}}-\underset{H}{\underset{|}{N}}-\overset{H}{\overset{|}{\underset{H}{\underset{|}{C}}}}\right]$$

25.101 a. **The more negative $\Delta H$ implies stronger alkane bonds; branching decreases the total bond enthalpy (and overall stability) of the alkane.**

b. **The least highly branched isomer (n-octane)** would have the greatest total bond energy and should thus produce the greatest heat. This prediction corresponds to that determined experimentally.